World Sustainability Series

Series Editor

Walter Leal Filho, European School of Sustainability Science and Research, Research and Transfer Centre "Sustainable Development and Climate Change Management", Hamburg University of Applied Sciences, Hamburg, Germany

Due to its scope and nature, sustainable development is a matter which is very interdisciplinary, and draws from knowledge and inputs from the social sciences and environmental sciences on the one hand, but also from physical sciences and arts on the other. As such, there is a perceived need to foster integrative approaches, whereby the combination of inputs from various fields may contribute to a better understanding of what sustainability is, and means to people. But despite the need for and the relevance of integrative approaches towards sustainable development, there is a paucity of literature which address matters related to sustainability in an integrated way.

Notes on the quality assurance and peer review of this publication

Prior to publication, the quality of the works published in this series is double blind reviewed by external referees appointed by the editor. The referees are not aware of the author's name when performing the review; the referees' names are not disclosed.

More information about this series at http://www.springer.com/series/13384

Walter Leal Filho
Editor

COVID-19: Paving the Way for a More Sustainable World

Springer

Editor
Walter Leal Filho
HAW Hamburg
Hamburg, Germany

Manchester Metropolitan University
Manchester, UK

ISSN 2199-7373 ISSN 2199-7381 (electronic)
World Sustainability Series
ISBN 978-3-030-69283-4 ISBN 978-3-030-69284-1 (eBook)
https://doi.org/10.1007/978-3-030-69284-1

This Springer imprint is published by the registered company Springer Nature Switzerland AG
The registered company address is: Gewerbestrasse 11, 6330 Cham, Switzerland

Preface

The COVID-19 pandemic has led to severe human suffering, and to substantial damages to economies round the globe, affecting both, rich countries and developing ones. The aftermath of the epidemic is also expected to be felt for sometime. This will also include a wide range of impacts in the ways sustainable development is perceived and how the principles of sustainability are practiced.

There is now a pressing need to generate new literature on the connections between COVID-19 and sustainability. This is so for two main reasons. Firstly, the world crisis triggered by COVID-19 has severely damaged the world economy, worsening poverty, causing hardships, and endangering livelihoods. Together, these impacts may negatively influence the implementation of sustainable development as a whole and of the UN Sustainable Development Goals in particular. These potential and expected impacts need to be better understood and quantified, hence providing a support basis for future recovery efforts.

Secondly, the lockdowns caused by COVID-19 have also been having a severe impact on teaching and research, especially—and not only—on matters related to sustainability. This may also open new opportunities (e.g. less travel, more internet-based learning), which should be explored further, especially in the case of future pandemics, a scenario which cannot be excluded.

This book *COVID-19: Paving the Way for a More Sustainable World* is a concrete effort to better understand the current and future impacts of the pandemic. It intends to gather and disseminate opinions viewpoints, studies, forecasts, and practical projects which illustrate the various pathways sustainability research and practice may follow in future, as the world recovers from the COVID-19 pandemic and prepares itself to the possibilities of having to cope with similar crisis. The book is structured around two main parts:

1. *COVID-19 and Its Societal Impacts*, describing a number of implications of the pandemic to various sectors, with their respective societal implications.
2. *COVID-19 and Sustainability Learning*, outlining the links between the pandemic and teaching and learning on matters related to sustainable development.

I want to thank all authors and reviewers for their contributions. I hope that the body of information and knowledge gathered on this volume will provide a timely support towards a greater understanding of the influences of COVID-19 to the cause of sustainable development and will help the on-going global efforts towards a recover from the pandemic.

Hamburg, Germany
Summer 2021

Walter Leal Filho

Introduction: COVID-19 and Its Influences on Sustainable Development

Abstract This introduction outlines the connections between COVID-19 and sustainable development and describes the influences of the pandemic on the future of sustainable development.

Keywords Sustainability, COVID-19, Goals, Support, Transformation

Introduction

As of 2015, many countries have made efforts to achieve the United Nations Sustainable Development Goals (SDGs). However, many factors have contributed to slowing down the progress in reaching these goals, which have caused several nations to fall behind when attempting to reach targets. The outbreak of the COVID-19 pandemic has caused countries worldwide to feel the financial burden brought about by the pandemic. This places further stress on finances allocated for the achievement of the SDGs, thus hindering the process of meeting the set targets by 2030 (Naidoo and Fisher 2020).

The most urgent SDGs that are affected by the COVID-19 pandemic include SDGs 1 (eradicating poverty), 2 (preventing hunger), 3 (attaining good health), 4 (education), 5 (addressing gender equality), 8 (ensuring employment), 10 (inequality reduction), and 16 (peace and justice). Therefore, it can be deduced that the effects of the pandemic are not only financial, but also include many social impacts Naidoo and Fisher 2020).

Apart from the health sector, the major impacts observed are probably related to those on the economy. The pandemic has caused the suspension of various economic activities, resulting in widespread job losses and pushing many families below the poverty line (Jain et al. 2020). This further delays the achievement of SDG1, SDG2, and SDG8. Additionally, the suspension of travel has hindered international supply chains, causing a reduction in the GDP of many countries (Gruszczynski 2020). The need for financial support has caused governments to redirect funds to aid people affected by the pandemic, thus reducing the funds that were intended for many sustainable development initiatives and projects (Barker and Russell 2020).

In addition to the financial impacts of the pandemic, the social impacts are equally pressing. Due to the lockdown imposed in many countries, several schools and universities were forced to close and to shift their learning to online platforms. Unfortunately, this form of e-learning is not accessible to all students due to socio-economic differences (Baticulon et al. 2020). This caused—or exacerbated—many inequalities during the pandemic between the students that were able to learn and have the means to do so and those that struggled. Therefore, it can be deduced that COVID-19 inhibited the progression of SDG4 and SDG10. Inequality was further increased between people who were able to retain their job (e.g. those working on essential services) and those that were unable to keep their jobs (non-essential workers) (Furceri et al. 2020).

The most obvious setback in sustainable development was in efforts towards the attainment of good health and well-being (SDG3). The COVID-19 pandemic placed significant strains on the healthcare systems of many countries, forcing governments to redirect finances and resources (Farrell et al. 2020; Khetrapal and Bhatia 2020). Additionally, not all civilians had equal access to medical care, and cases that were deemed non-urgent (such as elective surgeries or cancer cases) have been postponed (van de Haar et al. 2020). Many hospitals were forced to reprioritise their resources and medical personnel to ensure that they properly attended to COVID-19 cases (Farrell et al. 2020).

The imposition of social isolation has caused many mental health problems, which several healthcare systems are not properly equipped to handle. Such mental health problems were paralleled by increased incidences of gender-based violence in less developed countries, hindering SDG5 (Alon et al. 2020). This was followed by a reduction in women's economic empowerment and a reduced presence of women in leadership roles.

Overall, the inequalities, financial stress, and psychological impacts of the pandemic have caused a disruption in the peace and justice systems of several nations (SDG16). From the start of the pandemic to present, many countries have observed several protests that express opinions towards the pandemic. In some instances, the protests were peaceful, while in other cases, violence has occurred (Gibson et al. 2020; Forsyth 2020).

COVID-19 and Sustainability Efforts

With regard to sustainability and the overall efforts in this field, COVID-19 has created many problems. The negative impacts of the pandemic include the increases in the amounts of medical waste, the improper and hazardous discarding of disinfectants, and pollution created from disposable masks and gloves (PPE).

However, there are some positive environmental aspects. Following the onset of the pandemic, a significant reduction in greenhouse gases was observed due to decreased travel and subsequent lower use of fossil fuels. Furthermore, the lockdowns reduced levels of industrial activity and, to some extent, water pollution that previously occurred in great amounts. Reductions in the use of transportation (both air and land) have also caused decreases in noise pollution and improvements in air quality. Natural areas that were normally considered popular tourist destinations had the ability to be at least partly ecologically restored, due to closures during the pandemic. These positive impacts have contributed towards attaining environmental gains (Rume and Islam 2020), being advantageous to overall sustainability efforts.

Despite the major economic setbacks, the UN has called for countries to remain positive and develop new, innovative methods for the achievement of sustainable development and has encouraged countries to work together during these difficult times. Furthermore, the UN has committed to raising funds to combat COVID-19 in order to prevent any further repercussions of the pandemic such as famine (United Nations 2020 a, b).

In contrast to the delays, COVID-19 has caused in the achievement of the SDGs, it has been determined that prior progress of SDGs in developing countries may have reduced the effects of the pandemic in that particular country (Sakamoto et al. 2020).

A recent study described the need for post-pandemic strategies in developing countries to address the adversities caused by COVID-19 (Barbier and Burgess 2020). This is essential to reduce the influence that COVID-19 has had on achieving the SDGs created by the UN in 2015. New policies need to be designed to ensure that immediate action is taken to deal with the coronavirus as well as to allow for the progression of the 17 SDGs. In doing so, old policies need to be adjusted to accommodate for the changes required. Current and future policies should also highlight the need to uplift economies, deal with unemployment, ensure environmental sustainability, tackle poverty and hunger, and improve the overall health of people. Such regulations should be designed in a cost-effective manner with short-term and medium-term goals that are realistically achievable. Finally, more funds need to be reallocated in an appropriate manner to ensure that sustainability is pursued (Barbier and Burgess 2020). There is much at stake, since lack of action may undermine the progresses achieved to date.

Conclusions

As the various papers on this book show, the COVID-19 pandemic has impacted the world at different levels and has had an influence on the ways sustainability is perceived and practiced.

The coronavirus and the associated lockdown have put societies around the world in a difficult situation. Moving forward, we should also use the COVID-19 pandemic—and the experiences gathered from it—as an opportunity and a starting point for a global change towards greater environmental, economic, and social sustainability.

Moving forward, there is a perceived to add a sustainable development dimension to current and future recovery efforts. By doing so, we will be in a better position to prevent—and to handle—such crises in future.

Walter Leal Filho
European School of Sustainability Science
and Research
Hamburg University of Applied Sciences
Hamburg, Germany

References

Alon TM, Doepke M, Olmstead-Rumsey J, Tertilt M (2020) The impact of COVID-19 on gender equality. CRC TR 224 Discussion Paper Series, crctr224_2020_163 (Online ahead of print). https://doi.org/10.3386/w26947

Barbier EB, Burgess JC (2020) Sustainability and development after COVID-19. World Dev 135(1):105082. https://doi.org/10.1016/j.worlddev.2020.105082

Barker M, Russell J (2020) Feeding the food insecure in Britain: learning from the 2020 COVID-19 crisis. Food Secur 12(4):865–870. https://doi.org/10.1007/s12571-020-01080-5

Baticulon RE, Alberto NRI, Baron MBC, Mabulay REC, Rizada LGT, Sy JJ, Reyes JCB (2020) Barriers to online learning in the time of COVID-19: a national survey of medical students in the Philippines. medRxiv, Online ahead of print. https://doi.org/10.1101/2020.07.16.20155747

Farrell TW, Francis L, Brown T, Ferrante LE, Widera E, Rhodes R, Thothala N (2020) Rationing limited health care resources in the COVID-19 era and beyond: ethical considerations regarding older adults. J Am Geriatr Soc 68:1143–1149. https://doi.org/10.1111/jgs.16539

Forsyth DR (2020) Group-level resistance to health mandates during the COVID-19 pandemic: a groupthink approach. Group Dyn Theory Res Pract 24(3):139. http://dx.doi.org/10.1037/gdn0000132

Furceri D, Loungani P, Ostry JD, Pizzuto P (2020) Will Covid-19 affect inequality? evidence from past pandemics. Covid Econ 12(1):138–157. https://cepr.org.uk/sites/default/files/news/CovidEconomics12.pdf#page=143

Gibson AN, Chancellor RL, Cooke NA, Dahlen SP, Patin B, Shorish YL (2020) Struggling to breathe: COVID-19, protest and the LIS response. Equality Divers Incl Int J, Online ahead of print. https://doi.org/10.1108/EDI-07-2020-0178

Gruszczynski L (2020) The COVID-19 pandemic and international trade: temporary turbulence or paradigm shift? Eur J Risk Regul 11(2):337–342. https://doi.org/10.1017/err.2020.29

Jain R, Budlender J, Zizzamia R, Bassier I (2020) The labor market and poverty impacts of Covid-19 in South Africa. Southern Africa Labour Dev Res Unit, Online ahead of print. http://hdl.handle.net/11090/980

Khetrapal S, Bhatia R (2020) Impact of COVID-19 pandemic on health system and sustainable development goal 3. Indian J Med Res 151(5):395. https://doi.org/10.4103/ijmr.IJMR_1920_20

Naidoo R, Fisher B (2020) Reset sustainable development goals for a pandemic world. Nature, 583:198–201. https://www.nature.com/articles/d41586-020-01999-x

Rume T, Islam, SD-U (2020) Environmental effects of COVID-19 pandemic and potential strategies of sustainability. Heliyon 6(9):e04965. https://doi.org/10.1016/j.heliyon.2020.e04965

Sakamoto M, Begum S, Ahmed T (2020) Vulnerabilities to COVID-19 in Bangladesh and a reconsideration of sustainable development goals. Sustainability 12(13):5296. https://doi.org/10.3390/su12135296

United-Nations (2020a) Funding the fight against COVID-19 in the world's poorest countries. Retrieved 8 November, 2020, from https://www.un.org/en/un-coronavirus-communications-team/funding-fight-against-covid-19-world%E2%80%99s-poorest-countries

United-Nations (2020b). UN launches COVID-19 plan that could 'defeat the virus and build a better world'. Retrieved 2020, from https://news.un.org/en/story/2020/03/1060702
van de Haar J, Hoes LR, Coles CE, Seamon K, Fröhling S, Jäger D, Bergh J (2020) Caring for patients with cancer in the COVID-19 era. Nat Med 26(5):665–671. https://doi.org/10.1038/s41591-020-0874-8

Contents

COVID-19 and Its Societal Impacts

Sustainable Crises Management in Education During COVID-19 3
Dzintra Iliško, Madhavi Venkatesan, and Elizabeth Price

COVID-19 and Corporate Social Responsibility: A Canadian Perspective 23
David Billedeau and Jeffrey Wilson

COVID-19 and Pandemic Risk: The Link to SDG 13, Climate Change and the Finance Context 43
Federica Doni and Lara Johannsdottir

A Post COVID19 Blueprint for Sustainable Tourism in Small Island Developing States 61
Denise Galvin, K. Elliott, M. Handijaya Dewantara, L. Lewis, J. P. Platroz, W. Solomon, R. Tello Soberanes, and R. West-Gerald

Exploring Communication Framing Methods that Link Changes in Air Pollution Exposure and COVID-19 to Promote Post-pandemic Sustainability Policy 79
Lisa Jordan and Albert Tang

Achieving the UN's SDG 6 as a Means to Control the COVID-19 Pandemics in Brazil 109
Thiago Leite Cruz and Erleyvaldo Bispo dos Santos

Socioeconomic Pathways Toward the Sustainable Development Goals (SDGs) in Brazil During and Post-COVID-19 Pandemic 125
Thauan Santos, Luan Santos, and Ana Carolina Angelo

Sustainable Development Goals and Women: An Initial Reflection on Domestic Violence in Times of a Pandemic Crisis 143
Sonia Regina da Cal Seixas, João Luiz de Moraes Hoefel, Micheli Kowalczuk Machado, Gianlucca Consoli, and Waldo Emerson de Souza Nascimento

COVID-19 and Sustainable Tourism ... 163
Wendy M. Purcell, O'Shannon Burns, and Allison Voss

Fostering Innovation and Intercultural Exchange During a Global Pandemic: Lessons Learned from a Virtual Design Thinking Challenge in Nepal ... 185
Jens M. Unger, Aman Bhattarai, Bettina Maisch, Johannes M. Luetz, and James Obuhuma

Labor Market Sustainability and Corporate Governance—COVID-19 a Game-Changer ... 211
Diana-Mihaela Țîrcă, Laura Mariana Cismaș, and Cornelia Dumitru

Public Financing and Management for a Sustainable Healthcare Sector: Some Lessons from the Covid-19 Pandemic ... 233
Stella Tsani, Elena Riza, Panagiota Tsiamagka, and Margarita Nassi

COVID-19 and Sustainability Learning

Innovative and Sustainable Research-Based Learning & Community Services During Lockdown by COVID-19 ... 257
Ika Dewi Ana, Cahyono Agus, Hatma Suryatmojo, Wirastuti Widyatmanti, Irwan Endrayanto Aluicius, Sri Suning Kusumawardani, Mustofa, Mirwan Ushada, Ririn Tri Nurhayati, Adhy Kurniawan, Irfan Dwidya Prijambada, and Ambar Kusumandari

Challenging the Plague of Indifference: COVID-19 and Posthumanistic Education for Sustainability ... 281
Alison Sammel and Peter Blaze Corcoran

The Impact of COVID-19 on the Progress towards Sustainable Development Goal 4 in the Early Years: A Rapid Review ... 295
Jane Spiteri

COVID 19: Causal Loop Diagramming (CLD) of Social-Ecological Interactions for Teaching Sustainable Development ... 311
Gavin Melles, Stefan Lodewyckx, and Hariharan Thangatur Sukumar

Environmental Education as a Tool to Improve Sustainability and Promote Global Health: Lessons from the COVID-19 to Avoid Other Pandemics ... 331
Thiago Leite Cruz, Pedro Israel Mota Pinto, and Anderson Hélio Juwer Ferreira

Relationships Matter. New Paths for Tourism Beyond COVID-19 Pandemic. An Exploratory Research from Italy ... 349
Concetta Ferrara, Sabrina Tomasi, Chiara Aleffi, Annapia Ferrara, Giovanna Bertella, Gigliola Paviotti, and Alessio Cavicchi

Managing the Impact of COVID-19 on the Education Plans and Activities of South African Schools 371
Bongani Sibusiso Mchunu, Annatoria Zanele Ngcobo, and Thamsanqa Thulani Bhengu

COVID-19 and Education for Sustainable Development Role: An Assessment of Latin American Schools 389
Vanessa Tibola da Rocha, Bárbara Fritzen, and Luciana Londero Brandli

Using COVID-19 to Teach Sustainability Futures Thinking 411
Courtney E. Quinn and Matthew Cohen

Re-thinking Sustainability Teaching 427
Qudsia Kalsoom and Ayesha Jalil Shah

Women and Higher Education in COVID-19 Times: Challenges and Possibilities for Sustainable Development' 443
Tayyaba Tamim

Evolving Schools in a Post-pandemic Context 465
Dustin Eirdosh and Susan Hanisch

COVID-19 and Its Societal Impacts

Sustainable Crises Management in Education During COVID-19

Dzintra Iliško, Madhavi Venkatesan, and Elizabeth Price

Abstract COVID-19 has been a global disruptor. In many cases due to uncertainty and limited information, the pandemic has created chaos across business and private activity; education is no exception. The reality of the limited flexibility of existing systems of teaching was immediately acknowledged by the closure of physical facilities and the subsequent migration to online teaching. However, in spite of technologies to enable remote participation, it has become apparent that the capacity for learning in an autonomous environment is not homogeneous. From this perspective the pandemic created an opportunity to assess education infrastructure in real time as part of on-going crisis management. This paper explores aspects of crisis management in education as precipitated by the COVID-19 pandemic in Latvia and relates the mitigation and adaptation policies subsequently adopted to a holistic sustainability agenda for education. The research method employed in this study consists of semi structured interviews with the management team members from eight educational institutions. The interviews focus on the crisis management strategies and approaches implemented in their schools during the COVID-19 pandemic. The discussion concludes with a connection between crisis management and educational sustainability and sets an agenda for future research.

Keywords Sustainability · Crisis management · Education · COVID-19 · Complex adaptive systems

D. Iliško (✉)
Daugavpils University, Jātnieku 81-16, Daugavpils 5410, Latvia
e-mail: dzintra.ilisko@du.lv

M. Venkatesan
Department of Economics, Northeastern University, 360 Huntington Avenue, Boston, MA 02115, USA
e-mail: m.venkatesan@northeastern.edu

E. Price
Department of Natural Sciences, Manchester Metropolitan University, Chester Street, Manchester M1 5GD, UK
e-mail: e.price@mmu.ac.uk

W. Leal Filho (ed.), *COVID-19: Paving the Way for a More Sustainable World*,
World Sustainability Series, https://doi.org/10.1007/978-3-030-69284-1_1

1 Introduction

COVID-19 ushered in a new stage of globalization, pandemic. The pandemic was first recognized in March 2020; though there is now evidence to suggest that the roots of the contagion may have been in November 2019 (Bryner 2020). It was quickly understood, through the Italian experience, that facilities for the treatment of the disease along with the isolation of patients being treated, created inhumane conditions. Physicians and health care workers were forced to make decisions on who would be treated and who would not, given simply the lack of capacity for care and treatment at hospital facilities (Beall 2020). Further, the ability for production to mobilize to meet the demand for ventilation systems that were ultimately the treatment vehicle for the disease was severely limited (Kliff et al. 2020). In spite of technology, there was no capacity to meet unanticipated needs with the urgency required. The ongoing challenge of the disease has spread, while the survival rate and the lasting impacts on survivors of the disease has prompted a need for crisis management across society and its institutions (OECD 2020). The present lack of vaccine has provided an opportunity for pause and evaluation with respect to the relationship between the human system and the natural system.

However, it is important to highlight that the COVID-19 crisis did not happen in isolation, indeed it is consistent with a global and systemic behavioural relationship with our global commons. It fundamentally relates to the human relationship with the planet and the assignment by the former to the latter as being a "resource" to use for utilitarian gratification without responsibility for stewardship (Kalfagianni et al. 2020). COVID-19 represents the spread of disease from animals to humans (Campbell et al. 2020) and reflects the quality of care given to animal "resources"; the focus on market value; and fundamentally the treatment of externalities as market failures rather than moral ones. From this perspective it is directly relatable to Climate Change, inequalities, and demographics.

Given the holistic nature of the pandemic, on a global scale, all social, professional and government institutions were impacted and the majority of economic activity either ceased or was assigned a remote status, melding work and homelife for those fortunate enough to not be penalized by loss of employment.[1] Among the institutions affected by COVID-19, were schools. Across the globe, students were displaced as a result of COVID-19, heightening pre-existing inequities both between countries, as students in more developed countries retained access to education through remote channels, and within countries, as students with financial means were more likely to possess the technology to access educational programming and schooling remotely. From an aggregate perspective, the United Nations announced that the "COVID-19 pandemic has led to the largest disruption of education ever" (Guterres 2020).

[1]The level of employment loss varies across countries and in relation to the strictness of social distancing recommendations/regulations. Though not discussed in this paper, the loss of employment during this period also highlights the dependency of a given economy on non-essential institutions and correspondingly the reduction of the human footprint on the planet with the cessation of these activities.

In mid-July 2020, schools were closed in more than 160 countries, affecting over 1 billion students with at least 40 million children missing their critical pre-school year (Guterres 2020). School closures reflected mitigation of the spread of the disease, while arguably, the inability to seamlessly shift teaching to remote, autonomous learning surfaced a failure in managing the crisis. However, this failure has also provided an opportunity.

The Lebanese thinker Taleb (2014) in his book "*Antifragile: Things That Gain from Disorder*" wrote, that some things benefit from shocks and change—the elastic remains the same under the pressure, but the fragile becomes more durable. Taleb (2014) proposes that chaos and disorder provide the conditions to thrive and flourish. Considering his perception, to some degree, the educational system may benefit as a result of the present crisis. However, much of the challenge and opportunity is tied to crisis management.

As would be expected, COVID-19 prompted an increased focus on crisis management (Kouzmin 2008; Boin 2004) However, existing crisis management research has failed to provide a comprehensive interdisciplinary framework for action. Instead, the research has provided guideposts to integrate into a holistic agenda for crisis management. For example, Mikušova and Horvathov (2019) describe organizational crisis as a process of dealing with disruptive and unexpected events that threaten the functionality of organizations. Civelek et al. (2016) attribute the opportunity in crisis management as stepping outside what is comfortable. Fragouli and Ibidapo (2015) see crisis as an opportunity for growth. Kovoor-Misra (2019) addresses the complexity of crisis and role of organizational resilience in promoting individual action to be catalysts of transformative organizational change. Filho et al. (2020) affirms that every individual despite their position in an organization needs to take agency and become an initiator of change. The authors maintain that crisis needs to be perceived as a push to improve, to innovate, and to develop innovative solutions to an existing problem; and as an opportunity to learn and grow, and to build one's capacity to handle the crisis. Any organization needs to be seen as an open system that has interdependent relationships with the external environment. The external environment includes political and economic systems, and the risks they can cause for organizations. Organizations are also vulnerable to external threats, as in the case of COVID-19 (Winthrop 2020). Dedeurwaerdere (2013) argues that educational change needs to be incremental and supported by policy and ongoing training of all actors involved to incorporate unlearning of habitual ways of doing things. Ultimately, the goal of a crisis management outcome as a result of COVID-19 will be a transdisciplinary and holistic model for thinking and teaching.

In the sections to follow, we highlight crisis management actions that have been taken in Latvia specific to the education sector and COVID-19. We address how the Latvian crisis management process provides a country-level example of proactive engagement to promote sector-based sustainability objectives. The next section will address theories of crisis management; it is followed with a discussion of competencies required of educational leaders. We then provide a case study approach with respect to crisis management where Latvian educational leaders were surveyed on their perceptions of the opportunities and challenges resulting from COVID-19. Our discussion of the outcome of the interviews includes an assessment of how crisis management in education parallels needed action to ensure sustainability in the sector. We conclude with a research agenda for next steps.

2 Crisis Management in Education

According to a *World Development Report* (WDR) (2018), learning needs to be transformed from its narrow focus as a transmission of knowledge and skills to a process where learners are autonomous actors. In order to enable this, basic access issues need to be resolved. Most significantly, education needs to be inclusive and accessible to all children. This requires wider transformations in society and an adaptive framework with the following features: integrated (i.e. stakeholders, operations, communications), nonlinear, multi-dimensional, co-evolutionary, and nondeterministic (Anand et al. 2010; Fabac 2010; Prewitt et al. 2012). Non-linear systems have been the subject of discussion across disciplines. In Table 1 we reference five theories that have an application to crisis management in education.

Non-linear systems approach
Chaos theory relates to unpredictable courses of events. A chaotic system does not settle into a predictable pattern due to its nonlinear processes. Instead, it requires preparation for the unexpected. Applied to education chaos theory promotes adaptability and flexibility to deal with issues beyond human control (Resnicow and Page

Table 1 Non-linear systems theories

Theory	Description	Reference
Chaos theory	Any system is sensitive to pressures and changes from outside that might cause transformations in organizations	Van de Ven and Poole (1995)
Fractal theory	Similar patterns of behaviours and values on individuals' and group level	Zimmerman and Hurst (1990)
Autopoiesis	Feature of a system to self-generate and self-replicate	Drazin and Sandelands (1992), Zeleny (2005)
Self-organized criticality	A dynamic feature of an organization that supports introducing innovations for a better functioning of an organization	Stanley et al. (1996), Hoffmann and Payton (2018)
Synergetic	The feature that describes a synergetic behaviour of people working towards one goal	Zuijderhoudt (1990)

2008). Fractal theory emerges out of the Chaos theory and is built on principles of openness and networking that foster interdisciplinarity. Fractal theory refers to fractals as images of dynamic systems and offers insights for a new kind of pedagogy that requires one to rethink and to reflect how we teach students. Fractals have been used to model the evolutionary dynamics of complex systems that can be applied to the situation in education during remote teaching (Kokhanenko 2002). Flexible curricula can be described by a fractal metaphor dealing with complexity from a holistic and ecological perspective. Autopoiesis as a metaphorical perspective is rooted in nature studies and describes organizational phenomena of the twenty first century according to which organizations are seen as networks of interactions, reactions, and processes. Autopoiesis is related to living and autonomous systems that have their internal coherence resulting from the interconnectedness of system's inputs and outputs (Varela 1984). Autopoietic systems are described as operationally closed but at the same time maintain intense interactions with the environment and their ability to respond to bifurcations by maintaining autonomy (Zeleny 2005). Self-organized criticality refers to a dynamic feature of an organization that supports introducing innovations for its better functioning. The concept describes multiple interacting components in complex systems such as schools, universities or any other organizations. Self-organized criticality has been referenced as the special state between order and chaos in which schools were functioning during COVID-19 (Hoffmann and Payton 2018). Synergetic is a feature that describes a synergetic behavior of people working towards one goal in the organization.

In combination, these theories support the development of an adaptive system, which recognizes and builds on synergies between stakeholders. This type of framework will allow for a flexible model that is adaptive to chaos, thus allowing stressors from outside to encourage transformation and innovative practice within the organization. Adaptation would then also impact the school's culture, power distribution, and control system. Essentially, any transformation in the organization would ripple through all the functional areas of the entity (Uhl-Bien et al. 2007). Laszlo (1995) references quantum physics to describe a new form of leadership that is holistic, unifying, interconnecting, creative and transforming. Plsek and Greenhalgh (2001) describe a complex and adaptive system where groups of individual agents are free to act in a way that is not totally predictable but whose actions are aligned to one another. Cilliers (1998) provides a description of complex adaptive systems that includes the following: interaction of a large number of elements or stakeholders in a rich and nonlinear way with numerous enhancing and stimulating feedback loops that function in a constant process of change in response to a flow of energy from outside. The system is complex itself due to the patters of interactions between those elements. Goldstein (2000) argues that self-organization and emergence of a new order are significant characteristics of complex adaptive systems. The key components of complex systems are processes that are in constant co-evolution, adaptation on multiple levels, dynamic feedback, and mutually causal flows of knowledge across boundaries.

Crisis management competencies
Scharmer and Kaufer (2013) puts high demands on a leader to develop new competences during the major shifts in human consciousness fostered by crisis. From this perspective leaders need to have courage to be open-minded to new and unknown opportunities and transformations in education. According to this leadership model, education needs to be holistic and to engage multiple stakeholders in educating citizens to be able to solve the complex issues of a contemporary community. Schooling leadership requires a clear vision of available resources but also stakeholder engagement to motivate teachers and enable a smooth transition to a new model which can be characterized by its flexibility and adaptability to information delivery. Schools need to be seen as a complex adaptive system, made up of interdependent social partners situated in a dynamic world, as part of a larger complex ecosystem. As a fundamental quality, school leadership needs to have "a sense of evolutionary possibilities (and impossibilities) of the present moment by allowing safe experimentation" (Snowden 2011, 5).

Crisis management requires school management to adopt a new set of competencies to lead schools in the transition to a remote digital environment. Specifically, school leaders need to address and integrate sustainability competencies into their actions (Wiek et al. 2011a; Barth et al. 2007). Sustainability competencies comprise a set of knowledge, skills and attitudes that are needed for dealing with the real-world problems/issues. The key competencies include systems thinking, future thinking, strategic thinking, and value competence (Wiek et al. 2011b).

In times of crisis, it is particularly important to adopt systems thinking and to adopt a holistic view of the world. In engaging multiple stakeholders, and diverse partners an enterprise, in this case a school system, can develop resilience with respect to outside stressors. Anticipatory (future thinking) competence is necessary for building a new strategy in times of crisis, by developing a viable scenario for the efficient functioning of the school. Strategic competence is necessary to plan actions towards transformative change. Integrated problem-solving competence is essential for solving problems and offering viable solutions in developing, testing and implementing diverse solutions. Lambrechts et al. (2013) refer to such competencies as emotional intelligence, systems thinking and action taking (Fig. 1).

3 Latvian Education Experience

According to Latvia's prime minister, Krišjānis Kariņš, Latvia was quite successful in dealing with the spread of COVID-19 due to early steps of social distancing undertaken by the government (Latvian Public Broadcasting 2020). Latvia declared a state of emergency over COVID-19 on March 13, 2020 (Latvian Public Broadcasting 2020) and schools were closed immediately. On March 19, XinhuaNet (2020)

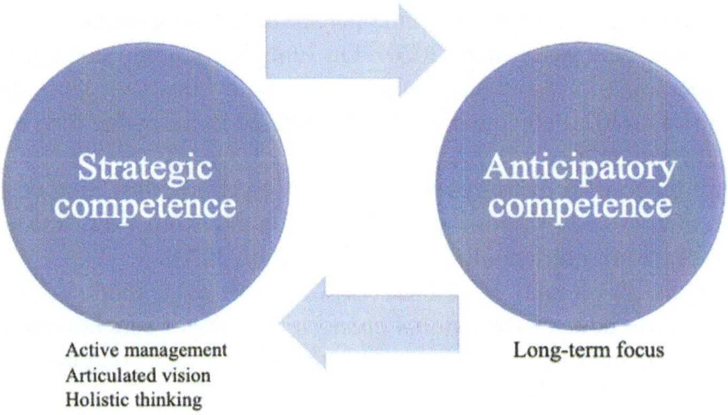

Fig. 1 Core competencies in crisis management

reported that "not all students in the Baltic country have the equipment necessary for distance learning" (XinhuaNet 2020). The Education and Science Ministry of Latvia stated that nearly 3 percent of students were without access to "personal computers, tablets, smartphones or internet connections available at home" and made an appeal to telecommunications providers (XinhuaNet 2020). Latvia Mobile Telephone and Bite Latvija pledged 5000 units and free calls and Internet access, "while other telecommunications companies also expressed willingness to help out" (XinhuaNet 2020). In May, the Ministry of Education and Science released a statement highlighting distance learning actions:

> There are several tools which are used during the period of distance learning. For example, E-klase (E-class) is the most widely used platform to access all important information, including guidelines for teachers for providing distance learning. Other platforms widely used are Uzdevumi.lv and Soma.lv, allowing to exercise online and to receive immediate feedback. Educational TV channel Tava Klase (Your Class) for pupils has been created in two weeks and has been on air since 6 April. It supports pupils, parents and teachers in the implementation of distance learning for 1st to 6th and upper grade pupils. More than 70 teachers are involved in the project, and a volunteering Parents Group is taking part in the project as well. At the higher education level and in vocational education and training, distance learning is provided in a decentralized way. Institutions provide such remote learning opportunities to their students as live online classes, seminars, discussion forums, and consultations via Skype, Webex, MS Teams, Zoom and other online tools (Ministry of Education and Science 2020).

The Ministry of Education and Science also affirmed its commitment to education by noting its monitoring of remote learning and establishment of a continuous improvement framework:

> In order to assess the situation more quickly and effectively, as well as to plan and implement improvements in the distance learning process, the Ministry together with the educational technology company Edurio is constantly carrying out surveys on the implementation of distance learning in the emergency situation. Over 60,000 respondents — teachers, pupils, parents, and school headmasters (99.6% of schools) — have been surveyed so far (Ministry of Education and Science 2020).

At the beginning of the pandemic, the percentage of infected people was around 2.7–2.9% (Rapid risk assessment 2020). On June 2, 2020, the number of positive tests reached a low of 1% and has remained at or below this level since then (Rapid risk assessment 2020). People were quite disciplined in following the requirements set by the government.

On June 9, emergency measures were softened. As of July, the China-CEE Institute reported in their weekly briefing that Latvian schools were being readied for the resumption of traditional schooling in the fall (Linde 2020).

The previous [pandemic] experience shows that the education system of Latvia has the capacity to adjust to the global changes in terms of education since overall students, parents, and teachers had valued positively the new conditions of distance learning. As the overall epidemiological situation in Latvia has improved it is planned that the upcoming schoolyear will happen on-site. However, there are still epidemiological risks that require schools to follow safety precautions. To ensure a safe on-site education process the educational institutions will have to establish new internal procedures in schools such as organizing flow of students and teachers, ensuring personal hygiene, monitoring of individual health conditions, particularly for persons at risk, frequent room cleaning algorithm, ventilation of rooms and others. Additionally, in accordance to the decision of the supervisor of the educational institution there will be an option for the supervisor to decide whether or not to ensure remote educational activities for students from 7th up to 12th grade, vocational institution students, and higher education students to promote distancing depending on the epidemiological situation in the country. To ensure not only epidemiological safety for students but to also improve the quality of education in Latvia the Ministry of Education and Science is developing a upgraded monitoring system that will look at the quality of education in a broader context, focusing on aspects such as quality learning, an inclusive environment, good governance and compliance with objectives (Linde 2020).

The Latvian response to COVID-19, specifically as it relates to the public education sector remains ongoing; however, the actions taken to date and the framework adopted provide a real-time perspective of crisis management that has both universal commonality and application.

The main issue that schools encountered as a result of sudden closure was a lack of a unified digital platform. Further, teachers had insufficient competencies to work with new technologies and experienced difficulties in designing online materials in a very compressed time frame. Additionally, teachers experienced difficulty in converting to an autonomous environment specific to organizing a self-directed learning process (Petere 2016). Learners encountered difficulties in organizing their learning routine from home since they were used to teacher facilitated instruction. Compounding this issue was the inconsistent accessibility of technology among students, which exacerbated pre-existing economic inequities, as students from disadvantaged households were most likely not to have the technology required for remote learning (Muižniece 2020). The private sector was responsive to these needs and as noted pledged resources to assist students.

Teachers faced the need to quickly adapt to a remote learning environment. Prior to COVID-19, given national educational reform through the *School*2030 initiative (EC 2019), they were required by the school administration to experiment with new tools to engage pupils in the learning process. However, as a result of COVID-19 and remote learning, there was an additional existential catalyst that sped up the reform processes in education to promote a student-centered and competency-based learning functionality.

The challenge for the administration of schools was how to lead schools in the uncertain and nonlinear situation. They had to learn to think holistically by developing a system view on issues.

4 Research Methodology

The methodology employed in this study consists of semi-structured interviews. The interviews were conducted with eight selected school leaders and included questions related to their strategies and approaches to crisis management in their schools as a result of COVID-19. The questions were informed by the literature. All selected leaders represented middle size city schools, all of which had in place the educational reforms mandated by the state, which pre-existed the COVID-19 pandemic by during the last three years. Selected school leaders were a representative sample for this study since they declared themselves as successful in leading the school during pandemic. The selected leaders were asked about the challenges and difficulties in dealing with the educational change in their institutions and about future challenges, post-pandemic. Ethical rules were observed while conducting interviews, such as confidentiality, privacy, informed consent, and avoidance of harm to the participants (Orb et al. 2001). Themes from the interviews emerged based on the emphasis the informants placed on an issue as prompted by the research questions (Bazeley 2009). The selected leaders were asked about the main difficulties and challenges encountered during the COVID-19 crisis in their schools along with the strategies to mitigate or adapt to these issues. They were questioned on sustainability competencies. The interviews were conducted online due to restrictions imposed by COVID-19. Though the study does provide insight into crisis management in schools, it is limited by small sample bias and the context defining "leader" selection, which may make findings not necessarily applicable across the Latvian education system.

5 Research Findings

All schools are unique. Perhaps the differences may be most easily observed in times of crisis as each school's engagement will differ based on resource access, facilities, teachers, students and other attributes and systems that define it. Table 2 summarizes the main challenges, difficulties and future prospects as reported by the

Table 2 Issues raised by the heads of the schools while overcoming the educational crisis caused by the COVID-19

Code	Description	Examples from the interviews
Opportunities caused by COVID-19	Selected school leaders discuss the need of a joint vision for the school and collaboration among all stakeholders and shared cases of good practice of creating interdisciplinary links in designing homework	"The main focus in our school was placed on establishing a joint vision among all stakeholders. Successful implementation of a remote mode of learning in our school occurred when all parties involved come to the agreement. Equilibrium and harmony then become equated with the success" "The pandemic made many teachers to cooperate in designing the tasks for their students with the respect of transdisciplinary that reduced the gap between subjects and allowed pupils to see a more holistic picture of a subject matter"
Challenges for the schools in turbulent times	Selected school leaders reflect on awareness of issues related to the technical equipment of schools and ensuring equal access to technologies	"By supporting digitalization of education, we need to further ensure equal opportunities for all without creating risks for pupils only due to lack of devices, poor internet connection, socio-economic status of parents or insufficient pedagogical tool for a digital mode of teaching"
Future prospects	Selected school leaders discuss the need for adopting flexible crisis management strategies in schools and discuss the functionality of schools as learning places	"There is a need to build a dialogue with the public and generate new ideas while waiting for the virus to end up and carry on in an old way" "No one can predict what awaits us in the future. There is always a possibility that schools will have to work remotely again, but no one knows when it happens. Therefore, schools need to adapt flexible planning strategies and practice and be ready to work in an unknown situation" "Maybe it will be the first push for the world to go even further, stopping to think about the school as a building with certain boundaries, but the school as a place of learning"

Table 2 (continued)

Code	Description	Examples from the interviews
Sustainability competencies for adaptive crisis management	Selected school leaders reflect on the competencies needed for the management of the crisis	"This was a completely new and unexpected situation when we had to design a strategy for dealing with a new situation that brings along sustainable changes" "Our management team needed to apply such sustainability competencies in the situation of pandemic as: complex problem solving, strategic thinking, future thinking competence, communicative competence, a competence to apply ICT technologies into teaching" "The most desirable competencies in time of crises are managing complexity, communicating vision to the staff members, challenging co-workers, introducing innovate solutions and thinking in a long term"

selected school leaders participating in this study. The following sections provide more granular detail with respect to the survey question responses.

5.1 Challenges for Schools in Turbulent Times of Pandemic

When questioned on the current challenges schools are facing during COVID-19, the selected school leaders mentioned the necessity and challenge to apply adaptive management strategies in a completely novel situation for them: involving multiple stakeholders, meeting the needs of children and their families, providing necessary feedback, making interconnectivity, adaptability, and establishing flexible management frameworks. One of the challenges mentioned was the transition from teaching in a physical environment to teaching in a completely digital environment. Participants in general commented that COVID-19 presented a good lesson for future challenges. As one of the school leader's reported:

> There is a need to build a dialogue with the public and generate new ideas while waiting for the virus to end up and carry on in an old way. No one can predict what awaits us in the future. There is always a possibility that schools will have to work remotely again, but on one knows when it happens. Therefore, schools need to adapt flexible planning strategies and practice.

Another participant commented that current experience shows that everyone is tired of intensive remote teaching and learning. One of the selected school leaders mentioned the possibility of introducing a blended mode of learning in schools to provide more autonomy for the teachers in future:

> There is a need to trust teachers and to organize the learning process flexibly, but the state and municipalities also need to be supportive by providing every child and teacher with the opportunities to work remotely.

A majority of teachers worked in the situation with existing patterns of behaviour, but the situation also taught the teachers to adopt more autonomy and independence in their teaching and to build trust in students' autonomous learning.

In order to meet the needs of a new generation the schools fostered a digitalization of the learning process with the new digital teaching tools and materials:

> Although many attempts have been made in the past to set up digital platforms in schools, the schools have failed due to various circumstances, and now, in the situation of crisis, appropriate decisions and immediate action should be taken.

The schools were forced to design relevant online materials in a very compressed time frame. Another challenge for schools was to create a sense of community among teachers and families by leading a remote learning process. Addressing common problems and sharing cases of good practice among teachers contributed to building a community of care and support. The selected school leaders were challenged to apply adaptive crisis management strategies and to implement required changes by creating a new strategic vision, communicating a vision to the staff members, empowering teachers to act upon this vision, thereby managing change.

5.2 *Difficulties the Schools Are Facing and Mechanisms of Overcoming Them*

The most topical issues discussed by both the government and the participants in this study are managing remote learning process and preparing pupils for the final exams in the remote mode. The concerns expressed by the selected school leaders include the following: ability of schools to meet the challenges of education for all, to entrust more autonomy to the learners in organizing self-directed learning process of pupils and developing teachers' competency of work in a digital environment. With respect to exams, survey participants believe that students should be allowed to attend classes in small groups and schools need to ensure this possibility by applying all social distancing precautions.

All interviewed participants discussed challenges in meeting the needs of all pupils, including those from disadvantaged families. For example:

> By supporting digitalization of educational process, we will be able to further ensure equal opportunities for education for all without creating risks of lack of devices, poor internet connection, socio-economic status of parents or insufficient pedagogical tool for a technological mode of teaching.

One of the difficulties for the teachers was to step out of their comfort zone, their habitual ways of doing and thinking. As one of the school leaders commented:

> From the beginning all educators experienced a greater or lesser shock, since the change in the environment and conditions came very rapidly and unexpectedly. Teachers were forced to step out of their comfort zone, which was not easy for them as a routine person. There was a great lack of emotional connection with the student, they had to cope with everything.

This was quite a stressful time for the teachers and affected both their emotional well-being and endurance.

5.3 *Opportunities for Schools Created by Covid-19*

Almost all the participants referred to crisis as time of opportunity that requires teachers' competence to adapt to new circumstances by offering diverse solutions, by choosing a diverse range of methods in teaching and by offering a wide variety of solutions for different subjects and pupils. The heads of the schools suggested numerous opportunities for schools caused by COVID-19, such as a cooperation among teachers, interdisciplinarity in teaching, speeding the digitalization of the learning process, making a transition to a more student cantered, self-directed learning mode, placing more trust on students in managing their learning process, updating teachers' digital skills and literacy.

The crisis calls for strengthening teachers' leadership. As one of the heads of the schools commented:

This requires developing such qualities as decision making in a situation of crisis, reflection and shared governance. This also requires setting clear objectives and building trust among all parties involved.

The pandemic fostered the digitalization of educational systems and acquisition of new Information and Communication Technology (ICT) tools for teaching that are relevant to the digital generation. It clear that today's children are a generation that can handle digital devices almost from birth (Gabre 2018). Survey participants reflected that digitalization of education will ensure equal opportunities for education for all, as long as the risk of accessibility is eliminated.

With respect to teachers' readiness in dealing with the crisis, one of the leaders commented:

It was easier to adapt to a new situation for those teachers who have had ICT before crisis. There were also those teachers who hoped that they would not experience digitalization in their lifetime, and that children and young people are likely to be satisfied with the educational process they offer so far. This situation of crisis was more complicated for them since they had to learn how to work in a new situation in a compressed time.

The selected school leaders admitted that the crisis reinforced the educational reforms to place higher emphasis on interdisciplinarity and organization of learning in a more holistic mode. During the first week of the "remote mode" of teaching, schools announced integrated interdisciplinary education design. This was supported by the Ministry of Education and Science, which issued new guidelines to support teachers' work in a remote learning mode by emphasizing the importance of interdisciplinary learning and cooperation among teachers in planning a learning process.

As one of the leaders commented on transfer to a more interdisciplinary learning mode:

In the first week of a remote teaching mode, my school gave up teaching separate subjects. Instead, teachers cooperated closely in creating interdisciplinary tasks for their students. Actually, remote teaching fostered a transition towards interdisciplinarity in teaching.

The Latvian Ministry of Education and Science has also emphasized the importance of cross-curricular links and cooperation among teachers in designing learning tasks, which is also one of the basic building blocks of *School 2030* (School 2020). This has fostered self-directed learning by increasing pupils' motivation to become more independent, self-disciplined, self-confident and purposeful.

One of the advantages of crisis was a transfer to a self-directed learning mode.

Self-directed learning helped the students to achieve better results, to develop their motivation, to become more independent, self-disciplined, self-confident and purposeful.

The leaders of the schools admitted that not all students are ready to make a transition to self-paced learning, particularly those who were taught in a teacher directed learning mode.

Another opportunity for schools caused by COVID-19 was developing durable relationships with the main stakeholders. COVID-19 caused the schools to build closer partnerships with families in developing a joint framework for remote teaching by providing all necessary support and a feedback for the pupils while and enabling a greater flexibility of the processes to endure pupils' well-being. The schools developed durable working relationships with families that increased value for both parties. The management team of schools developed partnership with families, thus helping them to understand, accept and to embrace needed changes. As one of the participants commented:

> The main focus in our school was placed on establishing a joint vision among all stakeholders. Successful implementation of a remote mode of learning in the school occurred when all parties involved come to the agreement. Equilibrium and harmony then become equated with the success.

5.4 Sustainability Competencies for Crisis Management in the Complex Adaptive Framework

Among the most required competencies during COVID-19, all leaders have mentioned the following sustainability competencies: complex problem solving, strategic thinking, future thinking competence, communicative competence and a competence to apply Information and Communication Technologies (ICT) into teaching. One of the leaders admitted that:

> This was a completely new and unexpected situation when all leaders had to adopt a new adaptive strategy for dealing with a new situation.

Almost all participants referred to a socio-emotional competence by ensuring well-being of teachers, pupils and their parents who need lots of emotional support during the emergency period.

Almost all leaders referred to the wisdom of learning from local traditions and practices and keeping a fresh and open view of the future to come. Wise leadership requires place-based knowledge and a courage to build a more sustainable school with a sustainability vision and agenda that is inclusive to all by achieving cooperation and solidarity of all stakeholders (Leal Filho and Brandli 2016). Leaders are required to adopt systemic and reflective ways of thinking and doing and synergy between all actors involved. This requires leader's ability to examine strategically the situation and to act upon it. For the educational changes to be efficient, they need to be adaptive, incremental, and dialogical (Stead and Stead 2014). Transformational change requires fundamental efforts to shift consciousness based on new values and ways of doing things. Therefore, strategic management becomes a spiralling and integrative process in a constantly changing co-evolving educational context (Stead and Stead 2014).

6 Conclusion

The COVID-19 pandemic caused disruption in education, depriving students of social interaction that has impacted their mental well-being (Nicola et al. 2020). It has also influenced the routine of families who were forced to adopt roles as coaches and mediators between school and home directly from their households (Huang et al. 2020). A smooth transition was possible only due to planned strategic efforts undertaken by the administration of schools in cooperation with motivated teachers who learned new ways of being and teaching. The crisis management literature asserts that crisis may have a damaging effect on productivity of an organization and therefore requires immediate attention and decisive actions (Civelek et al. 2016). It requires non-linear reaction and decision making by taking into account the contribution of all stakeholders involved. COVID19 as a global crisis has prompted crisis management in education on the global scale.

Among the challenges caused by crisis, which the selected school leaders interviewed in this study have mentioned, is how to make a smooth transfer from teaching in a physical environment to teaching in a completely digital environment. A transition "to a new normal," from the traditional classroom setting to a new digital environment reality also creates a need to guarantee a quality education by granting equal access to all parties involved (Tria 2020, p. 1) and to reinforce teachers' digital literacy.

Institutionally, schools had to overcome the following difficulties: unequal access to technologies for both teachers and the students, siloed approaches towards teaching, negotiating over the unified digital platform within one institution, limited access to the Internet for some students, and a lack of technological skills of some teachers and others. Among the psychological factors caused by the isolation resulting from physical school closures and restricted communication among students and teachers, was confusion, loneliness, fear of infection and boredom. As a result of COVID-19, survey participants noted that traditional classroom education needs to be modified to meet the ability and needs of contemporary learners. This highlighted the need to incorporate a joint digital platform for learning and readiness for blended forms of learning.

However, as reported by the survey participants, the crisis presented a number of opportunities for teachers. Among them, allowing students greater autonomy and agency as well as responsibility for their learning. Teachers had to adopt smart strategies to facilitate the learning process and allow their students to undertake autonomy over their learning. The remote learning mode strengthened a cooperation among teachers, interdisciplinarity in teaching, speeded up the digitalization of the learning process and made a smoother transition to a more student-centred, self-directed learning mode.

As reported by the leaders of schools, as a result of COVID-19, the most desirable competencies included the ability to actively manage complexity, communicate a strategic vision to staff members, challenge co-workers to think holistically, introduce innovative solutions and adopt anticipatory competence while maintaining a long-term focus.

From the initial recognition of the COVID-19 pandemic, the primary goal of the Latvian Ministry of Education and Science has been the safety of pupils and teachers and the maintenance of the educational process. The shift to online autonomous learning; the required rapid assimilation of institutions, teachers and learners, along with the commitment of the government and private sector to promote student learning in spite of the disruption of COVID-19 are factors that highlight the Latvian response to crisis management in education. However, given the scale of this study in conjunction with the work-in-progress of the transition to remote learning, there is benefit in monitoring the progress and success of online and future hybrid leaning adoption. Further, an understanding of the Latvian context, specific to cultural and national norms, as well as demographic and population attributes would add value to the application of this case study and future Latvia-based evaluations to other countries.

References

Anand M, Gonzalez A, Guichard F, Kolasa J, Parrott L (2010) Ecological systems as complex systems: challenges for an emerging science. Diversity 2:395–410

Barth M, Godemann J, Rieckmann M, Stoltenberg U (2007) Developing key competencies for sustainable development in higher education. Int J Sustain High Educ 8(4):416–430

Bazeley P (2009) Analysing qualitative data: More than 'identifying themes'. Malays J Qual Res 2(2):1–21

Beall A (2020, April 28) The heart-wrenching choice of who lives and dies. BBC. https://www.bbc.com/future/article/20200428-coronavirus-how-doctors-choose-who-lives-and-dies

Boin A (2004) Lessons from crisis research. Int Stud Rev 6(1):165–194

Bryner J (2020, March 14) 1st known case of coronavirus traced back to November in China. LiveScience. https://www.livescience.com/first-case-coronavirus-found.html

Campbell C, Yunnan Y, Park A (2020, July 23) Inside the Global Quest to Trace the Origins of COVID-19—and Predict Where It Will Go Next. Time. https://time.com/5870481/coronavirus-origins/CHED (2020). CHED Covid-19 Advisory No. 3. Retrieved from https://ched.gov.ph/wp-content/uploads/CHED-COVID-2019-Advisory-No.-3.pdf

Cilliers P (1998) Complexity and postmodernism: understanding complex systems. Routledge, London

Civelek ME, Cemberci M, Eralp NE (2016) The role of social media in crisis communication and crisis management. Int J Res Bus Soc Sci (2147-4478) 5(3):111–120. https://doi.org/10.20525/ijrbs.v5i3.279

Dedeurwaerdere T (2013) Transdisciplinary sustainability science at higher education institutions: Science policy tools for incremental institutional change. Sustainability 5:3783–3801

Drazin R, Sandelands L (1992) Autogenesis: a perspective on the process of organizing. Organ Sci 3:230–249

EC (2019) Izglītības un apmācības pārskats. Latvija. (Educational and teaching review. Latvia. Retrieved from https://ec.europa.eu/education/sites/education/files/document-library-docs/et-monitor-report-2019-latvia_lv.pdf

Fabac R (2010) Complexity in organizations and environment—adaptive changes and adaptive decision-making. Interdiscip Descr Complex Syst 8(1):34–48

Filho WL, Eustachio H, Caldana ACF, Will M, Salvia AL, Anholon RR, Platje J, Kovaleva M (2020) Sustainability leadership in higher education institutions: an overview of challenges. Sustainability 12:3761. https://doi.org/10.3390/su12093761

Fragouli E, Ibidapo B (2015) Leading in crisis: leading organizational change and business development. Int J Inf Bus Manage 7(3):71–90

Gabre A (2018) Zanda Rubene: Digitālie zīdaiņi mūsu vidū ir īstenība (Digital kids among is a reality). Retrieved from https://nra.lv/latvija/255323-zanda-rubene-digitalie-zidaini-musu-vidu-ir-isteniba.htm

Goldstein J (2000) Emergence: a construct amid a thicket of conceptual snares. Emergence 2(1):5–22

Guterres A (2020) Policy Brief: Education during COVID-19 and beyond. United Nations. Retrieved from https://www.un.org/sites/un2.un.org/files/sg_policy_brief_covid-19_and_education_august_2020.pdf

Hoffmann H, Payton DW (2018) Optimization by self-organized criticality. Sci Rep 8:2358. https://doi.org/10.1038/s41598-018-20275-7

Huang RH, Liu DJ, Amelina N, Yang JF, Zhuang RX, Chang TW, Cheng W (2020) Guidance on active learning at home during educational disruption: promoting student's self-regulation skills during COVID-19 outbreak. Smart Learning Institute of Beijing Normal University, Beijing

Kalfagianni A, Partzsch L, Beulting M (2020) Governance for global stewardship: can private certification move beyond commodification in fostering sustainability transformations? Agric Hum Values 37:65–81

Kliff S, Satariano A, Silver-Greenberg J, Kulish N (2020, March 26) There aren't enough ventilators to cope with the coronavirus. New York Times. https://www.nytimes.com/2020/03/18/business/coronavirus-ventilator-shortage.html

Kokhanenko IK (2002) Fractals in estimation of evolution of complex systems. Autom Remote Control 63:1255–1262

Kouzmin A (2008) Crisis management in crisis. Adm Theory Praxis 30(2):155–183

Kovoor-Misra S (2019) Crises management. Resilience and change. Sage, Thousand Oaks

Lambrechts W, Mulà I, Ceulemans K, Molderez I, Gaeremynck V (2013) The integration of competences for sustainable development in higher education: an analysis of bachelor programs in management. J Clean Prod 48:65–73

Laszlo E (1995) The interconnected universe: Conceptual foundations of a transdisciplinary unified theory. World Scientific, Singapore

Latvian Public Broadcasting (2020) Kariņš: first phase of COVID-19 response successful in Latvia. Retrieved from https://eng.lsm.lv/article/politics/politics/karins-first-phase-of-covid-19-response-successful-in-latvia.a363059/

Leal Filho W, Brandli L (2016) Engaging stakeholders for sustainable development. In: Leal Filho W, Brandli L (eds) Engaging stakeholders in education for sustainable development at university level. World Sustainability Series. Springer, Cham. https://doi.org/10.1007/978-3-319-26734-0_21

Linde N (2020, July) Latvia social briefing: Latvia's education system in times of Covid-19. China-CEE Institute. https://china-cee.eu/2020/07/29/latvia-social-briefing-latvias-education-system-in-times-of-covid-19/

Mikušova M, Horvathov P (2019) Prepared for a crisis? Basic elements of crisis management in an organisation. Econ Res 32(1):1844–1868

Ministry of Education and Science (2020, May 20) Situation caused by Covid-19 in Latvia. https://izm.gov.lv/en/highlights/3954-situation-caused-by-covid-19-in-latvia

Muižniece A (2020, April 15) Izglītība Covid 19 ēnā (Education in the shadow of COVID 19. *Ir*. Retrieved from https://ir.lv/2020/04/15/izglitiba-covid-19-ena/

Nicola M, Alsafi Z, Sohrabi C, Kerwan A, Al-Jabir A, Iosifidis C, Agha M, Agha R (2020) The socio-economic implications of the coronavirus pandemic (COVID-19): a review. Int J Surg 78:185–193

OECD (2020) Tackling Coronavirus (Covid 19). Contributing to a global effort. Retrived from http://www.oecd.org/coronavirus/en/

Orb A, Eisenhauer L, Wynaden D (2001) Ethics in qualitative research. J Nurs Scholarsh 33(1):93–96
Petere A (2016) Self-directed cross-curriculum teaching/learning process from the perspective of paradigm shift. In: Proceedings of the International Scientific Conference Society Integration Education, vol 2, pp 252–259, Rezekne, 27–28 May
Plsek P, Greenhalgh T (2001) The challenge of complexity in health care. BMJ 323:625–628
Prewitt JE, Weil R, McClure AQ (2012) A complex adaptive systems approach to strategic planning. Asian J Bus Manage Sci 1(11):94–99
Rapid Risk Assessment (2020) Coronavirus disease 2019 (COVID-19) in the EU/EEA and the UK—eleventh update: resurgence of cases. Retrieved from https://www.ecdc.europa.eu/sites/default/files/documents/covid-19-rapid-risk-assessment-20200810.pdf
Resnicow K, Page SE (2008) Embracing chaos and complexity: a quantum change for public health. Am J Public Health 98(8):1382–1389
Scharmer O, Kaufer K (2013) Leading from the emerging future. Berrett-Koehler, San Francisco
School 2030 (2020) Retrieved from https://www.skola2030.lv/lv/par-projektu
Snowden D (2011) Naturalizing Sense making. In: Informed by knowledge: expert performance in complex situations. Taylor and Francis Group
Stanley M, Amaral L, Buldyrev S, Havlin S, Leschhorn H, Maass P, Salinger M, Stanley H (1996) Scaling behaviour in the growth of companies. Nature 379:804–806
Stead JG, Stead WE (2014) Sustainable strategic management. ME Sharpe Inc., New York
Taleb NN (2014) Antifragile: things that gain from disorder. Penguin, Random House Trade
Uhl-Bien M, Marion R, McKelvey B (2007) Complexity leadership theory: shifting leadership from the industrial age to the knowledge era. Leadersh Q 18(4):298–318
Van de Ven A, Poole MS (1995) Explaining development and change in organizations. Acad Manage Rev 20:510–540
Varela FJ (1984) Two principles of self-organization. In: Ulrich H, Probst GJB (eds) Self-organization and management of social systems. Springer, New York
Whose learning crisis? Critical reflections on the world development report 2018. A briefing note on the World Bank's World Development Report 2018: Learning to realize education's promise. Global Education Network Europe. Retrieved from https://gene.eu/wp-content/uploads/Whose-Learning-Crisis-Briefing-on-WDR-2018.pdf
Wiek A, Withycombe L, Redman CL, Banas S (2011a) Moving forward on competence in sustainability research and problem solving. Environ Sci Policy Sustain Dev 53(2):3–12
Wiek A, Withycombe L, Redman CL (2011b) Key competencies in sustainability—a reference framework for academic program development. Sustain Sci 6(2):203–218
Winthrop R (2020, April 10) Top 10 risks and opportunities for education in the face of COVID-19. Brookings. https://www.brookings.edu/blog/education-plus-development/2020/04/10/top-10-risks-and-opportunities-for-education-in-the-face-of-covid-19/
World Development Report (WDR) Learning to realize education's promise (2018). The World Bank. Retrieved from https://www.worldbank.org/en/publication/wdr2018
XinhuaNet (2020, March 19) Latvian schools switch to distance learning during COVID-19 lockdown, call for devices. http://www.xinhuanet.com/english/2020-03/20/c_138896780.htm
Zeleny M (2005) Human systems management: integrating knowledge, management and systems. World Scientific Publishing, Singapore
Zimmerman B, Hurst D (1990) Breaking the boundaries: the fractal organization. J Manage Inq 2:334–354
Zuijderhoudt R (1990) Chaos and the dynamics of self-organization. Hum Syst Manage 9:225–238

Dzintra Iliško, Professor at Daugavpils University, Institute of Humanities and Social Sciences.

COVID-19 and Corporate Social Responsibility: A Canadian Perspective

David Billedeau and Jeffrey Wilson

Abstract Starting in June 2020—three months after the World Health Organization declared COVID-19 a pandemic—a series of interviews were started with representatives of large firms operating in Canada to address two primary questions: (1) How have corporations in Canada supported COVID-19 response efforts? (2) Will the COVID-19 pandemic result in increased responsibilities for, and societal expectations of, corporations? In response to the first query, it is concluded that many large firms have been successfully leveraging existing CSR programs to support COVID-19 response efforts. The companies assessed in this study are not indicative of corporate COVID-19 responses across Canada; however, the participants serve to demonstrate best practices for supporting pandemic relief and response efforts in tandem with continued operations. In response to the second query, it is concluded that the COVID-19 pandemic has created increased pressures for corporations within Canada to leverage their resources to advance and protect societal interests. Most notably, some industry leaders have recognized increasing requirements to maintain their social licence to operate and are thus seeking to expand CSR programs in a time of financial uncertainty brought about by the pandemic. As many civil society organizations continue to experience financial and operational pressures due to COVID-19, there is significant pressure on corporations to bolster their community advancement initiatives. This study provides insights into how leading firms in Canada have supported COVID-19 response as well as the future of CSR in Canada.

Keywords Corporate social responsibility · COVID-19 · Sustainability · Best practices · Philanthropy

D. Billedeau (✉) · J. Wilson
University of Waterloo, 200 University Ave W, Waterloo, ON N2L 3G1, Canada
e-mail: Dbbilledeau@uwaterloo.ca

J. Wilson
e-mail: Jeffrey.wilson@uwaterloo.ca

W. Leal Filho (ed.), *COVID-19: Paving the Way for a More Sustainable World*, World Sustainability Series, https://doi.org/10.1007/978-3-030-69284-1_2

1 Introduction

As of June 2020, there are over seven million confirmed cases of COVID-19 globally and over four hundred thousand COVID-19 related deaths—Canada accounts for over 95,000 cases and nearly 8000 deaths (World Health Organization 2020). The Government of Canada has responded to the COVID-19 pandemic by issuing a public health ethics framework, providing economic and financial supports, imposing travel restrictions and requirements, and collaborating with provincial governments to support coordination of responses and monitoring of pharmaceuticals and medical equipment (Government of Canada 2020). Across the Canadian provinces, there have been varying approaches to pandemic response—with an array of supports being created for income, housing, businesses, and utility costs (Lee and Hamidian 2020). National and sub-national administrations across the globe have been grappling with how best to protect their citizens and economies; however, governments—and the broader public sector—are not the only stakeholder involved in fighting the pandemic. Corporations—large, medium, and small—are contributing to the COVID-19 response and recovery efforts by contributing "their skills, networks and resources" (Clift and Court 2020) to support logistics, hospital construction, testing, and manufacturing of essential goods (e.g., hand sanitizer, gloves, and masks).

Not only are companies assisting with broad COVID-19 response efforts, they are resuming operations to support economic recovery efforts. Over the course of the summer of 2020, many companies have returned to the workplace—providing their employees with a changed workplace environment designed to restrict the spread of the virus. While governments and companies grapple with the logistics of reopening the economy, major corporations will continue to operate in a COVID-19 and post-COVID-19 world. However, many civil society organizations may crumble under the financial and logistical realities of operating during a pandemic. Not only will a prolonged pandemic limit government support for not-for-profits, many organizations will face challenges for raising funds and maintaining their core services at a time of social distancing and other transmission mitigation measures.

There is ongoing, international research focused on the impact COVID-19 is having on corporate social responsibility (CSR) programs and core operations more broadly. For example, there are studies that have examined how COVID-19 has changed office designs and health and safety programs (Parker 2020). Some researchers have argued that COVID-19 presents an opportunity for corporations to create more effective and meaningful CSR programs due to an increase in consumer ethical decision making (He and Harris 2020). There is also research indicating that corporations should tailor their CSR and COVID-19 response efforts to the strengths of their firm—in effect, corporations are being urged to avoid a "one size fits all" CSR framework (Aguinis et al. 2020). This study compliments existing literature on CSR in relation to COVID-19 by providing concrete examples of how corporations have modified their operations, leveraged and updated their CSR programs and core operations to support COVID-19 response efforts, and have engaged in CSR efforts that are aligned with their core operations and expertise.

In effect, this study provides insights into how Canadian corporations leveraged their core operations and CSR programs to support regional and national responses to the COVID-19 pandemic. Ten corporate representatives from diverse sectors operating out of Canada were interviewed to understand how corporations are responding to COVID-19 and how the virus has impacted CSR programs. The primary questions posed by this study are: (a) how have corporations in Canada supported COVID-19 response efforts; and, (b) will the COVID-19 pandemic result in increased responsibilities for, and societal expectations of, corporations? These questions are especially important given that many civil society organizations are struggling to operate, which is placing significant pressure on corporations to bolster their community advancement initiatives.

Prior to reviewing the results of those discussions, a brief examination of CSR will take place—with a focus on how and why corporations have been involved in past responses to largescale disasters. Subsequently, an overview of the methodology used to survey Canadian companies will be laid out—explaining why the companies included in this study were selected and how the interview process with participants was designed. Finally, the results of the interviews will be provided alongside analysis of the data and information provided by corporate participants prior to issuing concluding remarks.

2 Corporate Social Responsibility: An Overview

Corporate social responsibility is a widely studied and used term, with an equally expansive breadth of accompanying definitions, principles, and practices. Within the North American context, legislation pertaining to corporations requires board members to maintain and exercise fiduciary responsibilities in the financial interest of the firm. While there are a growing number of jurisdictions that expand board fiduciary duties beyond the scope of profit maximisation (Hiller 2013), CSR can simply be viewed as those corporate actions and operations that are not required/legislated and are designed to support wider community interests distinct—but not entirely disconnected—from a firm's financial interests. However, as society becomes more technologically and socially connected, the protection of a firm's reputation may very well rely on engagement in CSR activities. As noted by Paul Shrivastava:

> The lesson for corporations is that accepting corporate responsibility is not only an ethical matter but a matter of long-term survival. It is in the self-interest of companies to broadly conceptualize their responsibilities on safety, health, and environmental issues and fulfill them vigorously. By doing so, they are likely to act with more caution and more concern for human and environmental impacts of their activities. This caution and concern can reduce their crisis chances. (Shrivastava 1995, 225)

Therefore, the CSR programs maintained by firms protect against both risks to reputation and tangible risks to the viability of maintaining core business operations. But, what is CSR, exactly?

Broadly, corporate actions supporting the environment and the communities wherein firms operate are captured under the banner of CSR. In a review of 37 definitions of CSR, Alexander Dahlsrud captures five elements of CSR exhibited by firms: their actions pertaining to the environment, society, economic investment, and stakeholder relations, as well as voluntary actions not mandated by laws or regulations (Dahlsrud 2008). CSR, then, can be interpreted as a multifaceted concept that can be integrated into multiple aspects of a firm's strategic operating plan. Porter and Kramer explain: "Broadly speaking, proponents of CSR have used four arguments to make their case: moral obligation, sustainability, license to operate, and reputation" (Porter and Kramer 2006, 81). Most importantly to this review is how corporate support for CSR initiatives—inclusive of initiatives designed to mitigate negative operational impacts to the environment—can be supportive of the firm's reputation and long-term viability. Corporate investment in CSR can improve a firm's public reputation (Fombrun and Shanley 1990; Griskevicius et al. 2010). Further, CSR initiatives—and supporting public relations initiatives—can support closing the gap between corporate identity, a corporation's desired identity, and how the public perceives the corporation (Chun 2005). Using an index of CSR performance, studies have indicated that firms may yield financial returns for investment into CSR (Waddock and Graves 1997). CSR investments can support share prices (Frooman 1997) and the overall revenue of a corporation. As Liston-Heyes explains: "a firm that is socially responsible and responsive may be able to increase interpersonal trust between and among internal and external stakeholders, build social capital, lower transaction costs, and, therefore, ultimately reduce uncertainty about its financial performance" (Liston-Heyes and Ceton 2014, 391).

There is an abundant amount of studies linking CSR activity to a firm's financial performance (Kang et al. 2016; Zu 2009; Jo et al. 2015; Wang et al. 2015; Hasan et al. 2018). Importantly, corporate involvement in disaster response—when coupled with effective promotion of said involvement—has the ability to mitigate past damage to corporate reputation and create a positive image of the firm (Bodkin et al. 2015). This assertion is encapsulated in the stakeholder theory perspective on CSR, which "holds that CSR may benefit firms financially because various stakeholder groups may reward firms for their CSR activities" (Madsen and Rodgers 2015, 791). In terms of garnering social support for corporate operations, the greatest benefit derived from engaging in CSR practices stems from adequate advertising of corporate CSR activities (i.e., informing consumers about the good being done by the firm) and from partnering with entities outside of the private sector that have an innately higher degree of credibility than profit-driven organizations (Madsen and Rodgers 2015).

Typically, governments are the primary stakeholders in disaster preparedness and response; however, the private sector has become a vital component of coordinating an effective disaster response strategy. Bellesteros et al explains that "firms are being relied upon to adopt responsibilities that have traditionally fallen to governments, aid agencies, and nongovernmental organizations" (Ballesteros et al. 2017, 1682). Researchers have measured the heightened prevalence of private sector involvement in disaster response, noting that "private stakeholders represented only 9.8% (140 national organizations) of all organizations that joined forces during the September

11th response in 2001... and increased to 27% of all organizations that engaged in the Katrina Hurricane response in 2005" (Fontainha et al. 2016, 78). Further, within the United States, 85% of infrastructure critical for disaster response and recovery is owned and operated by private stakeholders—making their involvement in both disaster response necessary from a logistics standpoint, but also a requirement for their own financial viability (Fontainha et al. 2016, 78). Chen et al further explain the rationale for corporations to support disaster response efforts:

> [The corporation] undertakes its social responsibility strategically in natural disaster emergency management, which not only effectively eliminates the conflict between social responsibility and economic goals, but also combines them to realize the "win-win" between public welfare and corporate interests and contribute for the construction of a harmonious society. (Chen et al. 2012, 251)

Indeed, private sector support for disaster response is not entirely altruistic—firms may leverage involvement in disaster response and broader CSR initiatives to support more conventional business objectives (e.g., supporting profit, protecting brand image, etc.). Further, there is evidence that suggests corporate involvement in disaster response fosters greater employee satisfaction with their employer (Watkins et al. 2015).

But not all corporate responses to disasters result in positive returns (financial or reputational) for the firm. In their study focused on corporate responses to Hurricane Katrina, Muller and Kräussl note:

> If managers are interested in capturing value from CSR, they should consider curbing their firms' social irresponsibility rather than investing in prominent displays of corporate philanthropy. A firm's track record of minimizing its negative impacts appears to be a more genuine signal of trustworthiness that gives investors confidence in short-term recovery, not the accumulation of good deeds they do. (Muller and Kräussl 2010, 5)

Accordingly, leveraging CSR to support disaster response is not a sure-fire approach to garnering greater social support of a corporation's operations. Corporate responses to disasters must be genuine—if a firm has previously not supported any sort of community engagement, efforts to mitigate harm stemming from a disaster can readily be viewed by the public as self serving. To that end, while corporations engaging in CSR may seek to gain some return on their investment (in the form of reputational benefits), the primary motivation for engaging in disaster response initiatives should stem from a genuine desire to benefit the community wherein a firm operates.

There are also situations in which participation in disaster response is required—either by legislation or by the scope of the disaster. Ballesteros et al explain that the scope of large disasters may require an underfunded and underequipped public sector to seek private sector support:

> A typical large disaster causes a 20% reduction in national GDP... and the annual inflation-adjusted loss from even average disasters has grown from $54 billion in 1980 to more than $314 billion in 2015... Disasters are also underinsured, even in developed nations... As a result, there is a growing gap between the scale of disasters and the capacity of traditional aid providers, such as governments and multilateral agencies, to undertake effective responses. (Ballesteros et al. 2017, 1864)

Further, corporations need not directly involve themselves in disaster response—they can simply provide financial donations to organizations more directly involved in response efforts. Philanthropy is a central aspect of CSR programs, and evidence suggests that consumers respond favorably to corporations who engage and advertise their philanthropic initiatives (Patten 2008). Alternatively, firms may be required to support disaster response due to legislative action. In April 2020, United States President Donald Trump evoked the *Defense Production Act* to require companies (e.g., General Motors, General Electric, 3 M) to manufacture goods required for federal and state-run COVID-19 response efforts (Dzhanova 2020). Studies on consumer responses to being forced to support disaster response efforts are an area requiring further research.

In sum, consumer expectation of firms to support CSR and disaster response efforts are increasing—and firms are largely responding accordingly (Jordan et al. 2012). Corporations can no longer operate in a vacuum—there exist societal expectations in terms of CSR performance, and if corporations fail to meet societal expectations, their social license to operate may very well be revoked. However, increased communication of expectations between consumers and firms can result in tangible benefits for corporations—some of which have been articulated. Investment into CSR, then, results in the creation of "shared value" (Porter and Kramer 2006, 84) in the form of economic and social benefits. Weber and Feltmate note that shared value integrates CSR practices into core business strategies and can support corporate and consumer interests apart from, but inclusive of, financial returns (Weber 2016). Importantly, consumer expectations pertaining to CSR are not going away—and corporations should orient their operations accordingly. In other words, CSR "will remain as an essential part of business language and practice, because it is a vital underpinning to many of the other theories and is continually consistent with what the public expects of the business community today" (Carroll 1999, 292). Therefore, within Canada, the decision of many firms to support COVID-19 response efforts should not be viewed as abnormal—in fact, such actions should be understood as a new normal.

3 Methodology

To gain insights into why and how firms across Canada have leveraged their operations to support national COVID-19 response efforts—and to glean insights into the future of CSR operations in a post-COVID-19 world—this study leverages a series of interviews with corporate representatives from across Canada.

Companies operating in Canada that have been formally recognized as leading corporate citizens by third party sources (e.g., the Corporate Knights and the University of Waterloo's School of Environment, Enterprise, and Development) were contacted and asked to participate in this research. The first ten companies to respond to the request to participate were interviewed and included in our assessment. These companies were included in this research for two reasons. First, it is important to understand how CSR industry leaders are operating during the pandemic and how they intend to evolve their role in society in light of COVID-19. These leaders are setting the benchmark for how corporations will need to engage with society going forward. Second, it would not be beneficial to interview companies with limited CSR programs or engagement in community development practices. Such companies would provide limited value for outlining how corporations should evolve if civil society organizations fall. Instead, organizations with limited CSR engagements should review the conclusions of this study to prepare for increased societal, consumer, and investor expectations regarding their role within society.

The companies that were interviewed to support this research are as follows:

- **Financial sector (2)**: TD Bank, Vancity
- **Manufacturing sector (3)**: Bombardier, Celestica, Toyota
- **Natural resources sector (2)**: Iamgold Corporation, Teck Resources
- **Telecommunications sector (1)**: Telus
- **Transportation sector (1)**: Air Canada
- **Utility sector (1)**: Northland Power

In total, these corporate representatives were all asked the same 10 questions, which pertained to the past, current, and future states of their core operations and CSR programs. Interview questions were designed to explore the following areas:

1. Expectations of the corporation to support CSR and COVID-19 response/recovery;
2. Impact of COVID-19 on core business operations;
3. Impact of COVID-19 on CSR initiatives;
4. The interaction between government and business in responding to COVID-19;
5. How core business operations and CSR programs have been leveraged to support COVID-19 response efforts;
6. Whether companies utilized their COVID-19 response efforts in marketing campaigns;
7. The role of senior corporate leadership with respect to CSR and COVID-19 response; and,
8. The future of CSR programs.

The audio of all interviews was recorded, and companies were asked to verify information presented in this study prior to publication. To support data validation, all participants were provided with company-specific information included in this study (e.g., all text referencing their company, as well as the information noted in Table 1). No material changes to the text presented in this study were requested by corporate participants. Additionally, corporate participants were eligible to anonymize or completely withdraw from participating in this study. No corporation elected either of these options. To that end, the results and conclusions of this study are predicated on data that has been verified by corporate participants and the interview process aligns with research ethics requirements mandated by the University of Waterloo's Research Ethics Committee, which has reviewed and approved the interview and data collection process of this study.

Table 1 Results matrix

	Air Canada	Bombardier	Celestica	Iamgold	Northland Power	TD Bank	Teck Resources	Telus	Toyota	Vancity
Pre-existing CSR Program?	Yes	Yes	Yes	Yes	Yes	Yes	Yes	Yes	Yes	Yes*
What was the impact on core operations?	Significant	Significant	Moderate	Negative	Minimal	Significant	Moderate	Moderate	Significant	Moderate
What was the impact on CSR initiatives?	Neutral	Neutral	Neutral	Positive	Positive	Neutral	Neutral	Positive	Neutral	Neutral
Did the company coordinate with government on COVID-19 response?	Yes	Yes	Yes	Yes	Yes	Yes	Yes	Yes	Yes	Yes
Who initiated cooperation?	Mutual	Mutual	Corporation	Corporation	Government	Mutual	Mutual	Mutual	Corporation	Mutual
Was existing CSR program used to support COVID response?	Yes	Yes	Yes	Yes	Yes	Yes	Yes	Yes	Yes	Yes*
Were business operations modified to support COVID response efforts?	Yes	Yes	Yes	Yes	No	Yes	Yes	Yes	Yes	Yes
Does senior leadership support CSR/COVID-19 response?	Yes	Yes	Yes	Yes	Yes	Yes	Yes	Yes	Yes	Yes
Were there publicity campaigns noting COVID response?	Limited	Limited	Limited	Limited	Limited	Limited	Limited	Limited	Limited	Limited
What is the future of the CSR program?	Maintained	Expanded	Expanded	Expanded	Expanded	Maintained	Maintained	Expanded	Expanded	N/A*
Are there employee expectations to support CSR?	Yes	Yes	Yes	Yes	Yes	Yes	Yes	Yes	Yes	Yes
Are there employee expectations to support COVID response?	Yes	Yes	Yes	Yes	Yes	Yes	Yes	Yes	Yes	Yes
Are there customer expectation to support CSR?	Yes	Yes	Yes	Yes	Yes	Yes	Yes	Yes	Yes	Yes
Are there customer expectations to support COVID response?	Yes	Yes	Yes	Yes	Yes	Yes	Yes	Yes	Yes	Yes

*As a cooperative financial institution, Vancity does not operate a specific CSR program. Instead, environmental, social, and governance initiatives traditionally associated with CSR programs are woven throughout their core operations based on direction from membership

4 Results and Analysis

4.1 Expectations of the Corporation to Support CSR and COVID-19 Response/Recovery

All companies indicated that both employees and customers expect that they support ongoing CSR initiatives as well as specific actions focused on COVID-19. Importantly, companies indicated that both employee and customer support for CSR and COVID-19 response have factored into senior leadership decision making. Larger, successful, and profitable companies have a unique expectation to engage in community initiatives—and all companies interviewed recognize and have acted on this expectation in order to ensure their social license to operate remains intact. Employees had unique expectations of their employers to establish heightened occupational health and safety measures in order to ensure their wellbeing while at work.

The benefits of engaging in CSR and COVID-19 response efforts were also indicated by several respondents, who identified that both customers and investors align purchasing decisions with companies that share the same values as they do. Accordingly, it behooves corporations from a moral and financial perspective to exert leadership in CSR. Further, many corporations leverage CSR programs to support talent attraction and retention—as many potential employees (particularly younger generations) are looking to align with employers who share values.

As for responding to COVID-19, many companies supported joint efforts using corporate capital and employee desire to become involved. Manufacturing companies with many engineers, for example, allocated engineering manhours to support the deployment of manufacturing options for product health supplies and equipment. Other companies established donation drives, wherein the company would match the total dollar value of donations from employees to specific charities. It was noted that employees are significantly motivated to support their employer's efforts in addressing COVID-19, as the pandemic is a unifying issue that impacts every member of the company—and society at large.

Similarly, all companies indicated expectations from their customers in both supporting ongoing CSR initiatives and involvement with COVID-19 response efforts. Many companies noted that their relationships with customers have evolved during the pandemic, requiring the company to focus more efforts on customer engagement to ensure products and services are maintained and supported. Bombardier, for example, provided rail customers with increased maintenance and support services in order to install COVID-19 health and safety measures on rail lines and ensure the continued functioning of equipment. However, perhaps most interestingly was that many of the respondents categorized investors as the customer—and noted that CSR initiatives and COVID-19 responses were at the forefront of investors' minds. Not only do investors want to know how corporations have responded to COVID-19 from a financial and operational perspective, but they also want to ensure the continued delivery of corporate social responsibility programs, as they see clear value in meeting increased societal expectations.

In sum, both employees, customers, and investors have indicated a clear desire for increased transparency and action in CSR programs and COVID-19 response efforts.

4.2 *Impact of COVID-19 on Core Business Operations*

In terms of COVID-19's impact on core business operations, five companies (Air Canada, Bombardier, Iamgold Corp, TD Bank, Telus, and Toyota) noted a significant impact on core operations, three companies (Celestica, Teck Resources, and Vancity) noted a moderate impact, and Northland Power noted only a minimal impact. All corporations indicated that increased occupational health and safety measures have been rolled out, which impacts the financial bottom line of operations. Further, all companies, to varying degrees, have transitioned to a remote workplace—fundamentally changing the workplace dynamic and delivery of key products and services.

Since the pandemic was declared by the World Health Organization on March 11, 2020, it is clear that the aircraft industry was hardest hit. Air Canada's operations grounded to a near halt during the pandemic lockdown, which created serious impacts on operations. In the past quarter, Air Canada was operating at around 10–15% of normally capacity. Further, the company suffered significant layoffs (over 20,000). Notably, Air Canada was the first North American airline to cease operations out of China— following the federal government's advisory to avoid non-essential travel to the mainland due to the coronavirus epidemic. This decision was predicated on the company's best interest for employees, customers, and communities.

Bombardier has also been immensely affected by the pandemic. Many manufacturing locations have been shut down during lockdown, and global operations and supply chains have been negatively impacted. Accordingly, the company has laid off employees and has seen a temporary halt in some operations—which was also impacted by a ripple effect in the supply chain. Further, Bombardier has experienced issues with the delivery of products due to the closing of borders. Looking forward, the company is cognizant of the reduced demand for rail products from municipalities due to a public focus on social distancing measures over public transit considerations. Additionally, the toll the pandemic has taken on public coffers will mean a reduced ability to sell rail equipment to municipalities in the short-term future, although the overall outlook for the rail industry remains positive.

Iamgold Corp established a COVID-19 working group in February 2020, which consisted of representatives from corporate affairs, health and safety, and security, as well as site management leads from a range of disciplines. Despite significant planning for a COVID-19 outbreak at their operating sites, the scale of the pandemic significantly impacted operations. The company had to roll out safety equipment, testing facilities, and training programs across their global operations. Despite best efforts, numerous sites—namely Burkina Faso—experienced outbreaks that impacted operations and work schedules. In Quebec, provincial measures to combat COVID-19 resulted in a two-week closure of the company's provincial operations as mining was not considered an essential business.

TD Bank has experienced an enormous impact on its core operations due to COVID-19. The company was called upon by the federal government to assist in managing its Canadian Emergency Business Account loan program. Further, TD Bank has supported mortgage deferrals and interest rate forgiveness programs to assist Canadians experiencing financial hardship due to the pandemic. Close to half of TD Bank branches have been temporarily closed, and the company has heavily invested in technology to support working from home—with more than half of its 85,000 staff now operating remotely. Importantly, however, the CEO of TD Bank publicly stated that there would be no job loss in 2020 due to the pandemic, despite the heavy financial burdens placed on the organization.

Telus has seen a significant increase in demand for their telecommunications services, largely due to increased demand from home usage and because the federal government declared high speed internet access as an essential service. While Telus has been able to accommodate the roughly 175% increase in voice traffic on its network without disruption to service delivery, there has been significant movement towards remote working (95% of staff now work from home, with the exception of critical employees). Notably, Telus has not laid of a single employee during this time.

Toyota's Canadian manufacturing operations ceased production completely from March 19 through mid-May 2020, and they are only starting to return to normal production volume in the mid-summer. From a sales perspective, there have been varying responses from dealerships across Canada; however, there have been widespread closures throughout the country. With dealerships now open, many are supporting significant occupational health and safety and social distancing requirements, inclusive of a greater focus on web-based sales. This has required investment from dealerships in digital technologies to support online sales and distribution of vehicles.

The moderate impacts experienced by Celestica, Teck Resources, and Vancity mostly pertain to temporary closures of select operations, the introduction of increased health and safety measures to protect employees and customers, and a shift to remote working. Despite higher operating costs to carry out core operations, these companies have been able to readily adapt existing operations to the changing workplace environment brought about by the pandemic without significant modification.

The only company to report a minimal impact on core operations was Northland Power, who remained operational for the entirety of the pandemic (despite a brief closure at a Mexican location). The only impact on their operations has been an increased expenditure on occupational health and safety.

4.3 *Impact of COVID-19 on CSR Initiatives*

In addition to asking companies about how their core operations were impacted, they were asked about whether their existing CSR programs were reduced due to the pandemic. Eight companies (Air Canada, Bombardier, Celestica, Iamgold Corp, TD Bank, Teck Resources, Toyota, and Vancity) reported a neutral impact on their programs (meaning that funding or programing was not reduced; although, focus of CSR operations may have pivoted). Interestingly, two companies (Northland Power and Telus) reported an increase in CSR activity due to COVID-19.

Northland Power has a significant focus on the environmental performance of their operations. As all international business travel has stopped, their footprint (and thus climate targets) has been significantly improved. Further, charitable efforts (on top of pre-COVID-19 efforts) were introduced to help communities in eight countries that have been impacted economically and socially by COVID-19. The company also engaged in matching employee donations to specific charities, and—importantly— hired a Director of Sustainability from an employment competition that was launched during the lockdown period of the pandemic. This clearly demonstrates the company's dedication to investing financial and human capital into growing and refining its CSR initiatives.

Telus has also seen growth in its CSR initiatives. Prior to COVID-19, Telus operated two health services known as Babylon and Akira. Both of these applications allow Canadians to consult medical professionals remotely, and the services have seen exponential growth (roughly five times as many current users vs. pre pandemic users).

In all, not a single company interviewed noted that their CSR program would be scaled back to COVID-19. On the contrary, CSR programs are by and large continuing to operate—and in some cases expand—despite the pandemic's negative operational and financial impacts on corporate performance. Some companies did note that CSR programs could pivot or be reduced if prolonged financial hardship is experienced due to COVID-19; however, this will be determined in the years ahead, and it will be contingent on COVID-19 infection rates worsening or ameliorating.

4.4 *The Interaction Between Government and Business in Responding to COVID-19*

All corporations interviewed cooperated with provincial and government officials in coordinating COVID-19 response and relief efforts, and in some cases, companies worked with governments outside of Canada to support local community efforts close to their operations. Interestingly, three companies (Celestica, Iamgold Corp, and Toyota) reached out to government (federal or provincial) to offer support for COVID-19 response efforts, six companies (Air Canada, Bombardier, TD Bank, Teck Resources, Telus, and Vancity) reported an ongoing dialogue with government prior to the pandemic that resulted in organic strategic conversations, and one company (Northland Power) was contacted directly by government to support efforts. In all, the responses indicate that corporations maintaining active communications with government stakeholders prior to an emergency allows for corporations to synergize response efforts more readily. Additionally, there is a notable sense within Canada's business community—particularly among large firms—that their resources and role in society require them to act not only to preserve their operations but also to support the communities in which they operate.

4.5 How Core Business Operations and CSR Programs Have Been Leveraged to Support COVID-19 Response Efforts

The specific efforts of companies participating in COVID-19 response efforts vary greatly, but all efforts are uniquely attached to the core operations of the corporation. For example, Air Canada worked closely with government officials to organize rescue flights for over 200,000 Canadians traveling abroad during the initial stages of the lockdown. This required the company to work alongside federal officials to negotiate flights, schedules, and routes with members of the international community who had shut their borders. While organizing the biggest relief effort the company had ever conducted—with half of their staff furloughed at the time—the company modified passenger planes no longer in use to support cargo delivery to numerous communities, donated in-flight meals that would no longer be used to local community organizations, and piloted a drone cargo delivery program for remote northern communities. The company also leveraged its Aeroplan program, by allowing point holders to donate to charities engaged in COVID-19 response efforts.

Bombardier has donated personal protective equipment (PPE) to frontline workers across Canada and has designed and produced more than 40,000 protective visors to support essential workers across the country. They also helped manufacture ventilator equipment out of Thunder Bay, Ontario. Additionally, the company's foundation donated hundreds of thousands of dollars to support organizations such as the Red Cross in their relief efforts, and they have funded medical research with the Heart Institute Foundation on COVID-19.

Celestica has been producing components for air purifier respirator systems used by paramedics and first responders, who are often the first to come into contact with COVID-19 patients. Further, the company is building 7500 ventilators to help Canada's healthcare professionals prepare for the next wave of COVID-19 patients. In all, the company is involved in securing parts to support the manufacturing and eventual delivery of ventilators to hospitals. Additionally, the company is building ventilators at their facility in the Republic of Ireland.

Iamgold Corp mobilized to identify the needs of the communities wherein they operated, and have made cash and in-kind donations of over $1.5 million USD to numerous jurisdictions to secure PPE, testing and medical equipment, and to provide financial support to health ministries dealing with the pandemic. In effect, representatives from Iamgold Corp believe that the pandemic has accelerated societal trends mandating corporations to be proactively engaged with their communities by supporting a strong CSR program. Accordingly, the integration of environmental, social, and governance concerns—though already embedded into Iamgold Corp's operations—has since been bolstered.

TD Bank was called upon by the federal government to manage the Canada Emergency Business Account loan, which provides businesses with a $40,000 interest free loan to support operations during the pandemic. The company has also created programs for mortgage deferrals and interest rate forgiveness and has empowered local branches to support their customers in financial distress with increased supports. In terms of philanthropy, TD Bank has issued large short-term donations (e.g., $1 million to health centres across Canada, and another $1 million to a frontline workers support fund). TD Bank is also developing a $25 million dollar community resilience initiative to support the recovery phase of COVID-19.

Teck Resources procured over one million n95 respirator masks that are being distributed for widescale public use. Additionally, the company has created a $20 million COVID-19 response fund to support Canadian healthcare and social services. The company is also engaged in philanthropic COVID-19 efforts focused on supporting local community organizations and the Red Cross.

Toyota has donated PPE to local hospitals near their manufacturing operations, and they have also manufactured face shields. Further, the company has bolstered their longstanding partnership with Canadian Blood Services by donating unused advertising airtime to the organization. Moreover, Toyota provided their technical and engineering expertise associated with the Toyota Production System methodology to government and private sector organizations with the aim of increasing COVID-19 test processing, improving logistical networks from testing centres to labs, and manufacturing PPE and ventilator equipment.

Vancity is unique among the corporations assessed in the sense that it is a cooperative financial institution that does not incorporate a traditional CSR program or framework. Instead, the company incorporates CSR and sustainability into its core business operations—in effect, all work is centered around social and community justice. To that end, the company has been readily involved in supporting their customers and community during the pandemic. Vancity has pivoted some of its lending programs to support gig and contract workers that are otherwise ineligible to receive federal COVID-19 financial supports. Additionally, the company has modified its microfinance risk framework to allow more businesses to receive financial supports during the pandemic. Moreover, the company has dropped financial interest on credit cards, has renegotiated mortgage rates with customers, and has deferred payments on loans and mortgages for a period of six months.

4.6 Whether Companies Utilized Their COVID-19 Response Efforts in Marketing Campaigns

Despite the significant efforts to support their communities during the pandemic, not a single company assessed has invested financial capital into supporting a media campaign noting their CSR initiatives pertaining to COVID-19. All companies issued press statements on their activities in addition to supporting social media and web communications detailing their response efforts; however, there have been no dedicated advertising campaigns attempting to sway consumer behaviour with COVID-19 CSR initiatives. Some companies, namely financial institutions, did advertise new services and programs available to its customers; however, this was informational in nature. Additionally, other companies, such as Air Canada, have supported industry messaging that communicates new COVID-19 safeguards being put in place to support their operations and protect their customers and employees. However, no company attempted to leverage its efforts to assist in the pandemic response in order to garner increased sales or social capital.

4.7 The Role of Senior Corporate Leadership with Respect to CSR and COVID-19 Response

All companies reported that senior corporate leadership support both ongoing CSR initiatives and a specific response to COVID-19.

Air Canada noted that their goal to be a global sustainability leader existed before COVID-19—and their longstanding dedication to investors, communities, employees, and customers was the leading factor in their support for COVID-19 relief efforts. Other companies, such as Iamgold Corp, Northland Power, TD Bank, Teck Resources, and Telus, all noted that investors and customers both see expanding CSR initiatives as a requirement to maintain a social licence to operate. Similarly, Bombardier believes that the importance of CSR programs is only going to increase overtime due to a sharping focus on environmental, social, and governance indicators from customers and investors. Celestica, similarly, noted a long heritage of supporting CSR initiatives—and that their corporate leadership is not focused on whether they should support their communities but on how they can continuously improve community outreach initiatives.

4.8 The Future of CSR Programs

Seven companies (Bombardier, Celestica, Iamgold Corp, Northland Power, Telus, Toyota, and Vancity) noted an expectation for their CSR initiatives to increase in the coming years, albeit with some caveats. This potential growth in CSR programs is predicated on an increased focus of all stakeholders (e.g., customers, investors, communities, government) for corporations to have a more positive role in society. Additionally, many recovery plans will depend on corporate involvement. The European Union's recovery plan is highly supportive of a green economic transition. For this to take place, corporate support and investment will be required. Some companies, such as Celestica, speculated that expectations for sustainability reporting will increase, which will require firms to understand the environmental and social impact of all areas of their operations—inclusive of supply chain environmental impacts (an area often overlooked in integrated sustainability reporting). As previously mentioned, Northland Power hired a Director of Sustainability during the first COVID-19 lockdown; to them, the writing is on the wall: if operations are to continue, CSR initiatives must be bolstered and expanded. Teck Resources and TD Bank noted their expectation for their CSR programs to be maintained in the near future, but they could not provide affirmation on the likelihood of increased financial resources to bolster their CSR programs going forward. TD Bank, for example, noted that their philanthropic budget is based on a percentage of their net income before tax averaged over five years. Accordingly, if prolonged economic impacts from COVID-19 were to take place (over the period of years), funds available for TD Bank's philanthropy efforts could decrease—this is likely to be a reality for many companies.

5 Conclusion

This study addresses two primary questions:

1. *How have corporations in Canada supported COVID-19 response efforts*?
2. *Will the COVID-19 pandemic result in increased responsibilities for, and societal expectations of, corporations?*

In response to the first query, all corporations interviewed in this study have highlighted ways in which their existing CSR programs were leveraged to support the pandemic response. Although this cannot be extrapolated to *all* companies in Canada, the actions identified by companies in this study exemplify the gold standard for corporations engaging and supporting their community stakeholders during a significant public health crisis. Each of the companies leveraged their core operations in response to COVID-19 (e.g., changing manufacturing production to make ventilators, increasing health and safety measures to protect employees and local communities, supporting initiatives with government partners, etc.) due to a sense of duty—the role of a corporation in society is not simply to generate profit, but to actively support the communities wherein corporations operate. This indicates that companies with holistic CSR programs integrate corporate values across operations—and that companies with advanced sustainability programs are willing and able to shift business operations to support broader societal objectives in times of strife, despite the impact on the financial bottom line.

In response to the second query, it is important to first draw a stark contrast in philosophies. In 1970, Milton Friedman famously noted that there "is one and only one social responsibility of business–to use its resources and engage in activities designed to increase its profits so long as it stays within the rules of the game" (Friedman 1970). Clearly, this view is no longer tenable—and is not an accepted position by many firms within Canada. The role of corporations within Canadian society is moving away from the traditional model—wherein corporations provide jobs, goods, and services—to one wherein those corporations are key stakeholders in the protection and advancement of societal values.

Each of the companies interviewed for this study indicated that their operations have experienced varying degrees of negative impact due to the pandemic. Yet, despite instances of notable financial losses, all the companies have continued to support the communities wherein they operate through continued CSR programming and by creating specific COVID-19 response initiatives. These efforts range from increased workplace measures to protect employees, donations of equipment to frontline workers, organization of new philanthropic endeavours, and modification of core business operations to support provincial and national COVID-19 efforts (e.g., manufacturing of medical equipment, supporting rescue efforts, etc.).

Should a considerable number of civil society organizations fail due to operating pressures exerted by a prolonged pandemic, the expectations of corporations to support local communities will only increase. As this study has shown, this is a challenge that leading firms are ready to address. Although Canadian society will continue to be affected by the COVID-19 pandemic, the recovery effort will largely depend on the willing participation of corporations to leverage their operations and expertise to sustain and evolve the national economy and advance the interests of Canadians. Many of the corporate participants of this study noted that COVID-19 has increased investor, customer, and employee expectations for their corporations to support CSR initiatives and a broader role for their company in society (e.g., supporting pandemic response). This study concludes that COVID-19 has already changed the role and expectations of corporations in Canada—as indicated by industry leaders—and is ushering in a new approach to responsible enterprise that will require a sustained, long-term commitment to ethical behaviour and improved community engagement.

References

Aguinis H, Villamor I, Gabriel KP, Hall F (2020) Understanding employee responses to COVID-19: a behavioral corporate social responsibility perspective. Manage Res 1–30. https://doi.org/10.1108/MRJIAM-06-2020-1053

Ballesteros Luis, Useem Michael, Wry Tyler (2017) Masters of disasters? An empirical analysis of how societies benefit from corporate disaster aid. Acad Manage J 60(5):1682–1708. https://doi.org/10.5465/amj.2015.0765

Bodkin CD, Amato LH, Amato CH (2015) The influence of green advertising during a corporate disaster. Corp Commun 20(3):256–275. https://doi.org/10.1108/CCIJ-08-2014-0055

Carroll AB (1999) Evolution of a definitional construct. Bus Soc 38(3):268–295

Chen F, Da Q, Deng Y (2012) Realization of corporate social responsibility in natural disasters emergency management. In: Proceedings of the 2012 5th international conference on business intelligence and financial engineering, BIFE 2012, pp 247–251. https://doi.org/10.1109/BIFE.2012.60

Chun Rosa (2005) Corporate reputation: meaning and measurement. Int J Manage Rev 7(2):91–109. https://doi.org/10.1111/j.1468-2370.2005.00109.x

Clift K, Court A (2020) COVID-19: how companies are responding | world economic forum. World Economic Forum. 23 Mar 2020. https://www.weforum.org/agenda/2020/03/how-are-companies-responding-to-the-coronavirus-crisis-d15bed6137/

Dahlsrud Alexander (2008) How corporate social responsibility is defined: an analysis of 37 definitions. Corp Soc Responsib Environ Manag 15(1):1–13. https://doi.org/10.1002/csr.132

Dzhanova Y (2020) Coronavirus: trump used defense production act on these companies so far. CNBC News. 3 Apr 2020. https://www.cnbc.com/2020/04/03/coronavirus-trump-used-defense-production-act-on-these-companies-so-far.html

Fombrun Charles, Shanley Mark (1990) What's in a name? Reputation building and corporate strategy. Acad Manag J 33(2):233–258. https://doi.org/10.5465/256324

Fontainha TC, De Oliveira Melo P, Leiras A (2016) The role of private stakeholders in disaster and humanitarian operations. J Oper Supply Chain Manage 9(1):77. https://doi.org/10.12660/joscmv9n1p77-93

Friedman M (1970) The social responsibility of business is to increase its profits The New York Times Magazine

Frooman J (1997) Socially irresponsible and illegal behavior and shareholder wealth. Bus Soc 36(3):221–249

Government of Canada (2020) Canada's Response. June 5, 2020. https://www.canada.ca/en/public-health/services/diseases/2019-novel-coronavirus-infection/canadas-reponse.html

Griskevicius V, Tybur JM, Van den Bergh B (2010) Going green to be seen: status, reputation, and conspicuous conservation. J Pers Soc Psychol 98(3):392–404. https://doi.org/10.1037/a0017346

Hasan I, Kobeissi N, Liu L, Wang H (2018) Corporate social responsibility and firm financial performance: the mediating role of productivity. J Bus Ethics 149(3):671–688. https://doi.org/10.1007/s10551-016-3066-1

He H, Harris L (2020) The impact of Covid-19 pandemic on corporate social responsibility and marketing philosophy. J Bus Res 116(May):176–182. https://doi.org/10.1016/j.jbusres.2020.05.030

Hiller JS (2013) The benefit corporation and corporate social responsibility. J Bus Ethics 118(2):287–301. https://doi.org/10.1007/s10551-012-1580-3

Jo H, Kim H, Park K (2015) Corporate environmental responsibility and firm performance in the financial services sector. J Bus Ethics 131(2):257–284. https://doi.org/10.1007/s10551-014-2276-7

Jordan J, Diermeier DA, Galinsky AD (2012) The strategic samaritan: how effectiveness and proximity affect corporate responses to external crises. Bus Ethics Q 22(4):621–648. https://doi.org/10.5840/beq201222442

Kang C, Germann F, Grewal R (2016) Washing away your sins? Corporate social responsibility, corporate social irresponsibility, and firm performance. J Market 80(2):59–79. https://doi.org/10.1509/jm.15.0324

Lee M, Hamidian A (2020) Comparing provincial economic responses to COVID-19: policy note. Policy Note. April 20, 2020. https://www.policynote.ca/provincial-responses-covid/

Liston-Heyes C, Ceton GC (2014) Corporate social performance and politics. J Corp Citizsh 2007(25):95–108. https://doi.org/10.9774/gleaf.4700.2007.sp.00010

Madsen PM, Rodgers ZJ (2015) Looking good by doing good: the antecedents and consequences of stakeholder attention to corporate disaster relief. Strateg Manag J 36:776–794. https://doi.org/10.1002/smj.2246

Muller A, Kräussl R (2010) Social irresponsibility, firm value and philanthropy: the corporate response to hurricane katrina. Academy of Management 2010 Annual Meeting—Dare to Care: Passion and Compassion in Management Practice and Research, AOM 2010
Parker LD (2020) The COVID-19 office in transition: cost, efficiency and the social responsibility business case. Acc Audit Account J. https://doi.org/10.1108/AAAJ-06-2020-4609
Patten DM (2008) Does the market value corporate philanthropy? Evidence from the response to the 2004 Tsunami relief effort. J Bus Ethics 81(3):599–607. https://doi.org/10.1007/s10551-007-9534-x
Porter ME, Kramer MR (2006) Strategy and society: the link between competitive advantage and corporate social responsibility. Harvard Bus Rev 84(12):78
Shrivastava P (1995) Industrial/environmental crises and corporate social responsibility. J Socio-Econ 24(1):211–227. https://doi.org/10.1016/1053-5357(95)90036-5
Waddock SA, Graves SB (1997) The corporate social performance-financial performance link. Strateg Manage J 18(4):303–319. http://www.Jstor.Org/Stable/3088143
Wang DH-M, Chen P-H, Yu TH-K, Hsiao C-Y (2015) The effects of corporate social responsibility on brand equity and firm performance. J Bus Res 68(11):2232–2236. https://doi.org/10.1016/j.jbusres.2015.06.003
Watkins MB, Ren R, Umphress EE, Boswell WR, del Carmen M, Zardkoohi A (2015) Compassion organizing: employees' satisfaction with corporate philanthropic disaster response and reduced job strain. J Occup Organ Psychol 88(2):436–458. https://doi.org/10.1111/joop.12088
Weber O (2016) Sustainable banking: managing the social and environmental impact of financial institutions. In: Blair W (ed) Feltmate. business & sustainability. Toronto [Ontario] ; University of Toronto Press
World Health Organization (2020) Coronavirus Disease (COVID-19) Dashboard, 2020. https://covid19.who.int/?gclid=CjwKCAjw5vz2BRAtEiwAbcVIL9FDRUw_KNQHGfx36sP8GJHpNiTgC45yLi_VIaIDfbtE72CsZcqYZhoC7pYQAvD_BwE
Zu L (2009) Corporate social responsibility, corporate restructuring and firm's performance empirical evidence from Chinese enterprises, 1st edn., vol 20. Springer, Berlin. https://doi.org/10.1007/978-3-540-70896-4

D. Billedeau David Billedeau is a PhD student in the School of Environment, Enterprise and Development at the University of Waterloo. His research is focused on environment and trade, corporate social responsibility, and sustainable development. Professionally, David has significant experience in government relations and policy development. David is currently serving as a Research Fellow for the Energy Council of Canada, a non-partisan organization promoting sustainable energy leadership and transitions within and outside of Canada.

J. Wilson Jeffrey Wilson has worked at the interface of economic, environmental and well-being research and policy for over 17 years. Jeff specializes in economic and sustainability research, natural capital accounting, wellbeing research, indicator development, ecosystem goods and services valuation, and strategy and policy development for sustainability transitions. Jeff is an assistant professor in the School of Environment, Enterprise & Development, University of Waterloo, Canada and director of Terrametrics Research Lab, a thinktank that brings together academics and practitioners to advance low carbon pathways.

COVID-19 and Pandemic Risk: The Link to SDG 13, Climate Change and the Finance Context

Federica Doni and Lara Johannsdottir

Abstract This chapter aims to assess the impact of COVID-19 on Agenda 2030 specifically focusing on SDG 13, i.e. climate actions. Habitat loss, the creation of artificial environments, the manipulation and trade of wild animals and more generally the destruction of biodiversity are mainly affecting the dynamic balance of the biosphere, as is confirmed by the first world report on the world's ecosystems, i.e. the Millennium Ecosystem Assessment. The changes in the use of land and the destruction of natural habitats, such as tropical forests, may be the main origin of more than half of emergent zoonosis. Given the magnitude of the current health crisis and the potential of pandemic risks, the world needs to pay attention to climate change and the broader sustainability agenda at this time. As the current health crisis is turning worldwide interest on climate change there is an urgent need to assess the response to COVID-19 from the financial and insurance perspective. In doing this our analysis points out the relationship between pandemic risk and sustainable development by considering both negative and positive impacts on the achievement of the SDG 13 targets. The ongoing COVID-19 shows the urgent need to strengthen sustainability by reducing and managing climate and environmental risks that can be supported by radical solutions provided by sustainable finance, as the European Commission emphasized in its consultation on Renewed Sustainable Finance Strategy. From the insurance industry view, it is clear that potential actions will be needed to limit the impacts of extreme weather events and related effects on global supply chain.

Although the present chapter is the result of a collaborative analysis, Sects. 1 and 2 were written by Lara Johannsdottir, while Sects. 3 and 4 were written by Federica Doni.

F. Doni (✉)
Business Administration and Accounting, Department of Business and Law, University of Milano-Bicocca, Via Bicocca degli Arcimboldi 8, 20126 Milan, Italy
e-mail: federica.doni@unimib.it

L. Johannsdottir
Environment and Natural Resources Graduate Programme, Faculty of Social Sciences, University of Iceland, Gimli v/Saemundargotu, 102 Reykjavík, Iceland
e-mail: laraj@hi.is

W. Leal Filho (ed.), *COVID-19: Paving the Way for a More Sustainable World*, World Sustainability Series, https://doi.org/10.1007/978-3-030-69284-1_3

Keywords COVID-19 · Climate change · Finance · Insurance · Sustainable development goals

1 Introduction

> The virus has made it clear: the future is now
>
> (Bieri et al. 2020)

This is a statement proposed by four scientists who also state that the coronavirus crisis allows us "to perceive the state of the world with increased clarity" (Bieri et al. 2020). Furthermore, they explain how obsolete views can be transformed into forward-looking ideas do design capacity for addressing sustainable development challenges. The first topic is a socially-justice in the economy. In this case the transformation has to go from disgorgement logics, where social and ecological costs are externalized to the society, to the logic of value logic, where such costs are internalized and that nobody is left behind in the society. The second topic is global environmental change. In this case the transformation has to go from delay in actions until it is too late to respond to the consequences', to reshaping the people-nature relationship in the economies, cities and towns, and nutritional and energy systems. The third topic is global inequality. The transition is from prioritizing ones country first, free trade and development aid, to a cooperation and partnership so a shared vision of the future can be achieved. The fourth topic focuses on the role of science. This means a transformation from limited strength of experts, only able to voices their warnings, alongside those sharing doubts and fake news, to a situation where world-leaders and scientists develop joint solutions. The fifth, and the last, topic is further sustainability-related issues in areas such as happiness and prosperity, an inter-generational contract, participation, security, and social relationships (Bieri et al. 2020).

This context represents a significant opportunity to explore the Sustainable Development Goals (SDGs) from a new angle and how they can be achieved by transforming the global economy through a new pathway of transition. The United Nations Environment Programme (UNEP) has, consequently, identified four SDGs that can support global recovery in the wake of the COVID-19 crisis. These are goal 13, climate actions, goal 15, life on land, goal 14, life below water, and goal 12, responsible consumption and production (UNEP 2020). Addressing goal 13, requires urgent actions to battle climate change and its impacts, thus aligning actions with implementation of the Paris Agreement from 2015 (United Nations 2015). Implementation of SDG 13, furthermore, contributes to the fulfilment of the other 16 SDGs, but it requires increased investment and mobilization of billions of US$ annually to low-carbon development and adaptation to climate change (United Nations Development Programme 2020). However, the progress in this area has not been as expected so that capacities and access to finance has to scaled up at a much faster rate than since the ratification of the Paris Agreement, specifically for small island developing states and the least developed countries (United Nations 2019). This chapter aims to assess the impact of COVID-19 on Agenda 2030 specifically focusing on SDG 13, i.e. climate actions. Given the importance of finance, and thus investors, for implementation of SDG 13, the chapter intends to answer the following questions:

- What is the impact of COVID-19 on investors'?
- What is the role of investors in addressing SDG 13 (climate actions) in the wake of the COVID-19 crisis?
- What opportunities are there for investors and companies to address SDG 13 (climate actions)?

To address these questions the paper is structured around two focal points: (a) the impacts of COVID-19 on investors', and (b) the role of investors in addressing the SDGs and the opportunities this entails for investors and businesses. Previous studies have analysed the impact of the SDGs on businesses (Schramade 2017) by highlighting how Corporate Social Responsibilities (CSR) activities undertaken by companies are aligned with the SDGs (Poddar et al. 2019). Particularly, SDG 13 can produce relevant implications for finance and investors (Doni et al. 2020) by stimulating awareness on the climate change effects on the financial performance of private sector (Gallego-Álvarez et al. 2014; Gulluscio et al. 2020; Pizzi et al. 2020). Although some recent studies emphasized the interrelationship between sustainable development and global risks (Cernev and Fenner 2020; Pan and Zhang 2020), particularly, risk of emerging infection diseases (EIDs) such as COVID-19, has to be extensively analysed as it is a crucial component of the sustainable development planning (Di Marco et al. 2020). This risk may be relevant also in the light of the human rights framework (Fleetwood 2020) with relevant effects on the process of the Agenda 2030 implementation (Leal Filho et al. 2020). As Cernev and Fenner (2020) demonstrated, a set of targets, that includes SDG 13, can act as an important leverage point in preserving healthy human and environmental resources in the light of catastrophic and existential risks. Given the lack of academic literature on the link between SDG 13, climate actions and COVID-19, so far, the chapter relies on available information from relevant stakeholders, both international organizations, and investors alike. From a methodological point of view, this chapter uses a qualitative approach to give a first assessment on a relatively new field in order to stimulate a further development of this research. Methods used involve an analysis of both empirical and theoretical studies, a collection of comments on the main emerging issues and the current global trends (Leal Filho et al. 2020).

2 The Impact of COVID-19 on the Investors' Context

The COVID-19 pandemic situation highlights the issue of systemic risk (CERES 2020; Johannsdottir and Cook 2019; Schwarcz 2008; Thurm et al. 2018) and opportunities, that is if investors are willing to help countries and communities rebuild economies and societies in more sustainable and resilient way than before the crisis by investing in green technology, renewable energy, and new sustainable sectors

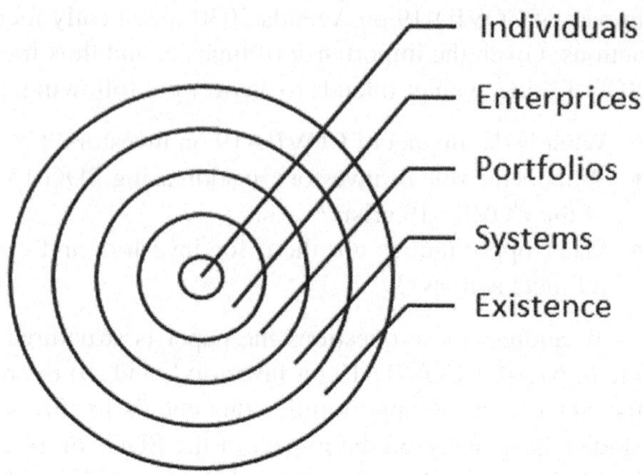

Fig. 1 Systemic risk. Model inspired by Johannsdottir and Cook (2019), Thurm et al. (2018)

that will help the global economy transitioning fast towards decarbonisation (UNEP 2020). Systemic risk includes risk to individuals, enterprises, portfolios, systems, and existance of species, such as in case of climate change (Johannsdottir and Cook 2019; Thurm et al. 2018), see Fig. 1. From a societal, economic and business prespective the society is a set of systems (Kennedy et al. 2017), where interactions between systems take place on micro, meso and macro levels going beyond time and place (Normann 2001). The COVID-19 pandemic demonstrates vulnerability of systems and the collective well-being of humans, thus bringing the discussion to the systemic level in case of health and fragility of systems (CERES 2020) that are increasingly interconnected (The Investment Integration Project 2020).

Systemic risk has, whether it is climate or pandemic-related, the ability to "destabilize capital markets and lead to serious negative consequences for financial institutions and the broader economy" (CERES 2020, p. 6). Therefore, investors are forced think in terms of different scales regarding risks and solutions and reconsider their consumptions about modern finance. To hedge against systemic risk investors' strategy is typically to diversify their investment portfolios. Previously, investors have mainly focused on enterprise level risks or risks in investment portfolios (meso level), although more recently they are starting to recognize systemic risk (macro level). The existential risk (cosmic level) is to a lesser degree recognized (Thurm et al. 2018), but understanding of systemic and existential risks is essential to deal with wicked problems (Batie 2008) such as the COVID-19 pandemic. Wicked problems differ from ordinary problems which is hard to deal with, as the challenges are unprecedented and the situation is constantly changing. The problem implicates many stakeholders, the roots of the issue are complicated and tangled, it is difficult to come to grips with the problem as it changes when attempts are made to address it, the challenge is unprecedented, and there are no indication of the right way to address the problem (Camillus 2008). Taming wicked problems requires systems thinking (Kennedy et al. 2017). The claim is that "COVID-19 has exposed how the financial sector has undervalued the importance of social impacts", and that "capital markets must better factor in the risk of future massive systemic failures". Furthermore, "[i]nvestors need to understand their role in ensuring that inevitable, future shocks to social systems are not as financially or economically catastrophic as COVID-19 has been" (The Investment Integration Project 2020).

Green fiscal stimulus packages, related to the COVID-19 crisis, on national, regional, and sub-regional levels, and green financing are encouraged and supported by United Nations Environmental Programme (UNEP). This should help prioritizing stimulus and finance towards sustainable jobs and income, investments in ecological and social foundations, support low-carbon production and consumption, public wealth, and steer responsible finance towards climate stability (UNEP 2020). Opportunities in relation to rebuilding a climate resilient and sustainable economy are in sectors such as food systems, building and construction, energy transition, mobility, and waste handling. It is furthermore suggested that policies put forth in so-called Global Green New Deal (GGND) following the economic crisis serve as lessons learned and can be valuable in rebuilding the world economy in the aftermath of the COVID-19 crisis (UNEP 2020). Comparably, GGND suggested prioritizing investments towards energy-efficiency, renewable energy, sustainable transportation, freshwater management and sanitation, food security, while reducing substantially subsides towards fossil fuels (United Nations Environment Programme 2009).

Sustainable finance has been defined as "any form of financial service integrating environmental, social and governance (ESG) criteria into the business or investment decisions for the lasting benefit of both clients and society at large", but sustainable finance "include sustainable funds, green bonds, impact investing, microfinance, active ownership, credits for sustainable projects and development of the whole financial system in a more sustainable way (Swiss Sustainable Finance n.d.). Another definition states that sustainable finance should create environmental, social and governance value that are sustainable over time (UNEP Finance Initiative 2016). As such, sustainable finance should support sustainable development and the implementation of the sustainable development goals, including Goal 13. More specifically, within the banking sector sustainable finance refers to the following business lines: (1) commerce and corporate lending, (2) project finance, (3) investment banking, and (4) trade finance. Energy efficiency and financing companies in their transition process towards sustainability are examples of commerce and corporate lending category, and sustainable energy infrastructure, such as renewable energy, energy efficiency, and smart meters, fall within the project finance category. The outcome should include, among other things, climate change mitigation and financial inclusion (UNEP Finance Initiative 2016). Some of the principles used to influence and mainstream commitment and actions of financial institutions regarding climate change and sustainability include the Principles for Positive Impact Finance (UNEP Finance Initiative n.d.-a), the Principles for Sustainable Insurance (UNEP Finance Initiative n.d.-c), the Principles for Responsible Banking (UNEP Finance Initiative n.d.-b), and the Principles for Responsible Investment (PRI n.d.). It is stated that signatories to the Principles for Responsible Investment (PRI) "should be supporting sustainable companies through this crisis – in the interests of public health and long-term economic performance – even if that limits short-term returns" (PRI 2020a).

In the interconnected financial system, financial institutions around the globe are dealing with the effects of the COVID-19 pandemic. Just as any other businesses, they need to plan their own recovery path. Health and well-being of employees comes first, followed by technological and productivity improvements, reconnection with customers, engagement with unconventional partners to develop new value propositions, stronger focus on ESG issues and considerations, and improved alertness and risk management (Liddy 2020a). In dealing with these issues leaders within the financial services focus on challenges in six key areas, namely employees, customers, liquidity, relationship with suppliers dependency on third-parties, communication and transparency, and scenario planning (Liddy 2020b). Although, financial institutions need to address these key areas they are also applicable for other types of industries. In the wake of the pandemic crisis huge level of growth is predicted in remote banking, given that banks and customers have had to change their behaviour by progressively moving to digital and cashless solutions (Caplain 2020). Other long-term impacts include a focus on cyber security, reduction of interest rates, diminished business activities', non-performing loans, and more (Deloitte 2020). For banks there are also opportunities related merger and acquisition (M&A) activities and restructuring of numerous industries (Deloitte 2020).

The COVID-19 pandemic has manifold impact on the insurance sector, including non-life (general) and life and health insurers, and consequently reinsurers. In case of non-life insurers pandemic risks may be excluded in insurance policies, but such exclusion clauses were in some cases strengthened after the SARS coronavirus outbreak in 2003 in case of travel insurance and business interruption. Instead, losses may be recorded in case of cancellation coverage as such policies may cover risk of pandemics. Other insurance types potentially affected are trade credit insurance, paid out when customers or supplies cannot pay their debts, or workers' compensation insurance in cases when they are not sufficiently protected against exposure to the coronavirus. Insurers offering life and health insurances may be affected by higher morbidity and mortality rates. Reinsurers are then affected indirectly through their insurance clients. Decline in interest rates and volatile markets will, additionally, impact the financial assets of insurers and their investment portfolios (Kölschbach 2020). During the COVID-19 outbreak some insurers have stepped up by developing insurance policies, either as standalone products or by extending terms in existing products, to protect their policyholders from the COVID-19 risk or by making donations to those on need, including hospitals and/or healthcare workers. The COVID-19 crisis is also pressuring insurers to further digitalize their daily operations, both in their interaction with customers and their agents. Consequently there is less need for physical offices, thus with implications' for the real estate market. The size and the scale of the COVID-19 crisis is such that business continuity plans of insurers need to be reviewed, both by the sector and regulators, and what is the future role of the insurance sector and how major risks could or should be mitigated and priced (Hay n.d.). Although premiums should reflect the risk policyholder are exposed to, they may become so high that they will exceed clients' ability or willingness to take out insurances covering pandemics.

The private equity firms and funds are mounting up cash according to analysts (Lachenmeier et al. 2020), suggesting investment opportunities. Post-COVID-19 opportunities include investments in firms others are afraid to invest in, namely businesses and economic sectors struggling hard in the wake of the pandemic crisis leading to short-term liquidly shortage (Lachenmeier et al. 2020; Menghi et al. 2020). Through flexible funding solutions private equity firms and funds can bring capital to the table, thus possible help preserve jobs, restructure debt, and help corporate managers steer through the crisis. Some of the equity firms are foreseeing how to carry this through. Alternatively, private equity firms may offer debt-financing in relation to small-business bailouts if they are putting their capital to work. The downside is that there is still lack of information on the impact of COVID-19 and how the virus will progress and contained (Menghi et al. 2020), and it is also recognized that equity funds and firms may earn immense returns through many small but messy deals (Ohrenstein n.d.), therefore not without caution and criticism if scrutinized through the lens of ESG and sustainable finance.

In the context of sustainable development and sustainable finance UNEP Finance Initiative has explained the link between climate, green and sustainable finance, but a framework proposed includes four categories, namely environment, social, economic and governance categories, thus adding economic to the previously ESG categories. The environmental categories can be subdivided into three categories, climate change mitigation, climate change adaptation, and other environmental issues. On the whole spectrum of the four categories climate change mitigation focuses on low-carbon finance, climate change mitigation and adaptation combined are categorised as climate finance, adding other environmental issues to these categories broadens the finance to green finance. The environmental category, with its subcategories, and the social category combined represent socio-environmental finance, and all the four main categories, ESG plus economics, symbolize the sustainable finance (European Commission 2017). This suggests that the environmental category is most important when implementing Sustainable Development Goal (SDG) 13 (climate change) keeping in mind that the implementation of SDG 13 contributes to the fulfilment of the other 16 SDGs (United Nations Development Programme 2020), see further discussion in the following section.

Green indexes, such as FTSE Russell, including FTSE Green Revenues which supports low carbon economy by recognizing eight industry sectors and 60 sub-sectors critical for energy transition, MSCI which uses Global Environment Indexes Methodology (GEIM) and Global Climate Index Methodology (GCIM) for screening different sectors, and NASDAQ offering a large number of environmental stock indexes, such as the Nasdaq Green Economy Global Benchmark Index (QGREEN) (European Commission 2017). The S&P Global Ratings Green Evaluation weighs, in their green evaluation, three categories. These are transparency, governance, and mitigation or adaptation. The mitigation criteria is the most important with regards to Goal 13, where "buildings, industrial efficiencies, energy infrastructure, transport, and water" are evaluated (Standard & Poor's Financial Services LLC 2017, p. 4). The net benefit ranking is based on key performance indicators (KPIs) such as carbon, waste or water. Then a specific Carbon Hierarchy is employed, followed by evaluation of environmental impact, finally resulting in a mitigation score. The Carbon Hierarchy has five categories, showing an increased contribution to avoidance of climate change (Standard & Poor's Financial Services LLC 2017):

- **Improvement of fossil fuel based activities' in regards to environmental efficiency and impact**. This includes a transition from coal to natural gas, clean fuel production, and clean use of coal.
- **Decarbonization technologies entailing environmental risks**. This includes nuclear technology and large hydro projects in tropical areas.
- **Decarbonization by lessening emissions in energy intensive industries**. This includes industrial efficiencies, green transport, and green buildings refurbishment, new build best standards, and energy-efficient products.
- **Significant decarbonization of main sectors through low-carbon solutions**. This includes green transport apart from hybrid/fuel efficient vehicles, green buildings, or new built best standards/net zero.
- **Systematic decarbonization of economies**. This includes wind and solar power technology, small and large hydro projects, in addition to energy management and control (Standard & Poor's Financial Services LLC 2017, p. 5).

The European Union (EU) Technical Expert Group (TEG) on Sustainable Finance presumes that the EU Green Bond Standard, Paris-Aligned and Climate Transition Benchmarks, and Sustainable Taxonomy may be useful for developing plans for the public and the private sector on how to recover from COVID-19 pandemic (EU Technical Expert Group on Sustainable Finance 2020a). Furthermore, EU's Roadmap to Recovery is intended to result in more resilience, fair and sustainable Europe. The principles guiding the Recovery Roadmap are solidarity, cohesion, convergence, inclusive, and co-owned, but the approach needs to be flexible and agile and evolve over time and grounded in EUs set of values and rights. In this regard the European Green Deal (European Commission 2020), but it aims for reaching the 2030 goal, and climate neutrality by 2050 (EU Technical Expert Group on Sustainable Finance 2020a).

To push EU's Sustainable Finance plans further the TEG suggests that "the European Commission creates a voluntary standard to enhance the effectiveness, transparency, accountability, comparability and credibility of the green bond market without disrupting the market, and to encourage bond issuers to issue their bonds as 'EU Green Bonds' (EU Technical Expert Group on Sustainable Finance 2019, p. 24)", but the core element of the standard are green projects, a Green Bond Framework (GBF), allocation and impact reporting, and verification of outcomes (EU Technical Expert Group on Sustainable Finance 2019). Climate benchmarks and benchmarks for ESG disclosure are "aimed at reallocating capital towards a low-carbon and climate resilient economy", one being EU Climate Transition Benchmark and the other one EU Paris-aligned Benchmark (EU Technical Expert Group on Sustainable Finance n.d., p. 2). The objectives of these benchmarks are similar, but the ambition level differ whereas the latter one is stricter in terms of minimum standards of disclosure (EU Technical Expert Group on Sustainable Finance n.d.). Additionally, the EU Taxonomy is put forth as a tool to guide investors, companies, issuers and promoters of projects with regards to low-carbon, resource-efficient, and resilient economy, where a performance threshold has to be reached, in terms of substantial contribution, in one of the following areas: (1) climate change mitigation, (2) climate change adaptation, (3) sustainable and protection of water and marine resources, (4) transition to a circular economy, (5) pollution prevention and control, (6) protection and restoration of biodiversity and ecosystems (EU Technical Expert Group on Sustainable Finance 2020b, p. 2).

3 SDG 13 Targets and Climate Change: COVID-19 as an Opportunity for Investors and Companies to Preserve Our Planet

The role of finance is addressed in the target 13.A of SDG 13, where the "*goal of mobilizing jointly $100 billion annually by 2020 from all sources to address the needs of developing countries in the context of meaningful mitigation actions and transparency on implementation and fully operationalize the Green Climate Fund through its capitalization as soon as possible*" (United Nations 2019), is specifically stated. Indicators for successful implementation of SDG 13, include 13.1.A, stating: "Mobilized amount of United States dollars per year starting in 2020 accountable towards the $100 billion commitment" and 13.1.B, where the role of finance is recognized as an important mean for supporting the least developed countries and small island developing states so capacities for planning and managing effectively climate change can be strengthen, such as by focusing on local and marginalized communities, women, and youth (United Nations 2019).

In discussing SDG 13 in the context of COVID-19 the United Nations Environment Programme (UNEP) states that the "*climate crisis may be seen as a slower moving crisis than the speed of this global pandemic, but it's the long-term effects are likely to be far more threatening*". This is due a situation where global temperature rise of the planet is on a track for a 3.2° warming, without increased commitment to decarbonisation (UNEP 2020). This situation increases the likelihood of extreme weather events, widespread destabilization of the global food system, flooding and droughts, pandemics, and negative economic and security issues. Furthermore, this will interrupt gains attained to address most of the other SDGs, as well as economic recovery in the wake of COVID-19 crisis (UNEP 2020).

"*An unplanned transition to a low-or-zero carbon economy can cripple key industries*", given that "*government policies, consumers sentiments, liability risks and technological innovation*" are ignored, potentially resulting in accumulative losses, and thus of great importance for the investment community given potential ripple effects on financial institutions (CERES 2020, p. vii). Even though markets have been slow to reflect the climate change as a risk factor, but are on the edge of restructuring finance fundamentally given a rapid change in the understanding of climate change physical impacts and the global economy. Investors are starting to understand climate risk as an investment risk, thus figuring out how investment portfolios need to be modified (Fink 2020). It is now considered a fiduciary of investors to help clients address these risk-factors (Fink 2020; Sullivan et al. 2015). Some major investors are therefore avoiding and selling-off shares in non-renewables, mainly coal production (Fink 2020; Lloyds of London & University of Oxford 2017), and transitioning to renewables even though such transition will take some decades. What is also brought forth that energy transition needs to be fair and just (Fink 2020), thus aligning with the implementation of the Sustainable Development Goals (United Nations Development Programme 2020). This requires improved disclosure for shareholders, and other stakeholders, on how companies manage climate and sustainability-related aspects of their business, even going so-far as in the case of BlackRock to *"vote against management and board of directors when companies are not making sufficient progress on sustainability-related disclosure and then business practices and plans underlying them"* and making sustainability as the new standard for investing (Fink 2020).

Role of the finance markets in response to climate risk and its socioeconomic impacts and physical hazards are through insurance availability, cost, terms and conditions (Dlugolecki 2009; Jóhannsdóttir et al. 2012), lending practices (Woetzel et al. 2020), just to name few. Decarbonisation is seen as a part of a climate change management approach in case of reducing and avoiding build-up of climate-related risks by scaling up solutions so that the goal of net-zero emissions can be reached. This means, more precisely, to remove carbon to a possible extent from future economic activities and from the atmosphere, highlighting the importance of decarbonizing investments supporting transition to renewable energy (Woetzel et al. 2020).

Given these reflections, decarbonisation can be used as an optimal opportunity to reallocate capitals and to foster investors' engagement in the transition to a lower carbon economy (Ceres 2020; Faioli and Natolli 2020) but it is also worthwhile to consider the social impacts that previously "had been neglected as externalities" (CERES 2020, p. 11). All businesses can be negatively affected by decabornisation, particularly some industries such as fossil fuel companies but also other sectors like cement, transportation, utilities, heavy industry, agriculture may suffer a significant decrease in their assets value or they can be exposed to serious climate risks and related legal liability (CERES 2020). The increasing focus on climate losses and the risk for companies to be considered responsible for climate damages can represent a serious weakness.

Nevertheless, it can be a strength because the pandemic crisis can exacerbate companies' responses to fight climate change by facilitating the achievement of some targets included in SDG 13. For example as a prompt reaction, a group of 155 large companies (IISD 2020) signed an agreement *"urging governments to align their COVID-19 economic aid and recovery efforts with the latest climate science"* (IISD, SDG Knowledge Hub 2020). As the current pandemic is a global disruption, climate change is *"as urgent as ever"* (IISD, SDG Knowledge Hub 2020). It is important to consider that all organizations have to fight an interconnected crisis that link the global climate crisis with relevant impacts on both human life and natural ecosystem. Given that, at micro.-level a number of 155 companies, that are part of the Science Based Targets initiative and its Business Ambition of 1.5 °C Campaign, reaffirmed their own science based climate commitments. This agreement identified three specific aims to recover better from COVID-19: (1) demonstrate that the best decisions and actions are grounded in science; (2) invest in recovery and resilience for a systemic socio-economic transformation; (3) work with Governments and scale-up the movements. This initiative emphasizes the importance to involve the public sector i.e. governments and policy makers (Doni et al. 2020), in the alignment of its recovery efforts to the companies' ambitious to reach net-zero emissions well before 2050 (UNGC 2020).

As mentioned before COVID-19 and climate emergency can be seen as both a systemic risk that is determining disruptive effects on financial markets, assets valuation and global economic stability (CERES 2020). By considering the global impact of this health crisis on all investors the engagement investors' process towards solidarity can represent an unusual opportunity to push investors towards green and sustainable finance. As PRI emphasized the signatories of PRI should influence companies during this crisis to focus on long-term economic performance rather than short-term returns (PRI 2020a). In this perspective, seven immediate investors' actions have been identified, among these actions, Action 1: engage companies that are failing in their crisis management; Action 3: re-prioritise engagement on other topics, Action 7: maintain a long-term focus in investment decision-making (PRI 2020a). As a general remark, PRI argued that this global crisis would determine radical changes in our financial system and governments should support companies and industries that can help to respond to climate emergency and inequality rather than those risk sharpening them (PRI 2020a). In this respect PRI commissioned the *Inevitable Policy Response*, a policy forecast to support investors anticipate and navigate transition risk (PRI 2020b). This pioneer project aims to prepare financial markets that have not adequately evaluated "the likely near term policy response to climate change" (PRI 2020c). As COVID—seems to be very similar to climate risk, we have to take into account that climate related risks continue to increase and it is much to learn from this health crisis and how to manage the related risks.

The COVID-19 virus and the economic crisis need to calibrate a global response to the SDGs. In doing this, the UN-led Inter-Agency Task Force on Financing for Development *"just called for in its new Financing for Development Report, a global and multilateral response is needed that attacks the virus and puts the global economy back on a path to achieve the Sustainable Development Goals (SDGs) and the Paris climate agreement"* (UN 2020b). That effort should start involving an effective action by the International Monetary Fund (IMF) and the World Bank to organize a radical emergency response *"to stem the crisis in the developing world in order to steer their economies toward the SDGs"* (UN 2020b). Developing countries are faced with significant devaluation of their exchange rates and a ballooning of debt. Such a situation *"is wreaking immediate havoc and derailing efforts to meet the SDGs"* (UN 2020b). Particularly, we should take into account SDG 14 and SDG 15 as they strictly related to biodiversity issues. Conservation efforts to fight illicit poaching and trafficking of wildlife, to protect animal breeds and to preserve rural areas in developing countries, should be enhanced and take in place (UN 2020a).

In this perspective, we have to point out some effects after COVID-19 that can determine both positive and negative effects in terms of the achievement of the SDG 13 targets. Regarding to positive effects we have to take into account some benefits in terms of pollution and human health. A relevant aspect is the reduction in pollution and carbon emissions during this period of lockdown. Some organizations, at local and international level, are checking the current benefits from the shutdown of factories and business activities on sea, land and air pollution. For example the European Environment Agency's (EEA) is assessing how coronavirus measures have influenced concentrations of air pollution by developing a viewer that tracks the weekly average concentrations of nitrogen dioxide (NO_2) and particular matter (PM10 and PM2.5). The drop in air pollution should be investigated in the view of forward-looking investments and ambitious policies to move towards a resilient and sustainable society (EEA 2020). Nevertheless, we cannot avoid some reflections on the further steps of the health crisis. *"Fall in COVID-linked carbon emissions won't halt climate change"* as UN weather agency chief said. The expected drop in gas emissions cannot be considered a stable and lasting consequence linked to the economic crisis after the COVID-19 emergency. It is "only short-term good news", the Head of the UN weather agency said recently (UN News 2020). The World Meteorological Organization (WMO) argues that the reboot of the global economy will cause the return of emissions to normal levels. Furthermore, *"there might even be a boost in emissions because some of the industries have been stopped"*, the WMO head cautioned (UN News 2020). In the last decade, the "One Health" approach has become increasingly popular at a global level, which recognizes that human health is closely linked to animal and environmental health. A strategic concept formally recognized by several United Nations bodies such as UNEP, UNDP, WHO, FAO, the World Organization for Animal Health (OiE), the European Commission, research institutes around the world, NGOs and others bodies. "One Health" identifies an holistic concept of health of people, animals, plants, living and working environments and ecosystems, promoting the application of a multidisciplinary and collaborative approach to address the potential or active risks that originate from the interface between living and working environment, animal populations and ecosystems. To make the "One Health" approach truly effective, it is necessary to establish a better and systematic interaction between professional groups with greater skills in this regard, in particular between doctors and veterinarians, epidemiologists, ecologists and wildlife experts, but also sociologists, economists, jurists. Only by acknowledging that our health and well-being are closely related to those of the nature that hosts us, we can protect our species from the most harmful effects of the pandemics. An effective and sustainable way of action should therefore ensure the natural functioning of the ecosystems and their careful management to regulate diseases, hinder their spread and thus reduce their impact on human health (WWF Italy 2020).

Particularly, the COVID-19 shock can cause negative impacts on the different kinds of capitals, such as financial capital (reduction of production capacity, fall in investment, etc.), human capital (unemployment and underemployment reduce the knowledge embedded in individuals) and social capital (reduction of social interactions).

Immediately after the crisis an Italian organization (ASVIS 2020) carried out a preliminary quantitative assessment of the COVID-19 likely impact on the SDGs by evaluating the predictable trend of the over 100 indicators that are used to elaborate the composite indices of the 17 goals. Evidences highlighted a large negative impact on SDG 1, 3, 4, 8, 9 and 10, on the other hand, positive effect is on SDG 13 and 16. Particularly, SDG 13 can take advantage from the overall improvement of climate indicators because of the lockdown.

As a final remark, coronavirus pandemic represents an excellent opportunity to tackle the SDGs in different ways as some recent studies confirmed. For example, Pan and Zhang (in press) argued that COVID-19 suggests a new critical journey of tackling the SDGs by developing the concept and practice of digital sustainability. Moreover, an interesting analysis demonstrated that the achievement of the foundational SDGs is essential to reduce global catastrophic and existential risks (Cernev and Fenner 2020). Taking into account the extensive interdependencies between the 17 SDGs this study demonstrated that SDG 13 with SDG 4, with other ones (2, 8, 12 and 16), is the most important to be used to reduce potential long-term global risks for humanity (Cernev and Fenner 2020). Given that, the interaction of all SDGs (Biehl and Thomson 2020) can also help to move the global system to desirable outcomes and to reduce the currently increasing of risks of disasters.

4 Conclusions

The chapter aimed to assess the impact of COVID-19 on Agenda 2030 specifically focusing on SDG 13, i.e. climate actions. Given the importance of finance, and thus investors, for implementation of SDG 13, the chapter intended to answer the following questions:

- What is the impact of COVID-19 on investors?
- What is the role of investors in addressing SDG 13 (climate actions) in the wake of the COVID-19 crisis?
- What opportunities are there for investors and companies to address SDGs 13 (climate actions)?

We have identified different issues to answer these questions and our analysis can contribute to shed light into the potential impact of COVID-19 on SDG 13 in different ways, despite its significant threat for human lives and livelihoods. A key lesson from this crisis is the profound revision of our fundamental assumptions in terms of achieving targets of human well-being and environmental protection (Kassouri and Altintas 2020). The achievement of the SDGs can also support the developing countries in their vulnerability to COVID-19 (Barbier and Burgess 2020). What the discussion also brings forth is an urgent need to revise and restructure the financial system in order to make it more linked to the social issues, like human needs and human rights, as an interesting study recently highlighted about the human rights implications from climate change (UNEP Inquiry 2016). The risk for finance is that the global recession forced investors to evaluate the rate of return and the risk of investment only. It is essential to promote investments towards the SDGs by reconsidering the optimal portfolio allocation by institutional investors (Yoshino et al. in press). Particularly, SDG 13 can offer useful insights to accelerate the transition to low-carbon economy. Some key lessons can be learned from the pandemic crisis by considering how an increase in investments in climate-resilient infrastructure can accelerate the transition to a lower carbon future and a resilient economy by causing a significant effect on near-term job creation (McKinsey Quarterly 2020). A further development of this study may investigate businesses' and investors' reactions to this emergency by carrying out an empirical analysis on climate actions and policies undertaken by different industries in the private sector, in Europe and in the world. As the European Union argued, "there is no place for business as usual" and "it is now time to chart the path towards a comprehensive recovery plan, with the ultimate objective of building a more resilient, sustainable and fair Europe" (European Commission 2020). This key statement is also valid at the global level and the process of implementation of the SDGs cannot be jeopardize by COVID-19 (Leal Filho et al. 2020).

References

ASVIS (2020) Politiche per fronteggiare la crisi da COVID-19 e realizzare l'Agenda 2030 (Policies to fight the COVID-19 crisis and achieve the Agenda 2030). Retrieved from https://asvis.it/public/asvis2/files/Pubblicazioni/RapportoASviSCovidAgenda2030.pdf

Barbier EB, Burgess JC (2020) Sustainability and development after COVID-19. World Dev 135(2020):105082 in press

Batie SS (2008) Wicked problems and applied economics. Am J Agr Econ 5(2008):1176–1191. https://doi.org/10.1111/j.1467-8276.2008.01202.x

Biehl C, Thomson I (2020) COVID 19 Interconnectedness: addressing the crisis as a 'Good Steward. Retrieved from https://www.global-solutions-initiative.org/wp-content/uploads/2020/04/COVID-19-Being-an-Active-Owner_305_1.pdf

Bieri S, Breu T, Heinimann A, Messerli P (2020) Corona Sustainability Compass. Retrieved from https://www.csc-blog.org/en/virus-has-made-it-clear-future-now

Camillus JC (2008) Strategy as a Wicked Problem. Harvard Bus Rev 1–15

Caplain J (2020) A catalyst for change for bank branches. Will the COVID-19 pandemic change the shape of bank branch networks? Retrieved from https://home.kpmg/xx/en/blogs/home/posts/2020/04/a-catalyst-for-change-for-bank-branches.html

CERES (2020) Addressing climate as a systemic risk: a call to action for U.S. financial regulators. Retrieved from https://www.ceres.org/resources/reports/addressing-climate-systemic-risk

Cernev T, Fenner R (2020) The importance of achieving foundational Sustainable Development Goals in reducing global risk. Futures 115:102492

Deloitte (2020) Understanding the sector impact of COVID-19. Banking & Capital Markets. Retrieved from file:///C:/Users/laraj/Downloads/COVID-19-Impact-Banking-Cap-Market-Sector.pdf

Di Marco M, Baker ML, Daszak P, De Barro P, Eskew EA, Cecile MG, Harwood TD, Herrero M, Hoskins AJ, Johnson E, Karesh WB, Machalaba C, Navarro Garcia J, Paini D, Pirzl R, Stafford

Smith M, Zambrana-Torrelio C, Ferrier S (2020) Opinion: sustainable development must account for pandemic risk. Proc Natl Acad Sci 117(8):3888–3892. https://doi.org/10.1073/pnas.2001655117
Dlugolecki A (ed) (2009) Coping with climate change: risks and opportunities for insurers. The Chartered Insurance Institute (CII), London
Doni F, Gasperini A, Torres Soares JF (2020) SDG13-climate action concise guides to the United Nations sustainable development goals series SDG13-climate action combating climate change and its impacts. Emerald Publishing Limited, UK
EU Technical Expert Group on Sustainable Finance (2019) Report on EU Green Bond Standard [Press release]
EU Technical Expert Group on Sustainable Finance (2020a) Sustainable recovery from the Covid-19 pandemic requires the right tools. Retrieved from https://ec.europa.eu/info/sites/info/files/business_economy_euro/banking_and_finance/documents/200426-sustainable-finance-teg-statement-recovery_en.pdf
EU Technical Expert Group on Sustainable Finance (2020b) Technical Report. Retrieved from Brussels
EU Technical Expert Group on Sustainable Finance (n.d.) Climate Benchmarks and Benchmarks' ESG Disclosures. Retrieved from https://ec.europa.eu/info/sites/info/files/business_economy_euro/events/documents/finance-events-190624-presentation-climate-benchmarks_en.pdf
European Commission (2017) Defining "green" in the context of green finance (final report). Retrieved from Brussels
European Commission (2020) A roadmap for recovery: towards a more resilient, sustainable and fair Europe. Retrieved from Brussels: https://www.consilium.europa.eu/media/43384/roadmap-for-recovery-final-21-04-2020.pdf
European Environment Agency (EEA) (2020) Air quality and COVID-19. Retrieved from https://www.eea.europa.eu/themes/air/air-quality-and-covid19
Faioli I, Natolli F (2020) The COVID-19 crisis and the future of the green economy transition, Covid-19 Notes, June 17, 2020, Banca d'Italia Eurosistema. Retrieved from https://www.bancaditalia.it/media/notizie/2020/2020.06.17-FaiellaNatoli_BdI_Covid_notes.pdf
Fink L (2020) A fundamental reshaping of Finance. Retrieved from https://www.blackrock.com/us/individual/larry-fink-ceo-letter
Fleetwood J (2020) Social justice, food loss, and the sustainable development goals in the era of COVID-19. Sustainability 12(12):5027. https://doi.org/10.3390/su12125027
Gallego-Álvarez I, García-Sánchez IM, da Silva Vieira C (2014) Climate change and financial performance in times of crisis. Bus Strategy Environ 23:361–374
Gulluscio C, Puntillo P, Luciani V, Huisingh D (2020) Climate change accounting and reporting: a systematic literature review. Sustainability 12:5455
Hay LJ (n.d.) COVID-19: the global insurance response. Global perspective on the various ways the insurance sector in affected countries have responded to COVID-19. Retrieved from https://home.kpmg/xx/en/home/insights/2020/04/covid-19-global-insurance-response.html
IISD SDG Knowledge Hub (2020) 155 Companies Urge Governments to Enact Net Zero Emissions Recovery from COVID-19. Retrieved from http://sdg.iisd.org/news/155-companies-urge-governments-to-enact-net-zero-emissions-recovery-from-covid-19/
UNEP Finance Initiative (2016) Guide to banking and sustainability, 2nd edn. Retrieved from Geneva
UNEP Inquiry (2016) Human rights and sustainable finance. Exploring the relationship. Retrieved from https://www.ihrb.org/uploads/reports/IHRB_UNEP_Human_Rights_Sustainable_Finance_Feb2016.pdf
Johannsdottir L, Cook D (2019) Systemic risk of maritime-related oil spills viewed from an Arctic and insurance perspective. Ocean Coast Manage 179(1):104853. https://doi.org/10.1016/j.ocecoaman.2019.104853
Jóhannsdóttir L, Wallace J, Jones A (2012) The primary insurance industry's role in managing climate change risks and opportunities. In: Stoner JAF, Wankel C (eds) Managing climate change

business risks and consequenses: leadership for global sustainability. Palgrave Macmillan, New York

Kassouri Y, Altıntaş H (2020) Human well-being versus ecological footprint in MENA countries: a trade-off? J Environ Manage 263:110405. https://doi.org/10.1016/j.jenvman.2020.110405

Kennedy A-M, Kapitan S, Bajaj N, Bakonyi A, Sands S (2017) Uncovering wicked problem's system structure: seeing the forest for the trees. J Soc Market 7(1):51–73. https://doi.org/10.1108/JSOCM-05-2016-0029

Kölschbach J (2020) What are the specific accounting implications for insurers? Retrieved from https://home.kpmg/xx/en/home/insights/2020/03/covid-19-insurers-11a.html

Lachenmeier SM, Dahl E, Juellund K (2020) Minority investments on the rise in the short-run as private companies seek funding to overcome the COVID-19 crisis. The COVID-19 liquidity gap. Retrieved from https://capitalmind.com/minority-equity-as-solution-to-covid-19-liquidity-gap/

Leal Filho W, Brandli LL, Lange Salvia A, Rayman-Bacchus L, Platje J (2020) COVID-19 and the UN sustainable development goals: threat to solidarity or an opportunity? Sustainability 12:5343

Liddy J (2020a) Five key priorities post COVID-19. Anticipate tomorrow, but deliver today. Retrieved from https://home.kpmg/xx/en/blogs/home/posts/2020/06/financial-service-priorities-post-covid-19.html

Liddy J (2020b) Six considerations in dealing with the impact of COVID-19. A view on the impact for financial institutions. Retrieved from https://home.kpmg/xx/en/blogs/home/posts/2020/03/six-considerations-for-financial-institutions-in-dealing-with-the-impact-of-covid-19.html

Lloyds of London, & University of Oxford (2017) Stranded assets: the transition to a low carbon economy: overview for the insurance industry. Retrieved from https://www.lloyds.com/news-and-insight/risk-insight/library/society-and-security/stranded-assets

Menghi J, Abrol B, Savoy E (2020) Opportunities for private equity post-COVID-19. How can private equity firms help reverse the economic damage? Retrieved from https://www2.deloitte.com/us/en/insights/economy/covid-19/private-equity-m-and-a-deal-activity-post-covid-tax-implications.html?id=us:2em:3pa:financial-services:eng:di:050420

Normann R (2001) Reframing business: when the map changes the landscape. Wiley, Chichester

Ohrenstein RI (n.d.) Differentiated diligence post COVID-19. Deal making just got a whole lot more complex for Private Equity players. Retrieved from https://home.kpmg/xx/en/home/insights/2020/05/differentiated-diligence-after-covid-19.html

Pan SL, Zhang S (2020) From fighting COVID-19 pandemic to tackling sustainable development goals: an opportunity for responsible information systems research. Int J Inf Manage 102196 (in press). https://doi.org/10.1016/j.ijinfomgt.2020.102196

Pinner D, Rogers M, Samandari H (2020) Addressing climate change in a post pandemic world. McKinsey Quarterly, April. Retrieved from https://www.mckinsey.com/~/media/McKinsey/Business%20Functions/Sustainability/Our%20Insights/Addressing%20climate%20change%20in%20a%20post%20pandemic%20world/Addressing-climate-change-in-a-post-pandemic-world-v3.pdf

Pizzi S, Rosati F, Venturelli A (2020) The determinants of business contribution to the 2030 Agenda: introducing the SDG reporting score. Bus Strategy Environ 1–18 (in press)

Poddar A, Narula SA, Zutshi A (2019) A study of corporate social responsibility practices of the top Bombay Stock Exchange 500 companies in India and their alignment with the Sustainable Development Goals. Corp Soc Responsib Environ Manag 26:1184–1205

PRI (2020a) How responsible investors should respond to the COVID-19 coronavirus crisis. Retrieved from https://www.unpri.org/covid-19-resources/how-responsible-investors-should-respond-to-the-covid-19-coronavirus-crisis/5627.article

PRI (2020b) A pandemic was near inevitable—so too is climate-driven disruption. Retrieved from https://www.unpri.org/pri-blog/a-pandemic-was-near-inevitable-so-too-is-climate-driven-disruption/5830.article

PRI (2020c) What is the Inevitably Policy response. Retrieved from https://www.unpri.org/inevitable-policy-response/what-is-the-inevitable-policy-response/4787.article

PRI (n.d.) Principles for responsible investment. Retrieved from https://www.unpri.org/

Schramade W (2017) Investing in the UN sustainable development goals: opportunities for companies and investors. J Appl Corp Financ 29:87–99
Schwarcz SL (2008) Systemic risk. Georgetown Law J 97(1):193–249
Standard & Poor's Financial Services LLC (2017) S&P Global Ratings Green Evaluation: Time to turn over a new leaf? Retrieved from New York
Sullivan R, Martindale W, Feller E, Bordon A (2015) Fiduciary duty in the 21st century. Retrieved from http://www.unepfi.org/fileadmin/documents/fiduciary_duty_21st_century.pdf
Swiss Sustainable Finance (n.d.) What is Sustainable Finance. Retrieved from https://www.sustainablefinance.ch/en/what-is-sustainable-finance-_content—1–1055.html
The Investment Integration Project (2020) TIIP and Moving the Market announce project to advance investor action on addressing systemic social risks in the wake of Covid-19. Retrieved from https://www.tiiproject.com/tiip-and-moving-the-market-announce-project-to-advance-investor-action-on-addressing-systemic-social-risks-in-the-wake-of-covid-19/
Thurm R, Baue B, Lugt CVD (2018) Blueprint 5. A Step-By-Step Approach to Organizational Thriveability and System Value Creation. Retrieved from Berlin
UN News (2020) Fall in COVID linked carbon emissions won't halt climate change UN weather agency chief. Retrieved from https://news.un.org/en/story/2020/04/1062332
UNEP (2020) COVID-19: four sustainable development goals that help future-proof global recovery. Retrieved from https://www.unenvironment.org/news-and-stories/story/covid-19-four-sustainable-development-goals-help-future-proof-global
UNGC (2020) Uniting business and governments to recover better. Retrieved from https://ungc-communications-assets.s3.amazonaws.com/docs/publications/recover-better-statement.pdf
United Nations (2015) The Paris Agreement. Retrieved from https://unfccc.int/files/essential_background/convention/application/pdf/english_paris_agreement.pdf
United Nations (2019) Sustainable development goal 13. Retrieved from https://sustainabledevelopment.un.org/sdg13
United Nations (2020a) Sustainable development goals. Retrieved from https://www.un.org/sustainabledevelopment/biodiversity/
United Nations (2020b) COVID-19 response. Retrieved from https://www.un.org/en/un-coronavirus-communications-team/calibrating-covid-19-crisis-response-sdgs
United Nations Development Programme (2020) Goal 13: climate action. Retrieved from https://www.undp.org/content/undp/en/home/sustainable-development-goals/goal-13-climate-action.html#:~:text=The%20goal%20aims%20to%20mobilize,also%20to%20the%20other%20SDGs
United Nations Environment Programme (2009) Global green new deal: policy brief. Retrieved from Geneva
Woetzel J, Pinner D, Samandari H, Engel H, Krishnan M, Boland B, Powis C (2020) Climate risk and response: physical hazards and socioeconomic impacts. Retrieved from Washington, DC
UNEP Finance Initiative (n.d.-a) Principles and tools for positive impact finance. Retrieved from https://www.unepfi.org/positive-impact/principles-for-positive-impact-finance/
UNEP Finance Initiative (n.d.-b) Principles for responsible banking. Retrieved from https://www.unepfi.org/banking/bankingprinciples/
UNEP Finance Initiative (n.d.-c) PSI principles for sustainable insurance. Retrieved from https://www.unepfi.org/psi/
WWF Italy (2020) Pandemie, l'effetto boomerang della distruzione degli ecosistemi. Tutelare la salute umana conservando la biodiversità (Pandemics, the boomerang effect of the destruction of ecosystems. Protecting human health by preserving biodiversity). https://d24qi7hsckwe9l.cloudfront.net/downloads/pandemie_e_distruzione_degli_ecosistemi.pdf
Yoshino N, Taghizadeh-Hesary F, Otsuka M (2020) Covid-19 and optimal portfolio selection for investment in sustainable development goals, Finance Research Letters, 101695, in press. Retrieved from https://doi.org/10.1016/j.frl.2020.101695

Federica Doni, PhD is a professor in Associate Professor in Business Administration and Accounting at the Department of Business and Law University of Milano-Bicocca. She is the scientific Co-Director Master's Degree First Level Sustainability in law Finance and Management (SiLFiM). She obtained the national qualification for full professor on the field 13 B/1 Business Administration session 2018. Her main research topics are Intangibles and Intellectual Capital, Business Valuation, Corporate and Integrated Reporting, Sustainability and CSR, Gender issues, Corporate Governance, Climate change and SDGs. Her papers have been published on Journal of Intellectual Capital, Social Responsibility Journal, Sustainability, Sustainability Science, Journal of Environmental Management, Journal of Cleaner Production, Environment, Development and Sustainability. She is the Associate Editor of W. Leal Filho, A. Azul, B., & P. Özuyar (eds.), Climate Action. Encyclopedia of the UN Sustainable Development Goals Springer, Cham. She published a book on SDG 13 in the Concise Guides to the United Nations Sustainable Development Goals, Emerald Publishing (2020). She is a member of the Research Group on Sustainable Development Goals (SDGs) organized by Manchester Metropolitan University, UK, coordinated by Prof. Walter Leal, MMU, UK and a member of the Executive Committee of the Special Interest Group on Corporate Governance della British Accounting and Finance Association (BAFASIGCG) coordinated by Prof. Jill Atkins. She is the Principal Investigator of the European project Jean Monnet Module in the Erasmus Plus Application No 611698-EPP-1-2019-1-IT-EPPJMO-MODULE entitled "Sustainability Disclosure in Corporate Reporting. Improvement and harmonization of best practices in European Union" that has been selected (acceptance rate 20%) for the support by EU of 30.000 euros covering the period 1/09/2019-30/08/2022.

Lara Johannsdottir, MBA, PhD is a professor of environment and natural resources at the University of Iceland (UIce) and a former academic director of the Environment and Natural Resources graduate programme (2019) and a Fulbright Arctic Initiative Scholar (2018–2019). She sits within the Faculty of Business Administration, University of Iceland (UIce). Her research focus is the business sector, in particular the insurance sector, and corporate social responsibility, corporate sustainability, sustainable business models, responsible investments, sustainability reporting, and Arctic business affairs. Her papers have been published by the Journal of Cleaner Production, Environmental Science and Policy, Ecological Indiscators, Renewable and Sustainable Energy Reviews, Polar Record, Local Economy, Ocean & Coastal Management, Sustainability, and others. Book chapters have been published by published by Jon Wiley & sons, Palgrave Macmillan and Springer. Dr. Johannsdottir has 14 years of working experience from the Icelandic insurance sector and was a board member of LSBÍ, an Icelandic occupational pension fund, for 8 years.

A Post COVID19 Blueprint for Sustainable Tourism in Small Island Developing States

Denise Galvin, K. Elliott, M. Handijaya Dewantara, L. Lewis, J. P. Platroz, W. Solomon, R. Tello Soberanes, and R. West-Gerald

Abstract The Covid-19 Pandemic has exposed the structural vulnerability of the global tourism industry affecting a vast chain of stakeholders, who whilst independent of each other are inextricably linked. The complexity and dynamism of the tourism industry render it particularly sensitive to internal and external random effects as witnessed with the rapid global expansion of this virulent pandemic and the responding severe quarantine and mobility measures imposed by national governments. Consequently, the fear generated by mass media and social media has led to successive border closures forcing overseas travellers to either cancel or indefinitely postpone travel plans. Could this be the opportunity to develop and implement a

---

D. Galvin (✉)
OGPI, Universidad de Alicante, 03690 Alicante, Spain
e-mail: denise.galvin@ua.es

K. Elliott
Kirk Elliott Limited, P.O. Box 672, Castries C04 101, Saint Lucia
e-mail: Kirk@KirkElliott.com

M. Handijaya Dewantara
Universitas Prasetiya Mulya, BSD City Kavling Edutown, I, 1, Jl, BSD Raya Utama, BSD City, Kec, Pagedangan, Tangerang, Banten 15339, Indonesia

L. Lewis
Rodney's Wellness Retreat Inc., Soufriere, St Mark, Dominica
e-mail: lucilla@rodneyswellness.com

J. P. Platroz
ADAMANDIA, Sustainable Tourism Solutions, 35, rue Docteur Rebatel, 69003 Lyon, France
e-mail: Jpp@adamandia.eu

W. Solomon
Mauritius Conscious Travel, 121 Av. Orchidées, Morc. Jhuboo, P.O. 22308, Trou aux Biches, Mauritius

R. Tello Soberanes
Montserrat Tourism Division, Hubert Buffonge Building, Brades MSR1110, Montserrat
e-mail: hello@mauritiusconscious.com

R. West-Gerald
Montserrat Tourism Division, P.O. Box 7, Brades MSR1110, Montserrat

W. Leal Filho (ed.), *COVID-19: Paving the Way for a More Sustainable World*, World Sustainability Series, https://doi.org/10.1007/978-3-030-69284-1_4

blueprint for adaptive and resilient tourism particularly in Small Island Developing States (SIDS)? The tourism sectors of these island states are highly vulnerable to external shocks the impacts of which are far reaching and potentially disastrous for long term national well-being. This pause provides a moment in time for the stakeholders of the Eastern Caribbean, Mauritius and Indonesia to ponder what robust and flexible strategic moves embrace multiple versions of a 'new normal' framed around well-being for local communities and visitors, sustainable economic growth and adequate mitigation measures to protect and conserve the environment.

Keywords Blueprint · Adaptative · Resilience · Well-being · New normal

1 Introduction

The world is dealing with a pandemic on a scale not experienced in recent times and it is affecting every aspect of life that involves movement including tourism. As tourism is an optional leisure activity whenever the sector is dealing with the negative impacts of an external shock tourists tend to postpone the desire to travel, particularly if their security and stability are to be placed into jeopardy. Up until now the sector has been able to bounce back largely due to continuous exponential growth in markets based on the realisation at national levels that the sector was responsible for significant inputs to the GDP of both developed and developing nations (UNWTO 2020a). Although, the sector has historically displayed structural vulnerability up until this point in time it has always recovered, with the aspiration of establishing new 'sustainable' states for tourism ecosystems to receive ever increasing numbers of tourists.

Despite a series of catastrophic events in the form of the economic crises created by two Gulf Wars in 1973 and 1979, the September 11, 2001 terrorist attack in the USA, the tsunamis in the Indian Ocean in 2004 and Japan in 2011, a series of catastrophic earthquakes in Iran, Haiti, Japan and Nepal, numerous plane hijackings, multiple terrorist attacks and other life threatening epidemics, the financial crisis of 2008, and the Arab Spring revolts in 2011, the global tourism sector has always been resilient enough to rebound maintaining a path of exponential growth. A notable exception however, was a 3.9% drop in 2009 in tourist numbers which was linked to global economic and environmental crises; and, the uncertainty arising from the A(H1N1) pandemic (UNWTO 2014).

Although 'mass tourism' did not emerge until the 1960s as a process of industrialisation (Christin 2017) this phenomenon has been largely responsible over the last two decades for the noticeable increase in tourist numbers. The Chinese outbound tourist market witnessed growth of +14% for international trips in the first half of 2019 (UNWTO 2019) has been one of the more recent drivers of mass tourism. Also, a new type of traveller commonly referred to as Insta tourists (associated with social media influencers) are used by the sector to market destinations to the ever-increasing numbers of younger generations of tourists across the globe (Hanan and Putit 2017).

Consequently, prior to the Covid-19 pandemic, tourism was at the forefront in terms of global service output, contributing 10.3% of world GDP and in at least twelve countries tourism input to GDP was running at 40% (WTTC 2019). The impact of the pandemic on this unprecedented rate of growth has exposed the structural vulnerability of the vast chain of service providers and stakeholders operating at different spatial and temporal scales. These actors while independent of each other, are so inextricably linked and consequently sensitive to both exogenous and endogenous random effects on tourism ecosystems.

Prior to the Covid-19 pandemic the industry standard adopted for *sustainability* in tourism ecosystems was the *Triple Bottom Line* as it supposedly aligns with the Sustainable Development Goals (SDGs). However, the low-cost airlines, cruise line sectors, and travel companies offering all-inclusive package vacations inadvertently created their own *race to the bottom* (Evans 2020) with Profit taking a place of priority over Planet and People such that business models became disengaged from the United Nations 2030 Sustainable Development Agenda. The sudden paralysation of mass tourism means the sector could retract by up to eighty percent this year against last year's figure of $US1.7tn. In addition, approximately one hundred and twenty million jobs could be lost or, one in ten jobs from the global tourism sector (World Economic Forum 2020).

On the positive side, international tourism promotes contact between different societies and cultures providing benefits to local communities in the form of expendable cash for the self-employed and SMEs. Consequently, the role tourism plays in the development of SIDS has been framed within a dialogue of favourable contributions towards economic, environmental and social sustainability. However, the immediate impact of this pandemic has been crippling to the tourism sectors of SIDS placing their long-term economic growth in jeopardy (UNCTD 2020). In the majority of SIDS macro-economic growth is inextricably linked to huge amounts of foreign investment which also brings negative impacts at the micro socio-economic level.

At the community level exponential growth in tourist numbers has created a tendency for local development to become distorted to support tourism with land being sold for hotels, water diverted to golf courses, and generally speaking transport, communications and other public infrastructure destined for the needs of tourists The fact that visitors needs take precedence over those of local communities and carrying capacities has just become a recently discussed and researched issue in the tourism literature (De Balleigue 2020).

In addition, mass tourism has given rise to the phenomenon of 'alien guests' who tend to have very limited contact with the locals (Candela and Figini 2012). Alien guests and their all-inclusive cruise and vacation packages tend to exacerbate the collapse of daily life at a particular destination (Ballester 2018) resulting in the erosion of environmental, economic and social well-being for local communities which is another topic which thus far has been largely overlooked in the tourism literature (Candela and Figini 2012). In fact, the sector has and to a certain extent is still clinging to the notion that the return to *business as usual* where Profit takes priority over Planet and People will become the new normal. The industry is not only floundering to find a unified dialogue with which to articulate the tenets of a new normal for *sustainable tourism* it is also hesitant to engage in substantial change despite widespread criticism, civil unrest and even protests (Higgins-Desbiolles 2008; Cassinger 2019).

The disconnect between the exponential economic growth of the sector at global level and the erosion of environmental, economic and social well-being at destinations has been and is still largely ignored by many influential tourism organisations that tend to focus attention on the economic benefits of tourism. For example, The World Travel and Tourism Council is currently focused on recovering jobs in the tourist sector (WTTC 2020a, b) while the UNWTO has launched a call for *better, more sustainable and resilient growth* (UNWTO 2020b, c, d) as the way forward with which to restore the industry to levels of sustainable growth whenever tourism resurges in the new Post Covid-19 normal.

This dichotomous dialogue at global level has been the framework in which tourism evolved pre Covid-19. While it is true that tourism can play a significant role in the economic growth and development of a country it can also have a 'parasitic effect' or in extreme circumstances produce no return for the destination economy owing to leakages to external actors in the supply chain. This is particularly the case in SIDS where food security expenditure is reliant on a high percentage of income from tourism and remittances from the related diasporas. Consequently, when tourist numbers are in decline the capacity to import food *is placed in jeopardy* (Schmidhuber and Qiao 2020).

Food Security is only one of the vital aspects of life in which SIDS need to urgently build resilience. This pause in proceedings offers SIDS destinations a unique opportunity to consider what robust and strategic moves are required to adapt to a new normal that strikes the equilibrium between Planet, Profit and People to deal with the structural vulnerability of the tourism sector. The Post Covid-19 Blueprint in Fig. 1 identifies the environmental, economic and social issues in which resilience building can commence. These issues equate with the continuing improvement of quality of life—or *well-being*—which coincides with the objectives embodied in the 17 SDGs (O'Brien 2020; Aksoy and Bayram Arli 2019). For SIDS this means identifying common vulnerabilities as they link to the achievement of the 2030 Sustainable Development Agenda and then taking the strategic and robust moves necessary to ensure future:

- Environmental adaptation and resilience building through attention to: Carrying Capacity, Climate Change Adaptation, Clean Energy
- Economic adaptation and resilience building through attention to: Food Security, Employment and SMEs, Diverse Economy
- Social adaptation and resilience through attention to: the overall *well-being* of Local Residents, Tourism Employees and Visitors.

This Post Covid-19 Blueprint for Sustainable Tourism in SIDS identifies those areas of life which will respond to immediate resilience building measures. Also, to complement and underpin any reactive measures taken at regional and national levels the next time the global tourism sector is dealing with unprecedented systemic

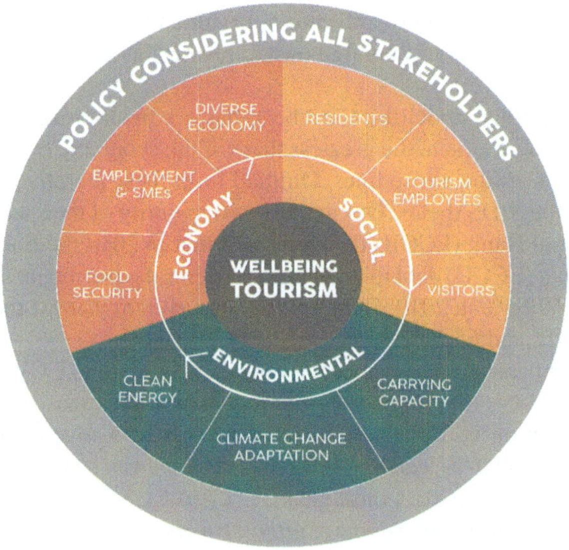

Fig. 1 Post Covid-19 blueprint for sustainable tourism in SIDS

shock. Following is a review of resilience building measures that can be adapted for the Caribbean which is still developing a regional response to the way forward, Indonesia where the central government has announced that the future of tourism will be tied to exponential increases in both domestic and international tourist numbers, and, Mauritius where decision makers appear to be locked into the pre Covid-19 tourism business model yet at the local level at least there is the recognition that any new business models must align with the SDGs.

2 Caribbean

For the entire twenty first century tourism has been a significant contributor to GDP in Caribbean SIDS. From Cuba with 22% of its total exports coming from international tourism revenue, to The Bahamas with 86% and Saint Lucia topping the list at 90% (UNWTO 2020b, c, d), the significance of the related economic contributions has been undeniable. Foreign Direct Investment is the primary economic engine of Caribbean tourism. Foreign investors in tourism are generally granted tax holidays which traditionally get extended as their expiration dates approach. While these hotels pay no corporate taxes and remit their profits back to their parent companies abroad, they provide employment for a significant percentage of the working population. Additionally, they collect guests' occupancy taxes as well as food and beverage taxes which are then remitted to government. From an administrative standpoint government satisfies two important objectives through this model: (i) employment for growing populations; and (ii) revenue for national treasuries. On the other side of this equation lies the largely foreign owned resorts that enjoy a lucrative business model.

In tandem with this model the Caribbean has a vibrant cruise sector. However, just as with stayover tourism, the greatest share of the economic benefits from cruise tourism remains with the cruise lines. Omitted from these lucrative tourism models cited above are the SIDS themselves and the vast majority of their nationals. Eastern Caribbean Central Bank (ECCB) data indicates that in 2018 the Eastern Caribbean Currency Union's (ECCU) debt to GDP ratio for central government stood at 59.78% in 2018, and at 57.78% in 2019. Debt to GDP was even higher for the public sector, at 68.05% in 2018 and 65.10% in 2019 (ECCB 2020).

Because Caribbean SIDS are so heavily tourism dependent, COVID-19 could not have come at a worse time impacting the lucrative winter season. The Caribbean Tourism Organisation (CTO) projects Caribbean growth reversal of as much as 25 years should COVID-19 lead to a reduction in Caribbean stayover arrivals beyond 50% year to date (CTO 2020). Consequently, *"Caribbean economies are likely to be scarred and left with permanent damage after the worst effects of COVID-19 subside"* (Ram 2020). Small to Medium Size Enterprises (SMEs) and the self-employed will be particularly vulnerable and many of their businesses are likely to close down.

2.1 Post Covid-19 Blueprint for the Caribbean Region

The almost complete shutdown of the region's economies has come at a heavy price, namely border closures and the imposition of heavy restrictions and protocols that have had a devastating effect on two key economic areas: Tourism and Food Security. In fact, the food import bill for fourteen member states of the Caribbean Community (CARICOM) rose from $US2.08 billion in 2000 to $US4.75 billion in 2018, and if the trend continues, it could increase to $US8-10 billion in 2020 (Ewing-Chow 2019a, b). Moreover, some of these foods are processed or over-processed, are low in nutritional value, and have health implications for local populations (CARICOM Today 2018).

The Food and Agriculture Organization (FAO 2013) notes that a high food import bill impacts a country's foreign exchange levels, budgets for social protection programmes, levels of chronic non-communicable diseases and displacement of local production. This leads to negative impacts including loss of employment, as well as loss of rural welfare and infrastructure. Because of COVID-19, reduced access to international food sources, coupled with zero revenue earnings by most of the Caribbean countries, paints a dire economic scenario in the short to medium term for many of these small island states.

While the modern issue of over-tourism may not apply to the region, mass tourism certainly does for many Caribbean islands, where the primary metrics used to determine success are the visitor arrival numbers. As previously mentioned, many Caribbean economies are organized and re-configured to provide goods and services to these tourists. Prior to the outbreak of the COVID-19 pandemic, travel to the Caribbean was on an upward trend, with total visitor arrivals (stayover and cruise) amounting to 61.7 million in 2019, a 3.9% increase over 2018. Total visitor expenditure in the Caribbean was an estimated $US40.6 billion for 2019, representing a 3.3% increase when compared to $US39.2 billion in 2018.

The pursuit of tourism-driven economic growth however, overlooks a significant environmental burden for the region. In smaller islands such as Dominica and St. Lucia *domestic* carbon emissions can be as high as 97% and 70% respectively (Gössling et al. 2013). Other negative environmental impacts on destination carrying capacities are in the form *of golf courses which use as much water as 60,000 rural residents and 1500 kg of chemical fertilisers, pesticides and herbicides each year, scientists and economists have suggested that there is a higher cost to the environment from developing tourism over other sectors* (Ewing-Chow 2019a, b).

Given the powerless of the Caribbean SIDS to reverse the prevailing situation either collectively or individually adaptation measure to build resilience will have to take place urgently at both regional and destination level. Through the implementation of the Post Covid-19 Blueprint (Fig. 1) both the region will be able pursue sustainable development options and strategies that promotes well-being for local destinations within the framework of the 2030 Agenda for Sustainable Development. Some initiatives take the form of resilience building in Health and Wellness Tourism, Agritourism and Virtual Tourism which are already underway at particular destinations and could be adapted to the regional level. Other regional initiatives such as the immediate support to facilitate intra-Caribbean travel require urgent consideration and implementation.

2.1.1 The Case for Intra-Caribbean Travel

The Covid-19 new normal presents an opportunity for the Caribbean to respond by strengthening its "domestic tourism" model and encouraging intra-Caribbean travel. This can be achieved in three ways:

1. by taking advantage of the region's proper management of the spread of the virus and creating a "*Caribbean Travel Corridor*" which exempts travellers from having to undergo testing and/or self-quarantining on arrival at the host island;
2. by reducing taxes, fees and charges (TFCs) on airline tickets to make travel within the region more affordable and accessible;
3. by enhancing air connectivity within the region.

TFCs and air connectivity have historically constrained the growth of intra-Caribbean travel. The International Air Transport Association (IATA) estimates that aviation supports over 1.6 million jobs and more than $US35.9 billion in GDP, equivalent to 14% of the total Caribbean economy. Unfortunately, it is generally considered an easy target for taxation by regional governments (IATA 2018). Studies undertaken indicate that in many cases, a share of these passenger-related charges is not intended to cover costs of aviation-related services and infrastructure. Rather, this tax income is a means to generate additional government revenue. It is therefore not uncommon to hear Caribbean citizens bemoan the fact that it's cheaper to visit North America than travelling from Trinidad to Montserrat.

In 2019, intra-Caribbean travel accounted for 2 million passenger arrivals, an increase of 7.4% over 2018 (CTO 2019). However, this represented just 6.4% of stayover visitors and an even more anaemic 3.3% of total arrivals to the region. The report notes that in the eastern and southern Caribbean, more than 50% of the total cost of a ticket was made up of taxes and charges which is an obvious disincentive to persons wishing to travel within the Caribbean, leaving other stakeholders out of the equation. Hence, by shifting the focus to investing in the development of intra-Caribbean vacations will equate with more attendant benefits and well-being for a greater number of stakeholders on different levels and building much required resilience at regional and destination level. Some of the strategic measures that could result from the opening up intra-Caribbean travel take the following form.

Build Environmental Resilience through:

- Lower carbon emissions due to the use of smaller aircraft types and more efficient travel between intra-regional destinations.
- Lower risk of community spread of COVID-19.
- A reduction in the Caribbean travel market footprint equates with a lesser impact on the carrying capacity of accommodation, sites and attractions, and transportation.

Build Economic Resilience through:

- Stimulation of the regional economy because leakage from the regional tourism sector will be diminished.
- Increased expenditure in local economies will account for increased tax revenue form Value-Added Tax (in most of the jurisdictions).
- Easier adoption of indigenous foods by visiting Caribbean nationals, thereby reducing the need to depend on imported food stuffs.
- Caribbean vacationers travel in the traditional low period for North American and European travellers, thus creating more sustainable year-round employment opportunities.
- Lower marketing costs for destinations targeting the Caribbean market.
- Accommodation providers in every category (from resort hotels to B&Bs) have the opportunity to capture the attention of a potential segment of travellers in their own backyard.

Build Social Resilience through:

- Greater inter-island integration and the creation of bonds between local communities.
- A more intimate knowledge of neighbouring Caribbean islands' culture, cuisine, history, people, etc. by Caribbean nationals.

- Connectivity growth from more routes and frequencies, resulting in more flexibility for passengers and shorter travel times.
- Opportunity for stronger intra-regional relationships (both business and leisure).
- Greater opportunity to re-open hotels, sites and attractions, allowing local tourist sector workforces to reintegrate into the economy.

2.1.2 A Case for Agritourism in Montserrat

The Food and Agriculture Organization (FAO 2015) indicates that a high food import bill impacts a country's foreign exchange levels, budgets for social protection programmes, levels of chronic non-communicable diseases and displacement of local production. All of which negatively impacts a destination's fiscal and social indicators. This points to a grim reality for SIDS during this pandemic. The Montserrat Agriculture Strategy and Marketing Plan is steering the sector which focuses on presenting a cohesive framework for a policy of food import substitution with a commitment to the exploration and development of niche market opportunities for selected commodities. Emphasis is being placed on horticulture including root crops, ornamental plants and fruits; livestock and eggs; fish; processed products; marketing, outreach and networking and institutional strengthening (CARICAD 2016). The shift of focus to Agritourism being implemented in Montserrat can be adapted across the region to suit individual SIDS contexts as a first and fundamental step in the diversification of local economies.

2.1.3 Health and Wellness Tourism

A Post Covid-19 new normal presents the opportunity to accelerate resource mobilisation to strengthen the health sector which is critical to the 2030 Agenda for Sustainable Development. In this new normal, one of the objectives of health policy could be to support sustainable tourism development. Investments in strengthening national health system augurs well for the health of residents in urban and rural areas, quality of the entire labour force, as well as health security for tourists visiting the region.

In fact, all of the tourism development plans and strategies in the Organisation of Eastern Caribbean States (OECS) espouse sustainability, and in this regard, include Health and Wellness as key market segments that can strengthen linkages among sectors and create job opportunities in rural areas. So far, the emphasis in these plans and strategies has been on the *wellness* component rather than the *health* component.

Pre Covid-19, the linkages between agriculture, education and creative industries and tourism were acknowledged. Each of those sectors generate outputs which are consumed by residents and are packaged into wellness products for consumption by visitors. Wellness tourism is expected to contribute to business viability, and potentially ensure other benefits to employees and the wider community. However, national health services have not been perceived to be sought after by tourists in the OECS. In fact, there is a tendency for non-residents and in the case of an emergency to some extent residents tend to make arrangements for air ambulance services to first class world health facilities in the US or Europe, and also Barbados. In practice, tourists to the Caribbean consume only routine basic services such as treating an insect sting, or emergency services as a gap to returning back home for comprehensive treatment.

With the exception of specialized cosmetic surgery offered in one or two islands, the Caribbean region does not stand out as a medical health destination for international tourists. However, some intra-regional medical health tourism exists with Barbados, Guyana, Jamaica and Trinidad and Tobago providing medical health services to patients mostly from the OECS sub-region. Traveller preferences now rank health and security as top priority for domestic and international tourism (UNWTO 2020b, c, d).

Hence, Post COVID 19 blueprints must adapt health sectors to respond to those concerns as well as to diversify local tourism ecosystems. In a regional and local sense, the intensity of and economic disruptions caused by hurricanes/tropical storms in the OECS islands place SIDS in the OECS as among the most vulnerable countries to climate change impacts. In this background the sub-region is taking the steps required to adapt and build resilience towards achievement of well-being as it relates to the Sustainable Development Goals.

2.1.4 Virtual Tourism

Virtual travel experiences can be created, packaged and sold to potential travellers unable or unwilling to leave home. In fact, *"Across industries, the coronavirus is inspiring innovation. In tourism, some operators are shifting from physical to virtual tours and using the opportunity to bring attention to noteworthy causes."* (Brown and Monitor 2020). The emergence of virtual tours or incorporating virtual and augmented reality combined with artificial intelligence has been shown to hold great promise even though it does not replace the authentic travel experience. However, for this to be realized Caribbean self-employed and SME owners must become creative thinkers and seize the opportunities the pandemic has afforded the region, rather than accepting a fate of doom and gloom.

An innovative initiative launched by an award-winning photography company in St Lucia takes the form of a virtual tourism experience. A free online photography learning experience was offered in exchange for participant feedback which has informed the design of a new virtual photography excursion to be featured exclusively online (available at https://IslandVR360.com). These virtual travel experiences hold potential as an important instrument for the promotion of sustainable tourism in a post-COVID-19 world by allowing a new type of exploration of travel destinations without a carbon footprint. These virtual experiences have the added benefit of keeping Caribbean destinations as priority for travellers with the potential to influence their travel decisions when safe international movement is once again possible.

2.2 A Post Covid-19 Blueprint for Indonesia

In 2019, tourism contributed 4.80% of Indonesia's total GDP (Lokadata 2020). As a consequence of the pandemic there has been a general loss IDR 85.3 trillion ($US5.85 billion) in the tourism industry for the first quarter of 2020. More than 2000 hotels and 8000 restaurants were closed in Indonesia, resulting in a revenue loss of IDR 30 trillion ($US2.05 billion) for hotels and IDR 40 trillion ($US2.74 billion) for restaurants. At the time of writing the airline industry has experienced a total loss of IDR 11.3 trillion ($US812 million). Tour operators also suffered losses of up to IDR 4 trillion ($US275 million). In addition, potential foreign exchange losses from January to April 2020 equate to $US4 billion (IDR 56 trillion) (Tirto-Id 2020).

Unlike the Caribbean where change is predicated on defining a new business model, the new normal for Indonesia will be to build upon existing tourism ecosystems and introduce the New 10 Bali Strategy in a hope of a more sustainable Post Covid-19 new normal. This return to new normal bears heavily on a significant resurgence in domestic and international tourism. Nevertheless, to cope with the current impact of the pandemic Indonesia has initiated some of the necessary resilience building measures to define its unique version of a Post Covid-19 Blueprint to promote environmental, economic and social well-being for its diverse tourism ecosystems.

Build Environmental Resilience through:

- The emergence of a discourse put forward by the National Development Agency (*Bappenas*) focused on determining carrying capacity at a particular location and based in an analysis of strategic environmental studies.
- Identifying the parameters that must be considered for carrying capacity limitations which include primary forests, forests on peatlands, endangered species habitats, coastal areas affected by climate change, disaster-prone areas, areas with minimum water availability, areas with minimum energy availability, areas with high level and intensity of greenhouse gas emissions.
- Engaging all of the relevant stakeholders in a reset of sustainable parameters, implementing a holistic strategy, and integrating various sectors by paying attention to the function of space and utilization of natural resources in tourism development at specific destinations.
- Prohibiting single use plastic in Jakarta Capital City.
- Promoting the use of private vehicles that are environmentally friendly, such as selling more motorbikes with electric fuel.
- Reducing electricity tariffs for some stakeholders.

Building Economic Resilience through:

- Protection of SME ecosystems through providing social assistance and reducing electricity tariffs over determined periods of time.
- Credit restructuring and relaxation, in collaboration with several banking institutions and financial technology applications.

- Provision of working capital to SMEs.
- Targeting six million female SMEs with programs to increase entrepreneurial capacity.
- Offering discounts in the hope of stimulating domestic tourism.
- With assistance from the Central Government, the provincial governments have been able to reduce hotel and restaurant tax rates by $US228 Million.
- A shift of focus towards agriculture in order to shore up resilience at local level in order to ensure food security.
- Incorporating more modern markets and online-based markets into supply chains.

Building Social Resilience through:

- Mechanisms that prevent termination for employees of the tourism sector.
- Providing training and capacity building in areas such as excellent service, information and technology, service standards during the pandemic, health and safety protocols, new trends for post-pandemic tourist activities, as well as entrepreneurial skills.
- Launching of virtual tour packages by most tour guides and dive centres to maintain contact with existing client bases.
- Popular tourist areas including Bali Island, Lombok Island, Komodo Island, and Yogyakarta region are taking the steps necessary to introduce new business models that embrace the agriculture/plantation (rice) sectors, fisheries, and handicrafts.

2.3 *A Post Covid-19 Blueprint for Mauritius*

The predicted contribution of tourism to Mauritius GDP was to be approximately 18.8% in 2020. While this percentage may indicate the country maintains a somewhat diverse economy historically inputs from tourism have been relatively high. It is worth noting that tourism employs more than 104,200 people, representing almost 19% of total employment in Mauritius (WTTC 2020a, b). Therefore, when tourism is impacted by negative external shocks, at least a fifth of the population, their consumption patterns and life styles are also affected. Consequently, the impact of the Covid-19 pandemic has exposed a vulnerable dependency on foreign visitors and their foreign currencies that facilitate trade particularly related to food security and the importation of pharmaceuticals.

The hundreds of hotel rooms built to cater for an ever-increasing number of tourists is testimony to a belief system held in Mauritius that the country would always be able to cater for an ongoing and increasing stream of tourism arrivals. This belief system based in *continuous and exponential growth* being an intrinsic element of sustainable tourism is hindering efforts for the country to implement a Post Covid-19 Blueprint that is framed around the type of robust and strategic adaptation measures that will help to shore up resilience for the tourism sector. However, this pause in tourism activity is providing destination managers with the breathing space in which to re-analyse carrying capacity without compromising the well-being of the natural environment or local communities over the short, medium and long term. Some of the initiatives that have been identified to facilitate adaptation and build resilience at the local level could be readily implemented with political level in order to place Mauritian tourism ecosystems on a path to future *well-being* for local communities.

Building Environmental Resilience through:

- Undertaking studies to establish carrying capacities for destinations and attractions to maintain their natural appeal. This is a particularly important and strategic move as a the National Long-Term Perspective Study: Vision 2020 (NLTPS), produced by the Ministry of Economic Development recommended that for Mauritius to sustain natural attributes over time, it would be necessary to cap the number of hotel rooms development to 9000 rooms and 600,000 visitors per year (Ministry of Economic Development 1997). Beyond this green ceiling, increased earnings were to come from high spending travellers and not from higher volumes of visitors. Nevertheless, by 2020 the number of hotel rooms has already surpassed the proposed ceiling by 100% to a total of 13,489 rooms (Statistics Mauritius 2019) and tourism arrivals for the year 2019 marked 1,383,488, more than doubling the recommended numbers contained in the NLTPS designed in line with the project *Maurice Ile Durable* (Mauritius, Sustainable Island).
- Climate Change mitigation measures are being facilitated through a program wherein accommodation providers and related SMEs can access fiscal incentives and employee training to consider rainwater harvesting as a way to provide a sustainable supply of water in a changing climate (Waite 2012).
- As the degradation of coral reefs due to increasing temperatures may lead to the loss of one of the main sources of food for Mauritians, food sufficiency for the tourism sector needs to be urgently developed around alternative sources such as private permaculture and free-range poultry. Both of which are included in some of the options that generate important value for customers in the form of farm to table experiences and meals, while generating relatively low CO_2 emissions.
- A Plan of Action that mitigates present and future impacts of climate change along the Mauritian coastline and the erosion of its world acclaimed beaches is urgently required.
- The creation of three new endemic forests and the maintenance of existing forests for hiking and mountain biking purposes would be an excellent initiative towards diversifying existing tourism ecosystems.
- The Mauritian government is subsidising the installation of solar panels to increase renewable energy usage by 40% (Padayachy 2020).
- The space for a stronger voice for the social collective *Arret Kokin Nou Laplaz* (Stop stealing our beaches) that has protested for years against the destruction of natural ecosystems for the construction of new hotels and IRS projects on the coast of Mauritius. This grass roots movement strives to maintain well-being for local populations and has the ability to influence future Land Use and Planning Management activity.

Building Economic Resilience through:

- Diversification into sustainable agriculture will represent a great milestone because in 2019 although there was an increase in cereal crop production there was below average production in other food crops (FAO 2020).
- A plan to optimise the use of agricultural land is to be developed by the Mauritian government, in order to develop a marketplace where supply and demand of land can meet in order to match unused land with those interested in pursuing small-scale farming (Padayachy 2020). To support this effort, the Development Bank of Mauritius aims to provide grants that may sponsor up to 90% of agricultural projects or loans.
- A solid strategy to improve the marketing, distribution, and integration between the tourism stakeholders that constitute the domestic market.
- Diversification to the Blue Economy (fisheries), reinforcement of the Circular Economy (waste management) both of which are initiatives driven by the Mauritius Investment Corporation (Padayachy 2020).

Building Social Resilience through:

- A proposal to shift to a model of community-based tourism to cater for the peak tourist season.
- Developing the necessary knowledge base and skills necessary for independent guest houses to provide the required quality and health standards in order to operate as licensed accommodation providers.
- Reforms to the labour market to secure the well-being of employees from the tourism sector. The disparities in pay rates between local employees and the foreign counterparts is an issue that has not been widely researched or published (Hampton and Jeyacheya 2013).
- Investment in the training of quality medical staff, infrastructure and equipment.
- Secure access to affordable pharmaceuticals.
- Develop effective health communications networks.

3 Conclusions

The paralysation of the transport sector and border closures have exposed the structural vulnerability of the global tourism sector. Contractions in the GDP of those countries with a heavy reliance of inputs from tourism will take years to reverse. This is particularly the case for SIDS whose economic growth is inextricably linked to foreign investment and income from foreign tourists. However, this pause in proceedings has also revealed the negative impacts that a heavy national reliance on tourism brings. These take the form of high food import bills (that need to be funded with foreign investment, remits from the related diaspora and to a certain extend foreign tourists), as well the erosion of local community values and access to basic infrastructure services.

In the Pre Covid-19 normal for tourism the exponential growth in tourist numbers was largely linked to the disconnection of the Triple Bottom Line from the SDGs because the race for ever increasing Profits overtook the sustainable development of the Planet and its People. The knock-on effect particularly for SIDS was that in many instances carrying capacities were exceeded resulting in negative impacts for the environmental, economic and social aspects of local tourist ecosystems. In the Post Covid-19 new normal many of these small island developing states are facing huge GDP to debt ratios, a significant loss of employment in the sector and the recognition that if indeed there is to be either a more sustainable revival of the sector, or, as in the case of Indonesia an expectation that Post Covid-19 will bring growth in domestic and international tourist numbers business models based on the Triple Bottom Line that focus on Profit must change. The focus must shift to building resilience in areas critical to the *well-being* of local communities and tourist ecosystems. The first draft of a Post Covid-19 Blueprint for Sustainable Tourism in SIDS has been developed based on the critical issues that affect the *well-being* of all actors involved in the sector as this equates with the Sustainable Development Agenda 2030.

Some measures that are currently being implemented and others that need to be considered by partnerships between stakeholders in the Caribbean, Indonesia and Mauritius have been presented here. The Caribbean region is desperately engaged in redefining the parameters for business models that ensure sustainable tourism for the region and its individual nation states, Mauritius is still somewhat trapped in the belief system that there could be a return to business as usual with the recognition that some fundamental changes are urgently required particularly in terms of reinforcing food security and providing access to viable pharmaceuticals for the entire population. Meanwhile, Indonesia is laying the Post Covid-19 groundwork necessary to grow the domestic tourism market and for a resurgence in international tourist numbers. Yet, these SIDS all share *well-being commonalties.*

In effect the only sustainable path forward is for SIDS to rise to the occasion and develop new business models that incorporate resilience building measures that will continue to adapt to the endemic structural vulnerability of the tourism sector. At the moment this seems to be the most intelligent way forward to overcome the dichotomous global industry body dialogues that are presenting two very different visions for the resurrection of a sustainable tourism sector. The concept of sustainability cannot be confined to a one-off heterogenous temporal or spatial concept nor can business models be framed to continually pursue Profit over the *well-being* of the Planet and its People. The caveat of course is that Post Covid-19 Blueprints for Sustainable Tourism must continually reinforce resilience building measures that align with the aspirations of the Sustainable Development Agenda 2030.

References

Aksoy F, Bayram Arli N (2019) Evaluation of sustainable happiness with Sustainable Development Goals: structural equation model approach, Sustainable Development, Wiley Online Library. https://doi.org/10.1002/sd.1985

Ballester P (2018) Téoros, Revue de Recherche en Tourisme, 2018 Vol.37 No. 2. Retrieved from https://journals.openedition.org/teoros/3367

Brown RL, CS Monitor (2020) The Dogs of Chernobyl: are virtual tours the future of tourism? Retrieved from https://www.csmonitor.com/World/Europe/2020/0605/The-Dogs-of-Chernobyl-Are-virtual-tours-the-future-of-tourism

Candela G, Figini P (2012) The economics of tourism destinations. Springer, Berlin

Caribbean Tourism Organization (CTO) (2019) Caribbean Tourism Performance Review 2019. Retrieved from onecaribbean.org

Caribbean Tourism Organization (CTO) (2020) The future of Caribbean tourism—key considerations for the COVID-19 recovery, p 9. Retrieved from https://documentcloud.adobe.com/link/track?uri=urn:aaid:scds:US:b9f9e241-0dea-468f-bd0c-92481169ee9b#pageNum=9

Caribbean Tourism Organization, Caribbean Tourism Performance Review 2019. Retrieved from onecaribbean.org

CARICAD (2016) Synopsis of an agricultural strategy and marketing plan (ASMP) For Montserrat. Retrieved from http://www.gov.ms/wp-content/uploads/2016/06/May-2016-Synopsis-of-Agricultural-Strategy-and-Marketing-Plan-Montserrat.pdf

CARICOM Today (2018) June 21, 2018. Retrieved from https://today.caricom.org/2018/06/21/caricom-and-its-food-security-agenda/

Cassinger C (2019) Communicating anti-tourism—movement, protest, phobia. 28th Nordic Symposium on Tourism and Hospitality Research, Roskilde University, Denmark, 23–25 Oct 2019

Christin R (2017) Manuel de l'Anti-Tourisme (Ecosociété)

De Balleigue C (2020) The end of tourism? The Guardian, Thursday 18 June, 2020. Retrieved from https://www.theguardian.com/travel/2020/jun/18/end-of-tourism-coronavirus-pandemic-travel-industry

ECCB (2020) Public sector debt—debt to GDP ratios (in percent of GDP). Retrieved from https://www.eccb-centralbank.org/statistics/debt-datas/comparative-report/2

Evans A (2020) The coronavirus will change how we travel. That will probably be good for us, nbcnews.com, think. Retrieved from https://www.nbcnews.com/think/opinion/coronavirus-will-change-how-we-travel-will-probably-be-good-ncna1186681

Ewing-Chow D (2019a) Five overlooked facts about Caribbean food security, Forbes. Retrieved from https://www.forbes.com/sites/daphneewingchow/2019/02/20/five-facts-about-caribbean-food-security/#65e6df8a5016

Ewing-Chow D (2019b) The environmental impact of tourism of Caribbean tourism undermines its economic benefits. Retrieved from https://www.forbes.com/sites/daphneewingchow/2019/11/26/the-carbon-footprint-of-caribbean-tourism-undermines-its-economic-benefit/#234edce23cb5

Export.gov, Mauritius Agricultural Sectors Retrieved from https://www.export.gov/apex/article2?id=Mauritius-Agricultural-Sectors

FAO (2013) Food and agriculture organization sub-regional office for the Caribbean, Issue Brief #5, October 2013: CARICOM Food Import Bill, Food Security and Nutrition. Retrieved from http://www.fao.org/3/a-ax740e.pdf

FAO (2015) State of Food Insecurity in the CARICOM Caribbean. Retrieved from http://www.fao.org/3/a-i5131e.pdf

FAO (2020) Global information and early warning system country brief Mauritius. Retrieved from http://www.fao.org/giews/countrybrief/country.jsp?code=MUS&lang=en

Gössling S, Scott D, Michael Hall C (2013) Challenges of tourism in low carbon economy, WIREs, Climate Change. Retrieved from https://onlinelibrary.wiley.com/doi/abs/10.1002/wcc.243

Hampton M, Jeyacheya J (2013) Tourism and inclusive growth in small island developing states. Hampshire, Hobbs the Printers, pp 9–11
Hanan H, Putit N (2017) Express marketing of tourism destinations using Instagram in social media networking. In: Sumarjan et al (eds) Hospitality and tourism. Taylor & Francis Group, London, pp 471–474. ISBN 978-1-00151-0
Higgins-Desbiolles F (2008) Justice tourism and alternative globalisation. J Sustain Tour 16(3):345–364. https://doi.org/10.1080/09669580802154132
IATA.org, June 29, 2018, Maximizing Aviation's Benefits in the Caribbean. Retrieved from https://www.iata.org/en/pressroom/pr/2018-06-29-01/)
Lokadata (2020) Kontribusi pariwisata terhadap PDB, 2010–2019. Retrieved from https://lokadata.id/data/kontribusi-pariwisata-terhadap-pdb-2010-2019-1582001327
Ministry of Economic Development and Regional Cooperation (1997) Vision 2020: The National Long-Term Perspective Study, Volume 2. Mauritius: Silvio M. Empeigne G.P.
O'Brien C (2020) Sustainable happiness and well-being: future directions for positive psychology. Psychology 2012 3(12A):1196–1201. Published Online December 2012 in SciRes, http://dx.doi.org/10.4236/psych.2012.312A177
Padayachy R (2020) Budget speech 2020–2021, Our new normal: the economy of life. Retrieved from the Ministry of Finance and Economic Development website: http://budget.mof.govmu.org
Ram J (2020) COVID-19 Policy Responses for the Caribbean. Retrieved from https://medium.com/@justinram/covid-19-policy-responses-for-the-caribbean-1-c88a37415399
Schmidhuber J, Qiao B (2020) Comparing crises. Great Lockdown versus Great Recession, Rome, Food and Agriculture Organization of the United Nations. https://doi.org/10.4060/ca8833en
Statistics Mauritius (2019). International travel and tourism, year 2019. Retrieved from http://statsmauritius.govmu.org
Tirto-ID (2020) Dampak Pandemi Corona Industri Pariwisata Keluhkan Kerugian Capai Rp85,3 T Akibat Corona. Quoted from General Chairperson of PHRI during discussion with National Parliament, Retrieved from https://tirto.id/industri-pariwisata-keluhkan-kerugian-capai-rp853-t-akibat-corona-fQSW
UNCTD (2020) July 2020, Covid-19 and Tourism Assessing The Economic Consequences. Retrieved from https://unctad.org/en/PublicationsLibrary/ditcinf2020d3_en.pdf
United Nations Sustainable Development Agenda 2030 Retrieved from https://www.un.org/sustainabledevelopment/sustainable-development-goals/
UNWTO, Rifai T (2014) Opening Speech by General Secretary, Mr. Taleb Rifai. Retrieved from http://www.unesco.org/new/fileadmin/MULTIMEDIA/HQ/CLT/pdf/taleb_rifai_may5.pdf
UNWTO (2019) International Tourism Up 4% In the First Half of 2019, World Tourism Organization Reports, PR No: PR 19051. Retrieved from https://www.unwto.org/global/press-release/2019-09-09/international-tourism-4-first-half-2019-world-tourism-organization-reports
UNWTO (2020a) Sustainability as the new normal. A Vision for the Future of Tourism, 5 June. Retrieved from https://www.unwto.org/covid-19-oneplanet-responsible-recovery
UNWTO (2020b) Covid-19 Response. Available at https://www.unwto.org/tourism-covid-19
UNWTO (2020c) Tourism in SIDS: the challenge of sustaining livelihoods in times of COVID-19, pp 8, 10. https://www.e-unwto.org/doi/pdf/10.18111/9789284421916
UNWTO (2020d) 2020, April 1 Supporting jobs and economies through travel and tourism. Retrieved from https://webunwto.s3.euwest1.amazonaws.com/s3fspublic/202004/COVID19_Recommendations_English_1.pdf
Waite M (2012) Climate-change mitigation and adaptation in small island developing states: the case of rainwater harvesting in Jamaica. Sustain: Sci Pract Policy 8(2):81–87. https://doi.org/10.1080/15487733.2012.11908101
Wong A (2015) Caribbean Island Tourism: Pathway to Continued Colonial Servitude. https://doi.org/10.4000/etudescaribeennes.7524
World Economic Forum (2020) May 2020, International Tourism is set to plunge by 80% this year—but some regions could recover more quickly. Retrieved from https://www.weforum.org/agenda/2020/05/tourism-industry-slump-recovery-coronavirus-lockdown/

World Travel & Tourism Council (2019) Economic Impact Reports. Retrieved from https://wttc.org/Research/Economic-Impact
World Travel & Tourism Council (2020a) Covid-19 Hub. Retrieved from https://wttc.org/COVID-19
World Travel & Tourism Council (2020b) Travel and Tourism Economic Impact 2020, Mauritius. Retrieved from https://wttc.org/

Denise Galvin Collaborative Senior Project Manager, OGPI, Universidad de Alicante, Alicante, Spain.

Exploring Communication Framing Methods that Link Changes in Air Pollution Exposure and COVID-19 to Promote Post-pandemic Sustainability Policy

Lisa Jordan and Albert Tang

Abstract This is a stock-taking piece on the role of COVID-19 social distancing measures in reducing urban air pollution and subsequent policy changes to reduce air toxics emissions from road traffic or energy production in dense urban areas. In particular, this chapter examines the use of multiple methods of communicating the benefits of reduced air pollution. Using a case study approach, the chapter identifies innovative sustainability programs championed by cities, in response to COVID-19. Creative and resourceful approaches to air pollution and COVID-19 by environmental justice organizations are also compared. By identifying and discussing intersectional work on air pollution and COVID-19, this chapter offers a diverse set of examples for how sustainability programming can emerge during and after the COVID-19 pandemic to reduce air pollution and improve wellbeing.

Keywords Air pollution · COVID-19 · Transportation · Environmental justice · Sustainability

1 Introduction

The disruption of COVID-19 to business as usual has produced the opportunity for major transitions in economies and societies toward more sustainable livelihoods (Vince 2020). Global air emissions declined substantially in the spring of 2020 in response to social distancing policies enacted worldwide (Harvey 2020c), with environmental groups and public leaders, backed by strong public opinion, proposing an infusion of sustainability goals with recovery efforts (Harvey 2020a; Ipsos 2020). An open letter from the global public health community to world leaders pointed to outdoor air pollution as a leading environmental health concern, potentially connected to disparities in COVID-19 risks, that should be a key focus for sustainability policy moving forward (Harvey 2020d; World Medical Association et al. 2020).

L. Jordan (✉) · A. Tang
Spatial Data Center, Drew University, 36 Madison Ave., Madison, NJ 07940, USA
e-mail: ljordan@drew.edu

W. Leal Filho (ed.), *COVID-19: Paving the Way for a More Sustainable World*, World Sustainability Series, https://doi.org/10.1007/978-3-030-69284-1_5

In the midst of the pandemic, a Pew survey found that two-thirds of Americans felt the federal government should act more aggressively to combat climate change (Dennis 2020; Tyson and Kennedy 2020). Relatedly, polls found that the vast majority of the British surveyed wanted the government to prioritize health and wellbeing over the economy (Harvey 2020b). A YouGov poll revealed that only 9% of Britons wanted to return to "normal" after the crisis, with 51% observing cleaner air as a benefit of social distancing measures (Binding 2020). A poll conducted by Ipsos estimated that 87% of people living in China saw climate change as serious a threat as COVID-19 (Rowlatt 2020a).

Not only is public opinion changing, behavior is changing too in response to the pandemic. Cycling is significantly increasing in many regions: bicycle counts are up 100% in the US Southwest, and ridership on Philadelphia, Pennsylvania and Arlington, Virginia trails doubled in early 2020, compared with 2019 (Bryant 2020). In Melbourne, ridership is up 79% (Landis-Hanley 2020), and bicycle counters saw increases all across Scotland (BBC News 2020). Use of bike shares increased by 150% in Beijing (ITDP 2020). These activities are encouraged in part by scholarship on the relationship between outdoor physical activity and health that have existed for over two millennia: walking was described as medicine by Hippocrates, the father of Western medicine, from whom the Hippocratic path gets its name (Stone and Roberts 2020). Non-peer reviewed, preliminary evidence found that transmission may happen indoors to a greater extent (Hirschauer 2020; Qian et al. 2020), encouraging some public health experts to prescribe increased time outside, as part of COVID and post-peak COVID responses (Makary 2020; National Association of City Transportation Officials (NACTO) 2020).

But as critics point out, most communities plan to return to business as usual, and if this is the case, we can expect any environmental benefits from diminished fuel combustion to be quickly erased (Gardiner 2020b; Goldfarb 2020). Several large countries have taken the opportunity of COVID-19 to advance even more developmentally aggressive policies, including regulation rollbacks in the US, deforestation in Brazil (Spring et al. 2020) and coal-based energy expansion in China (Gardiner 2020b). Scholars have also forecasted significant increases in traffic in major US cities, as large numbers of users quit public transportation, with San Francisco, New York, and Los Angeles expecting the greatest increases (Hu et al. 2020).

> Janette Sadik-Khan, a former transportation commissioner for New York City and principal with Bloomberg Associates, is working with Milan and other cities on their "transport recovery" programmes. "The pandemic challenges us, but it also offers a once-in-a-lifetime chance to change course and repair the damage from a century of car-focused streets," she says. "Cities that seize this moment to reallocate space on their streets to make it easier for people to walk, bike and take public transport will prosper after this pandemic and not simply recover from it." (Perry 2020)

Air pollution can be defined as "a mix of particles and gasses that can reach harmful concentrations" (Nunez 2019). As such, the transformation of urban environments to reduce exposure to air pollution is woven throughout the United Nations Sustainable Development Goals (SDGs). A global assessment of the impact of air pollution on human health by the World Health Organization identified air pollution in three

SDGs: essential for urban sustainable development (SDG 11, target 11.6), sustainable energy (SDG 7, target 7.1) and health and wellbeing (SDG 3, target 3.9) (World Health Organization 2016). Sources of air pollution and factors that influence air quality are numerous. Major human contributions to ambient air pollution include fuel combustion for vehicles, power generation, industrial processes, waste sites and incineration, and residential fuel use (World Health Organization 2020).

This chapter aims to provide a stock-taking summary of the role of COVID-19 social distancing measures in reducing urban air pollution, and subsequent policy changes to reduce air toxics emissions from road traffic or energy production in dense urban areas. In particular, this chapter examines the characteristics of multiple methods used in communicating the benefits of reduced air pollution, spanning both data-centric and humanistic approaches. Case studies for cities around the world are compared. We also study the web-based communications by US environmental justice organizations, and their practices of elevating COVID-19 concerns, air pollution concerns, and the two in combination. The effect of this work is to highlight human ingenuity and leadership in promoting diverse, sustainability-oriented programming.

2 Background

Historically, cataclysmic air pollution events in the UK and US prompted development, passage and enforcement of policies that regulate air pollution. The Great Smog of 1952 in London created blackout conditions that lasted for five days. Over four thousand people are believed to have died as a result of the event, which prompted long overdue passage of the Clean Air Act of 1956 (BBC News 2002). The Donora Smog event of 1948 is often credited for promoting the US federal Clean Air Act (Hamill 2008). After three days of smog, 20 people in the small Pennsylvania town had died.

In a few rare cases, the reverse situations can occur, where days of unusually clean air bring to light the consequences of urban air pollution. The 2008 Beijing Olympics was one example (Wang et al. 2010), where government interventions actively sought to temporarily reduce local air pollution. Beijing was again studied in 2014, during similar restrictions instituted for the Asian Pacific Economic Cooperation meetings, known as "APEC Blue" (Huang et al. 2015; Meng et al. 2015). Before that, epidemiologists found that childhood asthma hospitalizations were significantly reduced when Atlanta, Georgia restricted downtown traffic congestion for the 1996 Olympic Games (Friedman et al. 2001). Social and economic slowdowns in response to the COVID-19 pandemic offer a much larger scale natural experiment.

3 Methods for Reviewing Air Pollution Framing Approaches Applied During COVID-19

Some scientists have coined the slow down of human activity due to COVID-19 the "anthropause" (Rutz et al. 2020). The decline in economic activity following implementation of social distancing measures resulted in substantial reductions in air emissions, including CO_2 (Mufson 2020). Between March and April in the US, miles driven by Americans were cut by over half (Goldfarb 2020; Nguyen et al. 2020). These changes make for an important experiment on human health and air pollution (Phillips 2020), but have also been influential in moving forward urban sustainability policy in many cities around the world.

In order to review the framing approaches applied to urban air pollution, we chose to conduct a literature review of contemporary research and journalism that discusses connections between COVID-19 and air pollution. We summarize the communication framing that we observed. Two notable responses to COVID-19 emerged from our review. First, cities identified and promoted benefits to active transportation (walking and cycling) in reaction to COVID-19. Second, environmental activists called attention to the inequities of COVID-19 exposures as a means to problematize social divisions and unequal burdens from air pollution. We follow our summary of communication frames with a list of case studies involving cities that are making significant changes in urban transportation infrastructure in response to COVID-19. Similarly, we highlight case studies of environmental justice organizations who are working to demonstrate inequities in COVID-19 and environmental health.

The way that people perceive, communicate, and understand air pollution has changed significantly in recent years. Seven ways that reporters, scientists and activists share information about air pollution include: (1) the use of earth observation, satellite sensors, and air quality monitoring data, (2) shared photography and video illustrating visible changes from eye-level, (3) personal testimony about experiences of breathing cleaner air, (4) environmental health surveillance to quantify health impacts of air pollution, (5) identification of health disparities experienced by different groups, (6) economic benefits to reduced pollution exposure, and (7) co-benefits of declines in both ambient air pollution and greenhouse gas emissions. Examples and further explanation of each of these frames is shown next.

4 Communication Frames for Air Pollution

4.1 *Earth Observation*

A wide variety of scientific instrumentation is used to monitor air quality, locally and globally. Such forms of measurement help to quantify changes in air pollution intensity over time. During COVID-19, these methods were put to use to observe the impact of economic and social changes on our natural environment. This section lists the changes revealed by satellites, air quality monitors and seismographs.

Satellite Sensors

Satellite observation by the US National Aeronautics and Space Administration (NASA) revealed significant declines in nitrogen dioxide in March 2020 in comparison with the previous five years in the US Northeast (Blumberg 2020). Similarly, the US National Oceanic and Atmospheric Administration (NOAA) reported NO_2 declines in the US in March, but paid particular attention to the declines in emissions from less weekend travel (NOAA 2020a). *National Geographic* magazine published maps illustrating changes in air pollution over Europe, South Asia, North Africa, and the United States, derived from data obtained by the European satellite Sentinel (Gardiner 2020a). The *Washington Post* produced a stunning timelapse video mapping declines in NO_2 over Milan between January 1 and March 11 (Mooney et al. 2020).

Our reproduction of the NASA Aura Mission measurements of nitrogen dioxide emissions in Europe is shown in Fig. 1. Global data are readily accessible through the NASA Earth Observations website (NASA 2020b). Though a variety of factors can influence the comparability of air emissions, there are clearly substantive differences between April of 2019 and 2020, when the emission volumes are standardized across maps. The differences between the two time periods for the UK and Northern Italy are notable.

Peer-reviewed studies emerging on the association between COVID-19 responses and changes in air pollution frequently use satellite data. Studies using Europe Space Agency's Sentinel-5p TROPOMI instrument to estimate NO_2 have quantified changes in China during lock-down (Zhang et al. 2020). However, though often NO_2 concentrations are read as a proxy for overall air pollution, the reality of air pollution is more complicated. For example, satellite observations of NO_2 over China in January 2020, were at odds with ground-level observations of haze, PM2.5 and ozone (Le et al. 2020). Wind patterns and humidity, in combination with air pollution emissions from power plants, meant that ground-level pollution was high, even though emissions from transportation were significantly reduced.

A different instrument for observing changes by satellite is the NOAA/NASA Suomi NPP Day/Night Band (DNB), which is used to create maps of nighttime lights (NASA 2020a; NOAA 2020b). This sensor can reveal changes in electric generation, but it can also be used to monitor disparities in recovery efforts moving forward (Qiang et al. 2020). Essentially, access to electricity is visible from space at night.

Using satellite derived air pollution information helps both scientists and the public understand the geographic dispersion of urban air pollution and the measurable impact of changes in air pollution. Global monitoring sheds light on the sources, intensity and differences in emissions of nitrogen dioxide. The costs of economic

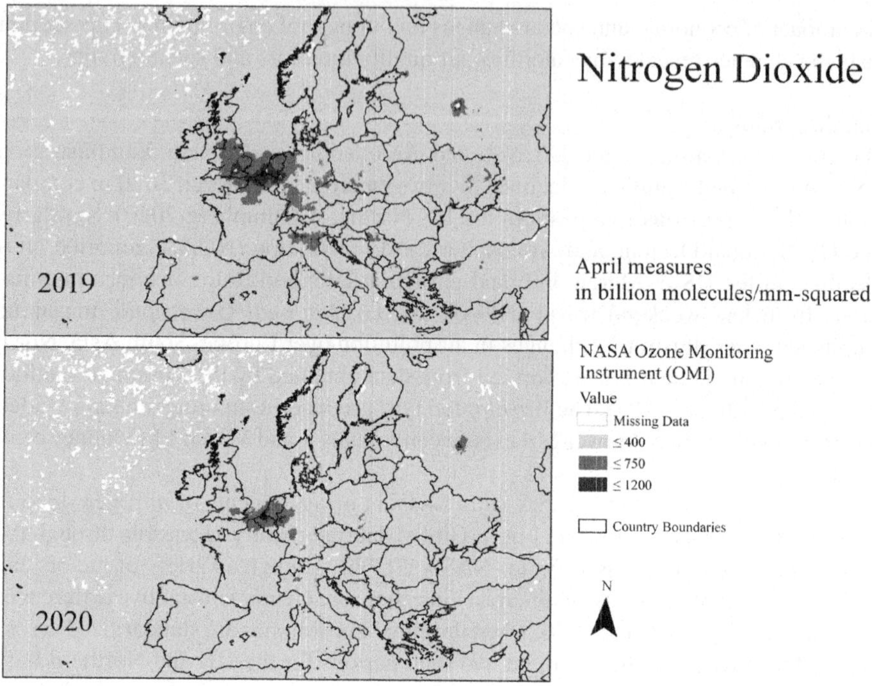

Fig. 1 Nitrogen dioxide concentrations derived from the ozone monitoring instrument (OMI) on NASA Aura satellite

and social activity to both the natural environment and bodies of urban dwellers can be measured, because of the availability of this information.

Air Quality Monitors

On-ground instrumentation also measures air quality. The New York City Department of Health and Department of Environmental Conservation took advantage of real time air quality monitors to demonstrate localized reductions in air pollutants as a result of less traffic congestion, due to COVID-19 social distancing measures (NYC Department of Health 2020). Similar analysis tools were used by National Public Radio in the US to map and share US Environmental Protection Agency (EPA) air quality monitoring data for criteria pollutants.

In Los Angeles, while anticipated reductions in air pollution because of COVID-19 distancing were noticeable, they were not as substantial as in other cities with less frequent truck traffic (Sommer et al. 2020). In Houston, despite declines in traffic by 40%, ozone declined only 11% due to continued emissions from refineries: in Pittsburgh, coal burning kept COVID-19 related declines in emissions smaller than other regions (Sommer et al. 2020). Similar conclusions were reached, regarding high levels of air pollutants observed even during lock-down in China: changes in electric generation, in addition to transforming transportation, are needed for substantially reducing urban air pollution (Le et al. 2020).

dB Sensors and Seismographs

Sensors like seismographs have revealed quieter cities, and while not directly a measure of air pollution, lowered levels of transit activity (cars, airplanes and trains) are also connected to reduced noise. A research team of scientists found that global noise dropped by up to fifty percent from March through May (Broad 2020; Lecocq et al. 2020). A quieter world seems to go hand in hand with higher air quality.

4.2 Photography and Video

Photography and video were instrumental to communicating the crises during the Great Smog in London and Donora, Pennsylvania. These tools have been used ever since to capture urban experiences of air pollution. However, COVID-19 distancing measures allowed photographers and videographers to capture the opposite: what cities can look like with clean air. Clear skies in New Delhi were featured in an article by Beth Gardiner in *National Geographic* (Gardiner 2020a), and a recent article by CBS, a US news agency, highlighted the major improvements in air quality by featuring the stark contrast in before and after photos of major urban cities including Jakarta, Los Angeles, New Delhi, and New York City (Lewis 2020).

4.3 Testimony

Many people have been able to observe the declines in air pollution due to reduced congestion and decreases in electric generation. For example, a Philadelphia cyclist Shari Hersh said,

> I don't remember it being clear like this day after day in my whole lifetime. You know how sometimes we have a high-pressure day, and it's really clear, it's like if it's rained the night before and there's no particulates, it's like it's washed it all away. But not day after day. (Phillips 2020)

In France, residents have reported similar improvements to air quality. French photographer William Daniels described,

> As I've roamed Paris, I've noticed that the air is much fresher—there's less pollution. And one day when I was shooting at the main entrance of Les Halles, one of the biggest commercial malls in Europe, I heard birds singing. I'd never realized there were birds at Les Halles of all places. It gave me hope. (Thiessen 2020)

In company with less pollution, many cities were observed as noticeably quieter. Reporter Sylvia Poggioli explained,

> Here in Rome, all of a sudden the city's usual chaotic, horn-honking soundtrack has been transformed. With hardly any traffic, you can actually hear the squeak of rusty door hinges. The chirping of birds, an early sign of spring, is almost too loud. (Poggioli 2020)

Stories about observed changes add an important humanity to improvements in our natural environment.

4.4 Environmental Health Surveillance

Historically, changes in deaths linked to air pollution events are powerful messages about the risks air pollution poses. Marshall Burke, a professor of Earth System Science at Stanford, calculated that declines in premature mortality from air pollution reductions for children under 5 and elders over 70 in China exceeded the deaths reported from COVID-19 (Burke 2020; Crist 2020). A similar study published in *The Lancet Planetary Health* reported similar findings, estimating 12,151 fewer pollution related deaths from February to March in China (Chen et al. 2020; Science 2020). A preprint study estimated 5300 fewer deaths in India, based on declining measures of NO_2, O_3, and PM2.5 (Venter et al. 2020). A working paper published by the National Bureau of Economic Research revealed that in the US, the reduction in air pollution has contributed to 25% fewer deaths, 360 fewer deaths per month, from illnesses such as asthma, lung disease, and heart disease (Davenport 2020).

A study by the Centre for Research on Energy and Clean Air estimated a decline of 11,000 deaths from air pollution in Europe, with the highest number of avoided deaths in Germany (2083), the UK (1752), and Italy (1490) (Myllyvirta 2020; Watts 2020a). Other reductions were found in work absences, cases of asthma in children, emergency room visits, and preterm births (Myllyvirta 2020). A survey of 14,000 people by the British Lung Foundation found that one out of six people with lung conditions claimed that they observed improvements in their health during COVID-19 lockdown and among UK citizens surveyed who suffer from asthma, a larger proportion, one out of four, reported an improvement in their symptoms (Carrington 2020). Public Health England data also shows the number of asthma-related visits to emergency departments in hospitals have been reduced by half during lockdown (Carrington 2020).

It is important to also note that an increasing body of research finds that air pollution exposure connects to a wider variety of health outcomes, including brain diseases. These early efforts have the potential to expand our assessment of the health harms attributable to air pollution (Gardiner 2019; Peeples 2020), which means that past measures of the health impacts of air pollution are likely to be underestimated.

4.5 Health Disparities and Environmental Justice

Research in health disparities focuses on the differences in health outcomes as they relate to different social groups, such as sex, race and ethnicity. Oftentimes the reason for health disparities can be explained by social determinants of health like racism, income, employment, or economic inequality. Environmental justice describes both research and advocacy related to the unequal share in exposure to environmental toxics or environmental benefits.

In the US, COVID-19 responses coincided with widespread social unrest following the murder of George Floyd by a police officer, igniting renewed interest in the Black Lives Matter movement and prompting a national reckoning with racism in police enforcement (Haseman et al. 2020). But a re-examination of policing was not the only area of US domestic policy that was deeply affected. Given that Floyd's dying words were "I can't breathe" (Oppel and Barker 2020), the relationship between racism and police violence were readily connected to differences in deaths experienced by Black people in the US from COVID-19, and from air pollution.

The widespread citation of a Harvard study connecting past pollution exposure to COVID-19 risks in the US demonstrates contemporary interest in health equity (Love 2020). *The Guardian* reported on disparities in pollution exposure and COVID-19 in Los Angeles, Detroit, Houston, and in Navajo county (Holden 2020). Frequently cited too are preliminary connections between pollution aerosols and COVID-19 (Setti et al. 2020), though scientists are uncertain about the role of pollution aerosol transmission and the risks from prior air pollution exposure on COVID-19 infection (Contini and Costabile 2020).

Evidence continues to build (Lerner 2020): studies of China (Zhu et al. 2020), Italy (Fattorini and Regoli 2020), and California (Bashir et al. 2020) also demonstrate strong connections between air pollution exposure and COVID-19 infections. Past air pollution exposure and elevated COVID-19 death rates were also linked (Fussell 2020; Wu et al. 2020). Similar preliminary studies from the UK are also considering housing quality, in addition to air pollution, in influencing SARS-COV-2 inflection (Soltan et al. 2020). These concerns further tie environmental justice and environmental health to urban design and history by demonstrating that cumulative impact of environmental burdens includes unequal infectious disease risks (Tabuchi and Fountain 2020).

Risk-based models now connect air pollution measurements to health outcomes: one was created for Pittsburgh, Pennsylvania, where air pollution risks were connected to elevated risks of coronary heart disease mortality in minority communities (Fabisiak et al. 2020). The new expectation that air pollution ties to COVID-19 was put on display when serious critiques of studies were levied against COVID-19 health impact assessments that excluded air pollution risks (The Guardian 2020).

4.6 Economic Justifications

Some groups have argued that renewable energy provides an opportunity to drive economic recovery following COVID-19 (Ambrose 2020a). A survey of senior officials in central banks worldwide revealed that most experts believe that rescue packages that benefit the climate are more likely than neutral or "brown" policy interventions to help the economy (Hepburn et al. 2020). Another recent study of different regions of the US identified a decoupling of carbon emissions and economic growth (Saha and Jaeger 2020).

4.7 Co-benefits: Putting the Evidence Together

Policies that reduce air pollution tend to produce a variety of co-benefits. For example, reductions in car oriented transportation are associated with decreases in traffic crashes, injuries and fatalities. COVID-19 reductions in car use were associated with a halving of traffic injuries and fatalities in California (Shilling and Waetjen 2020). In New York City, there were two months with zero pedestrian deaths (Manjoo 2020). These declines are tied to clear, quantifiable economic and health outcomes.

In contrast to active transportation, such as walking and cycling, and public transportation, where maintenance and environmental costs and benefits are shared, the individualized distribution of automobiles is also associated with an environmental impact across their life cycle from production to use and disposal. The widespread consumption and use of cars produces microplastic pollution, primarily from tire and brake wear, that is circulated globally, negatively affecting humans and wildlife, and entire ecosystems, including the arctic cryosphere (Evangeliou et al. 2020; Simon 2020).

While typically discussed in reverse direction, reductions in a variety of types of air pollution are often simultaneous with reductions in carbon emissions. Policies aimed at reducing greenhouse gases often enjoy the important and immediate co-benefit of reducing NO_2, ozone, and PM2.5. Analysis of the Regional Greenhouse Gas Initiative in the Northeast US quantified the health and financial impact of the policy in reduced asthma cases, preterm births, autism cases, and cases of low birthweight, modeled after declines in PM2.5 (Bienkowski 2020; Perera et al. 2020). Conversely, efforts to reduce air pollution lead to reductions in carbon emissions, which are driving global climate disruption both presently and long term.

The UK Department of Transport report "Gearing Up", is health oriented, listing health and wellbeing as primary motivators in expanding cycling and pedestrian access, but other benefits such as reduced traffic congestion, benefits to local businesses from foot traffic, environmental and air quality improvements, reducing greenhouse gas emissions, and economic benefits are also demonstrated (United Kingdom Department of Transport 2020b). The new UK program Active Travel England was granted a five year $2.5 billion budget, in response to COVID-19 and to increase active travel (O'Sullivan 2020b). In the introduction to the report, Prime Minister Boris Johnson wrote,

> The unprecedented pandemic has also shown many of us, myself very much included, that we need to think harder about our health. We need to think harder about how we can make lifestyle changes that keep us more active and fit - the way we travel is central to this. (United Kingdom Department of Transport 2020b)

4.8 Sustainability Policy and Advocacy Responses

Given the widespread availability and communication about the impacts of the anthropause on air pollution and active transportation, the COVID-19 pandemic responses can prompt accelerated changes in urban sustainability policy. The following sections examine case studies of cities and environmental organizations adapting to new information.

5 Case Studies in Sustainability Policy and Activism

While framing and communicating the deleterious effects of air pollution is important, outlining next steps to reduce present-day harms is essential. Two important groups exploring post-COVID-19 sustainable programs are cities and non-profit environmental organizations. City approaches are explored first, followed by methods adopted by environmental justice organizations.

5.1 Cases of Cities and Their Sustainable Transportation Projects

Initial motivations for street closures involved opening additional urban access to walkable spaces, as was the case in New York City (Hu and Schweber 2020; New York City Council 2020). But these initial actions cracked open a door to more permanent sustainability policy, where some streets stay permanently closed to motorized traffic, and where pedestrian and bicycle access are expanded (Rosa 2020). For example, as a result of COVID-19, European cities are opening up nearly 1500 km of cycling lanes (Weiss 2020). Table 1 identifies six cities, five within Europe, and one in Latin America, that responded to COVID-19 by expanding major sustainable transportation initiatives.

Temporary plans to improve pedestrian and cycling mobility have been instituted in New York, Budapest, Paris, and Dublin, and cities are exploring cycling incentives and subsidies in Paris and Rome. Paris saw a 70% decline in nitrous oxides in early 2020 (Atmo France 2020; Perry 2020). Major urban design changes were a central

Table 1 List of major sustainable transportation planning projects by city

City	Sustainable planning project	Key leadership
London	The UK announced emergency funding for active travel to promote walking and cycling, including a bike Tube, and plans to create "one of the largest car-free zones in any capital city in the world" (Holder 2020; Sweney 2020; United Kingdom Department of Transport 2020a)	Sadiq Khan, mayor
Milan	The Strade Aperte (open roads) plan operationalizes an expansion of pedestrian and cycling paths, affecting 35 km of streets (Kusmer 2020; Laker 2020)	Pierfrancesco Maran, Deputy Mayor for Urban Planning, Green Areas and Agriculture of the City of Milan
Bogata	80 km of "emergency" bike lanes will remain in place after the pandemic (Jaramillo 2020; Plosz 2020)	Claudia López, mayor
Berlin	Pop-up lanes were introduced, with efforts to keep the lanes post-pandemic (Plosz 2020; Weißenborn 2020). Kate Connolly, reporter for *The Guardian*, wrote that "In Berlin it can take up to a decade to create a new bike lane, but during the coronavirus crisis, 14 miles of pop-up bike lanes, separating from car lanes by traffic beacons, have been introduced in anything from three to 10 days. Most will be here to stay, city officials have said, arguing that increasing numbers of Berliners - at the last count 43% - have no car, and that bikes will help lessen the burden on public transport" (Connolly 2020)	Felix Weisbrich, Head of Friedrichshain-Kreuzberg Road and Green Area Office
Athens	Development of a four-mile "grand walkway" will connect archaeological sites, with traffic bans in the areas by the Acropolis (Connolly 2020). According to the mayor, Kostas Bakoyannis, "The goal is to liberate public space from cars and give it to people who want to walk and enjoy the city … Athens will be cleaner, greener and better lit" (Connolly 2020)	Kostas Bakoyannis, mayor
Brussels	In Brussels, a city built for commuters, where government positions subsidize car dependence, planners are creating 25 miles of bicycle lanes: some in the busiest areas of the city center (Rankin 2020)	Elke Van den Brandt, Minister of Government, Brussels-Capital Region

part of a mayoral election, with Ann Hidalgo releasing a "Manifesto for Paris" during her re-election campaign, involving permanent expansion of pedestrian and cycling lanes (O'Sullivan 2020a). Peru, Ecuador and Mexico are following in Colombia's footsteps: Lima promised to open 300 km of new bike lanes (Jaramillo 2020).

5.2 *Cases of Environmental Justice Organizations in Responding to Air Pollution and COVID-19*

In order to connect science communications to policy changes, many environmental activist organizations are working to communicate the significance of air pollution and COVID-19 through intersectional approaches. US environmental justice organizations provide helpful and diverse examples of methods for raising awareness about the importance of environmental health and equity programs. Table 2 is the result of a scan of 125 organizations, identified through searches of US EPA Environmental Justice grant awards (US EPA 2014), Guide Star listings (GuideStar 2020), and previous listings of environmental justice organizations (Newton 2009). We chose eight organizations to examine more closely to describe the different ways that advocacy groups are communicating about air pollution and COVID-19.

As Table 2 illustrates, advocacy organizations in the US are diverse in their approach to both air pollution and COVID-19. Post-pandemic policies can draw on the innovative approaches of these groups, as they work to address specific, local needs, while also advocating for systemic changes on the national level. Table 3 highlights the creative and intersectional work being pioneered by these groups to draw connections to air pollution, sustainability policy, COVID-19, and equity concerns.

In the web resources provided by the US Environmental Justice organizations outlined here, the many and varied practices of communicating about air pollution are demonstrated. While satellite-derived air pollution measures were rarely depicted, humanistic presentations (stories, photography, testimonials, video, and artwork) were frequently uplifted alongside maps, data on environmental health and health disparities. Each of the websites stored rich, updated resources and content for both advocacy and community-level health needs. The creativity, resourcefulness and innovation of environmental justice organizations can help people come up with new ways to tackle difficult problems.

6 Conclusion

How sustainability policy will move forward remains a key question. Some have begun to argue that the pandemic has expedited the way to peak CO_2 (Storrow 2020). A crash in oil prices in the US elevated the costs of coal power production, causing plants to close (Niiler 2020). Declines in coal use in India led to declines

Table 2 Examples of contemporary communications used by US environmental justice organizations to raise awareness of air pollution and COVID-19

Organization, location, and guide star information	Connection to air pollution	Connection to COVID-19
Environmental Health Coalition (EHC) Director: Alejandra Fen Location: San Diego, CA GuideStar: platinum Gross Receipts (last reported): $4,277,957	Advocates highlight the unequal exposure and environmental racism of diesel truck traffic in the community of Barrio Logan in San Diego, California (Environmental Health Coalition 2020b)	The organization assembled resources for the community related to COVID-19 and health care resources, employment relief, housing information, food access, and immigration support (Samad 2020)
Honor the Earth Director: Winona LaDuke Location: Minneapolis, MN GuideStar: N/A Gross Receipts (last reported): $2,883,204	The focus of Honor the Earth is on fossil fuel use, specifically pipeline expansion, and its impact on the earth and communities (Honor the Earth 2020a, b)	Honor the Earth is raising objections to the designation of pipeline workers as essential workers, highlighting that peak COVID-19 in July is coinciding with proposals to bring in 4200 workers, half from out of state (Honor the Earth 2020a, b)
WEACT for Environmental Justice Directors: Peggy Shepard, Vernice Miller-Travis Location: New York, NY GuideStar: silver Gross Receipts (last reported): $2,854,051	Organizers demonstrate the unequal burden of air pollution in New York: "In Northern Manhattan, residents continue to suffer disproportionately from the impacts of air pollution, particularly those living in East Harlem" (WEACT for Environmental Justice 2020a)	WEACT provides information on health, food, and economic resources for Northern Manhattan communities, with specific, local information on the New York State Rent Relief program, and COVID-19 testing locations (WEACT 2020b)
Asian Pacific Environmental Network (APEN) Director: Miya Yoshitani Location: Oakland, CA GuideStar: platinum Gross Receipts (last reported): $2,806,406	APEN began and continues to work to promote Laotian communities in Richmond, CA affected by air pollution by Chevron oil processing, though they have also expanded their outreach to other areas (APEN 2019)	APEN initially supplied emergency COVID-19 community stabilization funds, and suggested other sources, when funds were no longer available (APEN 2020)
Deep South Center for Environmental Justice (DSCEJ) Director: Beverly L. Wright Location: New Orleans, LA GuideStar: N/A Gross Receipts (last reported): $2,365,661	The current DSCEJ materials take a more holistic view to cumulative environmental burden, highlighting the combined impact of pollutants and disaster risks faced by the residents of the Mississippi River Chemical Corridor (DSCEJ 2020b)	DSCEJ provides regular COVID-19 essential and returning worker training, as well as disaster preparedness training (Deep South Center for Environmental Justice 2020)

(continued)

Table 2 (continued)

Organization, location, and guide star information	Connection to air pollution	Connection to COVID-19
Organization, Location, and Guide Star Information	Connection to Air Pollution	Connection to COVID-19
Detroiters Working for Environmental Justice (DWEJ) Director: Guy Williams Location: Detroit, MI GuideStar: N/A Gross Receipts (last reported): $1,628,409	DWEJ observed that truck traffic and toxic facilities disproportionately affect minority and low-income communities: prevalence of asthma among Detroit adults is 29% higher than Michigan as a whole (DWEJ 2020). DWEJ partners with the University of Michigan to work on strategies to improve air quality	Like many environmental justice organizations, DWEJ engages its community through active use of social media, such as Facebook, Twitter and Linked In. July posts on Facebook included uplifting art and stories about Detroit
Little Village Environmental Justice Organization (LVEJO) Director: Kimberly Wasserman Location: Chicago, IL GuideStar: gold Gross Receipts (last reported): $997,173	A current active campaign by LVEJO is #NoDieselLV. A recent expansion of Unilever Best Foods brought substantial increases in truck traffic and related air pollution. Residents are concerned about additional traffic due the conversion of an old coal plant to a logistics hub (LVEJO 2020)	LVEJO continued activist work during spring 2020, with virtual community meetings, and updates to the community about development actions planned by the city of Chicago
Texas Environmental Justice Advocacy Services (t.e.j.a.s) Director: Juan Parras Location: Houston, TX GuideStar: Bronze Gross Receipts (last reported): $369,983	t.e.j.a.s focuses on the Houston Ship Channel, "home to the largest petrochemical refinery complex in the western hemisphere" (t.e.j.a.s 2020a, b). Specific chemical exposures from air, water, and waste pollution in this region are identified	t.e.j.a.s maintains community watch status by posting through social media ways that the public can intervene in decisions related to environmental pollution, such as Valero Refining's request to increase hydrogen cyanide emissions in the Manchester neighborhood

(continued)

Table 2 (continued)

Organization, location, and guide star information	Connection to air pollution	Connection to COVID-19
La Union del Pueblo Entero (LUPE) Director: Juanita Valdez-Cox Location: San Juan, TX GuideStar: N/A Gross Receipts (last reported): N/A	LUPE is a community union that was founded by Cesar Chavez and Dolores Huerta, which fought to reduce farmworkers exposure to pesticides (LUPE 2015)	Janita Valdez-Cox shared a thoughtful reflection about how COVID-19 has affected LUPE members, and how LUPE can serve their members moving forward. Examples of recent work included regular health talks on Facebook Live, use of megaphones in colonias to encourage census response, and zoom bingo to teach about the US Census (Valdez-Cox 2020b)

Table 3 Examples of US environmental justice organizations and intersectional approaches

Organization	Intersectional approach
Environmental Health Coalition	EHC adopts community-specific messaging that highlights the increased vulnerability of people living in areas with high levels of air pollution during COVID-19. For example, during a major fire onboard a Navy ship, adjacent to Barrio Logan, the EHC asked for effectively communicated public guidance, air quality information, community input to disaster response (Environmental Health Coalition 2020a)
WEACT	Included in COVID-19 community specific resources is a statement about environmental racism and COVID-19, and many examples of staff outreach to the media to explain the connection between environmental injustice and COVID-19 deaths. In addition to local work, the organization provides examples of national testimony on elevated pollution concerns tied to COVID-19 (WEACT 2020a)
Honor the Earth	Honor the Earth takes an indigenous knowledge approach, connecting respect for indigenous knowledge with reduced use of fossil fuels, demilitarization and social transformation (Honor The Earth 2020a, b). Pipeline resistance and work with water protectors can be understood as an upstream solution to reduce fuel combustion and air pollution
Asian Pacific Environmental Network	APEN compiled an innovative and collaborative document "Reparations Now: Honoring the Movement for Black Lives in Richmond," which chronicled historical racism and a list of demands to support Black People in Richmond
Deep South Center for Environmental Justice	The DSCEJ publicizes Gulf Equity Consortium Updates, with weekly reports on county-specific information on: COVID-19 testing, prevalence, deaths, poverty, access to health insurance, age, ranking, and COVID results by race. Reports also include industrial toxic releases, stating "exposure to air pollution is linked to increased risk of death from COVID-19" (DSCEJ 2020)
Detroiters Working for Environmental Justice	Social media posts shared by DWEJ directed readers to articles and evidence on the associations between coronavirus and environmental inequities, as well as networks of environmental justice advocates seeking policy changes to increase protections to those burdened by pollution (Corbett 2020)

(continued)

Table 3 (continued)

Organization	Intersectional approach
Little Village Environmental Justice Organization	LVEJO sponsored virtual and in-person community meetings on Just Transition strategies, including presentation of "Let Us Breathe" campaign images (LVEJO 2018)
Texas Environmental Justice Advocacy Services	t.e.j.a.s shared social media posts with members from the Peak Plastic Foundation, to align with other organizations, demanding a Just COVID-19 Recovery (t.e.j.a.s. 2020a, b)
La Union del Pueblo Entero	Juanita Valdez-Cox wrote, "The current public health crisis reminds our nation how important it is to have hard-working farm workers making sure our grocery stores are stocked with produce and food for us to enjoy. … the Farm Workforce Modernization Act … would provide a path for farmworkers and their families to earn protected immigration status and a higher standard of living" (Valdez-Cox 2020a)

in carbon emissions for the first time in 37 years (Rowlatt 2020b). In the UK, the first quarter saw renewable energy make up 47% of the total electricity generated, a large improvement from the 39% quarterly record from the previous year (Ambrose 2020b). The UK also set a new coal-free record in electricity generation of over 67 days during lockdown (Ambrose 2020b). In the US, COVID-19 policy responses were problematized for their connections to fossil fuel subsidies (Gardiner 2020c; Holden 2020), and the Bank of England was criticized for bailing out oil companies as part of coronavirus stimulus (Jolly 2020a).

Major changes to the auto industry are anticipated, and there is some expectation that COVID-19 social transformation will open opportunities for a more swift transition to electric cars (Ewing 2020). During lockdown, while total car sales across all European markets decreased dramatically, battery electric and plug-in hybrid cars gained traction; in April, hybrid car sales increased by 7% and electric car sales fell by 28%, vastly outperforming traditional petrol and diesel cars, whose sales declined by 78% compared to the previous year (Jolly 2020b).

How we move forward as countries, and as communities matters, and can make an enormous difference. One important and hopeful example has been the bailout package for Air France. A $10.8 billion bailout package to Air France comes with strings attached, including ending short routes that compete with train routes, reducing carbon dioxide emissions by 50% relative to 2005 by 2050, and use 2% of its fuel from sustainable sources by 2025 (Irfan 2020; Watts 2020b). French finance minister Bruno Le Maire explains that Air France would have to become the "greenest airline in the world" (Irfan 2020; Watts 2020b).

In some countries, COVID-19 responses are having a positive impact on environmental policy. Due to South Korea's strong response to COVID-19, President Moon Jae-In's Democratic Party won the national legislature elections by a landslide in April, winning 180 out of 300 seats in the National Assembly. This victory allows for President Moon Jae-In to push for his Green New Deal, which will commit South Korea to reach net zero emissions by 2050 (Irfan 2020). Chile updated its national climate action plan:

> Announcing the new plan, Chilean Environment Minister Carolina Schmidt said once the country had overcome the coronavirus "crisis", "we will enter a rehabilitation phase which must be sustainable." (Bartlett 2020)

While there have been many instances of impactful decisions during the pandemic made by environmental justice organizations and governments alike, it is still uncertain whether their efforts will result in lasting change. Globally outdoor air pollution contributes to over 4 million premature deaths annually: while air pollution in Europe and North America has declined, levels in many regions remain consistently high (Shaddick et al. 2020). Likewise, evidence points to a return to business as usual post-COVID-19, as rapid declines in carbon emission are erased: global emissions for June show declines only 5% compared with 2019 (Le Quéré et al. 2020). Levels of air pollution in China have returned to pre-pandemic levels, and even exceeded past emissions in some places (Carrington and Kommenda 2020; CREA 2020).

Among the case studies on infrastructure expansion shared above, none problematize the material or methods used for expanding bicycle and pedestrian walkways. Concrete infrastructure has been increasingly criticized for its harmful impact on the environment. Sand has become an international criminal enterprise (with illegal extractions in Indonesia, North Africa, India and elsewhere, paving expansions in Singapore and major cities) (Beiser 2015; United Nations Environment Programme 2019). However, relative to cars and trucks, bicycle and pedestrian infrastructure costs less, fiscally and environmentally: "a citywide bikeway network can be built for about the costs of a single highway junction" (Saxe and MacAskill 2020).

All work on air pollution comes with the necessary disclaimer that humans do not have the monopoly on air pollution. Natural and cosmic forces over earth's long history have produced radically different atmospheres than the one we experience today. Arctic wildfires in the summer of 2020 emitted more CO_2 than Norway's annual human-made production (Sengupta 2020). While past anthropogenic pollution influences wildfires today, their increasing frequency and intensity demonstrate the potential setbacks to air quality even if significant changes to urban transportation are successfully implemented.

Moving forward in our research, we would like to expand our understanding of the communications methods that environmental justice organizations use in comparisons to mainstream environmental organizations and youth organizations, to help respond to problems around air pollution, COVID-19, and the connections between the two. A comparison of US organizations to international non-profits would also be a valuable contribution to our understanding of the intersectional approach adopted globally. It would be helpful to better understand how community organizations contribute to local and global media coverage, and subsequent sustainability programs.

To conclude, the objectives of the United National Sustainable Development Goals (SDGs) help identify the intersections among multiple policies, programs, and social groups, which are also consistent with our own findings of COVID-19 responses to outdoor air pollution. Goals of Good Health and Well-Being (Goal 3), Affordable and Clean Energy (Goal 7), Industry, Innovation and Infrastructure (Goal 9), Reduced Inequalities (Goal 10), Sustainable Cities and Communities (Goal 11), Climate Action (Goal 13), Peace, Justice and Strong Institutions (Goal 16), and Partnerships (Goal 17) all connect to clean air, as demonstrated through this chapter, even though Clean Air does not stand alone as one of the 17 overarching goals (United Nations 2020). Clean air, energy, transportation and equity all relate to one another. Perhaps being forced to slow-down, provides an opportunity to appreciate the rich diversity and complexity of environment-society relationships, allowing us to uncover the varied and innovative responses to improve both social and environmental wellbeing.

What we say, and how we said, can make a difference: we all have a part to play in building support for action on air pollution and climate change (Corner et al. 2018; FrameWorks Institute 2020). This chapter demonstrates different approaches taken by young and diverse leadership today, during COVID-19. By talking more often about air pollution and climate disruption, during and post-COVID-19, and by joining groups that speak to sustainability challenges that lie ahead, we can make a difference.

References

Ambrose J (2020a, April 20) Green energy could drive Covid-19 recovery with $100 tn boost. The Guardian. https://www.theguardian.com/environment/2020/apr/20/green-energy-could-drive-covid-19-recovery-international-renewable-energy-agency

Ambrose J (2020b, June 25) Renewable energy breaks UK record in first quarter of 2020. The Guardian. https://www.theguardian.com/business/2020/jun/25/renewable-energy-breaks-uk-record-in-first-quarter-of-2020

APEN (2019, Oct 8) Our history. Asian Pacific Environmental Network. https://apen4ej.org/our-history/

APEN (2020, July 20) COVID-19 emergency community stabilization fund. Asian Pacific Environmental Network. https://apen4ej.org/covid-form/

Atmo France (2020, Apr 21) COVID-19: Focus sur l'exposition des riverains à la pollution automobile près des grands axes avant/pendant le confinement. Atmo France. https://atmo-france.org/covid-19-focus-sur-lexposition-des-riverains-a-la-pollution-automobile-pres-des-grands-axes-avant-pendant-le-confinement/

Bartlett, J (2020, Apr 13) Chile charts path to greener, fairer future after coronavirus. Thomson Reuters Foundation News. https://news.trust.org/item/20200413182131-gtuoz/

Bashir MF, Ma BJ, Bilal, Komal B, Bashir MA, Farooq TH, Iqbal N, Bashir M (2020) Correlation between environmental pollution indicators and COVID-19 pandemic: a brief study in Californian context. Environ Res 187:109652. https://doi.org/10.1016/j.envres.2020.109652

BBC News (2002, Dec 5) Days of toxic darkness. BBC News. https://news.bbc.co.uk/2/hi/uk_news/2542315.stm

BBC News (2020, Apr 13) Cycling numbers jump during coronavirus lockdown. BBC News. https://www.bbc.com/news/uk-scotland-52269964

Beiser V (2015, Mar 26) The deadly global war for sand. Wired. https://www.wired.com/2015/03/illegal-sand-mining/

Bienkowski B (2020, July 29) A Northeast US climate initiative has had a major side benefit—healthier children. EHN. https://www.ehn.org/climate-change-plan-air-pollution-children-health--2646819170.html

Binding L (2020, Apr 17) Coronavirus: only 9% of Britons want life to return to "normal" once lockdown is over. Sky News. https://news.sky.com/story/coronavirus-only-9-of-britons-want-life-to-return-to-normal-once-lockdown-is-over-11974459

Blumberg S (2020, Apr 9) Data shows 30 percent drop in air pollution over Northeast U.S. [Text]. NASA. https://www.nasa.gov/feature/goddard/2020/drop-in-air-pollution-over-northeast

Broad WJ (2020, July 23) With Covid-19, a seismic quiet like no other. The New York Times. https://www.nytimes.com/2020/07/23/science/coronavirus-seismic-activity.html

Bryant M (2020, May 13) Cycling "explosion": coronavirus fuels surge in US bike ridership. The Guardian. https://www.theguardian.com/us-news/2020/may/13/coronavirus-cycling-bikes-american-boom

Burke M (2020, Mar 8) COVID-19 reduces economic activity, which reduces pollution, which saves lives. G-FEED. https://www.g-feed.com/2020/03/covid-19-reduces-economic-activity.html

Carrington D (2020, June 4) Cleaner air during UK lockdown relieves asthma for millions. The Guardian. https://www.theguardian.com/environment/2020/jun/04/cleaner-air-during-uk-lockdown-relieves-asthma-for-millions-lung-conditions-coronavirus

Carrington D, Kommenda N (2020, June 3) Air pollution in China back to pre-Covid levels and Europe may follow. The Guardian. https://www.theguardian.com/environment/2020/jun/03/air-pollution-in-china-back-to-pre-covid-levels-and-europe-may-follow

Chen K, Wang M, Huang C, Kinney PL, Anastas PT (2020) Air pollution reduction and mortality benefit during the COVID-19 outbreak in China. Lancet Planet Health 4(6):e210–e212. https://doi.org/10.1016/S2542-5196(20)30107-8

Connolly, K. (2020, May 18). "Cleaner and greener": Covid-19 prompts world's cities to free public space of cars. The Guardian. https://www.theguardian.com/world/2020/may/18/cleaner-and-greener-covid-19-prompts-worlds-cities-to-free-public-space-of-cars.

Contini D, Costabile F (2020) Does air pollution influence COVID-19 outbreaks? Atmosphere 11(4):377. https://doi.org/10.3390/atmos11040377

Corbett J (2020, July 7) 60+ Environmental justice advocates and groups issue coronavirus call to action demanding end to "Sacrifice Zones". Common Dreams. https://www.commondreams.org/news/2020/07/07/60-environmental-justice-advocates-and-groups-issue-coronavirus-call-action

Corner A, Shaw C, Clarke J (2018) Principles for effective communication and public engagement on climate change: a handbook for IPCC authors. Climate Outreach. https://climateoutreach.org/resources/ipcc-communications-handbook/

CREA (2020) China's air pollution overshoots pre-crisis levels for the first time. Centre for Research on Energy and Clean Air. https://energyandcleanair.org/wp/wp-content/uploads/2020/05/China-air-pollution-rebound-final.pdf

Crist M (2020, Mar 27) Opinion | what the coronavirus means for climate change. The New York Times. https://www.nytimes.com/2020/03/27/opinion/sunday/coronavirus-climate-change.html

Davenport C (2020, June 25) Pandemic's cleaner air could reshape what we know about the atmosphere. The New York Times. https://www.nytimes.com/2020/06/25/climate/coronavirus-clean-air.html

Dennis B (2020, June 23) Most Americans believe the government should do more to combat climate change, poll finds. Washington Post. https://www.washingtonpost.com/climate-environment/2020/06/23/climate-change-poll-pew/

DSCEJ (2020, July 30) HBCU-CBO Gulf Equity consortium updates—Deep South Center for Environmental Justice. Enmasse—Deep South Center for Environmental Justice (DSCEJ). Website https://www.dscej.org/be-prepared

DWEJ (2020, July 30) Environmental justice matters | Detroiters working for environmental justice. https://detroitenvironmentaljustice.org/environmental-justice-matters/

Environmental Health Coalition (2020a, July 13) San Diego community overwhelmed by smoke and toxins from navy ship fire. Environmental Health Coalition. https://www.environmentalhealth.org/index.php/en/media-center/blog-for-environmental-justice

Environmental Health Coalition (2020b, July 30) Environmental Health Coalition—home. Environmental Health Coalition. https://www.environmentalhealth.org/index.php/en/

Evangeliou N, Grythe H, Klimont Z, Heyes C, Eckhardt S, Lopez-Aparicio S, Stohl A (2020) Atmospheric transport is a major pathway of microplastics to remote regions. Nat Commun 11(1):3381. https://doi.org/10.1038/s41467-020-17201-9

Ewing J (2020, May 13) The pandemic will permanently change the auto industry. The New York Times. https://www.nytimes.com/2020/05/13/business/auto-industry-pandemic.html

Fabisiak JP, Jackson EM, Brink LL, Presto AA (2020) A risk-based model to assess environmental justice and coronary heart disease burden from traffic-related air pollutants. Environ Health 19(1):34. https://doi.org/10.1186/s12940-020-00584-z

Fattorini D, Regoli F (2020) Role of the chronic air pollution levels in the Covid-19 outbreak risk in Italy. Environ Pollut 264:114732. https://doi.org/10.1016/j.envpol.2020.114732

FrameWorks Institute (2020, May 1) Communicating about climate change in the time of COVID-19. https://www.frameworksinstitute.org/publication/communicating-about-climate-change-in-the-time-of-covid-19/

Friedman MS, Powell KE, Hutwagner L, Graham LM, Teague WG (2001) Impact of changes in transportation and commuting behaviors during the 1996 summer Olympic games in Atlanta on air quality and childhood asthma. JAMA 285(7):897–905. https://doi.org/10.1001/jama.285.7.897

Fussell S (2020, May 26) Covid-19 flares up in America's polluted 'Sacrifice Zones.' Wired. https://www.wired.com/story/covid-19-flares-americas-polluted-sacrifice-zones/

Gardiner B (2019) Choked: life and breath in the age of air pollution. University of Chicago Press.

Gardiner B (2020a, Apr 8) Pollution made COVID-19 worse. Now, lockdowns are clearing the air. National Geographic. https://www.nationalgeographic.com/science/2020/04/pollution-made-the-pandemic-worse-but-lockdowns-clean-the-sky/

Gardiner B (2020b, June 18) Why COVID-19 will end up harming the environment. National Geographic. https://www.nationalgeographic.com/science/2020/06/why-covid-19-will-end-up-harming-the-environment/

Gardiner B (2020c, June 23) In pandemic recovery efforts, polluting industries are winning big. Yale E360. https://e360.yale.edu/features/in-pandemic-recovery-efforts-polluting-industries-are-winning-big

Goldfarb B (2020, July 6) Lockdowns could be the 'Biggest Conservation Action' in a century. The Atlantic. https://www.theatlantic.com/science/archive/2020/07/pandemic-roadkill/613852/

GuideStar (2020 July 30) GuideStar nonprofit reports and Forms 990 for donors, grantmakers, and businesses. https://www.guidestar.org/

Hamill SD (2008, Nov 1) Unveiling a museum, a Pennsylvania town remembers the smog that killed 20. The New York Times. https://www.nytimes.com/2008/11/02/us/02smog.html

Harvey F (2020a, Mar 24) Covid-19 economic rescue plans must be green, say environmentalists. The Guardian. https://www.theguardian.com/environment/2020/mar/24/covid-19-economic-rescue-plans-must-be-green-say-environmentalists

Harvey F (2020b, May 10) Britons want quality of life indicators to take priority over economy. The Guardian. https://www.theguardian.com/society/2020/may/10/britons-want-quality-of-life-indicators-priority-over-economy-coronavirus

Harvey F (2020c, May 19) Lockdowns trigger dramatic fall in global carbon emissions. The Guardian. https://www.theguardian.com/environment/2020/may/19/lockdowns-trigger-dramatic-fall-global-carbon-emissions

Harvey F (2020d, May 26) World health leaders urge green recovery from coronavirus crisis. The Guardian. https://www.theguardian.com/environment/2020/may/26/world-health-leaders-urge-green-recovery-from-coronavirus-crisis

Haseman J, Zaiets K, Thorson M, Procell C, Petras G, Sullivan S (2020, June 18) Tracking protests across the USA in the wake of George Floyd's death. USA Today. https://www.usatoday.com/in-depth/graphics/2020/06/03/map-protests-wake-george-floyds-death/5310149002/

Hepburn C, O'Callaghan B, Stern N, Stiglitz J, Zenghelis D (2020) Will COVID-19 fiscal recovery packages accelerate or retard progress on climate change? Oxford Rev Econ Policy. https://doi.org/10.1093/oxrep/graa015

Hirschauer J (2020, Apr 15) Outdoor transmission of COVID. National Review. https://www.nationalreview.com/corner/coronavirus-transmission-chinese-study-shows-covid-more-likely-spread-indoors/

Holden E (2020, May 1) Revealed: US fossil fuel companies handed at least $50 m in coronavirus aid. The Guardian. https://www.theguardian.com/environment/2020/may/01/fossil-fuel-firms-coronavirus-package-aid

Holder, M. (2020, May 15). Large areas of London to go "car-free" as Mayor plots green Covid-19 recovery. BusinessGreen. https://www.businessgreen.com/news/4015297/london-car-free-mayor-plots-green-covid-19-recovery

Honor the Earth (2020a, July 30) No fossil fuels in the Great Lakes. Honor The Earth. https://www.honorearth.org/no_fossil_fuels_in_the_great_lakes

Honor the Earth (2020b, July 30) The green path: a just transition. Honor The Earth. https://www.honorearth.org/green-newdeal

Hu W, Schweber N (2020, Aug 10) Will cars rule the roads in post-pandemic New York? The New York Times. https://www.nytimes.com/2020/08/10/nyregion/nyc-streets-parking-dining-busways.html

Hu Y, Barbour W, Samaranayake S, Work D (2020, May 19) The rebound—how Covid-19 could lead to worse traffic. Medium. https://medium.com/@barbourww/the-rebound-how-covid-19-could-lead-to-worse-traffic-cb245a5b1da2

Huang K, Zhang X, Lin Y (2015) The "APEC Blue" phenomenon: regional emission control effects observed from space. Atmos Res 164–165:65–75. https://doi.org/10.1016/j.atmosres.2015.04.018

Ipsos (2020, Apr 22) Two thirds of citizens around the world agree climate change is as serious a crisis as Coronavirus. Ipsos. https://www.ipsos.com/en/two-thirds-citizens-around-world-agree-climate-change-serious-crisis-coronavirus

Irfan U (2020, June 4) How South Korea, France, and Italy are using the Covid-19 response to fight climate change. Vox. https://www.vox.com/2020/6/4/21276805/coronavirus-south-korea-france-italy-covid-19-stimulus-climate-change

ITDP (2020, Mar 26) Post-pandemic, Chinese cities gradually reopen transport networks. Institute for Transportation and Development Policy. https://www.itdp.org/2020/03/26/post-pandemic-chinese-cities-gradually-reopen-transport-networks/

Jaramillo A (2020, Aug 10) Bogotá is building its future around bikes. Bloomberg.Com. https://www.bloomberg.com/news/articles/2020-08-10/to-tame-traffic-bogot-bets-big-on-bike-lanes

Jolly J (2020a, Apr 16) Bank of England "failing climate" with Covid-19 stimulus programme. The Guardian. https://www.theguardian.com/business/2020/apr/16/bank-of-england-failing-climate-with-covid-19-stimulus-programme-oil-firms-debt-bond-governor

Jolly J (2020b, June 2) Electric cars gain market share in Europe despite Covid-19 crisis. The Guardian. https://www.theguardian.com/business/2020/jun/02/electric-carseurope-covid-19-emissions-reductions

Kusmer A (2020, June 9) After lockdown, Milan rolls out plan to open more streets to cyclists and pedestrians. The World from PRX. https://www.pri.org/stories/2020-06-09/after-lockdown-milan-rolls-out-plan-open-more-streets-cyclists-and-pedestrians

Laker L (2020, Apr 21) Milan announces ambitious scheme to reduce car use after lockdown. The Guardian. https://www.theguardian.com/world/2020/apr/21/milan-seeks-to-prevent-post-crisis-return-of-traffic-pollution

Landis-Hanley J (2020, Apr 22) "Bicycles are the new toilet paper": bike sales boom as coronavirus lockdown residents crave exercise. The Guardian. https://www.theguardian.com/lifeandstyle/2020/apr/22/bicycles-are-the-new-toilet-paper-bike-sales-boom-as-coronavirus-lockdown-residents-crave-exercise

Le T, Wang Y, Liu L, Yang J, Yung YL, Li G, Seinfeld JH (2020) Unexpected air pollution with marked emission reductions during the COVID-19 outbreak in China. Science 369(6504):702–706. https://doi.org/10.1126/science.abb7431

Lecocq T, Hicks SP, Noten KV, van Wijk K, Koelemeijer P, Plaen RSMD, Massin F, Hillers G, Anthony RE, Apoloner M-T, Arroyo-Solórzano M, Assink JD, Büyükakpınar P, Cannata A, Cannavo F, Carrasco S, Caudron C, Chaves EJ, Cornwell DG et al (2020) Global quieting of high-frequency seismic noise due to COVID-19 pandemic lockdown measures. Science. https://doi.org/10.1126/science.abd2438

Le Quéré C, Jackson RB, Jones MW, Smith AJP, Abernethy S, Andrew RM, De-Gol AJ, Willis DR, Shan Y, Canadell JG, Friedlingstein P, Creutzig F, Peters GP (2020) Temporary reduction in daily global CO_2 emissions during the COVID-19 forced confinement. Nat Clim Change. https://doi.org/10.1038/s41558-020-0797-x

Lerner S (2020, June 26) Scientists pin blame for some coronavirus deaths on air pollution, PFAS, and other chemicals. The Intercept. https://theintercept.com/2020/06/26/coronavirus-toxic-chemicals-pfas-bpa/

Lewis S (2020, Apr 22) Before-and-after photos show dramatic decline in air pollution around the world during coronavirus lockdown. CBS News. https://www.cbsnews.com/news/coronavirus-photos-decline-air-pollution-lockdown/
Love MC (2020) Pollution and Covid-19. America 222(13):50–50
LUPE (2015, May 26) Movement victories. LUPE. https://lupenet.org/organizing/la-union-del-pueblo-entero-victories/
LVEJO (2018, Mar 6) Current updates. LA VILLITA RESPIRA. https://lavillitarespira.com/updates/
LVEJO (2020, July 30) LA VILLITA RESPIRA. LA VILLITA RESPIRA. https://lavillitarespira.com/
Makary M (2020, May 14) Opinion | how to reopen America safely. The New York Times. https://www.nytimes.com/2020/05/14/opinion/reopen-america-coronavirus-lockdown.html
Manjoo F (2020, July 9) Opinion | I've seen a future without cars, and it's amazing. The New York Times. https://www.nytimes.com/2020/07/09/opinion/ban-cars-manhattan-cities.html
Meng R, Zhao FR, Sun K, Zhang R, Huang C, Yang J (2015) Analysis of the 2014 "APEC Blue" in Beijing using more than one decade of satellite observations: lessons learned from radical emission control measures. Remote Sens 7(11):15224–15243. https://doi.org/10.3390/rs71115224
Mooney C, Muyskens J, Dennis B, Freedman A (2020, Mar 13) Pollution is plummeting in Italy in the wake of coronavirus, emissions data show. Washington Post. https://www.washingtonpost.com/climate-environment/2020/03/13/italy-emissions-coronavirus/
Mufson S (2020, Apr 30) Coronavirus is driving down global carbon dioxide emissions to levels last seen 10 years ago, agency says. Washington Post. https://www.washingtonpost.com/climate-environment/2020/04/30/coronavirus-is-driving-down-global-carbon-dioxide-emissions-levels-last-seen-10-years-ago-agency-says/
Myllyvirta L (2020, Apr 30) 11,000 air pollution-related deaths avoided in Europe as coal, oil consumption plummet. Centre for Research on Energy and Clean Air. https://energyandcleanair.org/air-pollution-deaths-avoided-in-europe-as-coal-oil-plummet/
NASA (2020a, Mar 26) Nighttime images capture change in China [NASA Earth Observatory]. NASA Earth Observatory. https://earthobservatory.nasa.gov/images/146481/nighttime-images-capture-change-in-china
NASA (2020b, Aug 19) Nitrogen dioxide (1 month) | NASA [Text.Article]. Nitrogen dioxide (1 Month) | NASA; NASA Earth Observations (NEO). https://neo.sci.gsfc.nasa.gov/view.php?datasetId=AURA_NO2_M
National Association of City Transportation Officials (NACTO) (2020, June 25) Streets for pandemic response and recovery. National Association of City Transportation Officials. https://nacto.org/streets-for-pandemic-response-recovery/
Newton DE (2009) Environmental justice: a reference handbook, 2nd edn. ABC-CLIO, LLC. https://ebookcentral.proquest.com/lib/drew-ebooks/detail.action?docID=2051555
New York City Council (2020, Apr 17) Speaker Corey Johnson and council member Carlina Rivera to introduce legislation to open city streets during coronavirus/COVID-19 pandemic. Press. https://council.nyc.gov/press/2020/04/17/1939/
Nguyen T, Saleh M, Kyaw M-K, Trujillo G, Bejarano M, Tapia K, Interns REC (2020) Special report 4: impact of COVID-19 mitigation on wildlife-vehicle conflict. Road Ecology Center, p 10
Niiler E (2020, June 12) The Covid-19 economic slump is closing down coal plants. Wired. https://www.wired.com/story/the-covid-19-economic-slump-is-closing-down-coal-plants/
NOAA (2020a, May 4) NOAA's polar-orbiting satellites see drop in U.S. air pollution | NOAA National Environmental Satellite, Data, and Information Service (NESDIS). National Oceanic and Atmospheric Administration: Satellite and Information Service. https://www.nesdis.noaa.gov/content/noaa%E2%80%99s-polar-orbiting-satellites-see-drop-us-air-pollution
NOAA (2020b, May 11) Suomi NPP detects changes in nighttime lights in NYC metro | NOAA National Environmental Satellite, Data, and Information Service (NESDIS). National Oceanic

and Atmospheric Administration: Satellite and Information Service. https://www.nesdis.noaa.gov/content/suomi-npp-detects-changes-nighttime-lights-nyc-metro

Nunez C (2019, Feb 4) Climate 101: air pollution. National Geographic. https://www.nationalgeographic.com/environment/global-warming/pollution/

NYC Department of Health (2020) Air quality during coronavirus. Nyc.Gov. https://a816-dohbesp.nyc.gov/IndicatorPublic/Closerlook/covidair/

Oppel R Jr, Barker K (2020, July 8) New transcripts detail last moments for George Floyd—the New York Times. New York Times. https://www.nytimes.com/2020/07/08/us/george-floyd-body-camera-transcripts.html

O'Sullivan F (2020a, June 19) Paris speeds up its pursuit of a slower beltway. Bloomberg.Com. https://www.bloomberg.com/news/articles/2020-06-19/paris-is-plotting-a-greener-slower-beltway

O'Sullivan F (2020b, July 29) Britain is creating a government organization devoted to biking and walking. Bloomberg.Com. https://www.bloomberg.com/news/articles/2020-07-29/why-the-u-k-is-investing-big-in-cycling-and-walking

Peeples L (2020) News feature: how air pollution threatens brain health. Proc Natl Acad Sci 117(25):13856–13860. https://doi.org/10.1073/pnas.2008940117

Perera F, Cooley D, Berberian A, Mills D, Kinney P (2020) Co-benefits to children's health of the U.S. regional greenhouse gas initiative. Environ Health Perspect 128(7):077006. https://doi.org/10.1289/EHP6706

Perry F (2020, Apr 29) How cities are clamping down on cars. BBC: Future Planet. https://www.bbc.com/future/article/20200429-are-we-witnessing-the-death-of-the-car

Phillips S (2020, May 18) Has the pandemic cleaned up our air? Answers could lie on your doorstep. StateImpact Pennsylvania. https://stateimpact.npr.org/pennsylvania/2020/05/18/has-the-pandemic-cleaned-up-our-air-answers-could-lie-on-your-doorstep/

Plosz J (2020, June 22) How the Coronavirus recovery is changing cities. Bloomberg.Com. https://www.bloomberg.com/features/2020-city-in-recovery/

Poggioli S (2020, Mar 16) Reporter's notebook: what life is like in Rome under Coronavirus lockdown. NPR. https://www.npr.org/2020/03/16/816080263/reporters-notebook-rome-under-coronavirus-lockdown

Qian H, Miao T, Liu L, Zheng X, Luo D, Li Y (2020) Indoor transmission of SARS-CoV-2. MedRxiv, 2020.04.04.20053058. https://doi.org/10.1101/2020.04.04.20053058

Qiang Y, Huang Q, Xu J (2020) Observing community resilience from space: using nighttime lights to model economic disturbance and recovery pattern in natural disaster. Sustain Cities Soc 57:102115. https://doi.org/10.1016/j.scs.2020.102115

Rankin, J. (2020, May 11). Can coronavirus cure Brussels of its addiction to driving? The Guardian. https://www.theguardian.com/world/2020/may/11/slow-streetscan-coronavirus-cure-brussels-of-addiction-driving.

Rosa A (2020, Aug 11) Are more car-free streets in N.Y.C.'s future? The New York Times. https://www.nytimes.com/2020/08/11/nyregion/nyc-open-streets.html

Rowlatt J (2020a, May 31) How a green new deal really could go global. BBC News. https://www.bbc.com/news/science-environment-52848184

Rowlatt J (2020b, June 9) Could the coronavirus crisis finally finish off coal? BBC News. https://www.bbc.com/news/science-environment-52968716

Rutz C, Loretto M-C, Bates AE, Davidson SC, Duarte CM, Jetz W, Johnson M, Kato A, Kays R, Mueller T, Primack RB, Ropert-Coudert Y, Tucker MA, Wikelski M, Cagnacci F (2020) COVID-19 lockdown allows researchers to quantify the effects of human activity on wildlife. Nat Ecol Evol 1–4. https://doi.org/10.1038/s41559-020-1237-z

Saha D, Jaeger J (2020) America's new climate economy: a comprehensive guide to the economic benefits of climate policy in the United States. World Resources Institute. https://www.wri.org/publication/us-new-climate-economy

Samad B (2020, July 28) COVID-19: community information and resources. Environmental Health Coalition. https://www.environmentalhealth.org/index.php/en/media-center/blog-for-environmental-justice/1346-covid-19-community-information-and-resources
Saxe S, MacAskill K (2020, July 8) Opinion | stop building more roads. The New York Times. https://www.nytimes.com/2020/07/08/opinion/us-infrastructure-plan.html
Science (2020) China's halt cut pollution deaths. Science 368(6493):804–805
Sengupta S (2020, July 7) Intense Arctic wildfires set a pollution record. The New York Times. https://www.nytimes.com/2020/07/07/climate/climate-change-arctic-fires.html
Setti L, Passarini F, Gennaro GD, Baribieri P, Perrone MG, Borelli M, Palmisani J, Gilio AD, Torboli V, Pallavicini A, Ruscio M, Piscitelli P, Miani A (2020) SARS-Cov-2 RNA found on particulate matter of Bergamo in Northern Italy: first preliminary evidence. MedRxiv, 2020.04.15.20065995. https://doi.org/10.1101/2020.04.15.20065995
Shaddick G, Thomas ML, Mudu P, Ruggeri G, Gumy S (2020) Half the world's population are exposed to increasing air pollution. Npj Clim Atmos Sci 3(1):1–5. https://doi.org/10.1038/s41612-020-0124-2
Shilling F, Waetjen D (2020) Special report (update): impact of COVID19 mitigation on numbers and costs of California traffic crashes. Road Ecology Center, p 11. https://roadecology.ucdavis.edu/files/content/projects/COVID_CHIPs_Impacts_updated_415.pdf
Simon M (2020, July 14) Your car is spewing microplastics that blow around the World. Wired. https://www.wired.com/story/your-car-is-spewing-microplastics/
Soltan M, Crowley L, Melville C, Varney J, Cassidy S, Mahida R, Grudzinska F, Parekh D, Dosanjh D, Thickett D (2020) To what extent are social determinants of health, including household overcrowding, air pollution and housing quality deprivation, modulators of presentation, ITU admission and outcomes among patients with SARS-COV-2 infection in an urban catchment area in Birmingham, United Kingdom? BMC Public Health Preprint. https://doi.org/10.21203/rs.3.rs-35617/v1
Sommer, L., Hersher, R., Jingnan, H., & Benincasa, R. (2020, May 19). Traffic is way down because of lockdown, but air pollution? not so much. NPR.Org. https://www.npr.org/sections/health-shots/2020/05/19/854760999/traffic-is-way-down-due-to-lockdowns-but-air-pollution-not-so-much.
Spring J, McGeever J, Boadle A (2020, May 8) Deforestation in Brazil's Amazon surges, Bolsonaro readies troops. Reuters. https://www.reuters.com/article/us-brazil-environment-idUSKBN22K1U1
Stone EA, Roberts JD (2020) Park spaces and the user experience: reconsidering the body in park analysis tools. Nat Cult 15(2):123–133. https://doi.org/10.3167/nc.2020.150201
Storrow B (2020, June 1) PANDEMIC: global CO_2 has risen for a century. That appears to be over. E&E News. https://www.eenews.net/stories/1063286379
Sweney M (2020, May 15) TfL to raise congestion charge by 30% as part of £1.6 bn bailout deal. The Guardian. https://www.theguardian.com/uk-news/2020/may/15/transport-for-london-faces-3bn-funding-gap-despite-16bn-bailout-income-journeys-coronavirus
Tabuchi H, Fountain H (2020, May 20) Your ZIP code and your life expectancy. The New York Times. https://www.nytimes.com/2020/05/20/climate/nyt-climate-newsletter-pollution-coronavirus.html
t.e.j.a.s. (2020a, July 30) T.E.J.A.S. Advocacy. https://www.tejasbarrios.org
t.e.j.a.s. (2020b, July 30) Texas environmental justice advocacy service | Facebook. https://www.facebook.com/TejasBarrios/
The Guardian (2020, June 7) Omission of air pollution from report on Covid-19 and race 'astonishing'. The Guardian. https://www.theguardian.com/environment/2020/jun/07/omission-of-air-pollution-from-report-on-covid-19-and-race-astonishing
Thiessen T (2020, Apr 10) How clean air cities could outlast COVID-19 lockdowns. Forbes. https://www.forbes.com/sites/tamarathiessen/2020/04/10/how-clean-air-cities-could-outlast-covid-19-lockdowns/

Tyson A, Kennedy B (2020, June 23) Two-thirds of Americans think government should do more on climate. Pew Research Center Science & Society. https://www.pewresearch.org/science/2020/06/23/two-thirds-of-americans-think-government-should-do-more-on-climate/

United Kingdom Department of Transport (2020a, May 9) £2 billion package to create new era for cycling and walking. GOV.UK. https://www.gov.uk/government/news/2-billion-package-to-create-new-era-for-cycling-and-walking

United Kingdom Department of Transport (2020b, July 27) Gear change: a bold vision for cycling and walking. GOV.UK. https://www.gov.uk/government/publications/cycling-and-walking-plan-for-england

United Nations (2020, July 30) About the sustainable development goals. United Nations Sustainable Development. https://www.un.org/sustainabledevelopment/sustainable-development-goals/

United Nations Environment Programme (2019) Sand and sustainability: finding new solutions for environmental governance of global sand resources. https://wedocs.unep.org/handle/20.500.11822/28163

US EPA (2014, Nov 3) Environmental justice [Collections and lists]. US EPA. https://www.epa.gov/environmentaljustice

Valdez-Cox J (2020a, Apr 1) Cesar Chavez's legacy and farm worker struggle lives on today. LUPE. https://lupenet.org/2020/04/cesar-chavezs-legacy-and-farm-worker-struggle-lives-on-today/

Valdez-Cox J (2020b, Apr 3) A wall won't keep us healthy. LUPE. https://lupenet.org/2020/04/a-wall-wont-keep-us-healthy/

Venter Z, Aunan K, Chowdhury S, Lelieveld J (2020, Apr 14) COVID-19 lockdowns cause global air pollution declines with implications for public health risk | medRxiv. MedRxiv. https://www.medrxiv.org/content/10.1101/2020.04.10.20060673v1

Vince G (2020, May 17) After the Covid-19 crisis, will we get a greener world? The Observer. https://www.theguardian.com/environment/2020/may/17/after-the-covid-19-crisis-will-we-get-a-greener-world

Wang T, Nie W, Gao J, Xue LK, Gao XM, Wang XF, Qiu J, Poon CN, Meinardi S, Blake D, Wang SL, Ding AJ, Chai FH, Zhang QZ, Wang WX (2010) Air quality during the 2008 Beijing Olympics: secondary pollutants and regional impact. Atmos Chem Phys 10(16):7603–7615. https://doi.org/10.5194/acp-10-7603-2010

Watts J (2020a, Apr 30) Clean air in Europe during lockdown 'leads to 11,000 fewer deaths.' The Guardian. https://www.theguardian.com/environment/2020/apr/30/clean-air-in-europe-during-lockdown-leads-to-11000-fewer-deaths

Watts J (2020b, May 17) Is the Covid-19 crisis the catalyst for greening the world's airlines? The Guardian. https://www.theguardian.com/world/2020/may/17/is-covid-19-crisis-the-catalyst-for-the-greening-of-worlds-airlines

WEACT (2020a, June 10) Pollution and pandemics: COVID-19's disproportionate impact on environmental justice communities. WE ACT for Environmental Justice. https://www.weact.org/2020/06/pollution-and-pandemics-covid-19s-disproportionate-impact-on-environmental-justice-communities/

WEACT. (2020b, July 30). WE ACT Coronavirus Information & Updates. WE ACT for Environmental Justice. https://www.weact.org/coronavirus/.

Weißenborn S (2020, May 27) Berlin baut Pop-up-Radwege: "Wir stellen jetzt Flächengerechtigkeit her." Der Spiegel. https://www.spiegel.de/auto/berlin-baut-pop-up-radwege-planer-felix-weisbrich-im-interview-a-9215e2c4-7f91-4702-ba20-82b46b2bd07e

Weiss R (2020, July 4) Bicycles are pushing aside cars on Europe's city streets. Bloomberg.Com. https://www.bloomberg.com/news/articles/2020-07-04/bicycles-are-pushing-aside-cars-on-europe-s-city-streets

World Health Organization (2016) Ambient air pollution: A global assessment of exposure and burden of disease. World Health Organization. https://www.who.int/phe/publications/air-pollution-global-assessment/en/

World Health Organization (2020) Ambient air pollution: pollutants. WHO, World Health Organization. https://www.who.int/airpollution/ambient/pollutants/en/

World Medical Association, International Council of Nurses, ISDE International, World Federation of Public Health Associations, & World Organization of Family Doctors (WONCA) (2020) Letter to G20 leaders calling for a health recovery. https://healthyrecovery.net/

Wu X, Nethery RC, Sabath BM, Braun D, Dominici F (2020) COVID-19 PM2.5 (medRxiv 2020.04.05.20054502). Harvard T.H. Chan School of Public Health. https://projects.iq.harvard.edu/covid-pm/home

Zhang R, Zhang Y, Lin H, Feng X, Fu T-M, Wang Y (2020) NO_x emission reduction and recovery during COVID-19 in East China. Atmosphere 11(4):433. https://doi.org/10.3390/atmos11040433

Zhu Y, Xie J, Huang F, Cao L (2020) Association between short-term exposure to air pollution and COVID-19 infection: evidence from China. Sci Total Environ 727:138704. https://doi.org/10.1016/j.scitotenv.2020.138704

Lisa Jordan, is Director of the Spatial Data Center and Faculty Member in the Department of Biology at Drew University. She holds a PhD in Geography from the University Colorado at Boulder. Albert Tang is Research Associate for the Spatial Data Center at Drew University.

Achieving the UN's SDG 6 as a Means to Control the COVID-19 Pandemics in Brazil

Thiago Leite Cruz and Erleyvaldo Bispo dos Santos

Abstract Sanitation involves social, economic and environmental dimensions of Sustainable Development and influences the dissemination of diseases. Access to quality water and sanitation is fundamental for good standards of living, so it was defined by the United Nations as the Sustainable Development Goal 6, to be globally achieved by 2030. This goal is of special importance during the COVID-19 pandemic as recent studies identified long survival rates of the virus in human excrements and wastewater. However, data from the UN indicates that around 1.8 billion people worldwide consume fecally contaminated water, mostly in developing nations. Using publicly available data from scientific papers, national censuses and reports, the present research assessed the current access to quality water and sanitation in Brazil and related it to the spread of COVID-19. We conclude that the poor sanitary conditions in most developing nations enhance their vulnerability to the new coronavirus due to high contamination risks. We present and analyze some initiatives in Brazil that aim to address such issues. This study highlights the need for urgent and creative solutions to achieve the SDG 6, especially in developing countries, improving the quality of life and controlling the COVID-19 pandemic and other diseases.

Keyword Sanitation · Quality water · Sustainable development goals · COVID-19 pandemics · Brazil

1 Introduction

In recent debates on sustainable development, engagement with social issues is increasingly relevant (Sen 2011; Sachs 2002). Economic and natural resources should be adequately managed to produce short- and long-term social welfare, removing all

T. L. Cruz (✉)
Universidade Federal do Pará, Rua Augusto Corrêa, 01 – Guamá, Belém, Pará 66075-110, Brasil
e-mail: thiago.james@live.com

E. B. dos Santos
Institutos de FlorestasUniversidade Federal Rural do Rio de Janeiro, Rodovia BR 465, km 07 – Zona Rural, Seropédica, Rio de Janeiro 23890-000, Brasil

W. Leal Filho (ed.), *COVID-19: Paving the Way for a More Sustainable World*, World Sustainability Series, https://doi.org/10.1007/978-3-030-69284-1_6

barriers that prevent citizens from being free. Freedom, as defined by Sen (2011), means that problems such as poverty, oppression and unsanitary working conditions are extinguished, and access to sanitation, health, education and civil rights is granted to all.

Those global challenges are addressed in the Sustainable Development Goals (SDGs) established by the United Nations in 2015. Around 193 countries agreed to meet the 17 SDGs until 2030 to significantly reduce poverty and environmental degradation, thus improving living conditions on the planet for all people (UN 2019). This is especially relevant during global crises and pandemics when the most vulnerable social groups (low-income families, people with disabilities, etc.) are exposed to more risks (UN 2020b).

The novel coronavirus disease 19 (COVID-19) rapidly spread throughout the world. Even though it is a zoonotic disease, whose origins are connected to the mismanagement of wildlife, its transmission is intrinsically related to good hygiene practices and sanitation issues (WHO 2020a, b; UN 2020c). The United Nations and the World Health Organization intensely promote the importance of washing hands, considering it the basic and one of the most effective ways to contain the spread of infectious diseases, as it destroys contaminating agents (UN 2020c).

Further connections between previous coronaviruses, hygiene, clean water and sanitation have already been identified (Yeo et al. 2020). Earlier studies have found the presence of coronaviruses such as SARS-CoV (Chan et al. 2004) and MERS-CoV (van Doremalen et al. 2013; Corman et al. 2016) in the gastrointestinal system of patients (Fehr and Perlman 2015; Zhou et al. 2017) and sewage systems (Wang et al. 2005).

Viable RNA of the novel coronavirus, SARS-CoV-2, was also very recently found in human excrements and untreated water, with long survival rates (Wu et al. 2020). Although most preliminary studies have been conducted in China, viable samples of SARS-CoV-2 RNA have also been found in the sewage system of different countries, such as Australia, Netherlands and the USA (Ahmed et al. 2020; Holshue et al. 2020; Lodder and Husman 2020; RIVM 2020).

Thus, countries such as Brazil are exposed to more risks during pandemic times due to poor access to proper sanitation and clean water (Trata Brasil 2018). In May 2020, Brazil was declared by the World Health Organization a new global epicenter of the novel coronavirus pandemic (WBUR 2020; Andreoni 2020).

In this context, the SDG 6 (Ensure access to water and sanitation for all) is even more relevant as proper sanitation and hygiene facilities play a key role in improving the quality of life and controlling the spread of infectious diseases. However, around 33% of the world's population lacks access to clean water for consumption and 40% have no access to hand-washing facilities with soap and water (UN 2020a). Open defecation is still practiced by more than 670 million people, contaminating water that is consumed by around 1.8 billion people globally (UN 2020a).

Therefore, it is urgent to improve access to quality water and sanitation to effectively overcome the COVID-19 and avoid similar pandemics in the future. The present paper aims to analyze the current Brazilian status regarding SDG 6 and relating it to the COVID-19 pandemic. We also discuss some initiatives that address this issue and propose solutions to enhance advances on this matter.

Our methodology consisted of bibliographical reviews of previous studies that identified the presence of SARS-CoV-2 RNA in human excreta and the possibility of its enteric transmission. Google Scholar was used for data collection (Harzing and Vanderwal 2008). Other secondary data collected from official national censuses and research institutions (Trata Brasil 2018) regarding access to clean water, sanitation, and COVID-19 transmission and death rates in Brazilian regions. We analyzed these data to compare Brazilian regions regarding their different infrastructure and vulnerability to infectious diseases.

Finally, we selected and discussed the more prominent initiatives in Brazil that aim to improve the current sanitation and clean water issues, including the startup named Águas Resilientes. This project was chosen for a deeper analysis due to its importance and the intense participation of community members in the Rio de Janeiro metropolitan area. We conducted in-person meetings with the CEO and other participants of the startup to gather information about the initiative, and we present our findings in this study.

The present paper has four chapters, including this introduction. The second chapter discusses the important role played by quality water and sanitation in the dynamics of sustainable development and public health. We review some recent studies that relate poor sewage systems and untreated water consumption to COVID-19 transmission and assess data from official international and Brazilian institutions regarding sanitation and access to clean water.

In chapter three, we discuss proposals to improve social access to quality water and sanitation in Brazil that could contribute to the achievement of the SDG 6, improve public health and reinforce actions against the COVID-19 pandemic. We also present and discuss a successful initiative in Brazil, a startup named Águas Resilientes (Resilient Waters) in Rio de Janeiro. Finally, chapter four presents our main conclusions, acknowledging that the poor sanitary conditions in Brazil increase contamination risks and the vulnerability to the novel coronavirus. Thus, it is of ultimate importance to achieve SDG 6 in Brazil and other developing countries to promote more effective strategies to control pandemics such as the COVID-19.

2 COVID-19 and Sanitation

2.1 UN SDG 6: Quality Water and Sanitation in Pandemic Times

Even though COVID-19 has zoonotic origins, its transmission between humans is deeply related to sanitary issues. The WHO indicates that washing hands and other basic hygiene measures are the most effective ways to prevent contamination (WHO 2020a). Besides, recent studies identified long survival rates of the virus in human excrements and untreated water (Chen et al. 2020a; Lodder and Husman 2020; Wu et al. 2020; Yeo et al., 2020).

Understanding the ecology of the coronaviruses can help researchers better understand the functioning of the SARS-CoV-2 (Severe Acute Respiratory Syndrome Coronavirus 2) that causes COVID-19 (Chen et al. 2020b), as it is a mutation of previous coronaviruses such as SARS-CoV and MERS-CoV (Yeo et al. 2020). It is manifested by a range of symptoms that vary from asymptomatic or mild respiratory infection to fulminant pneumonia, which may lead to death (Guan et al. 2020; Wang et al. 2020). Notwithstanding, it has killed more than 450 thousand people worldwide by June 2020 (WHO 2020b).

It firstly appeared in Wuhan, Hubei Province, China, and rapidly spread to more than 200 countries by human-to-human transmission, mainly through airway exposure to respiratory droplets from infected people (Chen et al. 2020b; WHO 2020b). However, previous studies have identified the presence of multiple coronaviruses in the gastrointestinal system, regardless of the manifestation of gastrointestinal symptoms such as diarrhea (Wang et al. 2005; Fehr and Perlman 2015; Zhou et al. 2017).

In fact, Chan et al. (2004) detected the existence of SARS-CoV RNA in the excreta of patients during the international outbreak of 2002. On the 11th day of the disease's cycle, the proportion of viral RNA reached its peak in the stool specimens and remained present in smaller proportions for 30 days after the first symptoms were manifested. Wang et al. (2005) confirmed such findings, detecting SARS-CoV RNA in the sewage of hospitals that treated SARS patients and in their excreta in Beijing, China. Their study also seeded SARS-CoV into sewage water collected from such hospitals and found that the virus remained infectious for two days at 20 °C. When the temperature lowered to 4 °C, it remained infectious for 14 days.

Chan et al. (2011) confirmed the decrease in the viability of SARS-CoV in higher temperatures and relative humidity patterns. Whereas it could survive for a couple of weeks after drying, the virus remained viable for five days within 40–50% relative humidity and 22–25 °C. Its viability decreased significantly after being exposed to 38 °C and 80–90% relative humidity for 24 h. MERS-CoV was analyzed by Corman et al. (2016), who assessed stool samples from patients during the outbreak in 2012, finding that around 14–16% of the samples contained MERS-CoV RNA. Similarly to SARS-CoV, the viability of MERS-CoV decreases significantly in higher temperature and humidity conditions (van Doremalen et al. 2013).

Therefore, Yeo et al. (2020) stress that fecal excretion and environmental contamination might facilitate the fecal-oral transmission of SARS-CoV-2. Chen et al. (2020a) conducted further investigations on this matter. They indicated that patients with SARS-CoV-2 are discharged from hospitals after testing negative in at least two sequential respiratory-tract specimen collections according to the current Chinese Center for Disease Control and Prevention Guidelines. However, Chen et al. (2020a) assessed 42 COVID-19 patients in Wuhan, China, collecting pharyngeal swab and stool specimens, and testing each sample for SARS-CoV-2 RNA. It was found that 64.29% of the patients tested positive for viral RNA in stool specimens for 6–10 days after the pharyngeal swabs turned negative, regardless of gastrointestinal symptoms and COVID-19 severity.

Wu et al. (2020) confirmed such a trend, analyzing respiratory and fecal samples from 74 patients with COVID-19 at Zhuhai, China, from January to March 2020. After the first diagnosis, 55% of the observed patients tested positive for SARS-CoV-2 RNA in fecal samples. Among them, the pharyngeal swab samples remained positive for around 16 days, whereas fecal samples remained positive for SARS-CoV-2 for around 28 days after the first symptoms were manifested. Wu et al. (2020) highlight that fecal samples of four patients tested positive for more than 45 days after the first symptoms.

Xiao et al. (2020) came to the same conclusions when infectious SARS-CoV-2 was isolated from stool samples, confirming that the virus actively replicates and releases infectious virions in the gastrointestinal tract even after viral clearance in the respiratory tract. Such findings stress the importance of routinely testing for SARS-CoV-2 RNA in feces, besides the pharyngeal swab tests. In fact, it is currently compulsory to perform viral RNA detection in anal swabs in Shanghai and some other Chinese provinces to discharge COVID-19 patients from the hospital (Chen et al. 2020a).

Even though most studies assessing the relationship between the SARS-CoV-2 and the gastrointestinal system focused on China, the matter has been investigated in countries such as the Netherlands (Lodder and Husman 2020), Australia (Ahmed et al. 2020) and the USA. Holshue et al. (2020) described the case of the first patient who has been clinically diagnosed with COVID-19 in the USA, and SARS-CoV-2 RNA has also been identified in their fecal sample.

Lodder and Husman (2020) indicated that around four days after the first cases of COVID-19 had been confirmed in the Netherlands, SARV-CoV-2 RNA was identified in human wastewater samples at the Amsterdam Airport Schiphol (Haarlemmermeer, Netherlands). They also pointed out that viral RNA also tested positive in samples of human wastewater in different Dutch locations, less than a week after the first confirmed COVID-19 case in the country (RIVM 2020).

Liu et al. (2020) also noticed that SARS-CoV-2 spreads easily and sustainably in communities via many routes. Considering the significant presence of viable SARS-CoV-2 RNA in sewage and wastewater, such findings have serious implications for public health, especially in developing countries with poor sanitation systems (Lodder and Husman 2020; Yeo et al. 2020; Willemijn and Husman 2020). In fact, Lodder and Husman (2020) concluded that mismanagement of human wastewater and poor-hygiene conditions may play a key role in increasing virus concentration in the urban environment and, thus, enhancing the probability of COVID-19 enteric transmissions.

Therefore, we highlight the urgency to implement access to safe water and sanitation in all countries and regions within countries, especially the poorer ones. Providing adequate hygiene conditions, disinfecting sewage and adequately managing human wastewater in contained living spaces such as hospitals, residences and schools can protect people from COVID-19 and many other infectious diseases.

2.2 The SDG 6 in Developing Countries: The Case of Brazil

The UN emphasizes the crucial role played by water in social, economic and environmental dimensions of sustainable development. It is directly involved in food production and healthy ecosystems, and it is critical for human survival, in consumption and personal hygiene. UN also highlighted the importance of access to clean water for human dignity and safety, especially for women and girls, as it is essential to adequately manage matters such as menstruation and maternity chores (UN 2020a).

This matter has been subject to many UN debates throughout history. The UN Water Conference, held in Argentina in 1977, was a milestone in international debates on water resources. It had a holistic approach to water management, and developing nations actively participated in the first internationally coordinated discussions regarding water management. Although implementation schemes for an actual action plan to address this issue were not developed, the 80s were declared the International Drinking Water Supply and Sanitation Decade (UN 1977).

In 1992, the International Conference on Water and the Environment and the Earth Summit deeply discussed this important matter, but only in July 2010 the UN General Assembly recognized the access to water and sanitation as a basic human right. Every human being should have safe and affordable access to such an important resource (its costs should not exceed 3% of the household income), and water sources should be within 1000 m from homes, and its collection process should not exceed 30 min (UN 2020a).

The UN Sustainable Development Goals (SDGs) were implemented in 2015 as a set of objectives to be achieved by all UN Member States by 2030. The main purpose of the SDGs is to tackle growing poverty, empower women and girls and address climate change through 17 goals related to issues such as gender equality, quality education, clean and affordable energy, international cooperation, hunger, etc. The SDG 6 is to ensure access to water and sanitation for all (UN 2019).

All SDGs are interconnected. Clean water and sanitation access (SDG 6) contribute to good health and wellbeing (SDG 3) since contaminated water and poor sanitation are leading causes of child mortality due to diseases such as diarrhea, which causes 1.5 million child deaths each year. The SDG 6 is essential to decent work conditions (SDG 8), reduces inequalities (SDG 10), and is part of sustainable cities and communities (SDG 11). It is also a key-component of life below water (SDG 14) and balances life on land (SDG 15). In fact, the UN indicates that every US$1 invested in improved sanitation generates an average return of US$9, especially in poorer communities that lack such essential services (UN 2020b).

However, the UN agencies estimate that around 2.4 billion people worldwide do not have access to basic sanitation facilities, such as latrines or toilets, as at least 892 million people practice open defecation, polluting rivers and local water sources. Thus, at least 1.8 billion people drink water that is not protected against fecal contamination (UN 2020b, c). It was also noticed that around 40% of worldwide healthcare facilities lack basic water services and do not have soap or hand sanitizers.

Similarly, water and soap are unavailable to students in 50% of the world's schools (JMP 2020).

Thus, UN-Habitat stresses that the impacts of the SARS-CoV-2 could be considerably more devastating among the poorer social groups, especially the population living in urban slam areas. They live in overcrowded neighborhoods and lack basic sanitation and access to clean water, making measures to prevent the COVID-19 such as social distancing and quality hygiene practices very difficult to achieve (UN News 2020).

Brazil is the largest country in South America, stretching across around 2700 miles from north to south. It is also the most populated nation in the region, with 209,712,000 inhabitants who concentrate mostly in urban areas, and just 13.4% of the population lives in rural zones (Encyclopaedia Britannica 2020). The poorest population is the most vulnerable to the COVID-19 pandemic and other infectious and waterborne diseases in Brazil as they lack basic sanitation and hygiene infrastructure. According to official national data (Trata Brasil 2018), 33,129,083 people do not have access to clean water and 94,734,344 do not have access to adequate sewer systems. The distribution varies according to the region (Fig. 1). Likewise, each region was affected differently by waterborne diseases and the COVID-19 pandemic (Figs. 2, 3 and 4).

Brazil's North region hosts the largest portion of the Amazon rainforest and also hosts 73.6% of the country's total freshwater, mostly from the Amazon River Basin. It is also the least populated region in Brazil (ANA 2020; Encyclopaedia Britannica 2020). Paradoxically, it is the most vulnerable region as 43% (6,645,268 people) of the residents in the region have no access to clean water. Furthermore, only 10.5%

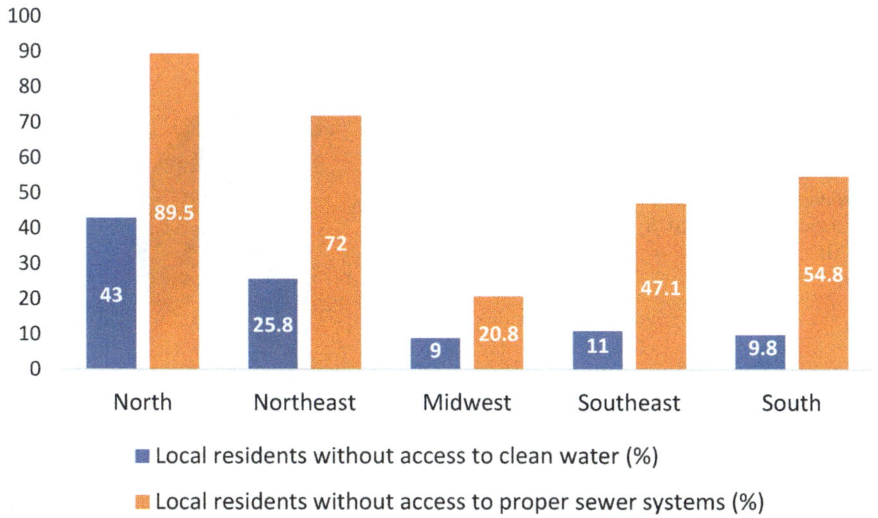

Fig. 1 Percentage of local inhabitants without access to water and basic sanitation in each Brazilian region. *Source* Created by the authors based on data collected from Brasil (2019)

Fig. 2 Hospitalizations due to waterborne diseases in each Brazilian region. *Source* Created by the authors based on data collected from Trata Brasil (2018)

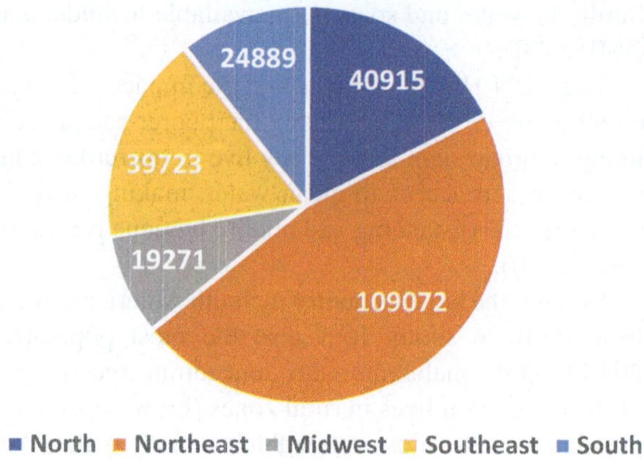

Fig. 3 Total COVID-19 cases per 100 thousand inhabitants in each Brazilian region. *Source* Created by the authors based on data collected from DATASUS (2020)

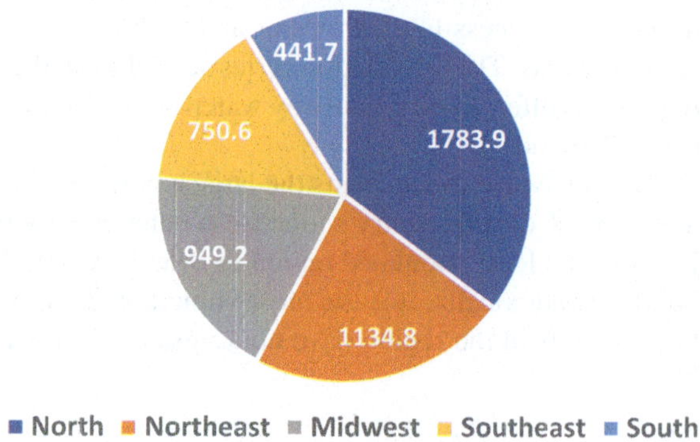

Fig. 4 Total deaths due to COVID-19 per 100 thousand inhabitants in each Brazilian region. *Source* Created by the authors based on data collected from DATASUS (2020)

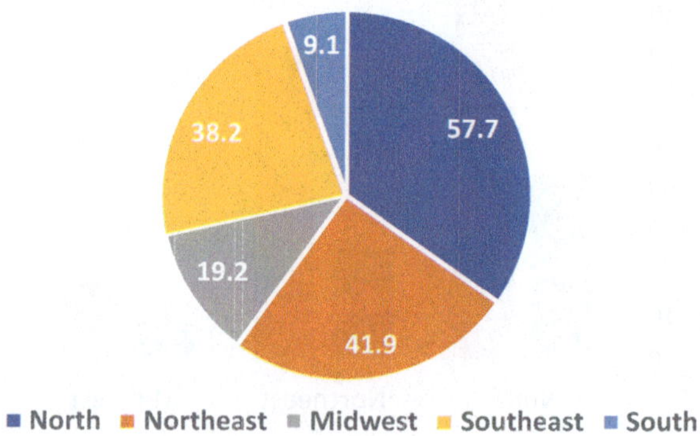

of the local population has proper sewer systems, mostly in urban centers (BRASIL 2019). In 2018, there were 40,915 cases of hospitalization due to waterborne diseases

(Trata Brasil 2018). Consequently, it was also the most affected region by the COVID-19 pandemic. Its incidence proportion was of 1,783.9 cases and 57.7 deaths per 100 thousand inhabitants by 15 July 2020 (DATASUS 2020).

Reduced rainfall in Brazil's Northeast is one factor that contributes to the lower availability of freshwater in this region (Pires et al. 2019). Among its citizens, 25.8% (14,047,934 people) are unable to consume clean water and 72% (39,219,767) of the residents have no adequate sewer systems at home (BRASIL 2019). Waterborne diseases contributed to 109,072 hospitalization cases in 2018 in the Northeast region (Trata Brasil 2018) and, per 100 thousand people, the novel coronavirus affected 1134.8 patients and caused 41.9 deaths by 15 July 2020 (DATASUS 2020).

In Brazil's Midwest, pesticides used for the expansion of agriculture seriously harm water quality (Pires et al. 2019). Around 11% (1,731,797 people) of the local population have no access to clean water and 47.1% (7,403,990) have no proper sanitation systems at home (BRASIL 2019). There were 19,271 cases of hospitalization because of waterborne diseases in 2018 (Trata Brasil 2018) and, by 15 July 2020, the incidence proportion of COVID-19 was 949.2 cases and 19.2 deaths per 100 thousand inhabitants (DATASUS 2020).

The Southeast region has the highest population density in the country and is also the most industrially developed region (IBGE 2020). There, challenges for sanitation and water access are related to high demands and the poor ecological status of local water bodies (Pires et al. 2019). However, only 9% (7,814,296 people) of its population suffer without access to clean water and 20.8% (18,102,430) lack adequate sewer systems at home (BRASIL 2019). In 2018, waterborne diseases caused 39,723 hospitalizations in the region (Trata Brasil 2018). The incidence proportion of the COVID-19 has been 750.6 cases and 38.2 deaths per 100 inhabitants by 15 July 2020 (DATASUS 2020).

Finally, in the South region, water quality is harmed by pollutants from irregular installations of wells and invasive species that interfere with local ecological dynamics (Pires et al. 2019). Among the local population, 9.8% (2,889,788 people) do not have access to clean water and 54.8% (16,158,707) suffer without basic sanitation (BRASIL 2019). In fact, 24,889 people have been hospitalized due to waterborne diseases (Trata Brasil 2018). Per 100 thousand inhabitants, the region had 441.7 cases of COVID-19 and 9.1 deaths by 15 July 2020 (DATASUS 2020).

Even though the enteric transmission of COVID-19 has not been fully confirmed yet, the viability of SARS-CoV-2 RNA in sewage and wastewater has been confirmed (Chen et al. 2020a; Yeo et al. 2020; Wu et al. 2020). Thus, it is alarming to notice that a large portion of Brazilian citizens does not have access to adequate wastewater-collection systems and clean water. The data analysis is clear when it indicates that the regions with less access to sanitation and clean water showed the highest incidence and mortality rates due to the novel coronavirus.

Such poor sanitary conditions increase the population vulnerability to contamination risks and highlight the urgency of achieving the SDG 6 in Brazil to improve the quality of life in the country. It urges government, businesses and civil society to join forces to be up to this challenge and combat the novel coronavirus and other infectious diseases (UN 2020b).

3 Tackling the Challenge and Changing Paradigms

UN-Water, a coalition of UN agencies, joins forces with national and local governments, non-government organizations and community leaders alike to tackle this issue globally. UN-Habitat is partnering with small-scale water and sanitation service providers, utilities and authorities to provide technical advice, online training and support to implement emergency safe drinking water and handwashing facilities in high-density key locations, such as community centers, markets and bus stations in slam areas. Community leaders are trained to properly manage such facilities and disseminate adequate information about the novel coronavirus (UN News 2020).

In Brazil, both public and private sectors are working towards improving people's access to sanitation and clean water. In fact, in June 2020, the Brazilian National Congress approved relevant changes in the legislation with the Law Project no. 4162/2019. This change aims to simplify procedures for the privatization of public firms in the sanitation business. The new companies must commit to guaranteeing clean water access and sanitation to all citizens until 2033, in line with the SDGs (Agência Senado 2020).

Nonetheless, NGOs, specialists and social associations complained about the lack of debates involving such an important matter. The Brazilian Association of Sanitation and Environmental Engineering (Associação Brasileira de Engenharia Sanitária e Ambiental) considers that this incentive to privatization highlights the government's failure to properly manage and allocate the appropriate resources to address water supply and sanitation issues (ABES 2020).

On the other hand, the Brazilian Council for Sustainable Development (Conselho Brasileiro de Desenvolvimento Sustentável) reckons that the aforementioned Law Project may be an opportunity to sustainably develop the national economy in a post-pandemic era, encouraging private business to address matters of social concern. It should also enable a quick response to the current lack of access to sanitation and clean water in Brazil, which is an inconvenient and serious nineteenth-century problem that still strikes many Brazilian households in the twenty-first century (CEBDS 2020).

It is clear that this Law Project is very important to provide a short-term improvement in the quality of life in Brazil. However, it transforms water from a public good into an economic good, disregarding its environmental and social value as a basic human right. Such a process is known as the commodification of water (Mitchell 1984). Therefore, different social segments must work in tandem to monitor the implementation of the Law Project n. 4162/2019 to avoid contraventions and effectively assure the meeting of its goals.

In fact, the importance of sanitation and water supply during the COVID-19 outbreak has already been identified by Brazilian research institutes and other government agencies. In the state of Minas Gerais (MG), the Agência Nacional de Águas (National Water Agency), the Instituto Nacional de Ciência e Tecnologia em Estações Sustentáveis de Tratamento de Esgoto (National Institute of Science and Technology in Sustainable Sewage Treatment Plants) and the Instituto Mineiro de Gestão das Águas (Minas Gerais' Water Management Institute) have partnered with the

Companhia de Saneamento de Minas Gerais (Sanitation Company of Minas Gerais) and the Secretaria de Estado da Saúde de Minas Gerais (Minas Gerais' State Secretary of Health) to monitor the spread of the novel coronavirus through the sewers of the cities of Belo Horizonte and Contagem. The Arruda and Onça river basins supply the water of the cities and are used by the sewage systems of around 2.2 million people, serving around 71% of the local urban population (ANA 2020).

The partnership between the institutions aims to detect and quantify the presence of SARS-CoV-2 RNA in different regions and neighborhoods of Belo Horizonte and Contagem. The research collects and analyzes samples from relevant locations in the area, including hospitals that treat COVID-19 patients, sewage treatment plants and some other spots in the Arruda and Onça river basins. The findings should guide decision-makers when dealing with social isolation and lockdown measures and assess the most vulnerable neighborhoods to improve the management of the COVID-19 outbreak in the region (ANA 2020).

Some important projects related to water access in Brazil are conducted by civil society organizations such as the startup named *Águas Resilientes* (Resilient Waters). It was founded in 2019 in the city of Seropédica, in the metropolitan area of Rio de Janeiro—Brazil. Its mission is to improve the access and adequate use of water resources by vulnerable social groups in low-income neighborhoods of the region, enhancing social awareness of environmental and sustainability issues.

The Águas Resilientes startup currently has four project managers that develop strategies and apply low-cost technologies to implement eco-friendly sewer systems and clean water facilities in households. They also contribute to improving the local management of water and other natural resources through Environmental Education, driving positive social changes as the local population is empowered. It raises people's awareness of environmental and health issues so that they may effectively influence and demand adequate investments from decision-makers in truly beneficial and sustainable infrastructure projects to improve local living conditions.

Currently, the startup has two main projects: *Águas Mil* (Thousand Waters) and *Água e Clima em Todos os Lugares* (Water and Climate Everywhere), which engage hundreds of community members and some private institutions. The former consists of constructing a low-cost rainwater-collecting system using reusable and recycled materials and implementing it in low-income households. This is a low-cost and eco-friendly solution to the lack of access to clean water in the region. The Water and Climate Everywhere Project is scheduled to be launched in August 2020 and consists of promoting Environmental Education. Using online platforms at first, it aims to share knowledge about climate, environment and their intersections with relevant social issues, using reliable and official data about water, pollution and sanitation in Brazil.

The members of the Águas Resilientes startup consider that knowledge is power. Therefore, their goal is to produce knowledge and share it with more than the 200 families that are already directly engaged with the organization. Since its establishment, the startup members have participated in two important international events: the Forum of the Countries of Latin America and the Caribbean on Sustainable Development, at the Economic Commission for Latin America and Caribbean's

(ECLAC) headquarters in Santiago—Chile from 22 to 26 April 2019; and the Water World Week in Stockholm—Sweden, from 25 to 30 August 2019. The members changed information and experiences with other international organizations of the civil society.

A similar initiative in Brazil's North region was launched in 2017. The startup named Amana Katu developed a rainwater catchment and purification system. Based in a circular economy approach, it is mainly made of reusable materials from food industry. The initiative was inspired by the paradox in the Brazilian Amazon region, where water is abundant, but most people do not have access to this important resource. Thus, one student founded the startup Amana Katu, to provide sustainable and safe access to clean water in the region. Young inhabitants in different communities are trained to produce, install and use the system units. Partnerships with local stores, industries and social movements are also established to provide financial support, and technical and logistical support. Currently, around 850 people use the system to have safe drinking water at home (UN 2020d).

Such initiatives highlight the importance of engaging different actors toward the solution of common problems in balance with the environment and adapting to their local realities. Thus, new and effective solutions can be developed and implemented to address important matters such as water, sanitation and hygiene, for benefit of all. It is also a call to action for other projects and social initiatives in different parts of the country and the world to tackle together the challenges for society to develop and succeed.

4 Conclusion

This paper discussed the ultimate importance of sanitation and access to clean water in improving people's quality of life and safety, especially in relation to the COVID-19 pandemic since viable SARS-CoV-2 RNA was confirmed to be present in the patients' fecal samples and wastewaters around the world. Good hygiene practices are among the most effective measures to control the spread of COVID-19. Thus, poor sanitary conditions intensify the vulnerability to infectious and waterborne diseases, as contamination risks increase.

Therefore, meeting the UN SDG 6, ensuring access to clean water and sanitation for everyone, is a crucial tool to control pandemics in countries where a significant part of the population lacks clean water and basic sewer systems. Our analysis focused on Brazil. Our findings have identified connections between sanitation, access to clean water and the COVID-19 in Brazil, highlighting the urgency to improve social development and build infrastructure for sanitation and water access for all.

We found that the North and the Northeast regions have the lowest rates of access to sanitation and clean water in the country, so they are the most vulnerable to COVID-19. In fact, their inhabitants are unable to maintain a safe and healthy hygiene routine and may be more exposed to the virus in the water they consume, as their incidence and mortality rates of COVID-19 cases per 100 thousand inhabitants have been the highest in Brazil. In contrast, regions where people have more access to safe water for consumption and basic sanitation showed lower incidence and mortality rates.

Future studies should replicate this analysis in other developing and developed countries and deepen the assessment by investigating the statistical correlation and influence of other factors in the transmission and mortality rates of COVID-19. This knowledge could guide decision-makers when dealing with similar pandemics in the future to apply more effective policies to control and avoid harm to the population. Furthermore, this study also highlighted the importance of adapting solution measures to the peculiarities of each region within countries, as their levels of development may vary.

We analyzed the legal initiative from the Brazilian Federal Government to promote the privatization of sanitation in Brazil. This measure may indeed accelerate the implementation of effective and necessary improvements in this field, but there are no quick responses to such a complex matter, especially considering the infrastructure and economic diversity between the different regions in the country. Besides, it commodifies water and sanitation, which should rather be valued as basic human rights.

In this context, the importance of social engagement through popular initiatives is highlighted. Projects such as the *Águas Resilientes* startup are critical to share knowledge and engage social actors to demand effective and fair policies from local governments. Such initiatives need to be encouraged, supported and expanded, as they also empower people to implement creative and sustainable solutions to achieve the SDG 6 in poorer areas. They contribute to improve local standards of living and should inspire and encourage social movements in different regions and countries to seek their own solutions, and engage and succeed as well.

References

ABES—Associação Brasileira de Engenharia Sanitária e Ambiental (2020) Nota da ABES sobre o PL 4162/2019. Available in: https://abes-dn.org.br/?p=35853. Accessed 20th June 2020

AGÊNCIA SENADO (2020) Senado aprova novo marco legal do saneamento básico. Available in: https://www12.senado.leg.br/noticias/materias/2020/06/24/senado-aprova-novo-marco-legal-do-saneamento-basico. Accessed 20th June 2020

Ahmed W, Angel N, Edson J et al (2020) First confirmed detection of SARS-CoV-2 in untreated wastewater in Australia: a proof of concept for the wastewater surveillance of COVID-19 in the community [published online ahead of print, 2020 Apr 18]. Sci Total Environ 728:138764. https://doi.org/10.1016/j.scitotenv.2020.138764

ANA—Agência Nacional de Águas (2020) Boletim de Acompanhamento: Monitoramento Covid Esgotos. Available in: https://www.ana.gov.br/noticias/monitoramento-covid-esgotos-constata-presenca-do-coronavirus-em-primeiras-coletas. Accessed 6th June 2020

Andreoni M (2020) Coronavirus in Brazil: what you need to know. Available in: https://www.nytimes.com/article/brazil-coronavirus-cases.html. Accessed 16th July 2020

BRASIL (2019) Ministério do Desenvolvimento Regional. Secretaria Nacional de Saneamento—SNS. Sistema Nacional de Informações sobre Saneamento: 24° Diagnóstico dos Serviços de Água e Esgotos, 2018. SNS/MDR, Brasília, 180p

CEBDS—Conselho Empresarial Brasileiro para o Desenvolvimento Sustentável (2020) Novo marco do saneamento é uma oportunidade para a retomada sustentável. Available in: https://cebds.org/novo-marco-do-saneamento-e-uma-oportunidade-para-a-retomada-sustentavel/#.XvXsJShKjIU. Accessed 20th June 2020

Chan KH, Poon LL, Cheng VC et al (2004) Detection of SARS coronavirus in patients with suspected SARS. Emerg Infect Dis 10:294–299. https://doi.org/10.3201/eid1002.030610

Chan KH, Peiris JS, Lam SY, Poon LL, Yuen KY, Seto WH (2011) The effects of temperature and relative humidity on the viability of the SARS coronavirus. Adv Virol 2011:734690. https://doi.org/10.1155/2011/734690

Chen N, Zhou M, Dong X et al (2020a) Epidemiological and clinical characteristics of 99 cases of 2019 novel coronavirus pneumonia in Wuhan, China: a descriptive study. Lancet 395:507–513. https://doi.org/10.1016/S0140-6736(20)30211-7

Chen Y, Chen L, Deng Q et al (2020b) The presence of SARS-CoV-2 RNA in feces of COVID-19 patients. J Med Virol 92(7):833–840. https://doi.org/10.1002/jmv.25825

Corman VM, Albarrak AM, Omrani AS et al (2016) Viral shedding and antibody response in 37 patients with Middle East respiratory syndrome coronavirus infection. Clin Infect Dis 62:477–483. https://doi.org/10.1093/cid/civ951

DATASUS—Departamento de Informática do SUS (2020) Painel coronavírus. Available in: https://covid.saude.gov.br/. Accessed 5th July 2020

Encyclopaedia Britannica (2020) Brazil. Available in: https://www.britannica.com/place/Brazil. Accessed 18th May 2020

Fehr AR, Perlman S (2015) Coronaviruses: an overview of their replication and pathogenesis. In: Maier HJ, Bickerton E, Britton P (eds) Coronaviruses: methods and protocols. Springer, New York, pp 1–23

Guan WJ, Ni ZY, Hu Y, Liang WH, Ou CQ, He JX et al (2020) Clinical characteristics of coronavirus disease 2019 in China. N Engl J Med. 382:1708–1720. https://doi.org/10.1056/NEJMoa2002032

Harzing A, Vanderwal R (2008) Google Scholar as a new source for citation analysis. Ethics Sci Environ Polit 8:61–73. https://doi.org/10.3354/esep00076

Holshue ML, Debolt C, Lindquist S et al (2020) First case of 2019 novel coronavirus in the United States. N Engl J Med 382:929–936. https://doi.org/10.1056/NEJMoa2001191

IBGE—Brazilian Institute of Geography and Statistics (2020) Population count, various years. Available in: https://sidra.ibge.gov.br/home/pnadcm. Accessed 15th May 2020

JMP (2020) Global estimates for basic hygiene services in households, schools and health care facilities are a call to action to prevent the spread of COVID-19. Available in: https://washdata.org/. Accessed 10th May 2020

Liu J, Liao X, Qian S et al (2020) Community transmission of severe acute respiratory syndrome coronavirus 2, Shenzhen, China, 2020. Emerg Infect Dis 26(6):1320–1323. https://doi.org/10.3201/eid2606.200239

Lodder W, Husman AMR (2020) SARS-CoV-2 in wastewater: potential health risk, but also data source. Lancet Gastroenterol Hepatol 5(6):533–534. https://doi.org/10.1016/S2468-1253(20)30087-X

Mitchell B (1984) The value of water as a commodity. Can Water Resour J 9(2):30–37. https://doi.org/10.4296/cwrj0902030

Pires APF, Farjalla VF, Faria BM, Rodriguez DA, Gomes EAT, Santos EC, Sodré FNGAS, Sabino J, Espécie MA, Pinheiro MRC, Ribeiro ML, Bozelli RL, Panosso RF, Mormul RP, Barthem R, Scofield V, Dib V (2019) Sumário para Tomadores de Decisão (STD) do Relatório Temático Água: biodiversidade, serviços ecossistêmicos e bem estar humano no Brasil. Editora Cubo, São Carlos. https://doi.org/10.4322/978-65-00-00068-9. ISBN:978-85-60064-84-7

RIVM (2020) Novel coronavirus found in wastewater. Available in: https://www.rivm.nl/en/news/novel-coronavirus-found-in-wastewater. Accessed 10 May 2020

Sachs I (2002) Caminhos para o desenvolvimento sustentável. Garamond, Rio de Janeiro. ISBN-10:858643535X

Sen A (2011) Development as freedom. Knopf Doubleday Publishing Group, Oxford. ISBN:9780307874290
Trata Brasil (2018) Painel Saneamento Brasil. Available in: https://www.painelsaneamento.org.br/. Accessed 2th June 2020
UN—United Nations (1977) Report of the United Nations conference on water, Mar del Plata, 1977, E/Conf. 70/29, Chapter I
UN—United Nations (2019) Sustainable Development Goal 6: ensure availability and sustainable management of water and sanitation for all. Available in: https://sustainabledevelopment.un.org/sdg6. Accessed 10th May 2020
UN—United Nations (2020a) Water. Available in: https://www.un.org/en/sections/issues-depth/water/index-2.html. Accessed 10th May 2020
UN—United Nations (2020b) Goal 6: ensure access to water and sanitation for all. Available in: https://www.un.org/sustainabledevelopment/water-and-sanitation/. Accessed 10th May 2020
UN—United Nations (2020c) Handwashing/hand hygiene. Available in: https://www.unwater.org/water-facts/handhygiene/. Accessed 10th May 2020
UN—United Nations (2020d) Amana Katu is a social business that works to solve the problem of social inaccessibility to drinking water in the Amazon, in order to reach the UN Sustainable Development Goal for 2030 number 06. Available in: https://sustainabledevelopment.un.org/partnership/?p=30346. Accessed 2 Aug 2020
UN News (2020) Water access critical to beating back COVID-19 spread in slum áreas. Available in: https://news.un.org/en/story/2020/03/1060042. Accessed 10th May 2020
van Doremalen N, Bushmaker T, Munster VJ (2013) Stability of Middle East respiratory syndrome coronavirus (MERS-CoV) under different environmental conditions. Eurosurveillance 18(38):20590. https://doi.org/10.2807/1560-7917.ES2013.18.38.20590
Wang XW, Guo T et al (2005) Concentration and detection of SARS coronavirus in sewage from Xiao Tang Shan Hospital and the 309th hospital. Water Sci Technol 52:213–221. https://doi.org/10.1016/j.jviromet.2005.03.022
Wang D, Hu B, Hu C, Zhu F, Liu X, Zhang J et al (2020) Clinical characteristics of 138 hospitalized patients with 2019 novel coronavirus-infected pneumonia in Wuhan, China. JAMA 323(11):1061–1069. https://doi.org/10.1001/jama.2020.1585
WBUR (2020) WHO declares Brazil new epicenter of coronavirus pandemic. Available in: https://www.wbur.org/hereandnow/2020/05/25/brazil-epicenter-coronavirus-pandemic. Accessed 16th July 2020
Willemijn L, Husman AMR (2020) SARS-CoV-2 in wastewater: potential health risk, but also data source. Lancet Gastroenterol Hepatol 5(6):533–534. https://doi.org/10.1016/S2468-1253(20)30087-X
World Health Organization—WHO (2020a) Coronavirus disease 2019 (COVID-2019): situation report 32. Available in: https://www.who.int/docs/default-source/coronaviruse/situation-reports/20200221-sitrep-32-covid-19.pdf?sfvrsn=4802d089_2. Accessed 5th Apr 2020
World Health Organization—WHO (2020b) Coronavirus disease (COVID-19) pandemic. Available in: https://www.who.int/emergencies/diseases/novel-coronavirus-2019. Accessed 20th May 2020
Wu Y, Guo C, Tang L, Hong Z, Zhou J, Dong X, Yin H, Xiao Q, Tang Y, Qu X, Kuang L, Fang X, Mishra N, Lu J, Shan H, Jiang G, Huang X (2020) Prolonged presence of SARS-CoV-2 viral RNA in faecal samples. Lancet Gastroenterol Hepatol 5(5):434–435. https://doi.org/10.1016/S2468-1253(20)30083-2
Xiao F, Tang M, Zheng X, Liu Y, Li X, Shan H (2020) Evidence for gastrointestinal infection of SARS-CoV-2. Gastroenterology 158(6):1831–1833. https://doi.org/10.1053/j.gastro.2020.02.055.3
Yeo C, Kaushal S, Yeo D (2020) Enteric involvement of coronaviruses: is faecal–oral transmission of SARS-CoV-2 possible? Lancet Gastroenterol Hepatol 5(4):335–337. https://doi.org/10.1016/s2468-1253(20)30048-0

Zhou J, Li C, Zhao G, Chu H, Wang D, Yan HH et al (2017) Human intestinal tract serves as an alternative infection route for Middle East respiratory syndrome coronavirus. Sci Adv 3(11). https://doi.org/10.1126/sciadv.aao4966

Thiago Leite Cruz has Bachelor's degrees in International Relations and Biology from Universidade da Amazônia (University of Amazon) and Universidade Federal do Pará (Federal University of Pará)—Brazil, and a Professional Master's degree in Sustainable Development. He has a postgraduate certificate in Business Management (London School of Business and Finance—UK). Currently, he is an specialist in Sustainability, Social License to Operate and Corporate Social Responsibility.

Erleyvaldo Bispo dos Santos has a Technician's degree in Agriculture and Livestock by the Instituto de EducaçãoCiência e Tecnologia de Sergipe (Institute of Education, Science and Technology of Sergipe)—Brazil. Currently, he is an undergraduate student of Forest Engineering at Universidade Federal Rural do Rio de Janeiro (Rural Federal University of Rio de Janeiro)—Brazil. He is an entrepreneur who produces research on water resources, technology and innovation.

Socioeconomic Pathways Toward the Sustainable Development Goals (SDGs) in Brazil During and Post-COVID-19 Pandemic

Thauan Santos, Luan Santos, and Ana Carolina Angelo

Abstract The outbreak of the new coronavirus (COVID-19) pandemic has caused instabilities in the world economy and significant changes in consumption patterns and production levels. These effects do not happen homogeneously on the population, varying according to income, gender, employment relationship and nature of work, in addition to other aspects related to urbanization and access to essential basic services. Although it is too early to predict the intensity and duration of these impacts, it is agreed that the COVID-19 has the potential to negatively influence the implementation of the UN Sustainable Development Goals (SDGs). However, it also presents an opportunity for considering new economic development pathways towards the 2030 Agenda. Ergo, this chapter seeks to discuss the impacts of COVID-19 in Brazil, by analyzing the main SDGs threatened, from the distinct indicators associated with its different goals. Besides, we look at the opportunities that this pandemic brings to the economy and society in terms of alternative development strategy and new ways of doing business, especially in the context of the circular economy. By doing so, we aim to associate the countercyclical economic policies for overcoming the crisis with the possible obstacles in reaching the SDGs, particularly looking at its socioeconomic effects.

Keywords Sustainable development goals · Circular economy · COVID-19 · Coronavirus · Brazil

---

T. Santos (✉)
Graduate Program in Maritime Studies, Brazilian Naval War College, Rio de Janeiro, Brazil
e-mail: thauan@marinha.mil.br

L. Santos
Production Engineering Program, Federal University of Rio de Janeiro, Rio de Janeiro, Brazil
e-mail: luan.santos@pep.ufrj.br

A. C. Angelo
Production Engineering Department, Fluminense Federal University, Volta Redonda, Brazil
e-mail: angeloana@id.uff.br

W. Leal Filho (ed.), *COVID-19: Paving the Way for a More Sustainable World*,
World Sustainability Series, https://doi.org/10.1007/978-3-030-69284-1_7

1 Introduction

The outbreak of the new coronavirus (COVID-19) has caused instabilities in the world economy and significant changes in consumption patterns and production levels. Such changes direct impact countries' energy demand and carbon footprint, at least in the short term. Naturally, these effects do not occur homogeneously on the population, varying according to income, gender, employment relationship and nature of work, in addition to other aspects related to urbanization and access to essential basic services (Santos et al. 2020).

COVID-19 is spreading human suffering, destabilizing the global economy and upending the lives of billions of people around the globe (Hakovirta and Denuwara 2020). The pandemic is an unprecedented wake-up call, laying bare deep inequalities and exposing precisely the failures that are addressed in the 2030 Agenda. Leveraging this moment of crisis, when usual policies and social norms have been disrupted, bold steps can steer the world back on track towards the Sustainable Development Goals (SDG).

It is too early to predict the intensity and duration of these impacts, but there is no doubt that the COVID-19 pandemic has the potential to structurally alter our personal and professional relationships, besides destroying or leveraging the SDGs. According to the United Nations (UN) Secretary-General António Guterres, "we need to turn the recovery into a real opportunity to do things right for the future". In this context, this chapter seeks to discuss socioeconomic pathways toward the SDGs' achievements in Brazil during and post-COVID-19 pandemic. One would ask "why Brazil?". Whether because it is the largest country in Latin America in terms of cases and deaths (and the second largest in the world, only behind the U.S.), or because the measures and policies proposed by the country have been the target of many questions and criticisms, nationally and internationally, the case of Brazil deserves to be analyzed in detail (Santos and Rodrigues 2020).

This chapter is structured as follows. This first section generally presents the topic to be discussed and the main purpose of the chapter. The next section will analyze the countercyclical economic policies that the country has created and/or expanded, mapping the main adverse socioeconomic impacts in the short and medium term. Initially, it focuses on the economic question, analyzing the evolution of selected macroeconomic variables in the period, besides the nature and amount of the Union's (federal) spending on COVID-19. Third section analysis the specific COVID-19 impacts on SDGs in Brazil. Section four looks at the opportunities that this pandemic brings to the economy and society in terms of alternative development strategy and new ways of doing business, especially in the context of the Circular Economy. Finally, we present our main conclusions and references.

2 Anti-cyclical Economic Policies and Socioeconomic Impacts in Brazil

Carvalho (2020) recently published a book in which she emphasizes the relevance of the State as a relevant economic actor in formulating countercyclical policies. The author highlights five (interrelated) roles of the State, namely: stabilizer of the economy, investor in physical and social infrastructure, protector of the most vulnerable, provider of services to the population, and entrepreneur. Therefore, it will be up to the State to use its economic, industrial and social policy instruments, through automatic stabilizers (taxes and unemployment insurance, for example) and a mix of other mechanisms (such as fiscal, monetary and income policy, for example) to mitigate the socioeconomic impact of the pandemic in Brazil.

Unlike other global crises, the case of the COVID-19 pandemic is particular, as it did not originate in the financial sector, consequently affecting the real sector of the economy; on the contrary, it comes from the health area. Hence, it will be necessary for the State to revitalize the economy by increasing public spending with multiplier effects, inducing households and business spending. Policies should come in two consecutive moments: at first, the focus of public policy should be on ensuring the survival of families and companies, transferring income; then, with the success of the drop in the number of cases and deaths, as well as the implementation of mass testing, economic recovery should be promoted (Carvalho 2020).

However, it is not what has been happening in Brazil. Following Santos (2020) and based on data from the Focus Report, published weekly by the Central Bank of Brazil (BCB, Portuguese acronym) with the summary of the statistics calculated considering the market expectations collected in the 30 calendar days until the reference date of the report, Table 1 presents and compiles all 27 published in 2020 (covering the period from January 3, 2020 to July 3, 2020).

Based on the evolution of the data, there is a drop in the Broad Consumer Price Index (IPCA), reaching 1.63%, a drop in the forecast for the growth rate of the Gross Domestic Product (GDP), totaling −6.5% and the dramatic drop in industrial production, reaching −8.10%, reflecting the drop in domestic demand and the country's economic dynamism. The drop in the current account deficit, reaching US$−11.75 billion, the devaluation of the exchange rate, reaching R$/US$5.20 at the end of the year, and the reduction in direct investment in the country, reaching only US$55 billion, show the drop in international flows to the country. A positive net result from the economic point of view is the expected increase in the trade balance, which will probably reach US$53.45 billion at the end of 2020.

In response to this fall in the (inter)national dynamics, there has been a continuous fall in the Selic rate, the basic interest rate of the Brazilian economy, reaching a level never seen before at 2% p.y. This movement, while making the country less attractive to foreign capital, reduces the cost of credit. In addition, the measures taken by the Brazilian government to combat COVID-19 and its socioeconomic impacts led to an increase in public sector net debt, reaching 67% of GDP; a drop in nominal result (which represents the difference between total revenues and expenses in the year),

Table 1 Evolution of selected macroeconomic variables, median (aggregate), per week

Median—aggregate	03 Jan	10 Jan	17 Jan	24 Jan	31 Jan	07 Feb	14 Feb	21 Feb	28 Feb	06 Mar	13 Mar	20 Mar	27 Mar
IPCA (%)	3.60	3.58	3.56	3.60	3.40	3.25	3.22	3.20	3.19	3.20	3.10	3.04	2.94
GDP (%)	2.30	2.30	2.31	2.31	2.30	2.30	2.30	2.20	2.17	1.99	1.68	1.48	−0.48
Exchange rate (R$/US$)	4.09	4.04	4.05	4.09	4.10	4.10	4.10	4.15	4.20	4.25	4.35	4.50	4.50
Selic rate target (% p.y.)	4.50	4.50	4.50	4.50	4.25	4.25	4.25	4.25	4.25	4.25	3.75	3.75	3.50
Industrial production (%)	2.19	2.10	2.19	2.19	2.10	2.33	2.33	2.33	2.41	2.00	1.63	1.00	0.85
Checking account (US$ bi)	−54.2	−54.2	−54.3	−54.2	−54.2	−54.2	−54.2	−55.8	−58.0	−58.7	−59.0	−56.5	−39.6
Trade balance (US$ bi)	38.2	37.3	37.4	37.2	37.3	36.4	35.4	37.0	36.7	36.4	36.1	35.3	37.7
Direct investment in the country (US$ bi)	80.0	80.0	80.0	80.0	80.0	80.0	80.2	80.0	80.0	80.0	80.0	80.0	72.0
Primary result (% do PIB)	−1.1	−1.1	−1.1	−1.1	−1.1	−1.1	−1.1	−1.1	−1.1	−1.1	−1.2	−1.2	−1.3
Nominal result (% do PIB)	−5.7	−5.7	−5.7	−5.7	−5.5	−5.5	−5.6	−5.5	−5.5	−5.5	−5.5	−5.6	−6.0

Median—aggregate	03 Apr	09 Apr	17 Apr	24 Apr	01 May	08 May	15 May	22 May	29 May	05 June	12 June	19 June	26 June	03 July
IPCA (%)	2.72	2.52	2.23	2.20	1.97	1.76	1.59	1.57	1.55	1.53	1.60	1.61	1.63	1.63
GDP (%)	−1.18	−1.96	−2.96	−3.34	−3.75	−4.11	−5.12	−5.89	−6.25	−6.48	−6.51	−6.50	−6.54	−6.50
Exchange rate (R$/US$)	4.50	4.60	4.80	4.80	5.00	5.00	5.28	5.40	5.40	5.40	5.40	5.20	5.20	5.20

(continued)

Table 1 (continued)

Median—aggregate	03 Apr	09 Apr	17 Apr	24 Apr	01 May	08 May	15 May	22 May	29 May	05 June	12 June	19 June	26 June	03 July
Selic rate target (% p.y.)	3.25	3.25	3.00	3.00	2.75	2.50	2.25	2.25	2.25	2.25	2.25	2.25	2.00	2.00
Industrial production (%)	0.50	−1.42	−2.25	−2.35	−2.75	−3.00	−3.68	−3.68	−3.59	−5.35	−5.44	−5.50	−6.00	−8.10
Checking account (US$ bi)	−52.3	−45.5	−40.8	−39.6	−38.0	−35.9	−34.1	−28.1	−28.1	−20.5	−14.0	−14.0	−13.5	−54.2
Trade balance (US$ bi)	34.1	35.0	36.1	37.7	42.0	42.5	43.4	45.5	45.5	47.8	52.5	52.5	53.0	38.2
Direct investment in the country (US$ bi)	76.5	73.0	71.0	72.0	70.0	70.8	65.0	65.1	64.0	60.0	60.0	60.0	57.5	80.0
Primary result (% do PIB)	−1.7	−4.1	−5.0	−6.2	−7.2	−7.5	−7.8	−8.0	−8.0	−8.0	−10.0	−10.1	−10.2	−10.5
Nominal result (% do PIB)	−6.9	−9.0	−10.0	−11.1	−11.3	−12.0	−12.0	−12.0	−12.0	−12.3	−14.3	−14.8	−14.9	−15.0

Source Own elaboration based on Focus Reports (BCB 2020a, b, c, d, e, f, g, h, i, j, l, m, n, o)

reaching −15% of GDP; and the fall in the primary result (which represents the difference between primary income and expenses—non-financial, that is, excluding the portion referring to nominal interest on net debt) in the year, reaching −10.5% of GDP.

Based on data from the Fiscal Policy Observatory of the Brazilian Institute of Economics of the Getúlio Vargas Foundation (OPF/IBRE/FGV, Portuguese acronym), on June 16 Brazil already reached 8.27% of GDP in government programs (exemptions, expansion of expenses, deferrals and prepayments) and 3.21% of GDP in tax credit (including financing for States and Municipalities, as well as operations with subsidies). The measures resulting from government programs are higher than what several countries are doing (U.S., UK, Germany), while the tax credit measures are far below what we see in other countries (such as the U.S., UK, Germany, France, Spain, Italy, Canada and Japan).

To face the challenges of COVID-19 on the Brazilian economy, the country has launched a series of specific programs, either to directly serve companies (small and medium-sized, as well as large ones), or to serve workers (especially the unemployed or dismissed in the period). Table 2 consolidates Union (federal) spending on COVID-19. Based on expenses paid by ministries, the Ministry of Citizenship (R$124.1 billion), Ministry of Economy (R$93.8 billion) and Ministry of Health (R$17.4 billion) stand out.

The Emergency Employment Support Program (Pese, Portuguese acronym) was created in early April, by Provisional Measure (MP, Portuguese acronym) 944/2020. According to the MP, Pese is aimed at micro and small companies that have, in the year of 2019, obtained annual gross revenue greater than R$360 thousand and equal to or less than R$10 million. In the approved text, it can count on up to R$20 billion, half of the amount initially forecast. The participating financial institutions must use their own resources to defray 15% of the value of each financing and the remaining 85% will be defrayed by the Federal Government. Being able to formalize contracts until June 30, 2020, they must present the following conditions: interest rate of 3.75% p.y., term of 36 months for payment and 6 months grace for beginning payment, with capitalization of interest during the grace period.

Until June 30, 2020 (latest data provided by the BCB), it supported 1,941,399 employees, financed 113,383 companies and totaled R$4.53 billion. Regarding its geographic distribution, the states of São Paulo (37.1%), Rio de Janeiro (10.1%), Minas Gerais (9.7%) and Paraná (7.8%) stand out. With regard to the distribution of the number of employees by sector of activity, stand out the sectors of services (25.9%), construction, wood and furniture (15.2%), health, sanitation and education (14.9), and media and leisure (10.7%). Analyzing the distribution by number of employees of the financed company, stand out companies in the range of 0 to 9 employees (44.3%), 10 to 29 employees (40.9%) and 30 to 59 employees (10.8%). When it comes to the stratification by minimum wages (SM, Portuguese acronym), there are employees who received up to 1 SM (30.7%), from 1 to 1.5 SM (28.6%) and from 2 to 5 SM (20.2%).

Focusing on informal workers, individual microentrepreneurs (MEI, Portuguese acronym), self-employed and unemployed, the Senate approved at the end of March

Table 2 Nature of union spending on COVID-19

Nature	Spending forecast (R$ bi)	Expenses paid (R$ bi)	Expenses paid (%)	Source
Emergency aid for people in vulnerable situations	254.24	121.79	47.9	MPs 937, 956, 970 and 988/2020
Expansion of the *Bolsa Família* program	3.04	0.37	12.2	MP 929/2020
Emergency employment and income maintenance benefit	51.64	15.98	30.9	MP 935/2020
Financial aid to states, municipalities and the federal district	79.19	39.94	50.4	MP 939 and 978/2020
Granting of payroll financing	34.00	17.00	50.0	MP 943/2020
Transfer to the energy development account	0.90	0.90	100.0	MP 950/2020
Quotas of credit and operations guarantee funds	35.90	20.90	58.2	MP 972 and 977/2020
Additional expenses to the ministry of health and other ministries	50.35	20.46	40.6	–
Total	509.26	237.34	46.6	–

Source Own elaboration based on Transparent National Treasury; updated on 7/17/2020

a bill that establishes the payment for three months of an emergency protection of R$600.00—originating Law 13,982/2020. Known as a "coronavoucher", it covered more than 65 million beneficiaries, totaling R$121.1 billion, with emphasis on the Brazilian Southeast region (36.9%) and Brazilian Northeast region (34.5%). Emergency assistance applies under the following conditions: being over 18 years of age, meeting family income criteria, not receiving social security benefits, unemployment insurance or participating in federal government cash transfer programs (with the exception of *Bolsa Família*). Women who are mothers and heads of families and are within the criteria, can receive R$1200.00 (two quotas) per month.

It is worth mentioning that the amount initially proposed for the voucher was only R$200.00 and was sanctioned by President Bolsonaro (April 1), with some vetoes: expansion of the Continuous Benefit Program (BPC, Portuguese acronym), reassessment of some criteria, and restriction to the bank account (Santos 2020). In addition, on June 30, 2020, the government announced an extension of the aid for another two additional months—the initial proposal was to extend it for another three months in a decreasing manner, to R$500.00, R$400.00 and R$300.00, respectively.

Therefore, it is quite evident that in the Brazilian case, although there is a strong movement in terms of public spending in favor of a robust package of anti-cyclical economic policies, they have a short-term objective to mitigate the negative impacts of the pandemic. The design of the policies and the economic package itself has not taken into account the commitments made by previous governments with the 2030 Agenda and the SDGs, making this narrative non-existent in the current Brazilian debate. Ergo, and in view of the worsening economic, social and political crisis that the country is in, experts fear that the economic situation imposes a limitation on the efforts of both agendas, including due to the current government's resistance to this discussion. Thus, different SDGs already have been affected in the country, which will be discussed in the next section.

3 COVID-19 Impacts on SDGs in Brazil

As previously mentioned, the pandemic we are facing has not only affected the physical health of hundreds of thousands of people. The global economic fallout from COVID-19 has been devastating, with people having lost jobs, and employers now having no means to support their employees while they are losing business and not able to pay their business's rent (Hakovirta and Denuwara 2020). Besides, a series of other social and economic impacts are on the table. All of them directly relates to the 2030 Agenda and the SDGs are translated into 17 goals, their 169 targets and 232 indicators.

We argue that it is a unique tool to help governments, businesses and NGOs to understand the full scope of complex policy challenges like COVID-19. This framework brings diverse aspects together—from poverty rates, to economic growth, to education, to health. The United Nations Department of Economic and Social Affairs (UN DESA) has conceptually mapped how COVID-19 affects each of the SDGs, from disruption to food supplies (SDG 2) to increased levels of violence against women (SDG 5) (UN 2020). This conceptual mapping (Fig. 1) shows the value of the SDGs as a framework for understanding the intersecting flow-on effects of COVID-19.

The UN Secretary-General, António Guterres, in addressing the eradication of poverty during a high-level meeting on July 30, warned that the impacts of the COVID-19 pandemic are falling disproportionately on the most vulnerable: people living in the poverty, poor workers, women and children, people with disabilities and other marginalized groups. In his speech, Guterres noted that the pandemic "exposed" some challenges such as structural inequalities, inadequate health care and lack of universal social protection (UN News 2020).

In regions with high inequalities, as in the case of Latin America, in the medium and long term, the impacts of COVID-19 can make explicit and increase existing

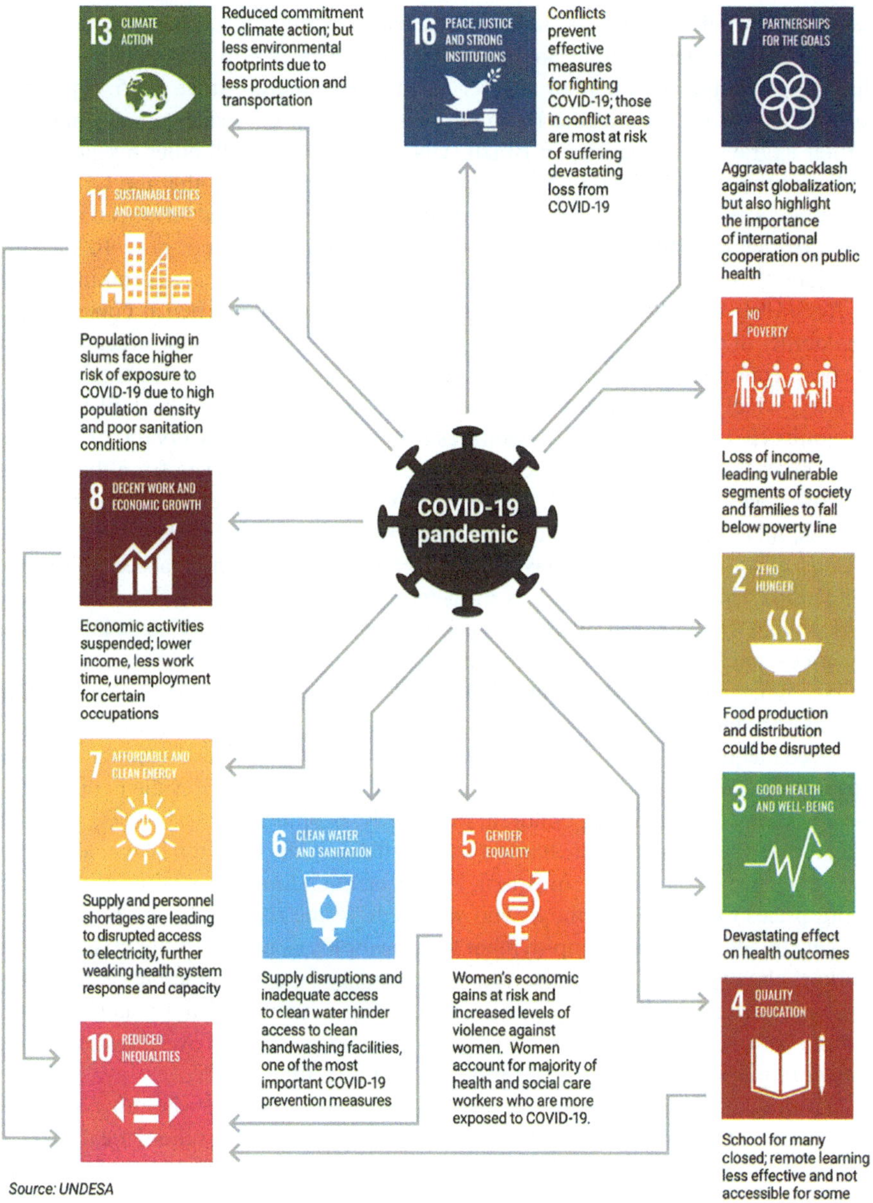

Fig. 1 Conceptual mapping on how COVID-19 affects the SDGs. *Source* UN (2020)

inequities, whether in income, access to services or the realization of basic rights. The Center for Metropolitan Studies at the University of Sao Paulo (USP) affirms that the pandemic and the poor recovery of the labor market since 2015, especially for the poorest sections of the Brazilian population, in the medium and long term further aggravate inequalities in Brazil (UN Brazil 2020). Still, other researchers at USP and the State University of Campinas (Unicamp) believe that the COVID-19 pandemic would have had less social impact if Agenda 2030 had advanced in recent years (Jornal da USP 2020). Northern Brazil has a greater need: only 10% of the population has access to solid waste treatment, and just over 50%, to drinking water.

According to these researchers, social inequality is clear when it comes to the availability of drinking water and sewage treatment in Brazil, and the precariousness of these conditions increases the rate of disease. The North and Northeast of the country, which are the most vulnerable regions and suffer from a lack of intensive care units (ICU) beds and health facilities to treat patients with COVID-19, are also those with the lowest rates of access to water and sanitation. Data from the Sanitation Information System (SNIS) of 2018 shows the discrepancy between Brazilian regions, including between wealthier and peripheral neighborhoods of some cities: in the Southeast, about 90.3% have access to water and 79.2% have access to sewage collection, while the urban area is even more privileged, with 97.7% and 72.1%, respectively. In the North of Brazil, just over half of Brazilians, 57.6%, have water resources and 10.5%, with waste treatment.

We agree with this study that highlights the Brazilian failure in forwarding actions to achieve SDGs 3 (good health and well-being), 4 (quality education), 10 (reduced inequality), 11 (sustainable cities and communities) and 13 (climate action). Regarding SDG 3 and 10, the UN report (UN 2020) shows that it is the poor and the most vulnerable—including children, the elderly, people with disabilities, migrants and refugees—who are hit the hardest by the effects of the COVID-19 pandemic. As seen in the previous section, in Brazil, loss of income, limited social protection and rising prices exposed even those who were safe to the risks of facing poverty and hunger. In addition, underemployment and unemployment due to the crisis expose workers already vulnerable in the informal economy to reduced income. Still, favela residents suffer from a lack of adequate housing, without running water at home, sharing bathrooms, with little or no garbage collection system, in overcrowded public transport and with limited access to formal health facilities.

With regard to SDG 4, the closing of schools kept an overwhelming share of students out of schools, causing them to be left without access to school meals, which in Brazil represents, in many families, children's daily meals. The lack of access to computers and the internet at home means that distance learning is out of reach for many. Particularly in relation to SDG 11 and 13, which are related to air pollution in cities and climate change, during quarantine, with fewer cars on the streets and a decrease in industrial activities, there was a fall in emissions of air pollutants. The Environmental Company of the State of São Paulo (Cetesb) registered in some metropolitan regions a maximum concentration of pollutants of 1.0 part per million (ppm), while the normal standard on other days was 9.0 parts per million (ppm) (CETESB 2020). However, for the post-pandemic economic recovery, it would be essential to have discussions with various social actors on the generation of jobs and the transition to the production of clean and renewable energy, to the detriment of the use of fossil fuels (which is not expected to happen).

In order to reduce the weakening of the 2030 Agenda in Brazil, guaranteeing the reach of the SDGs in the country, some interesting initiatives have been developed and have sought support from the private sector. For example, the Global Compact Brazil Network, together with other institutions, is leading Covid Radar, a collective of companies and organizations that operate in various sectors of the Brazilian economy and articulate joint fronts to assist Brazilian society in confronting COVID-19. Through the mobilization of business leaders, alignment between the public, private and civil society sectors, project development and an online platform, the coalition seeks to minimize the impacts of the pandemic in Brazil (Global Compact 2020). It is noteworthy that Covid Radar provides information for decision making by aggregating data, statistics and technologies made available by companies and partner organizations. In addition, it promotes support and partnerships between the business sector, identifying emergency needs across the country, putting donors in touch with institutions that need help, and offering the population a self-diagnosis tool.

Furthermore, in the international context, a new online and free corporate impact management tool was launched globally. Available in five languages, SDG Action Manager enables companies of all sizes, worldwide, to take meaningful measures and track their progress towards global goals (ODS Brasil 2020). The tool was developed by B Lab and the United Nations Global Compact and brings together the B Lab's Impact Assessment, the Ten Principles of the Global Compact and the SDGs with the intention of enabling companies to take meaningful actions through dynamic self-assessment, benchmarking and improvement.

4 Circular Economy for the Brazilian SDG Achievement

The global economy is only 8.6% circular, relying on a deeply interconnected and long supply chains that involves a material consumption about more than 100 billion tons every year, and thus high rates of resource extraction, high volume of stock of materials and products, besides low levels of recycling (Circle Economy 2020). The COVID-19 crisis has exacerbated the fragility of this system, evidencing the need of a fundamental change (Ellen Macarthur Foundation 2020). Therefore, we argue that this pandemic period can bring some interesting opportunities to the economy and society in terms of alternative development strategy and new ways of doing business globally and in Brazil.

The Circular Economy (CE) has emerged as a new paradigm of production and consumption of goods and services, suggesting a deep rethink and redesign of the current linear model based on the take-make-waste production system (Ellen Macarthur Foundation 2015). Due to its characteristic of being more tangible to organize economies and societies (Geissdoerfer et al. 2018), the CE can operate at micro, meso and macro levels. For instance, it can deal with products, companies and consumers at micro level, eco-industrial parks at meso level and with cities, region, nation and global at macro level (Kirchherr et al. 2017).

Based on three principles—the preservation and enhancement of natural capital, by controlling the stock of finite resources and balancing renewable resource flows, the optimization of resource and products yields and the effectiveness of the system by designing out waste and pollution (Ellen Macarthur Foundation 2015), the CE can be seen as a restorative and regenerative system that represents a systemic shift to resilient in supply chains, business and economic opportunities, thus providing positive environment and society wide benefits through multiple value-creation mechanisms (Ellen MacArthur Foundation 2017a, b; Geissdoerfer et al. 2017, 2018).

From a practical point of view, the CE comprises several strategies such as eco-design of products and processes, sustainable supply, energy and resource efficiency and industrial symbiosis on a perspective of supply from economic stakeholders; responsible consumption, reuse and repair on a perspective of consumer demand and behavior; and recycling materials (closed loops) on a waste management perspective (SOeS 2017), that can be applied for large and small business, for organizations and individuals, globally and locally, therefore gaining attention from multinational companies (Lacy et al. 2014) and policy makers mainly in industrialized countries (EC 2015) in the last years.

Concerning the CE in the pandemic context, the COVID-19 crisis demonstrates the potential of circular practices especially in the healthcare sector and in public sector in general. Bottom-up and decentralized initiatives such as several private companies implementing closed material loops (e.g. Breweries produced disinfection alcohol from waste flows) and individual citizens recycling locally available resources (e.g. Makers' movements produced masks for textile leftovers) clear represent the potential of the CE to deal with shortages of resources, reducing import dependency (Wuyts et al. 2020), the use of sterilization agents to decontaminate N95 masks extending their useful lives (Potter 2020), the partnership of laboratories that are working together to develop innovations of prototype designs and processes for medical use (Whyman 2020), packaging companies seeking to use a more recyclable and reusable packaging due to the increase of grocery shopping and home delivery (Poole 2020) are examples of circular practices amid COVID-19 that increase the society's resilience and therefore help to reach the Sustainable Development.

In this sense, the CE can potentially help achieve the SDGs. A qualitative research carried out by Schroeder et al. (2018) discussed the relevance of CE for the implementation of the SDGs by analyzing the linkages between CE practices, the 17 SDGs and their 169 targets. Their findings reveal strong contribution of the CE to several SDGs beyond the SDG 12 (Sustainable Consumption and Production), that is the most closely SDG related to the concept of CE. This occurs due to the transversal characteristic of CE strategies, practices and related business models.

The strongest relationships between CE practices and SDGs lie on SDS 6 (Clean Water and Sanitation), SDG 7 (Affordable and Clean Energy), SDG 8 (Decent Work and Economic Growth), SDG 12 (Sustainable Consumption and Production) and SDG 15 (Life in Land), representing a direct contribution to 21 targets and indirect contribution to an additional 28 targets. Besides the contribution of CE for implementing several SDG targets, the authors also identified the relevance of SDGs for the promotion of CE practices despite they are not directly related to CE. For instance, SDG 4 (Quality education) and SDG 16 (Peace, Justice and Strong Institutions) help to develop capacity and the skills required for scaling-up and replicating CE strategies, SDG 9 (Industry, Innovation and Infrastructure) and SDG 17 (Partnerships for the goals) are crucial for building resilient and closed-loop supply chains, industrial symbiosis and circular business models (Schroeder et al. 2018).

A consult in the main scientific databases (Scopus and Web of Science) (Chadegani et al. 2013; Boyle and Sherman 2006) taking "Circular Economy and SDG" as search string has resulted in 34 documents, including articles, conference papers and book chapters published between 2018 and July, 2020. By analyzing the most frequent words of this collection, represented by the WordCloud in Fig. 2, it is possible to observe planning, agriculture, waste water, drinking water, human development, poverty, waste disposal, electronic waste, recycling, climate change as related subjects besides the most frequent ones (circular economy, environmental sustainability, economics and sustainability).

Regarding these subjects in Brazil, the National Solid Waste Policy (PNRS), the National Policy on Climate Change and the Energy Policy are those most closely associated with CE practices. In addition, Brazilian companies have also an important role towards the CE and many circular strategies have already been adopted before the emerging of the CE concept, including practices related processes optimization, energy and resource efficiency, reducing waste and material losses (CNI

Fig. 2 WordCloud with the 50 most frequent words of CE and SDG. *Source* Own elaboration based on Biblioshiny Software

2020). However, policies, actions and scientific studies related to the CE on a national level in Brazil can be still considered restrict (Oliveira et al. 2018).

There are several opportunities of the CE in Brazil that can be divided into three focus areas: agriculture and biodiversity assets, building and construction, and electrical and electronic equipment (EEE). Regarding agriculture and biodiversity assets, the opportunities consist to expand the existing efforts in regenerative business models in agriculture and biodiversity assets in order to restore the reserve of natural capital, boost biological diversity, close nutrient loops and increase nutrient content in food; to stimulate the development of bio-intelligence; and to use technology in digitized supply chains for increase resource efficiency and transparency. In terms of buildings and construction, the opportunities are on the investment in new real estate to avoid linear lock-in; on the increase accessibility in the built environment; and on the harness of the digital technology to create new value in the sector. Finally, the opportunities of EEE comprise the leverage of EEE market dynamics, the integration of informal and formal economies into the sector in a win-win model, the development of new business models to increase access to EEE products, and the creation of mechanisms to inform design processes to other players in the sector (Ellen Macarthur Foundation 2017a, b).

5 Conclusion

As discussed in this chapter, the COVID-19 pandemic has a strong impact on Agenda 2030. It increases poverty and social inequality, limits the availability of medical services and generates damage to the health of the world population. By July, Reuters (2020) showed more than 400,000 people had died worldwide from COVID-19, with the spread far from contained in many parts of the world. Other regions are already experiencing a second wave of infections. The International Labour Organization (ILO 2020) estimates nearly half of the world's workforce—1.6 billion people—risk losing their livelihoods due to the 'great lockdown'. According to the IMF's latest World Economic Outlook, the cumulative loss to global GDP over 2020 and 2021 could be about US$9 trillion (IMF 2020).

Although Brazil has spent a significant portion of its GDP mainly on government programs to support workers (including the unemployed or who lost their jobs during the COVID-19 pandemic) and (small, medium, and large) companies, the government has been severely criticized for the late payments. The fact that the country started its number of cases in the large coastal cities and delayed making isolation and lockdown decisions made the virus spread throughout the country, making Brazil the second largest country in terms of cases and deaths. Thus, from a socioeconomic point of view, the data and forecasts point to a worsening of the pre-existing economic crisis, a significant increase in public debt, as well as a deepening of social inequality in the country. In its policy package, there is no mention of the 2030 Agenda or the SDGs, which makes specialists in Brazil fear that previously signed commitments will be set aside due to the pandemic.

It must be highlighted that the SDGs are not only useful for understanding COVID-19, but they are also a valuable tool to guide the national and international recovery effort. The pandemic has exposed weaknesses in the (inter)national system that need to be addressed as part of the 'build back better' agenda. The past couple of months have shown how poverty (SDG 1), weak health systems (SDG 3), inadequate water and sanitation (SDG 6), and substandard international cooperation (SDG 17) have exacerbated the COVID-19 crisis. In the SDGs we have a comprehensive framework to address these risk factors holistically to build resilience to current and future shocks, whether they be health, economic or environmental shocks.

Undoubtedly, while the COVID-19 crisis has demonstrated the urgent need of a fundamental change in the current production and consumption model, based on the linear economic model of take-make-waste, it has also demonstrated the potential of the CE as a constructive change driver to increase societies' and supply chains' resilience. As a result, CE would support the SDGs' achievements. The transformation to the CE needs leaving the linear mindset to think circularly, which implies to think from simple to complex, from predictive to adaptive and from competition to cooperation. The CE is not an abstract concept and, during the COVID-19 pandemic, several circular practices were adopted worldwide, and the European Circular Economy Action Plan (EU 2020) is a well representation of how countries' economy and environment can both prosper post COVID-19. Thus, we can identify several opportunities to facilitate the achievement of the SDGs through the CE in Brazil, mitigating the adverse impacts resulting from the COVID-19 pandemic.

References

BCB Brasil (2020a) Focus – Relatório de Mercado, 3 de julho de 2020. BCB, Brasília

BCB (2020b) Focus – Relatório de Mercado, 19 de junho de 2020. BCB, Brasília

BCB (2020c) Focus – Relatório de Mercado, 5 de junho de 2020. BCB, Brasília

BCB (2020d) Focus – Relatório de Mercado, 22 de maio de 2020. BCB, Brasília

BCB (2020e) Focus – Relatório de Mercado, 8 de maio de 2020. BCB, Brasília

BCB (2020f) Focus – Relatório de Mercado, 24 de abril de 2020. BCB, Brasília

BCB (2020g) Focus – Relatório de Mercado, 10 de abril de 2020. BCB, Brasília

BCB (2020h) Focus – Relatório de Mercado, 27 de março de 2020. BCB, Brasília

BCB (2020i) Focus – Relatório de Mercado, 13 de março de 2020. BCB, Brasília

BCB (2020j) Focus – Relatório de Mercado, 28 de fevereiro de 2020. BCB, Brasília

BCB (2020l) Focus – Relatório de Mercado, 14 de fevereiro de 2020. BCB, Brasília

BCB (2020m) Focus – Relatório de Mercado, 31 de janeiro de 2020. BCB, Brasília

BCB (2020n) Focus – Relatório de Mercado, 17 de janeiro de 2020. BCB, Brasília

BCB (2020o) Focus – Relatório de Mercado, 3 de janeiro de 2020. BCB, Brasília

Boyle F, Sherman D (2006) Scopus: The product and its development. The Serials Librarian 49(3):147–153.

Carvalho L (2020) Curto-Circuito: O vírus e a volta do Estado. Todavia, São Paulo

CETESB (2020) COVID-19: Cetesb constata diminuição da poluição em SP durante a quarentena. Retrieved from https://www.saopaulo.sp.gov.br/ultimas-noticias/covid-19-cetesb-constata-diminuicao-da-poluicao-em-sp-durante-a-quarentena/

Chadegani AA, Salehi H, Yunus MM, Farhadi H, Fooladi M, Farhadi M, Ebrahim NA. (2013) A comparison between two main academic literature collections: Web of Science and Scopus databases. Asian Social Science 9(5).

Circle Economy (2020) The circularity gap report 2020. Retrieved from https://www.circularity-gap.world/

CNI (2020) Circular economy: strategic path for Brazilian industry. Retrieved from https://bucket-gw-cni-static-cms-si.s3.amazonaws.com/media/filer_public/3a/a6/3aa67ad0-4e7c-46c8-8b10-0c6d097f35cf/circular_economy_-_interativo.pdf

EC (2015) Closing the loop: an EU action plan for the circular economy. European Commission, Brussels

Ellen MacArthur Foundation (2015) Growth within: a circular economy vision for a competitive Europe. Retrieved from https://www.ellenmacarthurfoundation.org/

Ellen MacArthur Foundation (2017a) The concept of a circular economy. Retrieved from https://www.ellenmacarthurfoundation.org/circular-economy/concept

Ellen MacArthur Foundation (2017b) A circular economy in Brazil: an initial exploration. Retrieved from https://www.ellenmacarthurfoundation.org/assets/downloads/A-Circular-Economy-in-Brazil-An-initial-exploration.pdf

Ellen MacArthur Foundation (2020) The Covid-19 recovery requires a resilient circular economy. Retrieved from https://medium.com/circulatenews/the-covid-19-recovery-requires-a-resilient-circular-economy-e385a3690037

EU (2020) European circular economy stakeholder platform: a joint initiative by the European Commission and the European Economic and Social Committee. Retrieved from https://circulareconomy.europa.eu/platform/

Geissdoerfer M, Savaget P, Bocken NMP, Hultink EJ (2017) The circular economy—a new sustainability paradigm? J Cleaner Prod 143:757–768

Geissdoerfer M, Morioka SN, Carvalho MM, Evans S (2018) Business models and supply chains for the circular economy. J Cleaner Prod 190:712–721

Global Compact (2020) Pacto contra a COVID-19. Retrieved from https://www.pactoglobal.org.br/pg/pacto-contra-covid-19.

Hakovirta M, Denuwara N (2020) How COVID-19 redefines the concept of sustainability. Sustainability 12:3727

ILO (2020) COVID-19: stimulating the economy and employment. As job losses escalate, nearly half of the global workforce is at risk of losing livelihoods. Retrieved from https://www.ilo.org/global/about-the-ilo/newsroom/news/WCMS_743036/lang--en/index.htm

IMF (2020) World economic outlook, April 2020: the great lockdown, April 2020. Retrieved from https://www.imf.org/en/Publications/WEO/Issues/2020/04/14/weo-april-2020

Kirchherr J, Reike D, Hekkert M (2017) Conceptualizing the circular economy: an analysis of 114 definitions. Resour Conserv Recycl 127:221–232

Lacy P, Keeble J, McNamara R, Rutqvist J, Haglund T, Cui M, Cooper A (2014) Circular advantage: innovative business models and technologies to create value in a world without limits to growth. Accenture, Chicago, IL, USA

ODS Brasil (2020) Nova ferramenta ajuda empresas a adotar medidas para alcançar os ODS. Retrieved from https://odsbrasil.gov.br/Home/Noticia?id=65

Oliveira FR, França SLB, Rangel LAD (2018) Challenges and opportunities for circular economy in a furniture local productive arrangement in Brazil. Resour Conserv Recycl 135:202–209

Poole J (2020) Chicken or the egg? Recyclable packaging and infrastructure must advance together, says Amcor VP. Packaging Insights. Retrieved from https://www.packaginginsights.com/news/chicken-or-the-egg-recyclable-packaging-and-infrastructure-must-advance-together-says-amcor-vp-sustainability.html

Potter S (2020) Circular economy principles are alleviating COVID-19 shortages. GreenBiz. Retrieved from https://www.greenbiz.com/article/circular-economy-principles-are-alleviating-covid-19-shortages

Reuters (2020) Global coronavirus deaths top 400,000 as outbreak grows in Brazil, India: Reuters tally. Retrieved from https://www.reuters.com/article/us-health-coronavirus-casualties/global-coronavirus-deaths-top-400000-as-outbreak-grows-in-brazil-india-reuters-tally-idUSKBN23E0PK
Santos T (2020) Socio-economic impacts and counter-cyclical policies to face coronavirus in Brazil. E-Int Relat 1:1–11
Santos T, Rodrigues BR (2020) Laissez-faire, laissez- l'État: o panorama sul-americano e a economia brasileira na pandemia da COVID-19. Sul Global 2(3)
Santos L, Weiss M, Santos T, Caldas L, González E, Grottera C (2020) Novo coronavírus (COVID-19) sob a ótica da demanda de energia e dos impactos sociais: o que podemos esperar e como nos preparar? Article and Opinion EPBR. Retrieved from https://epbr.com.br/artigo-novo-coronavirus-covid-19-sob-a-otica-da-demanda-de-energia-e-dos-impactos-sociais-o-que-podemos-esperar-e-como-nos-preparar/
Schroeder P, Anggraeni K, Weber U (2018) The relevance of circular economy practices to the sustainable development goals. Res Anal 23(1):77–95
SOeS, The Monitoring and Statistics Directorate (2017) 10 Key Indicators for Monitoring the Circular Economy
UN (2020) Shared responsibility, global solidarity: responding to the socio-economic impacts of COVID-19. Mar 2020. Retrieved from https://www.un.org/sites/un2.un.org/files/sg_report_socio-economic_impact_of_covid19.pdf
UN Brazil (2020). Impactos socioeconômicos da COVID-19 são mais intensos entre população mais pobre no Brasil. Retrieved from https://nacoesunidas.org/impactos-socioeconomicos-da-covid-19-sao-mais-intensos-entre-populacao-mais-pobre-no-brasil/
UN News (2020) Impacts of COVID-19 disproportionately affect poor and vulnerable: UN chief. Retrieved from https://news.un.org/en/story/2020/06/1067502
Whyman E (2020) COVID-19 ecosystems and the circular economy. Reflow Project. Retrieved from https://reflowproject.eu/blog/covid-19-ecosystems-and-the-circular-economy/
Wuyts W, Marin J, Brusselaers J, Vrancken K (2020) Circular economy as a COVID-19 Cure? Resour Conserv Recycl 162(May):105016

Thauan Santos, is an Assistant Professor at the Graduate Programme in Maritime Studies at the Brazilian Naval War College (EGN), Coordinator of the Economy of the Sea Group (GEM), and Researcher at the Latin American Council of Social Sciences (CLACSO). His main research area covers blue economy, ocean governance, sustainable development, energy, and regional integration.

Luan Santos, is Assistant Professor of the Production Engineering Program (PEP/COPPE/UFRJ) and Industrial Engineering (UFRJ-Macaé), Head of Research of the Brazilian Business Alliance for Sustainable Finance and Investment (BRASFI), and Member of the Laboratory of Financial Innovation (LAB). His main research area covers sustainable finance, climate policy, carbon pricing, and green bonds.

Ana Carolina Angelo, is an Adjunct Professor at the Production Engineering Department in Fluminense Federal University (EEIMVR/UFF) and Coordinator of the Industrial Symbiosis and Sustainable Production Group (PSSI). Her main research area covers circular economy, life cycle analysis, and sustainable decision-making.

Sustainable Development Goals and Women: An Initial Reflection on Domestic Violence in Times of a Pandemic Crisis

Sonia Regina da Cal Seixas, João Luiz de Moraes Hoefel, Micheli Kowalczuk Machado, Gianlucca Consoli, and Waldo Emerson de Souza Nascimento

Abstract This paper reports on a research (FAPESP 2019/08044-3) on the importance of sustainability, through the 17 Sustainable Developments Goals (SDGs) defined by Agenda 2030 with a particular emphasis to SDG 5 which deals with Gender Equity, and the importance of preventing and combating violence against women and girls. Since WHO's declared the COVID 19 pandemic in March 2020, there has been a significant increase in domestic violence rates. Social isolation and hygiene practices are the most effective measures to prevent the spread and containment of the virus. Several feminist, national and international NGOs and societies have warned through their publications and social networks about these significant increases in this period. This chapter addresses this fact, through a systematic survey of these publications (e.g. UN Women, Patrícia Galvão Institute) and papers published in some relevant Journals (e.g. The Lancet). This reflection aims not only to contribute to the ongoing research, but also to present strategies related to SDG 5, and to offer proposals and approaches to face the challenges related to Gender Equity.

Keywords Sustainability · Sustainable Development Goals · Agenda 2030 · SDG5 · Gender Equity · Violence against women and girls · COVID 19 pandemic · Domestic violence rates · Social isolation and hygiene practices · Social relations

---

S. R. da Cal Seixas · J. L. de Moraes Hoefel (✉) · G. Consoli · W. E. de Souza Nascimento
Center for Environmental Studies and Research, NEPAM, State University of Campinas, UNICAMP, Rua dos Flamboyants, 155, Cidade Universitária, Campinas, São Paulo 13083-867, Brazil

S. R. da Cal Seixas
e-mail: srcal@unicamp.br

J. L. de Moraes Hoefel · M. K. Machado · G. Consoli · W. E. de Souza Nascimento
Center for Sustainability and Cultural Studies, UNIFAAT University Center (NESC/CEPE/UNIFAAT), Estrada Municipal Juca Sanches 1050, Atibaia, São Paulo CEP 12954-070, Brazil

W. Leal Filho (ed.), *COVID-19: Paving the Way for a More Sustainable World*, World Sustainability Series, https://doi.org/10.1007/978-3-030-69284-1_8

1 Introduction

A series of pneumonia cases of unknown origin in Wuhan China caused concern amongst health officials in late December 2019. On December 31, the Wuhan Municipal Health Commission issued an alert, the Chinese Centre for Disease Control and Prevention (China CDC) then sent a team to Wuhan and the World Health Organization was notified (WHO). Causes like Influenza, avian influenza, coronavirus (SARS-CoV) and coronavirus of the Middle East (MERS-CoV) were eventually ruled out. The epidemiological investigation implicated Wuhan's Huanan Seafood Wholesale Market and active case tracing was begun and vigorously pursued (Salzberger et al. 2020; Srivastava et al. 2020; Zhang 2020).

On January 7, 2020 the causative pathogen was identified as a novel coronavirus and given the name COVID-19 (Zhang 2020). Respiratory droplets were found to be the primary mode of transmission of COVID-19 virus from human to human (Allaerts 2020; Bandyopadhyay 2020; Srivastava et al. 2020).

Although the transfer route of the virus had not yet been determined Arora and Mishra (2020) are certain that the reservoir of the coronavirus is in a wild animal. Civets, bats or pangolins are thought to be the source of transmission of this deadly virus to humans.

In order to prevent the spread of COVID-19 the main strategy has been personal hygiene, physical distancing, and respiratory hygiene (Bandyopadhyay 2020). In comparison to outbreaks of other viruses COVID-19 is highly contagious. The number of primary infections reproduced can be lessened through various containment and mitigation strategies and it has been estimated that there was a significant reduction of the reproductive number of initial infections through travel restriction and physical distancing.

Governments across the world are taking steps to reduce the spread of the virus. The most common steps have been increased physical distancing, the shutting down of factories, offices, educational institutions and by prohibiting crowding. Today these actions and restrictions are wreaking havoc on society by drastically changing social behaviours, economic activities, and effecting environmental issues (Allaerts 2020; Srivastava et al. 2020).

The World Health Organization (WHO) declared SARS-CoV-2 a pandemic caused by the novel COVID-19 on March 11, 2020. SARS-CoV-2 is a novel strain of corona virus not previously identified in humans. Infected individuals have reported symptoms that include fever, cough, muscle aches, and shortness of breath, as well as standard or decreased leukocyte counts (Saxena et al. 2020). Serious cases of infection cause pneumonia, severe acute respiratory syndrome, kidney failure, and death.

The ongoing spread of the COVID-19 pandemic is having profound effects on both private and professional lives, the world economy, and the social fabric of communities. We are also aware that the shape of the world we live in will be profoundly altered by this pandemic (Filetti 2020). The current situation calls for enforcement of strict laws to help brake the further spread of COVID-19 (Srivastava et al. 2020).

The first confirmed case of CoVid19 in Brazil occurred on February 26, 2020. On April 1, confirmed cases totalled 5868 with 203 officially reported deaths. On May 1, confirmed cases had reached 91,589 and the number of deaths was at 6329. On June 1, 2020 Brazil surpassed 500 thousand confirmed cases of the novel coronavirus with 514,849 people having been infected since the beginning of the pandemic and the country saw 29,314 deaths. On June 1, the official world number of cases was around 6,100,500 cases of COVID-19 with around 375,000 deaths (SanarMed 2020).

For Stang et al. (2020) the infection rate is substantially underestimated because a significant proportion of COVID-19 patients have not been confirmed by polymerase chain reaction (PCR) tests and many patients with symptoms are not tested at all. It has been speculated that the total number of COVID-infected people is about five times higher than the official statistic. This is something that should be considered when interpreting any COVID19 statistic.

Another important factor is that humans are disturbing the habitats of wild animals and the normal cycle of pathogens and their hosts, as mentioned by Arora and Mishra (2020) causing humans to become more and more vulnerable to new diseases. The corona virus is not yet fully understood and there may be several other strains of this virus in nature or in their wild animal reservoir hosts which can become a matter of global health security in the future as well.

The COVID19 pandemic raises questions about our ongoing efforts to improve the earth's environment and there are key lessons to be learned from COVID-19 in relation to our preparedness, survival and our responsibility towards nature. Whether we learn these key lessons or not will determine our ability to control future pandemics. According to Arora and Mishra (2020) the way to tackle such pandemics in advance is to go all out and achieve the targets of environmental sustainability.

Leal Filho et al. (2020) observe that current trends suggest that the process of implementing the SDGs may be delayed due to socio-economic pressures and setbacks that are lowering the level of priority they are given. However, for the authors (Leal Filho et al. 2020) the global crises triggered by COVID-19 shows that pursuing and implementing the SDGs is more critical now than ever because they represent ways which quality of life can be restored and how social and environmental problems can be addressed.

For many people the current pandemic evokes an experience of deeply rooted existential anxiety, threatens their accustomed identity and their sense of place in the world. Many suffer from uncertainty, fear of infection, moral distress and grief, and this is often experienced alone. There is increasing concern about how to cope with the resulting anxiety as well as with its long-term individual and collective impacts (Peteet 2020).

Both men and women are affected by COVID-19, but biology and gender norms are shaping the burden the disease brings to people. Women are being affected differently than men with inequities disproportionately affecting their wellbeing and economic resilience during lockdowns. The responsibilities of housework, childcare and the care of the elderly typically fall upon women and with households under strain concerns over increased domestic violence are growing (Editorial 2020).

During pandemics women may worry about their physical safety or experience additional mental or emotional distress making it even more challenging to create the necessary space at work, in relationships and to mitigate immediate risks of violence. Women working in humanitarian efforts may be at even higher risk and their protection depends on specific organization policies and procedures (Sharma et al. 2020). The COVID-19 outbreak may make existing protective strategies identified by women and girls—such as moving in groups or ensuring aid workers are accompanied when visiting refugee households, more challenging to implement. At the same time shortages in goods mean women and girls face more pressure to access these items for themselves and for their homes (Peterman et al. 2020).

1.1 Sustainability, SDG 05 and Domestic Violence Against Women

Arruza et al. (2019) are accurate when they emphasize that gender violence takes many forms, but it is always entangled in capitalist social relations, and reflects the contradictory dynamics of family and personal life in a capitalist society. These dynamics are based on the systems unmistakable division between production and profit, family and work. The authors also emphasize that historically domination had a political character but that this domination has recently become more private, informal and psychological, rather than rational and controlled. Gender violence has existed in every country, class and in all ethnic and racial groups during every period of capitalist development and is stimulated by outside factors such as exacerbated alcohol consumption, shame in regard to status and anxiety about maintaining domination. Moreover, domestic violence increases and spreads during times of crisis and gender vulnerabilities and becomes more apparent where there is economic precariousness, anxiety, and political uncertainty.

In relation to the objectives of sustainable development stated in Agenda 2030 the violence against women and girls in contemporary capitalist society has been an impediment to building sustainability at the global level for some time (UN 2015). With this in mind research carried out by Seixas and Hoefel (2020) urges us to contribute to the reflection on this very unique moment of global crisis.

From the article cited above we were able to highlight the importance of gender equity and emphasize that in order to be practical, political actions for sustainability must correct disproportionate differences and impacts on women and girls and consider the three dimensions of sustainable development: economic, social and environmental. They also recognized the distortions that the predatory capitalist development model causes in social relations, to the environment and to everyday life far exceeds the political and private dimensions.

Historically, the role of women in the process of building sustainable development has been critical not only because of the suffering caused by conflicts and inequality but also because of the increasing rates of violence (Seixas and Hoefel 2020). Because of the immense inequalities in Brazil the difficulties for women and girls are even greater and for black women and girls in particular this situation is aggravated for besides being women they are black.

The data exists to prove this harsh reality and according to Oxfam (2017), 88.3% of black women living in rural areas earn less than a minimum wage in contrast to 42.8% of urban white men. Black Brazilian women are overrepresented among the poorest with almost 20% of them being among the poorest 10% in contrast to 5% of white men. In Brazil the homicide rate of women is the fifth-highest in the world and the death of black women increased by 54% between 2003 and 2013 (Oxfam 2017).

It is possible to state with certainty that the limitations imposed on sustainable development are directly related to the incomplete creation of equal rights and opportunities for the female population. The importance of putting SDG 5 Gender Equality and Equity into effect across the board and recognize women as an essential part of society and the economic system, combat ignorance, give value to women's contribution to both home and community and seek to build inclusive and peaceful societies is clear (Seixas and Hoefel 2020).

With this in mind it is important to highlight that women and girls are the most likely to suffer from the physical and psychological violence that is a consequence of economic crises and conflicts. The present plight of women suppresses their creative potential and possibilities of leadership and it is essential to offer women social and economic security in their homes and within society and to guarantee their dignity. Without this it is impossible to imagine a culture of peace for humanity. Achieving gender equity is a fundamental part of the 2030 Agenda and of Human Rights.

Providing quality education for girls and women is vital to achieving this goal and will have positive effects such as employability, reduction of forced early marriage, less possibility of becoming victims of violence and improved health and well-being.

Human development requires strong institutions that have the capacity and willingness to face discrimination and guarantee the protection of rights, overcome prejudice and abuse, strengthen the quality of education in the lives of women and girls, and build a strong legal structure that can defend human rights.

Seixas and Hoefel (2020) considered the hypothesis that without the gender equity postulated by the objectives of SDG 05 and especially without combating one of the greatest scourges to humanity, the violence against women and girls, it would be impossible to build a society based on values of sustainability. Therefore, in the search for the real meaning of gender equity and the empowerment of all women and girls it is necessary to consolidate collective action against discrimination against women and provide education, employment and wages that are equal to those of men.

1.2 *Being a Woman in the Context of the Pandemic*

It is important to recognize that Brazil's immense economic, social, sexual, ethnic and cultural inequalities present tremendous difficulties to implementing the SDGs particularly Objective 05. It is, however, possible to point out a few advances such as the Maria da Penha Law (Law nº 11,340.07, of August 2006), and 9 years later the inclusion of article 121 to the Brazilian Penal Code. This article deals with homicide, specifically the classification of femicide (the intentional murder of women because they are women) Law of Femicide, Law No. 13.104, which came into force in 2015.

These advances do not mean that violence against women has decreased. We are still a long way from this and in addition to this problem Brazilian women and girls are significantly less likely to participate in the educational process and especially if they are black. Table 1 and Fig. 1 presents this data (Brasil/IBGE 2010).

It is important to mention that the data is out of date because the importance of the demographic census that would have been applied to the population in 2020 was questioned by the current president when he assumed power in 2019. All funds for health, education and other social policies had their resources frozen despite

Table 1 Proportion of people by gender, color and education level (%), Brazil, 2010

Education level	White man	Black man	White woman	Black woman
No instruction and incomplete elementary school	42.4	59	41	55.2
Complete elementary school and unfinished high school	15.1	14.8	14.3	14.5
Entire high school and incomplete higher education	26.9	21.4	26.7	23.2
Complete higher education	15.3	4.55	17.7	6.71

Source Brasil/IBGE, Censo Demográfico (2010)

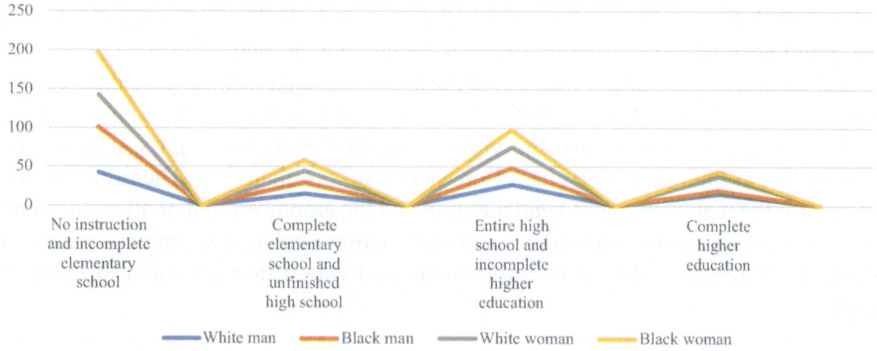

Fig. 1 Proportion of people by gender, color and education level (%), Brazil, 2010. *Source* Brasil/IBGE, Censo Demográfico (2010)

numerous protests from the community and civil society. These manifestations led to partial maintenance with significant decreases. Currently, due to the pandemic, the census has been postponed to 2021.

In Table 1 and Fig. 1 we can see the significant disadvantages for women regarding education, especially when the color factor is considered. Black women suffer the most from this inequality. The outlook for the country in 2010 was unequal for women in general and especially for black women.

The gap presented by Table 1 and Fig. 1 places women in a significant situation of disadvantage concerning men in terms of education if we look at the percentage data and the position of the lines. It indicates that objective conditions of employment, access to better salaries, housing conditions and access to public goods, and better living conditions, cannot be directly addressed. The pandemic of COVID 19 further worsens a situation of vulnerability already present in the lives of these women. Not to mention that women are the ones most subject to care of the home, care of children and care of the elderly and often have to live directly with partners who could already be their abusers.

It is worth noting that this situation is not exclusive to Brazil, as Arruza et al. (2019) point out, in global terms, more than one in three women experienced some form of gender violence in their lifetimes. Intimate partners are responsible for 38% of the murders of women. The authors also point out that physical, emotional, sexual abuse or all of them together committed by intimate partners happens in all forms in a capitalist society and should not be considered accidental but seen as rooted in the basic institutional structure of this model of society.

1.2.1 The Pandemic Crisis and Violence Against Women

Times of crisis and unrest have been associated with an increased incidence of violence against women and children. In this context, pandemics are no exception. The regional or global nature and the fear and uncertainty associated with pandemics provide an environment that can help exacerbate or unleash various forms of violence against women and children (Peterman et al. 2020).

According to Global rapid gender analysis for COVID-19 (2020) on a global scale, more than 35% of women experience gender-based violence (Gender-based Violence 2020) during their lifetime. During humanitarian crises combined factors such as increased levels of anxiety and stress and economic difficulties can further increase rates of gender-based violence.

As crucial as mandatory lockdowns, quarantine and self-isolation may be these measures can have detrimental effects to those already living in violent situations. Individuals, especially women, are intrinsically tied to the aggressor without physical relief from the abusive relationship. The attacker can also use the virus to further isolate his victim from family, friends and social networks, as well as from the services that could support them (Global rapid gender analysis for COVID-19 2020).

Although there is a shortage of rigorous studies to estimate increases in data on violence against women and children, the media frequently reports it. For example, when the Ebola outbreak hit West Africa an "epidemic" of "rape, sexual assault and violence against women and girls" was reported to have been widely documented as collateral damage. Currently during the coronavirus outbreak (COVID-19) in mid-March 2020 there have been reports from Australia, Brazil, China and the United States suggesting an increase in violence against women and children (Peterman et al. 2020).

Kofman and Garfin (2020) mentioned that current pandemics share some critical similarities with previous disasters in history and present a unique and distressing paradox for victims. If a victim decides or is forced by their partner to stay at home, they run the risk of suffering or being susceptible to increased violence. On the other hand, if they leave, they are at risk of being exposed to a highly infectious and dangerous virus.

There are some worrisome indicators about domestic and family violence against women from the social isolation imposed by the COVID-19 pandemic. Organizations addressing domestic violence have seen an increase due to forced coexistence, economic stress and fears about the coronavirus (Vieira et al. 2020).

According to Sigal et al. (2020) social isolation in Latin America has brought with it a sad consequence. There has been an increase in requests for help that also demonstrate an increase in domestic violence in a region where almost 20 million women and girls suffer from physical and sexual abuse every year. The authors also state that the fear is that violence against women which was already prevalent in the region is being further exacerbated (Sigal et al. 2020).

According to the authors, the Argentine emergency call line 137 for victims of abuse that is supported by the Justice Department saw a 67% increase in calls in April compared to a year earlier after a national lockdown was imposed on March 20. In Chile, the Women's Ministry stated that calls to domestic helplines increased by 70% in the first quarantine weekend. Also, in Chile, the mayor of the district of Providencia, in Santiago stated that requests for help from a local office that provides legal, psychological and social assistance jumped 500% under the shutdown (Sigal et al. 2020).

In Colombia daily domestic violence calls to the national women's hotline increased by almost 130% during the first 18 days of the country's quarantine according to government data (Sigal et al. 2020).

According to the Brazilian Public Safety Forum (2020) analysis showed that in 12 States the number of registered femicides grew by 22.2% jumping from 117 victims in March/April 2019 to 143 victims in March/April 2020. There was also an increase of 27% in complaints registered at the Ligue-180 Central of Assistance to Women in Situations of Violence help line (Ligue 180) and in April after all Brazilian states had adopted social isolation the increase in complaints was 37.6%.

A study carried out by the Brazilian Department of Justice has shown that according to data made available by the Courts of Justice the number of emergency protective measures concessions decreased 31.2% in the Acre 8.2% in Pará 14.4% in São Paulo and 28.7% in Rio de Janeiro.

Campbell (2020) mentions that reports of rising rates of domestic violence are beginning to appear around the world. In China domestic violence is reported to have tripled during the quarantine. France indicated a 30% increase in reports of domestic violence. Brazil estimates that reports of domestic violence jumped 40–50% and Italy also showed reports of rising domestic violence. From the viewpoint of the author, a global trend of data showing increasing domestic violence is likely to be seen throughout the pandemic and may represent only the tip of the iceberg because many of the victims remain trapped by their abuser and are unable to report the abuse (Campbell 2020).

According to Campbell (2020), in the United States agencies across the country are also reporting an increase in domestic violence. In addition to the risk of physical harm, victims are also at high risk of emotional abuse. The author mentions that there have been reports from the USA about perpetrators of domestic violence using Covid-19 as a weapon by prohibiting hand washing in an attempt to increase the victim's fear of contracting the virus and threatening to ban medical treatment if they contracted the virus.

Therefore this chapter has two mains goals: (1) to contribute to the reflection on the increase in data on violence against women in times of the Covid-19 pandemic and (2) to suggest ways in which society and institutions can collaborate to lessen the vulnerability and create better conditions for this social group.

1.3 Methodology: A Brief Description of the Tools

The methodology applied in this work utilizes an exploratory study and bibliographic research. Qualitative and quantitative data were collected from publications and websites that analyze and publish data on violence against women and girls and violence against women and girls in the COVID 19 pandemic (Gil 2019). The information on violence against women and girls during COVID 19 presented and analyzed in this work followed a timeline that began in December 2019 and ended in June 2020.

Information on the violence against women and girls during the COVID 19 pandemic happening internationally was collected from the UN Women, Women's Safety NSW and Women's Aid websites.

Data on Brazil was gathered from the UN Women website (https://www.onumulheres.org.br/covid-19), at the Patrícia Galvão Institute, on the Information Portal of the Ministry of Women, Family and Human Rights (MWFHR), from the Federal Government on its communication channel National Human Rights Ombudsman (NHRO) and from the Brazilian Public Safety Forum.

UN Women (www.unwomen.org/en) is the United Nations entity dedicated to gender equality and the empowerment of women and was established to accelerate progress in meeting women's needs worldwide. UN Women supports UN Member States as they set global standards for achieving gender equality and work with governments and civil society to design laws, policies, programs and services that are needed to ensure that the standards are effectively implemented and truly benefit women and girls worldwide. It works globally to make the vision of the Sustainable

Development Goals a reality for women and girls and stands behind women's equal participation in all aspects of life.

Women's Safety NSW (https://www.womenssafetynsw.org.au) is a state-wide body for women's specialist services advocating for women's safety in the context of #DFV in New South Wales. It advocates for systemic reform to increase women's safety, justice and wellbeing in the context of domestic and family violence.

Women's Aid (https://www.womensaid.org.uk) is a grassroots federation working together to provide life-saving services in England and build a future where domestic abuse is not tolerated.

UN Women (https://www.onumulheres.org.br) is the body that represents UN Women in Brazil and works in the area of gender equality and the empowerment of women.

Patrícia Galvão Institute (PGI)—Media and Rights (https://agenciapatriciagalvao.org.br) is a feminist social organization that works in the fields of women's rights and communication.

The Federal Government's Portal of the Ministry of Women, Family and Human Rights (MWFHR) is a channel of access to Information on the Federal Government that aims to provide data and public information about all the organs of the Federal Government including data on violence against women and girls. The information made available by the ministries is gathered on the Portal and made available digitally to all citizens (Brazil 2020).

The Brazilian Public Safety Forum—BPSF (https://forumseguranca.org.br) is a non-profit organization that acts as a permanent and innovative space for debate, articulation and technical cooperation for public safety in Brazil particularly in the area of violence against women and girls (BPSF 2020).

2 Results Obtained

2.1 Covid-19: Confinement and Domestic Violence in the Global Context

The relationship between increasing orders of social confinement and the rise in the number of requests for help found in the records of hotlines and shelters for victims of domestic violence worldwide is almost directly proportional. According to the executive director of UN Women in a statement to the Organization's website government officials and women's rights activists from developed countries; Canada, France, the United Kingdom, Spain and the United States have denounced an uncontrollable increase in the number of reports of domestic violence during the Covid-19 pandemic.

France is the country in Europe with one of the highest rates of domestic violence. Every year about 219,000 women aged between 18 and 75 face physical or sexual abuse by current or former partners but only 20% report it. On March 17, 2020 the government made lockdown mandatory and descriptions of domestic violence up until April 04 alone recorded an increase of 30% across the country; with the capital registering an increase of 36%. The French government has announced that it will finance hotel rooms to house victims of domestic violence as a possible outlet for many women who are being raped in their homes (Euronews 2020).

In Australia, Women's Safety NSW (Women's Safety New South Wales—Australian) reports that "Sydney's wealthiest suburbs are experiencing an 'alarming' increase in cases of domestic violence with increases above 30% which is well above the average in the State" (June 03, 2020). As of March 2020, Baulkham Hills, a neighbourhood in New South Wales, and Hawkesbury, a neighboring city, have reported a 38% increase in registered attacks related to domestic violence and Sutherland has registered 31%. Australia now has 7249 confirmed cases of CoVid19 and 102 deaths.

An "initial" survey conducted in April by Women's Aid, an English institution that works to investigate and reduce domestic abuse against women and children, found that in England, of the victims who said they are currently in an abusive relationship, 67.4% reported that the abuse has gotten worse since Covid-19. The survey also states that "the lockdown meant that more than three-quarters of the survivors (76.1%) would have to spend more time with the aggressor". Unsurprisingly, many survivors said they felt trapped with nowhere to run. Abusers are using Covid-19 as a tool for abuse.

According to Spain's government statistics the number of calls made by victims of domestic violence to protection services up until April of this year are considerably higher than those of last year. In January, February, March and April 2019 the number of calls made by the victims were 3482, 3310, 3794 and 5396 respectively and for the same months in 2020 3654, 3417, 4254 and 8692 calls were registered (Espanha 2020a, b).

Updates from the "Monthly Statistical Bulletin of April—2020" made available by the Spanish Government Delegation against gender-based violence states that from January 1 to April 30, 2020, 25,642 "pertinent" calls were received. This was 20.6% more calls than in the same period of 2019 (21,269 "pertinent" calls) and that during April 2020 alone 8692 pertinent requests were received. This is the highest number in the last three years and the second-highest in the entire series and represents a 61.1% increase in comparison to April 2019 (5396 pertinent calls). According to the Bulletin average daily calls during April 2020 reached 290 (Spain 2020).

Domestic violence has long been regarded as one of the most significant violations of human rights. According to UN Women data, 243 million women and girls aged 15–49 worldwide were subjected to physical or sexual violence by an intimate partner in the last 12 months before the Pandemic (UN Women 2020).

The Director also added that underreporting is a direct challenge to the ability to respond to this category of violence and is also a constant in this barbaric universe. Less than 40% of women who suffer abuse seek help of any kind or denounce the crime and less than 10% of these women who seek help go to the police. It is not surprising that the current scenario is a factor that further impedes access to aid. Public services such as police, justice and even social services have totally or partially shut down and this creates an extremely favorable environment for impunity.

It is hoped that the current situation will at least give rise to a more significant concern for women and girls worldwide. It is important to mention that 1 in 4 countries do not have laws that specifically protect women from domestic violence. Times of crisis are fertile ground for substantial change. All civil societies and governments must understand that even in this scenario this increase in cases of abuse is a cry for a change.

2.2 Covid-19: Confinement and Domestic Violence in the Brazilian National Context

The National Human Rights Ombudsman (NHRO) receives, collects and organizes data from calls made by victims of domestic violence and makes this data available on the Federal Government website for information purposes.

The following data corresponds to the period from January thru April 2020—the beginning and rise of the CoVid-19 health crisis and the crisis of increasing domestic violence against women and children that ensued.

The Federal Government website states that information is collected for later publication through the service's networks and the NHRO. Through Dial 100, Ligue-180, the website and its app the Ombudsman receives, examines and forwards complaints and specific complaints about human rights violations.

There was an average increase of 14.1% in the number of complaints made to Ligue-180 during the first four months of 2020 compared to last year. The total number of complaints registered was 32.9 thousand between January and April 2019 compared to 37.5 thousand in the same period this year. April alone showed an increase of 37.6% in the comparison between the two years (NHRO 2020). Figure 2 illustrates this exponential growth.

In Brazil, the states with the highest number of cases of violence against women as of March 1, 2020 are São Paulo with an average of 102 new cases per day, Rio de Janeiro with an average of 70 new cases per day, Minas Gerais with an average of 30 new cases per day and Pará, with an average of 25 new cases per day. These cases include all kinds of violence against women. The most common kinds of violence seen are physical, psychological, moral harassment, emotional neglect, shame and aggression. According to data gathered by the MWFHR, across the country there was an average of 352 new cases of physical violence and 273 cases of psychological abuse reported per day. Figure 3 shows the daily evolution of new cases for each type of registered violence.

The average age among women who suffer the most from some form of violence is 35–39 years; there were 672 recorded cases of violence against women of that age.

Figure 4 shows the map of Brazil with information on the distribution of complaints by the Federative Unit.

The Patricia Galvão Institute (PGI) surveyed high-circulation national newspapers on March 27, 2020 and based on data from Ligue 180 found that there had been an

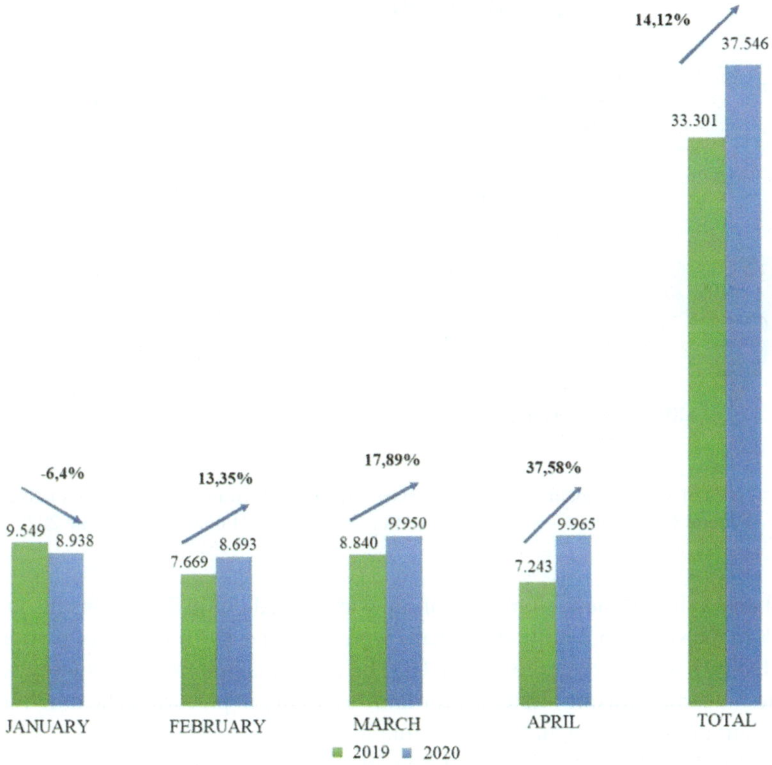

Fig. 2 Domestic violence in comparative data in the first four months of 2019 and 2020. *Source* Systematized from Brasil—Federal Government—Ministry of Women, Family and Human Rights, June 2020

8.5% increase in cases of domestic violence since social isolation was imposed to contain the spread of Covid-19. This significant increase in domestic violence confirmed what many experts had already warned and proved that social isolation forces victims and their aggressors to live in the same environment.

Through the same source the data analyzed from the NHRO showed that between March 1 and March 16 of this year 3045 calls were received and 829 complaints were registered. Between March 17 and 25 there were 3303 calls for help and 978 complaints were registered. This was equivalent to an 8.5% increase.

On March 31 the PGI pointed out another fact that further aggravated this situation. Many women had to abandon their jobs in order to care for others and to carry out domestic tasks. This situation demonstrates the already unequal pay of women compared to men had tended to increase in this period. It also demonstrates how gender roles are defined and the immense burden placed upon women.

On April 2, 2020 the PGI presented the reality of Brazilian slums after the arrival of the corona virus and sought to analyze the living conditions of the 13.5 million

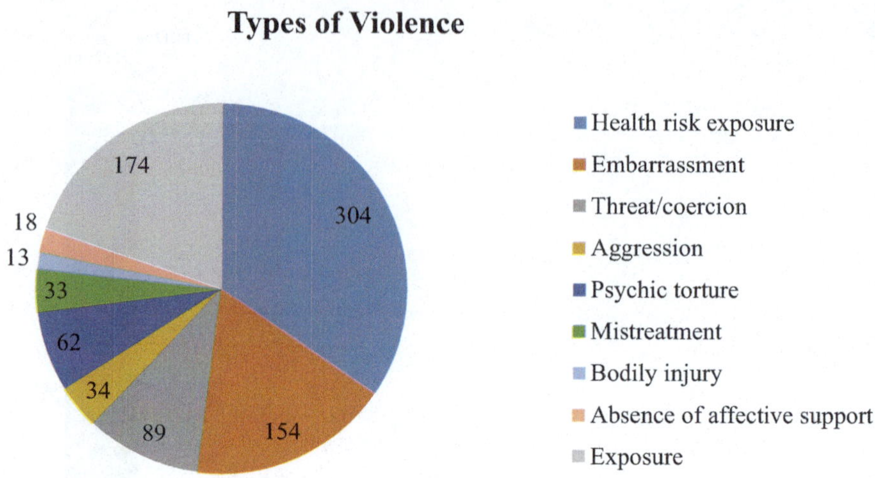

Fig. 3 Types of violence against women, January–April 2020, Brazil. *Source* Systematized from Brasil—Federal Government—Ministry of Women, Family and Human Rights, June 2020

Brazilians living in these slums. A survey was carried out in 260 slums throughout Brazil. In the survey 5.2 million residents were mothers. The universe of participants included 621 women over the age of 16. The data obtained from the survey will be used to boost fundraising strategies for families living in precarious health, hygiene, basic sanitation, housing and income situations.

Results show that 92% of mothers had difficulty buying food after the first month without income, 72% stated that the lack of income affected the families' diet, 73% said they have no financial reserve to meet their needs, and 76% reported that expenses increased because children cannot go to school. All of this implies significant difficulty in maintaining social isolation to combat the spread of COVID-19.

The Institute draws attention to the need for the pandemic crisis to reinforce the debate on gender equity at work. According to what was reported by the Brazilian press on June 18, 2020 women's routines were much more affected by the pandemic because of the need to care for the home and family as others went back to work. At the same time children have not returned to school and according to research this situation impacts not only the family's economy but the country's economy as well. In the current model of division of domestic work women dedicate more hours than men and their labor often amounts to a double workday.

The International Monetary Fund (IMF) estimates that the Global Gross Domestic Product (GGDP) would increase by at least 4% if unpaid work were better distributed. The Federal University of Ceará together with the Maria da Penha Institute released a study that estimates a loss of R$1 billion a year in Brazil from domestic violence that causes employee absenteeism and impacts to health. According to the IBGE men earn 47% more than women with equal training and while white men earn 55% more than black women.

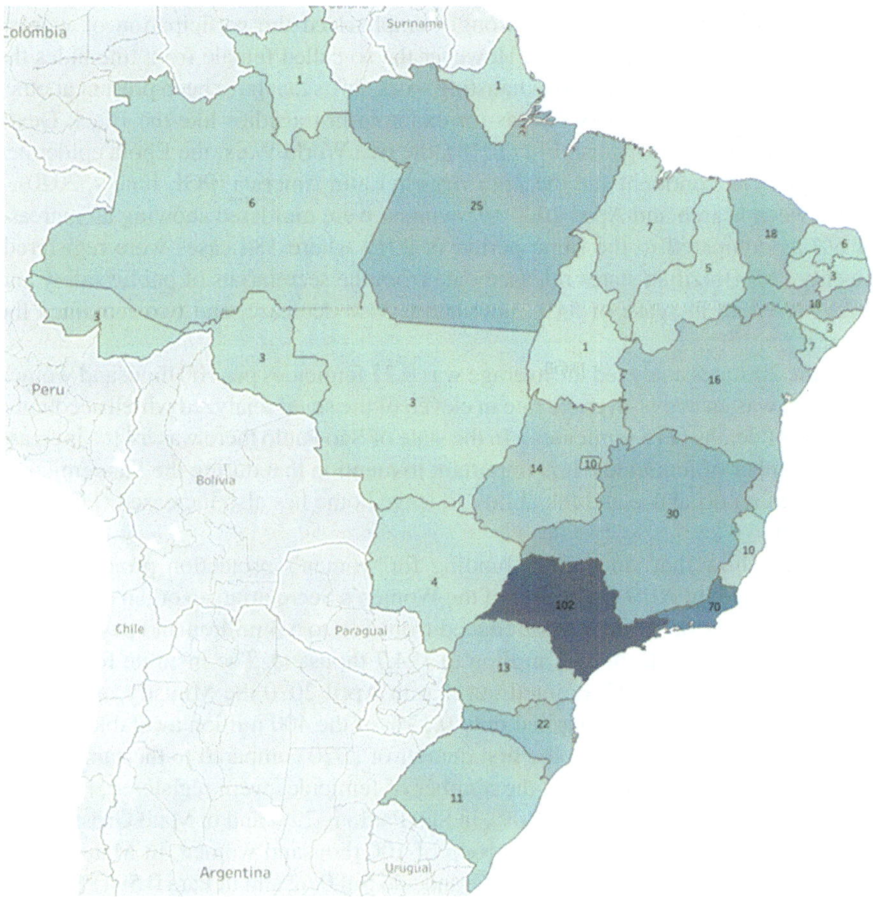

Fig. 4 Distribution of the number of complaints by Federative Unit, Brazil, January–April 2020. *Source* Brasil—Federal Government—Ministry of Women, Family and Human Rights. Available at: https://ouvidoria.mdh.gov.br/portal/indicadores. Accessed: June 24, 2020

Latin America has come to be considered the new epicentre of the Covid-19 pandemic. It has more than one million cases and countries like Mexico and Brazil have faced increasing rates of violence against women during the same time period (PGI, June 12, 2020).

In Mexico the situation shows an increase superior to 50% in the number of attacks on women during the first four months of this year as compared to the same period last year. In Brazil the complaints of violence against women received by Ligue-180 grew by almost 38% in April 2020 alone. Femicides increased by 22% in March and April in twelve Brazilian states. According to a report from The Brazilian ministry of Public Safety in the state of Ceará this number was double and in Acre there was an increase of 300%.

The media and society have strongly emphasized the participation of women in the fight against the pandemic. However the so called female front line hides the many sacrifices that are behind exhausting work shifts that have been present at other times in the history of the world as for example in tragedies like the Black Death (fourteenth century), Spanish Flu (1918), the two World Wars, the Ebola epidemics on the African continent and the Zika virus in Latin America (PGI, June 9, 2020).

Between March and April 2020 195 women were murdered showing an increase of 5% as compared to the same period in 2019 where 186 cases were registered. Twenty (20) Brazilian states released data from the secretariats of public safety and 9 registered an increase of 54%, nine had a 34% decrease, and two remained the same.

In the 20 states analyzed the average was 0.21 femicides per 100 thousand women and there was an above-average rate in eleven of the states analyzed which accounted for 59% of deaths (115 femicides). In the state of São Paulo there was a 41% increase in the number of femicides. It is important to mention that during the Pandemic it is difficult in report abuse and the ability to leave home has also increased (PGI, June 18, 2020).

Studies show that since 2015 funding for women's protection programs have been decreased. In 2019 the budget of the Women's Secretariat, an organ of the now-defunct Ministry of Women, was reduced from 119 to 5.3 million and payments for victim care dropped from 34.7 million to 194.7 thousand. The Institute for Socioeconomic Studies (INESC) pointed out that in April 2020 the Ministry of Women, Family and Human Rights applied only 0.13% of the 400 million available.

In the State of Para during the first quarter of 2020 compared to the same period of the previous year three times the number of femicides were registered, in Rio de Janeiro there was an increase of 70%, in São Paulo is 29% and in Mato Grosso 40%. In Acre there were 1.32 cases per group of 100 thousand women, in Mato Grosso there were 1.26, Sergipe 0.67, in Rio Grande do Sul 0.62 and in Pará 0.59 (PGI, June 18, 2020).

3 Conclusions

The COVID-19 pandemic has caused society to fear the unknown, contagion and life and death. These fears along with social isolation and other restrictive measures to prevent disease from the virus have caused an increase in domestic violence and especially for groups of the most vulnerable women.

As pointed out by Queiroz (2020) this serious problem is not specific to Brazil and that on every continent many countries have registered an increase in numbers involving some type of violence against women during the Covid-19 pandemic and the consequent need for social distancing.

The data shows that in many countries around the world including Brazil, between March and June 2020 the data jumped the same time as the contamination began and the need for social isolation became present.

Although the reality we have sought to present has been restricted to the collection of data from official organs, from associations for the protection of women and from international agencies it is still indicative of a frightening and extremely worrying panorama that compels us to conclude that we must achieve the 17 SDGs to attain a sustainable development model. Women must be strengthened through education, improved living conditions and better employment for women and girls. Many today live in a situation of dependence on men (father or husband/partner) for their subsistence and often that same man is their abuser. If we do not face this situation, we will not achieve sustainability much less offer a safe place in society.

Queiroz (2020) drew attention that in Brazil between March and April cases of femicide grew 22.2% in 12 states as compared to the same period last year. States like Acre showed a 300% increase in these crimes followed by 166.6% in Maranhão and 150% in Mato Grosso. The current health crisis facilitates increasing crimes that are already happening inside the home. With the lockdown women and children living with violent relatives have ended up more vulnerable.

Data like this together with poor schooling and the precarious conditions of employment and life that follow will not allow for a project of global sustainability, gender equity and the recognition of human rights to be achieved.

With this in consideration, we present below some measures as recommendations to confront this situation:

1. Create quality education for women and girls;
2. Build training projects for women so they may achieve better terms of employment;
3. Organize committed secular institutions that will direct efforts and support to women at risk and victims of violence;
4. Create Public and multidisciplinary health services so that women can overcome the trauma caused by abuse;
5. Create safe places for women and girls to face situations of social isolation that happen in situations like pandemics or extreme weather events where they won't be victims of violence.

It is our hope that these recommendations will collaborate with the reduction of violence against Women and Girls and to the implementation of the Sustainable Development Goals and that once the SDG's are implemented that they are monitored, and further studies carried out to verify their results.

Acknowledgements We would like to thank Sao Paulo State Foundation for Research Support (FAPESP, Brazil) for financial support for the research that led to this article (FAPESP n. 2019/08044-3) and for the scholarships to the fourth and fifth authors (FAPESP n. 2019/26490-0; 2020/03603-1), and CNPq for the Fellowship granted to the first author.

We would like also to thank Debra Barbarick, from Sierra Water Protectors, California, USA, for revising and editing the English version of the Chapter.

References

Allaerts W (2020) How could this happen? Acta Biotheor. https://doi.org/10.1007/s10441-020-09382-z

Arora NK, Mishra J (2020) COVID-19 and importance of environmental sustainability. Environ Sustain. https://doi.org/10.1007/s42398-020-00107-z

Arruza C, Bhattacharya T, Frazer N (2019) Feminismo para os 99%: um manifesto. Boitempo, São Paulo, Brasil

Bandyopadhyay S (2020) Coronavirus disease 2019 (COVID-19): we shall overcome. Clean Technol Environ Policy 22:545–546. https://doi.org/10.1007/s10098-020-01843-w

BPSF—Brazilian Forum and Public Security (Fórum Brasileiro e Segurança Pública) (2020) Violência doméstica durante a pandemia de Covid-19—ed. 2 29 de maio de 2020. Nota Técnica. Available at: https://forumseguranca.org.br/wp-content/uploads/2020/06/violencia-domestica-covid-19-ed02-v5.pdf. Accessed 05 jun 2020

Brasil. Federal Government—Ministry of Women, Family and Human Rights (Governo Federal—Ministério da Mulher, da Família e dos Direitos Humanos) (2020) Denúncias registradas pelo Ligue 180 aumentam nos quatro primeiros meses de 2020. Available at: https://www.gov.br/mdh/pt-br/assuntos/noticias/2020-2/maio/denuncias-registradas-pelo-ligue-180-aumentam-nos-quatro-primeiros-meses-de-2020. Accessed 24 jun 2020

Brasil. Instituto Brasileiro de Geografia e Estatística-IBGE (2010) Censo Demográfico, 2010. IBGE, Brasília. Available at: https://censo2010.ibge.gov.br. Accessed 05 jun 2020

Campbell AM (2020) An increasing risk of family violence during the Covid-19 pandemic: strengthening community collaborations to save lives. Forensic Sci Int Rep

Editorial—The gendered dimensions of COVID-19 (2020, Apr 11) Lancet 395(10231):1168. https://doi.org/10.1016/S0140-6736(20)30823-0

Espanha (2020a) Ministerio de Igualdad. Principales datos sobre violencia de género. Espanha, Madrid, Abril 2020. Available at: https://violenciagenero.igualdad.gob.es/violenciaEnCifras/boletines/boletinMensual/2020/docs/Principales_Datos_Abril_2020.pdf. Accessed: 05 jun 2020

Espanha (2020b) Portal Estatístico—Delegação do governo contra a violência de gênero. Espanha, Madrid. Available at: https://estadisticasviolenciagenero.igualdad.mpr.gob.es/. Accessed 06 jun 2020

Euronews (2020) Casos de violência doméstica aumentam 30% durante confinamento na França. Última atualização: 28/03/2020. França, Paris. Available at: https://www.euronews.com/2020/03/28/domestic-violence-cases-jump-30-during-lockdown-in-france. Accessed 04 jun 2020

Filetti S (2020) The COVID-19 pandemic requires a unified global response. Endocrine 68:1. https://doi.org/10.1007/s12020-020-02293-6

Gil AC (2019) Métodos e técnicas de pesquisa social. Atlas, São Paulo

Global Rapid Gender Analysis for COVID-19 (2020) Care International/International Rescue Committee, Suíça. https://www.care-international.org/files/files/Global_RGA_COVID_RDM_3_31_20_FINAL.pdf

Kofman YB, Garfin DR (2020, June 1) Home is not always a haven: the domestic violence crisis amid the COVID-19 pandemic. Psychol. Trauma Theory Res Pract Policy. Advance on line publication. https://psycnet.apa.org/fulltext/2020-37317-001.pdf

Leal Filho W, Brandli LL, Lange Salvia A, Rayman-Bacchus L, Platje J (2020) COVID-19 and the UN sustainable development goals: threat to solidarity or an opportunity? Sustainability 12(13):5343. https://doi.org/10.3390/su12135343

Oxfam Brasil/INESC/Center for Economic and Social Rights (2017) Brasil. Direitos humanos em tempos de austeridade. Available at: https://oxfam.org.br/publicacao/direitos-humanos-em-tempos-de-austeridade

Peteet JR (2020) COVID-19 anxiety. J Relig Health. https://doi.org/10.1007/s10943-020-01041-4

Peterman A, Potts A, O'Donnell M, Thompson K, Shah N, Oertelt-Prigione S, van Gelder N (2020) Pandemics and violence against women and children. CGD working paper 528. Center for Global

Development, Washington, DC. https://www.cgdev.org/publication/pandemics-and-violence-against-women-and-children

PGI-Patrícia Galvão Institute. Available at: https://agenciapatriciagalvao.org.br/quem-somos. Accessed 02, 5, 7, 18, 20, 23 and 24 junho 2020

Queiroz C (2020) No contexto da pandemia, agressões contra mulheres crescem no Brasil. São Paulo, Pesquisa FAPESP 293:54–59, julho

Salzberger B, Glück T, Ehrenstein B (2020) Successful containment of COVID-19: the WHO-report on the COVID-19 outbreak in China. Infection 48:151–153. https://doi.org/10.1007/s15010-020-01409-4

SanarMed (2020) Linha do tempo do Coronavírus no Brasil. Available at: https://www.sanarmed.com/linha-do-tempo-do-coronavirus-no-brasil. Accessed 1st June 2020

Saxena SK, Kumar S, Maurya VK, Sharma R, Dandu HR, Bhatt MLB (2020) Current insight into the novel coronavirus disease 2019 (COVID-19). In: Saxena S (eds) Coronavirus disease 2019 (COVID-19). Medical virology: from pathogenesis to disease control. Springer, Singapore

Seixas SRC, Hoefel JLM (2020) Human rights and gender equity: building sustainable. In: Leal Filho W, Azul A, Brandli L, Özuyar P, Wall T (eds) Gender equality. Encyclopedia of the UN sustainable development goals. Springer, Cham. Available at: https://doi.org/10.1007/978-3-319-70060-1

Sharma V, Scott J, Kelly J, VanRooyen MJ (2020) Prioritizing vulnerable populations and women on the frontlines: COVID-19 in humanitarian contexts. Int J Equity Health 19(66):1–3. https://doi.org/10.1186/s12939-020-01186-4

Sigal L, Miranda NAR, Martinez AI, Machicao M (2020) 'Another pandemic': in Latin America, domestic abuse rises amid lockdown, April 27. Reuters. Available at: https://www.reuters.com/article/us-health-coronavirus-latam-domesticviol/another-pandemic-in-latin-america-domestic-abuse-rises-amid-lockdown-idUSKCN2291JS

Srivastava N, Baxi P, Ratho RK, Saxena SK (2020) Global trends in epidemiology of coronavirus disease 2019 (COVID-19). In: Saxena S (ed.) Coronavirus disease 2019 (COVID-19). Medical virology: from pathogenesis to disease control. Springer, Singapore

Stang A, Standl F, Jöckel K (2020) Characteristics of COVID-19 pandemic and public health consequences. Herz. https://doi.org/10.1007/s00059-020-04932-0

UN-United Nations (2015) Transforming our world: the 2030 agenda for sustainable development, resolution adopted by the general assembly on 25 Sept 2015, New York, United Nations

UN Women (2020) Violência contra mulheres e meninas: a pandemia das sombras: Declaração de Phumzile Mlambo-Ngcuka, Diretora Executiva da ONU Mulheres, 6 de abril de 2020. EUA, Nova Iorque. Available at: https://www.unwomen.org/en/news/stories/2020/4/statement-ed-phumzile-violence-against-women-during-pandemic#notes. Accessed 04 jun 2020

Vieira PR, Garcia LP, Maciel ELN (2020) Isolamento social e o aumento da violência doméstica: o que isso nos revela? Rev Bras Epidemiol 23:e200033. Epub 22 Apr 2020 https://www.scielo.br/pdf/rbepid/v23/1980-5497-rbepid-23-e200033.pdf

Women's Aid—Until Women and Children are Safe (2020) The impact of Covid-19 on survivors: findings from Women's Aid's initial Survivor Survey, Abril 2020. Reino Unido, Bristol. Available at: https://www.womensaid.org.uk/research-and-publications/evidence-briefings-the-impact-of-covid-19-on-survivors-and-services/. Accessed 05 jun 2020

Women's Safety NSW (2020) Subúrbios mais ricos de Sydney relatam violência doméstica 'muito alarmante': BOCSAR. Publicado: 3 de junho de 2020. Nova Gales do Sul, Austrália. Available at: https://www.womenssafetynsw.org.au/impact/article/wealthier-sydney-suburbs-reporting-very-alarming-domestic-violence-increases-bocsar/. Accessed 04 jun 2020

Zhang Y (2020) The epidemiological characteristics of an outbreak of 2019 novel coronavirus diseases (COVID-19)—China, 2020. Chinese Centre for Disease Control and Prevention—CCDC Weekly 2(8):113–122

Sônia Regina da Cal Seixas has a Ph.D. in Social Sciences, from Campinas State University, UNICAMP, Brazil and Post-doctoral in University of Reading/UK. Nowadays she is researcher at the Centre for Environmental Studies and Research, NEPAM, State University of Campinas, UNICAMP. She is a professor of Ph.D. Program Environment & Society (NEPAM-IFCH-UNICAMP) and of Postgraduate Program Energy Systems Planning (FEM-UNICAM). She has a productivity research fellowship CNPq-Brazil's National Council for Scientific and Technological Development and Group Leader-Directory CNPq: Lab of Environmental changes studies, Quality of life and subjectivity.

João Luiz de Moraes Hoefel has a Ph.D. in Social Sciences at Campinas State University (IFCH/UNICAMP), São Paulo, Brazil, with a focus on environmental issues and developed a Post-Doctoral research at the Center for Environmental Education and Public Policies at São Paulo University (ESALQ/USP). He teaches at UNIFAAT – University Centre, Atibaia, São Paulo, where he also coordinates the Centre for Sustainability and Cultural Studies (NESC/CEPE/UNIFAAT) developing research projects on Natural Resources Conservation, Environmental Education and Environmental Planning. He also develops research activities at NEPAM/UNICAMP on Global Environmental Change and Quality of Life as a Collaborator Researcher.

Micheli Kowalczuk Machado has a Master's Degree and Ph.D. on Applied Ecology from ESALQ/USP, Piracicaba, São Paulo, Brazil. She teaches at UNIFAAT – University Centre, Atibaia, São Paulo, Brazil, where she also develops research projects on Natural Resources Conservation, Environmental Education and Environmental Planning.

Gianlucca Consoli has a graduation degree in Environmental Management at UNIFAAT – University Centre, Atibaia, São Paulo, Brazil, and a Technical Training Scholarship Level III (TTIII) from The São Paulo Research Foundation – FAPESP.

Waldo Emerson de Souza Nascimento is a Law undergraduate student at UNIFAAT – University Centre, Atibaia, São Paulo, Brazil, and has a Technical Training Scholarship Level I (TTI) from The São Paulo Research Foundation – FAPESP.

COVID-19 and Sustainable Tourism

Wendy M. Purcell, O'Shannon Burns, and Allison Voss

Abstract The dual health and economic crises of the COVD-19 pandemic have thrown the disruptive forces acting on the travel and tourism (T&T) sector into sharp relief, drawing attention to the interconnected and hyper-dependent nature of sustainability, health and business. Lockdowns and social distancing strategies effectively closed and could ultimately decimate the sector. With the pandemic affording people and planet some short-term relief from T&T's impact on communities, destinations and the environment, this is the time to re-imagine the sector. Looking ahead, a range of approaches to advance sustainable T&T could help it 'build back better', attenuating its negative impacts and advancing the contribution T&T makes to global citizenship and to a more balanced economy and equitable society. T&T business leaders are widening their view of sustainability beyond immediate operational impact to consider the broader systems in which they operate, adopting sustainability leadership practices for the twenty-first century and beyond. COVID-19 represents a 'teachable moment' for the T&T sector to accelerate sustainability, paying greater attention to the trade-offs and dilemmas presented by its activities. Indeed, T&T has enormous potential to educate the traveler and drive fulfilment of the Sustainable Development Goals. We propose that adopting the 'Culture of Health' framework could serve to fast track the sector's move to sustainable T&T, supported by conscious consumerism and greater governmental oversight.

Keywords Travel and tourism · Sustainability · Culture of Health · Sustainable tourism · Disruption · Transformational change

---

W. M. Purcell (✉)
Harvard T.H. Chan School of Public Health, Harvard University, 401 Park Drive, 4th Floor West, Suite 406, PO Box 15677, Boston, MA 02215, USA
e-mail: wpurcell@hsph.harvard.edu

O. Burns
Harvard Division of Continuing Education, Harvard University, Cambridge, USA

A. Voss
Harvard Kennedy School, Harvard University, Cambridge, USA
e-mail: allison_voss@student.hks.harvard.edu

W. Leal Filho (ed.), *COVID-19: Paving the Way for a More Sustainable World*,
World Sustainability Series, https://doi.org/10.1007/978-3-030-69284-1_9

1 The World's Fastest Growing Sector Comes to a Halt: Travel & Tourism Macro-Trends and COVID-19's Impact

> Travel is fatal to prejudice, bigotry, and narrow-mindedness, and many of our people need it sorely on these accounts. Broad, wholesome, charitable views of men and things cannot be acquired by vegetating in one little corner of the earth all one's lifetime. Mark Twain[1]

The United Nations World Tourism Organization (UNWTO), defines tourism as "*a social, cultural and economic phenomenon which entails the movement of people to countries or places outside their usual environment for personal or business/professional purposes*" (UNWTO 2010). Before the pandemic struck, travel and tourism (T&T) was one of the world's largest and fastest growing economic sectors, with last decade marking the first time that more than one billion people crossed an international border as a visitor in a single year (McCaul 2020). The T&T sector is comprised of a range of disparate and inter-connected industries, with small and medium enterprises (SMEs) representing 80% of the sector (Pololikashvili 2020), involving interactions among residents, visitors, administration of the destinations and the companies that integrate and operate in them (Azcarate et al. 2019). This complex value chain comprises six key segments that drive overall performance:

- **Aviation and Airports**: New and expanded airports and airline routes fueled growth in tourism, opening up new destinations. Low-cost airlines (e.g., Ryanair, easyJet) made travel cheaper and T&T more accessible (Johnson 2018), with 4.4 billion passengers estimated to have traveled by air in 2018 (IATA 2019). Greenhouse gas (GHG) emissions from aviation were estimated to triple by 2050, representing 25% of the global carbon budget (Graver et al. 2019).
- **Cruise Lines**: Some 28.5 million passengers traveled with cruise companies in 2018, and a record number of new berths were added to the global fleet in 2019 (Cruise Lines International Association 2019).
- **Destinations**: Destinations represent the critical environmental and cultural assets that attract travelers, with Destination Management Organizations (DMOs) working with national, state and/or local governments on marketing. Issues linked to improper waste management, depletion of water resources and 'overtourism'—the situation in which a place exceeds its carrying capacity (UNWTO 2018a, b)—have all come to the fore with destinations needing to master plan around place and consider local impacts of T&T (Epler-Wood et al. 2019).
- **Hotels and Accommodation**: Two primary business models make up the hotel sector, namely branded franchise hotels, and small independent hotels, the latter representing some 60–70% of the industry worldwide (Stringam and Partlow 2016). Airbnb, as a platform business, has served to disrupt the accommodation offer in the sector—now with 150 million users and 7 million listings (iProperty Management 2020).

[1] In *The Innocents Abroad* by Mark Twain (1869). London: Collins Clear-type Press.

- **Sales and Booking Platforms**: Technology has facilitated the development of online travel agencies and booking platforms that are transforming T&T, allowing airlines, accommodation providers, and small tourism businesses to sell direct to the consumer. Technology increases the complexity of the travel supply chain, making collective sustainability efforts more challenging.
- **Tour Operators**: Tour operators curate package and bespoke T&T experiences, bringing together accommodation, food and beverage, transportation and local transfers, excursions and activities, and guides.

Issues relating to COVID-19 are affecting the traveler experience across the T&T value chain. Deciding whether to leave home at all now includes a careful consideration of health risks, safety, affordability and other factors as they relate to the journey, stay and experiences at the destination. We have therefore focused our efforts on exploring COVID-19 and sustainable T&T as it pertains to transport, accommodation and destinations; these parts of the T&T ecosystem also present the most significant opportunities to advance sustainability in the sector. We undertook an extensive literature review, which was critically evaluated and emergent themes identified. We also drew on our extensive network of global T&T leaders to ensure the literature was situated in context.

Over the past decade and before the crisis, growth of the T&T sector was championed as a "*key driver for socio-economic progress*" (Euractiv 2019), because of its significant contribution to the development of local economies and its sizeable contribution to GDP in many countries. The World Travel and Tourism Council (WTTC), which represents private T&T companies, highlighted the sector's input to global GDP at 10.3% in 2019 (WTTC 2020). The sector contributed over US$8.9 trillion to the global economy in 2019, supporting one in 10 jobs (330 million) worldwide and one in five new jobs over the last five years, with 3.5% growth in 2019 compared to the global economy at 2.5% (WTTC 2020). The sector represented nearly 30% of total service exports and was core to the economy of many nations. In 2019, domestic tourism represented 71.3% of total tourism spend, business travel 21.4% and leisure travel 78.6%, with a high proportion of women in employment and a high dependency on natural and cultural resources (WTTC 2020). With 1.5 billion international tourist arrivals in 2019 and an additional one billion people forecast to join the global middle class by 2030, international travel was predicted to grow by some 35% (Lonely Planet 2016), with 1.8 billion international arrivals by 2030 and domestic tourism up to four times this figure (UNWTO 2016, 2017). The T&T sector has seen six decades of consistent growth (OECD 2020a, b, c), with tourism outpacing the United Nations (UN) growth projections over the period 2010–2019 and 45% of international travel arrivals to emerging economies[2] in 2017 (UNWTO 2018a). Late 2019 forecasts predicted that these trends would continue, with tourism arrivals forecast to grow 3–4% globally in 2020, despite a number of expected economic, political, and health disruptions (UNWTO 2019a, b). These international tourism statistics only capture a small part of the picture—a growing number of domestic tourists dominate

[2]Classification based on International Monetary Fund criteria.

travel destinations like China and the United States (US), for example, just 3.3% of person-trips[3] in the US are international arrivals (U.S. Travel Association 2020). With budget airlines (e.g. WizzAir, Spirit) lowering the costs of travel, disruptive technology creating platforms for accommodation (e.g. Airbnb), and travel searches making it easier to plan a trip (e.g., Google Travel), the T&T growth curve looked set to continue unabated before the pandemic.

Against these confident T&T global growth figures, there were however mounting concerns about the extractive nature of the sector and its detrimental impact on people and planet. T&T macro-trends hide the dependency of some countries and destinations on its income and the effects that a poorly designed and managed sector can have on place and local communities. With the infrastructure supporting T&T linked to other economic sectors, from food and farming to energy and public health, the sector can strain natural and societal resources. The risk of infectious diseases, water scarcity, environmental degradation, carbon emissions, waste, worker health, deforestation, deregulation, decentralization, privatization, and the fragmentation of power are just some of the global sustainability issues in the T&T sector (Sustainability Leaders Project 2019). In recent years, new terms have been coined to draw attention to these issues, including 'flygskam' or 'flight shaming' (Quick 2019) and 'overtourism' (Rafat 2018). While travelers explore all areas of the globe, the international arrivals of just 20 countries combined exceeds that of the rest of the world (McKinsey and WTTC 2017). Overtourism has severe consequences in some destinations, for example, Boracay island in the Philippines was closed to tourists in an effort to minimize harm (Alexander 2019), while citizens in Amsterdam and Venice protested to draw attention to their concerns (Henley 2020). The over-popularity of certain destinations, from Thailand's Koh Khai islands and Peru's Machu Picchu to various European and US cities amplified the negative environmental and social impacts of T&T. While destinations and T&T businesses responded by adopting traveler dispersion policies and moves to shift traveler behavior by targeting higher-value overnight travelers rather than day-trippers (McKinsey and WTTC 2017), a more concerted approach is needed.

The detrimental impacts of the T&T sector have prompted many stakeholders to warn of worsening conditions if appropriate steps are not taken. The WTTC's report 'Connecting Global Climate Action' (WTTC 2015) drew attention to existing research on the environmental footprint of T&T. It proposed the sector take a lead, stating the next 20 years "*...will be characterized by the sector fully integrating climate change and related issues into business strategy, supporting the global transition to a low carbon economy, and strengthening resilience at a local level against climate risks.*" (WTTC 2015). WTTC went on to make public commitments with its members towards carbon reduction and agreed a common agenda for climate action in T&T in accord with the United Nations Framework on Climate Change (WTTC 2015). The United Nations Environment Program 2050 forecast the impact

[3]Person-trip is defined as one person on a trip away from home overnight in paid accommodations or on a day or overnight trip to places 50 miles or more [one-way] away from home (Source: US Travel Association).

of tourism, highlighting the requirement for T&T to adopt sustainability practices, with energy consumption predicted to increase by 154%, GHG by 131%, water consumption by 152% and solid waste by 251% by 2050 (United Nations Environment Program 2050 (UNEP) 2017). Private T&T businesses and governments play an important role in addressing these challenges and driving sustainable T&T; however, their efforts have historically been limited in scope. While some are making concerted efforts to integrate sustainable practices into their operations, performance and reporting it typically remains confined to the corporate social responsibility agenda and is not mainstreamed into either business strategy or the traveler's experience. Similarly, national sustainability or low carbon strategies are lacking, with just 11% of national T&T objectives related to sustainability (UNWTO 2019a).

There are many perceived barriers to the pursuit of strategic sustainability in business, such as low consumer demand, policies failing to drive enough market incentives, short-termism of financial markets, among others. However, it is essential that the T&T sector adopts sustainability across all its domains. Given Brundtland's definition of sustainability states, "*... development that meets the needs of the present without compromising the ability of future generations to meet their own needs.*" (World Commission on Environment and Development 1987), T&T needs to adopt sustainability wholeheartedly across:

- the environment: making optimal use of resources, maintaining essential ecological processes and helping to conserve natural heritage and biodiversity;
- the socio-cultural axes: respecting the values and customs of host communities, conserving their built and living heritage and traditional values, and contributing to inter-cultural understanding and tolerance; and
- the economy: ensuring viable, long-term economic operations, providing benefits equitably to all stakeholders, including employment and contributing to poverty alleviation (UNEP & World Tourism Organization 2005).

Historically, some positive steps to establish sustainability governance in the sector have been taken. In 2008, the United Nations Foundation, UNWTO, UNEP, and the Rainforest Alliance developed the Global Sustainable Tourism Criteria (GSTC) to establish the "*minimum requirements that any tourism business or public destination management authorities should aspire to reach to protect and sustain the world's natural and cultural resources while ensuring tourism meets its potential as a tool for poverty alleviation*" (GSTC 2020). However, there is still no universally accepted sustainability standard or certification for T&T, perhaps reflecting the complexity of the sector, and the adoption of sustainability has been slow with only a few companies and destinations embracing it strategically and holistically. Conscious consumers have pushed the industry forward on some issues, such as single-use plastics (UNEP 2020), but sectoral motivations have largely been reactive. Although the importance of sustainable T&T and destination management has come into sharp focus, the sector as a whole still appears to lack meaningful commitment to sustainable tourism development across its various domains (Mullis 2017).

2 React and/or Adapt: Travel & Tourism's Responses to the Pandemic and Re-Opening Efforts

> Sustainability is essential for tourism to be compatible with the fragile environment… while providing benefits and opportunities to communities, which gives a key role to innovation to harmonize these objectives. Manuel Butler, UNWTO Executive Director[4]

In January 2020, the first cases of COVID-19 were reported; by mid-March, as the virus spread, borders began closing across the world. Governments implemented quarantine measures—with 91% of the world's population living in countries that limited or forbid entry to noncitizens or non-residents (Connor 2020). International T&T effectively stopped as COVID-19 cases rose worldwide. By May 2020, a UNWTO report noted that every destination in the world had implemented travel restrictions, and 75% of those restrictions completely halted international tourism UNWTO (2020a). By July 2020, many governments continue to require quarantine, with international tourism arrivals forecast to decrease 60–80% in 2020 (UNWTO 2020a), making T&T one of the sectors most impacted by COVID-19. In the US alone, travel spending in 2020 will likely drop by US$300 billion, translating into a loss of nearly a trillion dollars in economic output (Oxford Economics 2020). COVID-19 highlighted the impact of the T&T sector on communities, in both positive and negative ways. While grounded flights resulted in improved air quality and locals were again able to enjoy their neighborhoods free of tour buses and day-trippers, this also meant no income for millions around the world who rely on T&T for their livelihoods. Lockdown measures, designed to preserve health and health systems, led to some temporary environmental benefits, including global carbon dioxide emission levels down by 8% in March and April (Le Quéré et al. 2020).

The T&T sector responded to the pandemic in various ways, with government-led stimulus packages and worker furlough support, and individuals taking personal action to protect their health. Global hotels were at 29% occupancy compared to 72% in 2019 (Dalrymple et al. 2020), and some chains leaned into help. For example, the 'Hospitality for Hope' initiative saw more than 15,000 hotels sign up for the American Hotel and Lodging Association campaign to match hotels with government agencies in need, offering temporary housing for emergency and health care workers amid the pandemic (Simon 2020). Hilton and Marriott donated 1 M rooms to front-line workers (Hilton 2020; Clabaugh 2020), Melia hotels transformed one of its properties into a hospital for mild COVID patients (Majorca Daily Bulletin 2020), and Indian Hotels Company provided millions of meals to healthcare and migrant workers (IHCL 2020). Many airports closed entirely, others shut one or more terminals and airlines suspended operations or cancelled a significant proportion of flights, with seat miles for US airlines down by 71% in April 2020 (Curley et al. 2020; Dalrymple et al. 2020). To adapt to the COVID-19 crisis, airlines have shut down and/or altered routes and frequency, with the number of seats offered by airlines in 2020 expected

---

[4]https://www.unwto.org/snow-and-mountain-tourism-industry-professionals-to-discuss-links-between-tourism-sustainability-and-innovation, February 2020.

to be 42–52% less than originally planned (ICAO 2020). Most airlines laid off staff, with Air Canada letting go of 50% of its staff (CBC 2020), Lufthansa shedding some 22,000 employees (Reuters 2020), and United Airlines announcing plans to lay off 36,000 employees by October 2020 (Josephs 2020). Meeting future needs for more flexibility, for example in airline ticketing, could cost 4% of 2019 revenues at US$40 billion (Dalrymple et al. 2020). The resultant decline in world passenger traffic is forecast to be seven to 17 times larger than during the global financial crisis of 2008 (Boin et al. 2020).

Many governments, recognizing the economic impact of the T&T sector, were quick to announce bailout and stimulus packages, for example, Hong Kong bailed out Cathay Pacific at US$5 billion (BBC 2020), and the US passenger airlines called for US$50 billion to survive the crisis (Financial Times 2020). Typically bailouts did not come with any 'green recovery' requirements, except in a few countries, for example, France tied their bailout package to emission reduction targets for Air France requiring the airline to cut back on 40% of its domestic flights (Dunn 2020). However, early analysis already suggests that these 'green' measures might not have the lasting impact on emissions that is needed to reduce climate change (Keating 2020). Governments also implemented a myriad of reopening strategies to attract tourists to return, from superior hygiene to stipends to encourage travel. For example, the SG Clean plan was launched by Singapore's National Environment Agency in February 2020 (National Environment Agency 2020) to send a strong signal that tourism businesses should take cleanliness and hygiene very seriously for locals and visitors alike. Elsewhere, the use of data was identified as a core strategy, for example, Iceland did not initiate a lockdown, opting instead for extreme tracing and quarantine measures using biotech firm deCODE for contact tracing; the official tracing app was in use by 40% of the population in Iceland by June 2020 (Kolbert 2020). Others adopted strict quarantine measures, for example, Hawaii's 14-day quarantine came with severe penalties including jail for transgressors (Sampson 2020). Some offered to take care of travelers who got sick, for example the Cyprus government committed to look after travelers who tested positive during their stay, as well as their families and close contacts (Hadjicostis 2020). Other nations are extending payments to entice travelers, for example, Sicily announced in April 2020 that it would pay a portion of travelers' trip costs (Wilson 2020), and Switzerland offered a 200 euro voucher for citizens to spend while exploring the country (Broom 2020).

Recovery of the T&T sector from the pandemic's impact will vary based on governmental, company and individual responses to both the health and economic crises, the perceptions of risk and behavioral changes due to business and leisure practices adopted in the lockdown. Some regions are implementing 'travel bubbles' between places with low prevalence of COVID-19, where people may travel across borders without quarantine (Mzezewa 2020). The European Union was the first to implement such an approach (Palmer 2020); the US, which has struggled to accurately track and contain the pandemic, was notably excluded from the bubble. A trans-Tasman bubble between Australia and New Zealand has been debated (Derwin 2020; Locker 2020), and countries in Asia and South America are exploring similar policies

(Andina 2020; Pinandita 2020). Some forecasts predict that group and volume travel will be the last to rebound after domestic and regional tourism (E-Tourism Frontiers 2020). McKinsey & Company (Krishnan et al. 2020) predicts normal hospitality operations and demand levels will not return to pre-COVID-19 levels until 2023 at the earliest. Additional waves of the virus could halt or curtail reopening strategies, bringing additional disruptions to the sector. As the sector grapples with adopting stringent health and safety protocols, the future may see borders open and close as the number and geography of COVID-19 cases fluctuate. For example, a June 12, 2020 report (Enger et al. 2020) signaled China's travel recovery, but just three days later, a massive spike in COVID-19 cases in Beijing effectively halted traveler plans. These fluctuations, as well as enhanced contact tracing and social distancing protocols, will demand greater flexibility and rapid response within the T&T sector and may demand greater regulatory oversight. Overall, these recovery strategies largely focus on traveler health and safety rather than efforts to reimagine a more sustainable future for T&T (Osbourne 2020).

The uncertainty and economic consequences brought on by the pandemic have put at risk 100–120 million T&T jobs around the world (UNWTO 2020a). This loss is particularly devastating in destinations where the sector contributes significantly to GDP. A recent UN report highlighted the exposure of Small Island Developing States given their deep reliance on tourism, noting that impacts would fall disproportionately on women and informal workers (Coke-Hamilton 2020). For example, T&T represents over 90% of total GDP in Macau, 74% in Aruba and over 50% in the Maldives and the British and US Virgin Islands (United Nations Conference on Trade and Development 2020). The T&T sector contributes 8.5% of Africa's total GDP and the pandemic will likely result in the loss of 2 million direct and indirect tourism jobs (Ighobor 2020). In Asia, the UN estimates that 15.3 million workers' jobs and livelihoods are at risk, and that 75% of these workers hold informal jobs that are especially vulnerable (International Labour Organization 2020). Thus, as the T&T sector plays a critical role for millions around the world, governments, private sector, local communities, and travelers have a critical role to play in shaping the future of T&T (Higgins-Desbiolles 2019). Re-framing sustainable T&T will therefore require a delicate balance of people, planet and prosperity.

3 Building Back Better: Reimagining Sustainable Travel and Tourism

> From a sustainability point of view, all political and economic systems have failed. But humanity has not yet failed. Greta Thunberg[5]

[5]Thunberg, Greta. "No 'green deal' will be ambitious enough to save the planet." *Time* 20 Jul. 2020: 67. Print.

A recent OECD report on COVID-19 recovery (OECD 2020a) warned against policies and programs that return the world to 'business as usual' and instead recommends the recovery, whatever shape or duration, be leveraged as a mechanism to 'build back better'. If ever there was a crisis that taught the world about health and the fragility of our economic model, it was COVID-19 as we collectively repressed the economy to save lives. The impact of the pandemic demands we consider new ways to sustain business and health. We propose that adopting the 'Culture of Health'[6] framework (Quelch and Boudreau 2016), which places well-being as a strategic business priority, could accelerate sustainable T&T. Quelch's argument that "*Every company, knowingly or unknowingly, impacts public health…*" (Quelch and Boudreau 2016) is ostensibly an agenda for sustainability and connects health and business across fours domains:

- Environment: how businesses impacts the environment;
- Community: how much business invests in the health of its communities;
- Employees, including supply chain: how business treats its employees and supply chain workers, and
- Consumers: how businesses deliver products and services to consumers.

Environmental, Social and Governance (ESG) frameworks focus primarily on metrics related to the environment, with any consideration of health typically limited to employee health; by adding Consumer and Community as equally important pillars (Quelch and Boudreau 2016) the framework widens the lens of how business might envision more sustainable practices. In a world post the COVID-19 pandemic, those businesses that take the health of people and the natural world as core strategic agendas are poised to pivot more powerfully and emerge not just more environmentally and socially sustainable, but also more economically viable (Paun 2020).

By focusing on sustainability as well-being, inclusivity, emissions reductions, slowing biodiversity loss, creating circular supply chains, and increasing resiliency in addition to recovering economies and livelihoods (OECD 2020a), T&T has the opportunity to re-imagine itself. Positive trends that support a more responsible future include localism, with travelers exploring closer to home. Destinations too are adopting social distancing measures, with fewer people allowed in any one place, for example, Machu Pichu stated it will reduce numbers by 75% when it reopens in July 2020 (Merco Press 2020). Interestingly, virtual tourism is predicted to increase with those in lockdown enjoying online experiences (Chen 2020), but whether this then translates to seeking a real-life travel experience is as yet unknown. Overall, governments and destinations may end up embracing sustainability by default through a reduced volume, higher value strategy to manage T&T as they reopen. On the other hand, the negative side effects of exclusivity, inequity and higher carbon footprint

---

[6]Culture of Health https://www.hsph.harvard.edu/news/hsph-in-the-news/companies-culture-of-health/ is a study supported by the Robert Wood Johnson Foundation under the grant No. 74275 'Building a Culture of Health: A Business Leadership Imperative' and is a joint initiative between the Harvard Chan School of Public Health and the Harvard Business School.

may take root in the sector. The likelihood of people avoiding shared mobility solutions, such as buses and ride shares, turning instead to their own cars may worsen air quality. T&T may again become the reserve of the only those with sufficient disposable income to afford socially distanced T&T and who can cover the associated costs of increased health risks. While private jet flights decreased by some 70% in April 2020, with places such as the Seychelles and the Maldives open only to such flights (Skirka 2020; Thani 2020), levels had recovered to near pre-COVID levels by July 2020 (Powley and Bushey 2020). Tradeoffs within T&T need to accommodate the impact on people, planet and prosperity and balance the desire to travel with that of health and safety, seeking to secure economic interests against preserving the quality of life in a destination and globally. Figure 1 captures some predicted near-term (1–3 years) outcomes of COVID-19 across the T&T sector mapped against the four pillars of the 'Culture of Health' framework (Quelch and Boudreau, 2016).

Within **the environment pillar**, pricing should reflect the true impact of T&T recognizing the positive contribution the sector makes to conservation efforts against the negative impacts, including carbon emissions, freshwater supply, waste management and overtourism on communities, cultural and heritage artefacts and ways of living (UNWTO 2018a, b). Many biodiversity and conservation efforts have been realized because of T&T activities and revenue. Although ecotourism emerged in the 1990s as a niche alternative to mainstream T&T, today it is seen as a key driver of biodiversity and ecosystem health, contributing some US$120 billion to global GDP in 2018 (WTTC 2019). Some travel companies have expanded their purview beyond

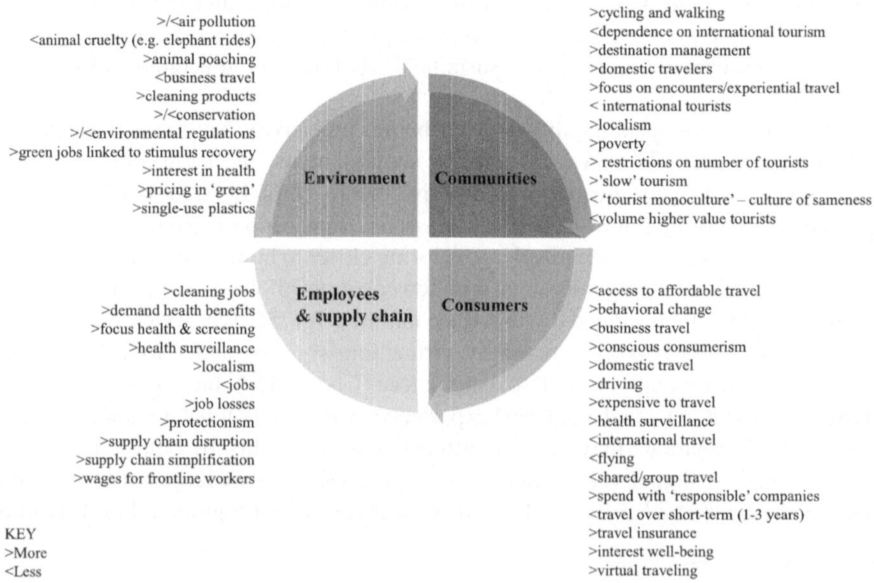

Fig. 1 Forecasting the impact (1–3 years) of COVID-19 on the Travel & Tourism sector using the 'Culture of Health' framework

simply packaging tourism experiences and are actively managing the ecosystems upon which their businesses rely. For example, African safari company Wilderness Safaris has helped to conserve 2.3 million hectares of wildlife areas across six African countries, committing to its preservation despite the cessation of T&T due to COVID-19 (Wilderness Safaris 2017; Austin et al. 2020). As countries develop strategies to recover the T&T sector, those companies and regions critical to maintaining biodiversity hotspots should be prioritized. However, we may see mitigating strategies aimed at dealing with COVID-19, from increased use of personal cars for local road trips, to more widespread usage of single-use plastics and toxic chemical cleaning products, leading to greater environmental damage.

Prior to the pandemic, 'flight shaming' brought significant attention to aviation and the T&T sector's contributions to GHG emissions (Quick 2019). Simultaneously, the UN's International Civil Aviation Organization—ICAO (ICAO 2020) made significant progress toward establishing global emissions reduction targets for commercial aviation, identifying technologies and operational efficiencies, launching the CORSIA carbon offsetting and reduction scheme, and engaging states to report and reduce emissions. Upholding these frameworks and commitments is critical as the world tackles climate change (Bremner 2020; Carrington 2020). Efforts to de-carbonize T&T must be accelerated, with government bailout funds linked to green economy requirements including hotel retrofits, circularity, renewable energy projects and clean transport. In the last recession, environmental regulations were relaxed to get the economy back on track; given the economic impact of COVID-19 is predicted to be harsher than the 2008 crash (The World Bank 2020), this is a threat. Both China and the US have rolled back environmental regulations and, coupled with the lobbying effects of the aviation and automotive industries, have retreated from CO_2 emission targets (Carrington 2020). In contrast, the European Commission plans to recover the EU economy by funding sectors that will also help the region tackle climate change (Carrington 2020). The T&T industry should leverage recovery policy and financing mechanisms and instruments to accelerate sustainability across the sector. Early signs are promising, for example, the UNWTO partnered with the International Finance Corporation to promote green finance and infrastructure improvements within the T&T sector (UNWTO 2020b). These are good first steps for a broader shift to sustainable T&T, but much more is needed if global carbon targets are to be met. Overall, there is a significant need for multinational and global coordination, with frameworks and policies aimed at creating a more holistic sustainability approach for the T&T sector. Further research is needed to develop a 'true' cost for T&T, for example using impact weighted accounting principles (Serafeim et al. 2020) or total impact measurements (PWC 2020) that seek to extend ESG reporting and, we propose here, adding in health as a key metric.

Within **the community pillar**, the T&T sector could see changes that seek to address those concerns relating to overtourism, prioritizing the needs of local residents and the importance of preserving local history, culture and traditions, with fewer tourists and/or better management of visitor flow. Abrupt lockdowns and slow reopening of borders have highlighted how important it is for countries to diversify their economies beyond a reliance on T&T revenue, in particular those destinations heavily reliant on international tourism. Because of increased travel times, due to stricter health and safety compliance, social distancing measures, and border restrictions, international travel will be slower to rebound. As such, governments and T&T operators will look to develop and spur demand among domestic travelers. For example, Costa Rica's President Carlos Alvarado announced in July an increased number of long weekends through 2024 in an attempt to spur domestic travel (The Tico Times 2020). A return to localism, allowing citizens to explore their own countries and its different communities, can serve to reduce not only the carbon footprint of air travel, but also fight against what has been called a 'tourist monoculture', one in which all tourist experiences become the same globally (Perelli 2007). Moreover, as some tourists may have an increased ability to work from home in the future, there could be a rise in 'slow' tourism,[7] one in which the average trip length increases, with travelers looking to spend more time getting to know the community of their destination and seeking out more local experiences (Callot 2013). Leaning into this potential trend, Barbados' Prime Minister Mia Amor Mottley announced in July 2020 a new visa that allows visitors to stay up to 12-months and work remotely (Holcombe 2020). Overall, communities living in tourist destinations will look to restrict tourist numbers, encourage domestic and regional travel and create experiences that honor and respect local culture.

Within **the employees and supply chain pillar**, COVID-19 has highlighted the nature of T&T jobs and the lack of protection and benefits for some who work in the sector. The pandemic has also brought to the forefront the importance of health benefits and protection for workers in T&T who put their lives at risk to be in contact with domestic and international travelers. Job losses will likely continue as the sector shifts and makes changes to reopen; however, there may be new job opportunities in cleaning, and health and wellness, as tourists prioritize cleanliness as well as trips that offer peace and tranquility after this time of unexpected change. The next months and years will be challenging for workers in T&T as roles pivot and significant cuts are made to hiring. The pandemic has had an outsized effect on SMEs that support tourism destinations (OECD 2020b), with supply chain engagement and investment now key to rebuilding T&T. This could help reduce environmental impacts, with increased attention to local sourcing and shorter more resilient supply chains that empower vulnerable groups.

Within **the consumer pillar**, tourists themselves make decisions about where to travel, when, why and how—all of which will change depending on their risk tolerance, affordability, and particular needs and desires. With many destinations, like Machu Pichu taking this moment to restrict visitor numbers, the cost of travel could rise as destinations lean into a lower volume, higher price model. A shift in this direction could mean that travelers may take fewer trips and/or travel less

---

[7] Slow tourism "*is about slowing down the rate of tourism and a guarantee of rediscovering oneself (the physiological and the psychological); it is about low greenhouse gas emissions and it is a synonym of patience, peace of mind, deeper experiences, improved cultural understanding and knowledge*" (Callot 2013).

internationally, focusing more locally and within their region. Given the attention paid to the 'relief' destinations experienced during COVID-19, travelers may look to spend their T&T dollars with 'responsible' companies and embrace more conscious consumerism. While virtual travel can provide one outlet for responsible 'travel', the enduring desire for T&T—to connect with people and places—is unlikely to diminish. A McKinsey survey conducted shortly after China reopened showed growth in domestic tourism numbers and a desire to travel in the next few months (Enger et al. 2020). Overall, tourists as consumers can play a much larger role in making T&T more sustainable, being more conscious of their 'footprint' and examining the sustainability credentials of T&T companies.

A key player that impacts across **all four pillars** of the 'Culture of Health' are DMOs. To date, tourism has largely been managed by DMOs that focus on increasing tourist volumes, tourism employment, and the amount of money spent in the destination. Going forward, DMOs should broaden their purview to a more holistic focus on destination management that prioritizes and advances sustainable T&T. Master planning must be undertaken to guard against overtourism and ensure the invisible burden of the sector—the additional cost of protecting and managing destination assets—is priced in and/or ameliorated (Epler-Wood et al. 2019). Additionally, COVID-19 has highlighted the importance of economic resilience and the need to address tourism dependency such that local industries, including important craft and artisanal business, are not overtaken by T&T as the sole source of income over the long-term. DMOs will need to engage multi-actor stakeholder groups to support strategic planning across the different industries and businesses that together create a traveler experience. Technology offers a significant advantage to better aggregate, monitor, and report data that is essential to destination planning, for example, geographic information systems can be used to help local communities, and governments create dynamic sustainable tourism plans (Epler-Wood et al. 2019). Moreover, recent initiatives to use 5G to track and trace health needs could also be utilized to manage tourist flows more effectively if appropriate privacy is safeguarded; this may become even more important as domestic T&T increases as predicted. Resultant destination stewardship plans should seek to capture the long-term environmental, social and economic viability of place and can be a powerful means to progress more sustainable T&T. For example, Visit California was the first state and largest DMO to develop a destination stewardship plan (Farr 2019). In addition to examining how tourism could effectively contribute to California's economy and preserve natural and cultural assets, Visit California seeks to enhance the visitor experience and the quality of life for local residents (Farr 2019).

Finally, **sustainability leadership** plays a critical role in shaping, carrying out, and communicating sustainable T&T changes. Reframing sustainability as a goal for today was a focus for T&T leaders in 2019, who with WTTC focused on Sustainability Leadership (WTTC & Harvard University 2019). Emergent company-wide sustainability initiatives illustrate the approaches and frameworks that could be adopted more widely across the sector. For example, the 'Make Travel Matter' agenda, created by the TreadRight Foundation of The Travel Corporation (Treadright 2020), represents an integrated sustainability strategy across 40 different brands in the group focused on people, planet and wildlife. Another example is Intrepid Group, which is the largest travel company to achieve B Corp certification (Intrepid 2020); this required the assessment of 23 associated companies across the globe and independent analysis of the company's sustainability practices. The UNWTO has called on the sector to balance business survival through the COVID-19 pandemic with the longer-term goals of protecting ecosystems, and to making a fuller contribution to climate change (UNWTO 2020a, b, c). This transition to a more sustainable T&T sector will depend heavily upon strategic partnerships along the value chain and must prioritize inclusive participatory approaches to implement recovery plans.

4 Conclusions

The global health crisis of the COVID-19 pandemic illustrates how human health and the health of our planet are connected; planetary health, which refers to the human health impacts of human-caused disruptions to the Earth's natural systems, must be at the forefront of our agenda for sustainable T&T. Safeguarding fragile communities and ecosystems and creating a healthier future and economy for all parts of society to thrive must be a shared purpose within the sector and its global stakeholder community. While a welcome respite, savoring clean air over Beijing, clean water in Venice or traffic free roads in London, imposing human inactivity is not an answer to progressing a more sustainable T&T sector. We need to clean up that activity and 'degrow' T&T (Higgins-Desbiolles et al. 2019) as part of economies based on well-being over solely profit-driven goals. The collective global urgency to tackle COVID-19 might be contrasted unfavorably with that needed to address climate change, but the pandemic shows us that we have an inherent leadership and change capacity once we gather humanity around the problem and work collectively with a global mindset.

What might it take for T&T to rebound in a more sustainable manner? As the pandemic served to all but close the sector (de Bellaigue 2020), we have been afforded a unique opportunity to advance sustainability with deliberate re-opening strategies that seek to attenuate the negative and extractive effects of T&T activities and advance its positive impacts. The desire for T&T is not going away, and we are already seeing demand begin to recover (Boin et al. 2020), but the sector may look very different in the coming decades as health of customers, employees and those in the supply chain, community and the environment come to the fore. With business travelers learning the opportunities afforded by digital connectivity for relationship building and trade, as well as the benefits to mental and physical well-being of reduced international travel, leisure travelers look to hyper-local experiences to break the monotony of lockdown life, setting the stage for a lower-carbon footprint, more sustainable sector. In the run up to COVID-19, there were already mounting pressures on the sector, with place-based interventions from closing destinations to allow for environmental recovery to the so-called 'Greta effect'—overall, highlighting the fact that the pre-COVID-19 'normal' for T&T was not sustainable.

Can the climate emergency and the health crisis of the pandemic collide to drive more sustainable T&T? The need to re-open the sector while reducing carbon emissions is a difficult dichotomy as we seek to balance environmental, social and economic domains. There is an opportunity now to reset and refocus, with T&T prioritizing the United Nations 2030 Agenda (United Nations General Assembly (UNGA) 2015) both to assure long-term resilience of the sector as well as to make its fullest contribution to sustainable development. Leadership at all levels, but especially among CEOs and global T&T organizations, can help guide the attention of the sector's actions toward the next normal (Levy et al. 2020). Given the challenges of the T&T sector, 'building back better' is essential but may not be enough to save our planet, our communities, and ourselves. Therefore, we must reimagine and reinvent new ways for sustainable T&T, a future where travelers stay closer to home, pay for the true cost of their environmental and social footprint and act with a greater awareness of the people and planet around them.

Embracing a '**Culture of Health**' (Quelch and Boudreau 2016), is proposed as a way for the T&T sector to make important strides in sustainability as well as position itself for a world after the COVID-19 pandemic. A move away from crowded places, to explore the great outdoors in parks and trails, offers a way for the positive impact of nature on well-being to be championed, which serves to reinforce the need to protect our planet. The T&T sector can be a force for good, a means to promote deep cross-cultural understanding, global citizenship and personal development—revealing our shared humanity as well as the beauty and fragility of our world. With nature and culture T&T's most valuable assets, careful stewardship involving all stakeholders is needed, working in concert with appropriate governmental oversight and conscious personal choices and behaviors. The T&T sector itself recognizes both its important contribution to the global economy and the need to accelerate its sustainability plans in response to growing concerns among conscious consumers, locals subject to the ravages of unplanned destination growth, and macro-interests related to climate change and equity. T&T enterprises must now embrace rigorous sustainability frameworks used in other sectors, integrating triple-bottom line accounting and sustainability management systems that ensure companies track and report results transparently. Sustainability should define the foundation for the next normal in T&T post the COVID-19 crisis, as travelers are more aware of the impact their choices and behaviors have on our world. Purpose-driven T&T was already emerging as travelers intentionally seek to grow and engage with the world in new ways, with rises in immersive, experiential and community-based experiences for education and learning (Daniel 2018; Warren 2019).

COVID-19 is a moment of crisis for T&T, re-shaping the sector in real-time and placing unprecedented demands on all stakeholders. Accelerating pre-COVID trends in the sector, in particular a focus on sustainability and well-being, the pandemic represents an opportunity to make T&T better. The sector went into the crisis with spectacular growth figures and strong forecasts. However, this was accompanied by growing public unease about its impact on destinations, the over-reliance of some countries on T&T receipts and major concerns about its environmental impact.

Seeking to navigate these volatile, uncertain, complex and ambiguous times—so-called VUCA conditions—sustainability can be positioned as a major driver of T&T's future. As one of the largest sectors in the world, embracing the 'Culture of Health' framework as a lens to view the various domains of T&T could accelerate sustainable T&T and help the sector reimagine itself in line with the SDGs. The T&T sector is not a monolithic sector, nor are the nations involved uniformly eager to add more regulations to international bodies or public health organizations. However, sustainable T&T demands more coordination, oversight, surveillance, and planning than ever before. This is a time for a new social contract to be forged with the T&T sector, with sustainability a driver of radical transformation. By acting to protect people and planet over the long-term, the T&T sector can create and sustain shared value, making its fullest contribution to fulfilling the SDGs and a world where "*no-one will be left behind*" (UNGA 2015).

References

Alexander B (2019) Closure of Boracay Island. Conservation Guide. January 11, 2019. Retrieved from https://www.conservationguide.org/news/closure-of-boracay-island

Andina (2020) Peru: Cusco proposes cross-border travel to resume tourism in Machu Picchu. Andina Agencia Peruana de Noticias. Retrieved from https://andina.pe/ingles/noticia-peru-cusco-proposes-crossborder-travel-to-resume-tourism-in-machu-picchu-801390.aspx

Austin JE, Epler-Wood M, Leonard HB (2020) Wilderness Safaris: responses to the COVID-19 crisis. HBS Case Study, July 17 2020. Retrieved from https://hbsp.harvard.edu/cases/

Azcarate T, Benaya J, Nerilli G, Juster A (2019) A guide to sustainable tourism. RED. Madrid. ISBN: 978-84-09-19768-2

BBC (2020) Coronavirus: Cathay Pacific gets $5bn state-backed bailout. Retrieved from https://www.bbc.com/news/business-52960684

de Bellaigue C (2020) The end of tourism? Guardian, June 18, 2020. Retrieved from https://www.theguardian.com/travel/2020/jun/18/end-of-tourism-coronavirus-pandemic-travel-industry

Boin R, Cosmas A, Dichter A, Wittkamp N (2020) A new approach in tracking travel demand. McKinsey & Company Travel, Transport & Logistics, April 2020. Retrieved from https://www.mckinsey.com/industries/travel-logistics-and-transport-infrastructure/our-insights/a-new-approach-in-tracking-travel-demand#:~:text=The%20traditional%20booking%20funnel%2C%20in,much%20shorter%20timelines%20than%20before

Broom D (2020). hese countries are looking to subsidize holidays after COVID-19. World Economic Forum, June 4, 2020. Retrieved from https://www.weforum.org/agenda/2020/06/coronavirus-tourism-government-subsidies-sicily-japan/

Bremner C (2020) Travel 2040: climate emergency to force a revolution in the industry. Euromonitor Int. June 2020. Retrieved from https://go.euromonitor.com/white-paper-travel-2020-travel2040.html

Callot P (2013) Slow Tourism. In: Idowu SO, Capaldi N, Zu L, Gupta AD (eds) Encyclopedia of corporate social responsibility. Springer, Berlin, Heidelberg

Carrington D (2020) EU green recovery package sets a marker for the world. The Guardian. May 28, 2020. Retrieved from https://www.theguardian.com/environment/2020/may/28/eu-green-recovery-package-sets-a-marker-for-the-world

CBC (2020) Air Canada to lay off 20,000 workers as pandemic collapses travel industry. Retrieved from https://www.cbc.ca/news/business/air-canada-layoffs-1.5572596

Chen A (2020) Is virtual travel here to stay, even after the pandemic subsides? April 20, 2020. Retrieved from https://www.nationalgeographic.com/travel/2020/04/can-virtual-reality-replace-real-tourism-during-pandemic-and-beyond/

Clabaugh J (2020) Marriott to join Hilton in offering rooms to front-line health care workers during coronavirus pandemic. WTOP News. Retrieved from https://wtop.com/coronavirus/2020/04/marriott-joins-hilton-in-offering-rooms-to-front-line-healthcare-workers/

Coke-Hamilton P (2020) Impact of COVID-19 on tourism in small island developing states. United Nations Conference on Trade and Development. Retrieved from https://unctad.org/en/pages/newsdetails.aspx?OriginalVersionID=2341

Cruise Lines International Association Inc. (2019) 2019 Cruise trends & industry outlook. Cruise Lines International Association. Retrieved from https://cruising.org/-/media/research-updates/research/clia-2019-state-of-the-industry-presentation-(1).pdf

Connor P (2020) More than nine-in-ten people worldwide live in countries with travel restrictions amid COVID-19. Pew Research Center (blog). April 1, 2020. Retrieved from https://www.pewresearch.org/fact-tank/2020/04/01/more-than-nine-in-ten-people-worldwide-live-in-countries-with-travel-restrictions-amid-covid-19/

Curley A, Krishnan V, Riedel R, Saxon S (2020) Coronavirus: Airlines brace for severe turbulence. McKinsey & Company Travel, Transport & Logistics, April 2020. Retrieved from https://www.mckinsey.com/industries/travel-logistics-and-transport-infrastructure/our-insights/coronavirus-airlines-brace-for-severe-turbulence

Dalrymple M, Mann R, Peters M, Seltzman N (2020) Make it better, not just safer: the opportunity to reinvent travel. McKinsey & Company Travel, Logistics & Transport Infrastructure Practice. June 2020. Retrieved from https://www.mckinsey.com/industries/travel-logistics-and-transport-infrastructure/our-insights/make-it-better-not-just-safer-the-opportunity-to-reinvent-travel

Daniel D (2018) Transformative travel' is the industry's latest twist on making vacations more meaningful. Washington Post, July 12, 2018. Retrieved from https://www.washingtonpost.com/lifestyle/travel/why-just-go-on-vacation-when-you-can-have-a-life-changing-experience/2018/07/11/a2886b6c-7eee-11e8-bb6b-c1cb691f1402_story.html

Derwin J (2020) Jacinda Ardern says New Zealand 'can't risk' a travel bubble with Australia just yet, as Victoria's COVID-19 outbreak continues. Business Insider. July 6, 2020. Retrieved from https://www.businessinsider.com.au/australia-new-zealand-travel-bubble-september-victoria-covid19-2020-7

Dunn K (2020) Climate conditions on airline government bailouts are rare—and the coronavirus likely won't be an exception. Fortune, June 26, 2020. Retrieved from https://fortune.com/2020/06/26/airline-bailouts-climate-conditions-coronavirus/

Enger W, Saxon S, Suo P, Yu J (2020) China's travel recovery gains steam: How families are planning their summer vacations. McKinsey & Company, June 2020. Retrieved from https://www.mckinsey.com/featured-insights/china/chinas-travel-recovery-gains-steam-how-families-are-planning-their-summer-vacations July 6, 2020

Epler-Wood M, Milstein M, Ahamed-Broadhurst K (2019) Destinations at risk: the invisible burden of tourism. The Travel Foundation. Retrieved from https://www.thetravelfoundation.org.uk/invisible-burden/

E-Tourism Frontiers (2020) Tourism phases of recovery diagram. Retrieved from http://www.e-tourismfrontiers.com/what-we-do/research/

Euractiv (2019) Tourism—a key driver for socio-economic progress. Retrieved from https://events.euractiv.com/event/info/tourism-a-key-driver-for-socio-economic-progress

Farr LJM (2019) Visit California spearheads DMO trend toward destination stewardship. Retrieved from https://thetravelvertical.com/2019/09/03/visit-california-spearheads-dmo-trend-toward-destination-stewardship/

Financial Times (2020) US airlines call for $50bn in emergency support to survive crisis. Retrieved from https://www.ft.com/content/b775ce72-67ac-11ea-800d-da70cff6e4d3

Global Sustainable Tourism Council (2020) What is the GSTC? Retrieved from https://www.gstcouncil.org/about/about-us/

Graver B, Zhang K, Rutherford D (2019) CO_2 emissions from commercial aviation, 2018. The International Council on Clean Transportation. Retrieved from https://theicct.org/sites/default/files/publications/ICCT_CO2-commercl-aviation-2018_20190918.pdf
Hadjicostis M (2020) Cyprus pledges to cover costs for virus-hit tourists. May 27, 2020. Retrieved from https://apnews.com/fad64eb5fcc1ea79ad94770936b2dc9a
Henley J (2020) Overtourism in Europe's historic cities sparks backlash. The Guardian, January 25, 2020, sec. World news. Retrieved from https://www.theguardian.com/world/2020/jan/25/overtourism-in-europe-historic-cities-sparks-backlash
Higgins-Desbiolles F (2019) Has tourism had its day? RSA J 2:39. Retrieved from https://medium.com/rsa-journal/has-tourism-had-its-day-4c534b2b5e58
Higgins-Desbiolles F, Carnicelli S, Krolikowski C, Wijesinghe G, Boluk K (2019) Degrowing tourism: rethinking tourism. J Sustain Tour. https://doi.org/10.1080/09669582.2019.1601732to Published online: 22 Apr 2019. Retrieved from https://www.researchgate.net/publication/332565985_Degrowing_tourism_rethinking_tourism
Hilton (2020) Hilton and American Express to donate up to 1 Million rooms to frontline medical professionals during COVID-19 crisis. Hilton Press Center. April 6, 2020. Retrieved from https://newsroom.hilton.com/corporate/news/hilton-american-express-team-up-to-donate-rooms
Holcombe M (2020) Working from home can soon mean working in Barbados for up to a year. CNN, July 14, 2020. Retrieved from https://www.cnn.com/travel/article/barbados-coronavirus-work-from-home-trnd/index.html
IATA (2019) WATS: 4.4bn traveled by air in 2018. IATA, August 1, 2019. Retrieved from https://airlines.iata.org/news/wats-44bn-traveled-by-air-in-2018%C2%A0
ICAO (2020) Effects of novel coronavirus (COVID-19) on civil aviation: economic impact analysis. Retrieved from https://www.icao.int/sustainability/Documents/COVID-19/ICAO_Coronavirus_Econ_Impact.pdf
Ighobor K (2020) AU study: COVID-19 could cost Africa $500 billion, damage tourism and aviation sectors. United Nations: Africa Renewal, April 15 2020. Retrieved from https://www.un.org/africarenewal/magazine/au-study-covid-19-could-cost-africa-500-billion-damage-tourism-and-aviation-sectors
IHCL (2020) IHCL exceeds 1 Million meals offered to healthcare providers and migrant workers. Taj. April 30, 2020. Retrieved from https://www.ihcltata.com/ihcl/press-room/IHCL-exceeds-1-million-meals-offered-to-healthcare-providers-and-migrant-workers/
iProperty Management (2020) Airbnb Statistics. iPropertyManagement, March 2020. Retrieved from https://ipropertymanagement.com/research/airbnb-statistics
International Labour Organization (2020) COVID-19 and employment in the tourism sector: impact and response in Asia and the Pacific. International Labour Organization, April 24, 2020. Retrieved from https://www.ilo.org/asia/publications/issue-briefs/WCMS_742664/lang--en/index.htm
Intrepid (2020) B corp. Retrieved from https://www.intrepidtravel.com/us/b-corp
Johnson K (2018) How low-cost airlines are changing the face of air travel. Business Insider. Retrieved from https://www.businessinsider.com/how-low-cost-airlines-are-changing-the-face-of-air-travel-2018-7
Josephs L (2020) United warns 36,000 employees of potential job cuts as pandemic roils travel demand. CNBC. Retrieved from https://www.cnbc.com/2020/07/08/coronavirus-united-braces-36000-employees-for-job-cuts-as-pandemic-roils-travel-demand.html
Kolbert E (2020) How Iceland beat the coronavirus. The New Yorker. June 1, 2020. Retrieved from https://www.newyorker.com/magazine/2020/06/08/how-iceland-beat-the-coronavirus
Keating D (2020) Has Covid-19 extended aviation's free pass to pollute? Power Technology Energy News and Market Analysis. July 8, 2020. Retrieved from https://www.power-technology.com/features/has-covid-19-extended-aviations-free-pass-to-pollute/
Krishnan V, Mann R, Seitzman N, Wittkamp N (2020) Hospitality and COVID-19: how long until 'no vacancy' for US hotels? McKinsey & Company Travel, Transport & Logistics, April 2020. Retrieved from https://www.mckinsey.com/industries/travel-logistics-and-transport-infrastructure/our-insights/hospitality-and-covid-19-how-long-until-no-vacancy-for-us-hotels

Le Quéré C, Jackson RB, Jones MW (2020) Temporary reduction in daily global CO_2 emissions during the COVID-19 forced confinement. Nat Clim Chang 10:647–653. Retrieved from https://doi-org.ezp-prod1.hul.harvard.edu/10.1038/s41558-020-0797-x

Levy C, Mieszala J-C, Mysore M, Samandari H (2020) Coronavirus: 15 emerging themes for boards and executive teams. McKinsey & Company Risk, June 2020. Retrieved from https://www.mckinsey.com/business-functions/risk/our-insights/coronavirus-15-emerging-themes-for-boards-and-executive-teams

Locker M (2020) Five to know about travel bubbles. Smithsonian Mag, May 28, 2020. Retrieved from https://www.smithsonianmag.com/travel/five-things-know-about-travel-bubbles-180974983/

Lonely Planet (2016) International travel set to increase by 35% over the next decade. Lonely Planet. June 2016. Retrieved from https://www.lonelyplanet.com/articles/international-travel-increase-next-decade

Majorca Daily Bulletin (2020) Meliá Palma Bay hotel becomes a hospital. Majorca Daily Bulletin April 24, 2020. Retrieved from https://www.majorcadailybulletin.com/news/local/2020/03/24/64561/melia-palma-bay-hotel-becomes-hospital.html

McCaul J (2020) 5 things that will define tourism in 2020. Resonance. Retrieved from https://resonanceco.com/insights/5-things-that-will-define-tourism-in-2020/

McKinsey & Company and the World Travel and Tourism Council (2017) Coping with success: managing overcrowding in tourism destinations. Retrieved from https://www.mckinsey.com/industries/travel-logistics-and-transport-infrastructure/our-insights/coping-with-success-managing-overcrowding-in-tourism-destinations

Merco Press (2020) Machu Picchu reopens in July but with a sharp reduction in daily visitors. Merco-Press June 15, 2020. Retrieved from https://en.mercopress.com/2020/06/15/machu-picchu-reopens-in-july-but-with-a-sharp-reduction-in-daily-visitors

Mullis B (2017) The growth paradox: can tourism ever be sustainable? World Economic Forum. Retrieved from https://www.weforum.org/agenda/2017/08/the-growth-paradox-can-tourism-ever-be-sustainable/

Mzezewa T (2020) 3 Baltic States announced a 'travel bubble.' What is it and could it work in the U.S.? New York Times. Retrieved from https://www.nytimes.com/2020/04/29/travel/coronavirus-travel-bubble.html

National Environment Agency (2020) "SG Clean" campaign launched to rally public and businesses to work together to keep Singapore clean. Retrieved from https://www.nea.gov.sg/media/news/news/index/sg-clean-campaign-launched-to-rally-public-and-businesses-to-work-together-to-keep-singapore-clean

OECD (2020) OECD tourism trends and policies 2020. OECD, March 4, 2020. Retrieved from https://www.oecd.org/cfe/tourism/oecd-tourism-trends-and-policies-20767773.htm

OECD (2020a) Building back better: a sustainable, resilient recovery after COVID-19. OECD, June 5, 2020. Retrieved from http://www.oecd.org/coronavirus/policy-responses/building-back-better-a-sustainable-resilient-recovery-after-covid-19-52b869f5/

OECD (2020b) Tourism policy responses to the coronavirus (COVID-19). OECD, June 2, 2020. Retrieved from https://www.oecd.org/coronavirus/policy-responses/tourism-policy-responses-to-the-coronavirus-covid-19-6466aa20/

Osbourne J (2020) The future of travel: how the industry will change after the pandemic. New York Times, May 2020. Retrieved from https://www.nytimes.com/interactive/2020/05/06/travel/coronavirus-travel-questions.html

Oxford Economics (2020) The economic impact of the coronavirus due to travel losses: 2020 Analysis. Retrieved from https://www.ustravel.org/sites/default/files/media_root/document/Coronavirus_2020_Impacts_WEB.pdf

Palmer E (2020) For much of Europe, reopening day is here. New York Times, June 12, 2020. Retrieved from https://www.nytimes.com/2020/06/12/travel/reopening-europe-caribbean-virus.html

Paun A (2020) ESG Stocks Did Best in COVID-19. HSBC Insights, March 27, 2020. Retrieved from https://www.gbm.hsbc.com/insights/global-research/esg-stocks-did-best-in-corona-slump
Perelli C (2007) Governing tourism monoculture: Mediterranean mass tourism destinations and governance networks. Tour Polit. Retrieved from https://www.academia.edu/23403475/Governing_Tourism_Monoculture_Mediterranean_Mass_Tourism_Destinations_and_Governance_Networks
Pinandita A (2020) Indonesia proposes regional travel corridor at ASEAN summit. Retrieved from The Jakarta Post, June 26 2020. Retrieved from https://www.thejakartapost.com/seasia/2020/06/26/indonesia-proposes-regional-travel-corridor-at-asean-summit.html
Pololikashvili Z (2020) COVID-19 statement from Secretary-General of World Tourism Organization. UN World Tourism Organization, March 17, 2020. Retrieved from https://www.unwto.org/news/covid-19-statement-zurab-pololikashvili
Powley T, Bushey C (2020) Wealthy switch to private jets to avoid coronavirus. Financ Times, July 25, 2020. Retrieved from https://www.ft.com/content/063785d0-ed9a-4eae-b4ba-8a56cf59e905
PWC (2020) Total impact measurement & management. Retrieved from https://www.pwc.com/gx/en/services/sustainability/total-impact-measurement-management.html
Quelch JA, Boudreau EC (2016) Building a culture of health: a new imperative for business. Springer International Publishing AG, Basel
Quick M (2019) Flygskam. Retrieved from https://www.bbc.com/worklife/article/20190718-flygskam
Rafat A (2018) The genesis of overtourism: why we came up with the term and what's happened since. Skift, August 14, 2018. Retrieved from https://skift.com/2018/08/14/the-genesis-of-overtourism-why-we-came-up-with-the-term-and-whats-happened-since/
Reuters (2020) Lufthansa seeking cuts equivalent to 22,000 positions. Retrieved from https://www.reuters.com/article/us-lufthansa-restructuring/lufthansa-seeking-cuts-equivalent-to-22000-positions-idUSKBN23M2M1
Sampson H (2020) Hawaii isn't messing around when it comes to enforcing tourist quarantines. Washington Post, May 20, 2020. Retrieved from https://www.washingtonpost.com/travel/2020/05/20/hawaii-isnt-messing-around-when-it-comes-enforcing-tourist-quarantines/
Serafeim G, Trinh K, Zochowski R (2020) A framework for product impact weighted accounts. Harvard Business School, February 18, 2020. Retrieved from https://www.hbs.edu/faculty/Publication%20Files/20-076_2e82d088-8dbd-4a4d-844c-60b6617a2799.pdf
Simon E (2020) 15,000 hotels sign up for AH&LA's hospitality for hope drive. Hotel Manag, April 6, 2020. Retrieved from https://www.hotelmanagement.net/operate/15-000-hotels-sign-up-for-ah-la-s-hospitality-for-hope-drive
Skirka H (2020) The Seychelles has reopened to tourists—but only to those travelling by private jet. The Natl, June 3, 2020. Retrieved from https://www.thenational.ae/lifestyle/travel/the-seychelles-has-reopened-to-tourists-but-only-to-those-travelling-by-private-jet-1.1028188
Stringam BB, Partlow CG (2016) A profile of the hospitality industry. Business Expert Press LLC, New York
Sustainability Leaders Project (2019) Sustainable tourism trends and challenges in 2019: expert panel. Retrieved from https://sustainability-leaders.com/sustainable-tourism-trends-challenges-2019/
Thani D (2020) India's wealthy are still planning their travels—and they're taking to private jets. Condé Nast Traveller India. July 1, 2020. Retrieved from https://www.cntraveller.in/story/india-wealthy-planning-travels-private-jets-cost-booking-holiday-maldives-rajasthan-kerala/
The Tico Times (2020) Costa Rica passes law creating more long weekends through 2024. Retrieved from https://ticotimes.net/2020/07/17/costa-rica-passes-law-creating-more-long-weekends-through-2024
The World Bank (2020) COVID-19 to plunge global economy into worst recession since World War II. Retrieved from https://www.worldbank.org/en/news/press-release/2020/06/08/covid-19-to-plunge-global-economy-into-worst-recession-since-world-war-ii
Treadright (2020) Make travel matter. Retrieved from https://www.treadright.org/

United Nations Conference on Trade and Development (2020) COVID-19 and Tourism: assessing the economic consequences. Retrieved from https://unctad.org/en/pages/PublicationWebflyer.aspx?publicationid=2810
United Nations Environment Programme & World Tourism Organization (2005) Making tourism more sustainable: a guide for policy makers. URI http://hdl.handle.net/20.500.11822/8741. Retrieved from https://www.unwto.org/sustainable-development and http://wedocs.unep.org/handle/20.500.11822/8741
United Nations General Assembly (2015) Transforming our world: the 2030 agenda for sustainable development. UN General Assembly Retrieved from https://www.refworld.org/docid/57b6e3e44.html
UNEP (2017) Assessing global resource use. Retrieved from https://www.unenvironment.org/news-and-stories/press-release/resource-use-expected-double-2050-better-natural-resource-use and https://www.resourcepanel.org/reports/assessing-global-resource-use
UNEP (2020) Tourism to tackle plastic pollution with new commitment. Retrieved from https://www.unenvironment.org/news-and-stories/story/tourism-tackle-plastic-pollution-new-commitment
United Nations World Tourism Organization (2010) Retrieved from https://www.unwto.org/glossary-tourism-terms
UNWTO (2016) UNWTO tourism highlights: 2016 Edition. Retrieved from https://www.e-unwto.org/doi/pdf/10.18111/9789284418145
UNWTO (2017) World could see 1.8 billion tourists by 2030—UN Agency. UN News, December 27, 2017. Retrieved from https://news.un.org/en/story/2017/12/640512-world-could-see-18-billion-tourists-2030-un-agency
UNWTO (2018) Overtourism? Understanding and managing urban tourism growth beyond perceptions. Retrieved from https://www.e-unwto.org/doi/pdf/10.18111/9789284420070
UNWTO (2018a) UNWTO tourism highlights: 2018 Edition. Retrieved from https://www.e-unwto.org/doi/pdf/10.18111/9789284419876
UNWTO (2019) International tourism growth continues to outpace the global economy. Retrieved from: https://www.unwto.org/international-tourism-growth-continues-to-outpace-the-economy#:~:text=1.5%20billion%20international%20tourist%20arrivals%20were%20recorded%20in%202019%2C%20globally.&text=Looking%20ahead%2C%20growth%20of%203,the%20same%20level%20of%202019
UNWTO (2019a) Transforming tourism for climate action. Retrieved from https://www.unwto.org/one-planet-stp-cop25
UNWTO (2020) Covid-19: putting people first. Retrieved from https://www.unwto.org/tourism-covid-19
UNWTO (2020a) International tourist numbers could fall 60–80% in 2020. Retrieved from https://www.unwto.org/news/covid-19-international-tourist-numbers-could-fall-60-80-in-2020
UNWTO (2020b) Green investments for sustainable tourism. Retrieved from https://www.unwto.org/green-investments-for-sustainable-tourism
U.S. Travel Association (2020) U.S. Travel Answer Sheet. U.S. Travel Association. Retrieved from https://www.ustravel.org/answersheet
Warren K (2019) Living like a local was the big travel trend 5 years ago. Now, travelers want their experiences to be life-changing, and it's ushering in a new era of 'transformational travel'. *Business Insider,* September 1, 2019. Retrieved from https://www.businessinsider.com/transformational-travel-experiential-luxury-safari-trends-2019-8
Wilderness Safaris (2017) Wilderness safaris—biodiversity conservation 2017. May 22, 2017. Retrieved from https://wilderness-safaris.com/blog/posts/wilderness-safaris-biodiversity-conservation-2017
Wilson A (2020) Sicily to subsidise post-Covid holidays as Italy ponders reopening to tourists. The Guardian, May 5, 2020. Retrieved from https://www.theguardian.com/travel/2020/may/05/sicily-to-subsidise-post-covid-holidays-as-italy-considers-reopening-to-tourists

World Commission on Environment and Development (1987) Our common future. Oxford University Press, Oxford
World Travel and Tourism Council (2015) Travel and tourism 2015: connecting global climate action. WTTC, London
WTTC & Harvard University (2019) Responsible travel and tourism with Harvard University. Retrieved from https://wttc.org/News-Article/WTTC-ushers-in-new-critical-collaboration-with-Harvard-T-H-Chan-School-of-Public-Health
WTTC (2019) Economic impact reports. Retrieved from https://wttc.org/Research/Economic-Impact
WTTC (2020) Global Economic Impact & Trends 2020. Retrieved from https://wttc.org/Research/Economic-Impact

Dr. **Wendy Purcell** PhD FRSA. Harvard T.H. Chan School of Public Health, Harvard University. Leads research on global leadership and governance for sustainable transformation in higher education and business sectors, including Travel & Tourism, the Auto industry, and the Aviation sector; she directs the Sustainable Development Solutions Group. Wendy is Emeritus Professor and President Vice-Chancellor Emerita of a UK university (2007–2015), taking it to the top 1% of world universities and 1st 'green' UK university. She is a Council Member United Nations University, Member Association of American Universities and Colleges, was a Non-Executive Director UK Government's Department of Business, Innovation and Skills and is Expert Adviser and Board member to global Boards and charities. Wendy's purpose lies in transforming lives through education and research in pursuit of social equity, the knowledge economy and sustainable development. She has authored over 100 academic publications, book chapters, books and patent filings. Editor Sustainability special issue 'Sustainable Higher Education and Leadership', co-editor textbook 'Handbook on Sustainability in Higher Education: an Agenda for Transformational Change' and series editor 'Higher Education and the Sustainable Development Goals'

Ms. **O'Shannon Burns**. Harvard Division of Continuing Education. Sustainability and tourism professional with experience in program management and development. Practitioner of science-based models for sustainable tourism development. Curator of travel experiences that foster authentic cultural interaction, support local economies, and innovate sustainable tourism. Developed conservation-, storytelling-, and education-focused programming. Driven by the belief that travel can transform our understanding of the world, build community, and help create a more sustainable, equitable, and resilient planet.

Ms. **Allison Voss**. Harvard Kennedy School. Finance professional with experience working in sustainable land use and agriculture investments, anti-money laundering and combating the financing of terrorism, small and medium-sized enterprise (SME) finance, and investment banking in the Middle East, Latin America, Africa and the U.S.

Fostering Innovation and Intercultural Exchange During a Global Pandemic: Lessons Learned from a Virtual Design Thinking Challenge in Nepal

Jens M. Unger, Aman Bhattarai, Bettina Maisch, Johannes M. Luetz, and James Obuhuma

Abstract Countries in the majority world are widely argued to be most severely affected by the corona pandemic. To fight the outbreak of the virus and reduce risks of economic hardships and starvation these countries need an ability to innovate and quickly develop feasible solutions in areas such as health, education, and business. In response to this crisis, the Impact Week (IW), a global non-profit organization run by a community of professional volunteers, developed a 3-day virtual format to quickly develop and test solutions using human-centric innovation techniques. Based on a case study approach we present the format, lessons learned and impact of a pioneer project conducted to support people in Nepal called Nepal-vs-COVID-19 with 54 students and 21 professionals. Despite severe technical problems like connectivity and power cuts, overall the concept proved to be highly effective. More

---

J. M. Unger (✉)
Lufthansa Group, Lufthansa Aviation Center, Airportring, 60546 Frankfurt, Germany

A. Bhattarai
zeroG GmbH, Hugo-Eckener-Ring 1, FAC, Building 234, D7.01, 60549 Frankfurt, Germany
e-mail: aman.bhattarai@zerog.aero

B. Maisch
Hochschule München/Strascheg Center for Entrepreneurship, Heßstraße 89, 80797 Munich, Germany
e-mail: bettina.maisch@sce.de

J. M. Luetz
School of Social Sciences, Christian Heritage College (CHC), Carindale/Brisbane, QLD 4152, Australia

School of Law and Society, University of the Sunshine Coast (USC), Maroochydore, QLD 4558, Australia

School of Social Sciences, University of New South Wales, Sydney, NSW 2052, Australia

J. M. Luetz
e-mail: jluetz@chc.edu.au; jluetz@usc.edu.au; j.luetz@unsw.edu.au

J. Obuhuma
School of Science & Technology, Africa Nazarene University, P.O. Box 53067-00200, Nairobi, Kenya
e-mail: jobuhuma@anu.ac.ke

W. Leal Filho (ed.), *COVID-19: Paving the Way for a More Sustainable World*,
World Sustainability Series, https://doi.org/10.1007/978-3-030-69284-1_10

specifically, the quality of solutions generated during this virtual event was perceived to be even higher than in previous physical events. For participants, the event created an opportunity to learn digital skills and to connect with others in a meaningful and productive way, despite global lockdown. The virtual Design Thinking (DT) challenge subsequently sparked events in a number of other countries (Kenya, Nigeria, Germany, India, and South Africa). Our study adds to the sustainability literature by presenting a format that is flexible, scalable, cost efficient, environmentally friendly and can easily be applied in contexts of crisis management, business and education. Furthermore, it provides a concrete example of successful collaboration and knowledge exchange beyond differences and national boundaries and thereby constitutes a blueprint to tackle future global challenges.

Keywords Impact week · Social impact · Design thinking · Online collaboration · Intercultural exchange · Sustainability · Innovation

1 Introduction

The corona pandemic has led to an unprecedented crisis in modern history, severely affecting countries worldwide. While immediate problems in Europe and America concerned capacity of health systems including the availability of intensive care units, ventilators and test capacities, (Goldin 2020), it had been clear from the onset of the pandemic that countries in the majority world would likely be hit even harder (Goodman et al. 2020; Walker 2020). In many of these countries, medical resources are scarce or hardly existent, precautionary measures are difficult if not impossible to implement and lockdown effects are felt almost immediately for the extreme poor who depend on jobs and mobility to cover most basic needs and to nourish their families (Leal Filho et al. 2020; Goldin 2020). Similarly, negative effects in education are likely to be accentuated in less advanced countries resulting in learning loss, reduction of human capital and may even negatively impact future labor market outcomes (Carmichael and Darko 2020). These countries therefore are the ones that are most in need of immediate solutions in this crisis and that will benefit the most from a mind and skill set to mitigate the effects of the pandemic, preferably in a sustainable way that is in harmony with the local socioeconomic and cultural context.

Before the crisis, the Impact Week (IW), a global non-profit organization originating in Germany, has been supporting countries in the majority world fostering innovation and entrepreneurship to solve local problems by teaching and applying human-centered design (Unger and Luetz 2019). After one physical IW took place in Kenya in February 2020, all other scheduled events for the year had to be canceled due to health risks and lockdowns. Even so, within 2 days a small team of the IW community developed a virtual format in order to be able to also help countries in the majority world in the crisis. This concept was tested 10 days later in a pilot project called Nepal-vs-COVID-19. Our goal for contributing to this book is to make our program available to policy makers, decision makers in the educational systems or any other stakeholders interested in supporting and enabling people in the majority world to develop solutions to their most pressing problems, in particular in emergency situations like the coronavirus crisis. We describe how the virtual format was developed, what it looks like and how it works, present outcomes of the event, and discuss learnings as well as shortcomings. In doing so we aim to contribute to the sustainability literature so that our analysis may foster a better understanding of virtual formats, and more specifically, virtual Design Thinking (DT) challenges as a quick, scalable, cost efficient and environmentally friendly response to similar crisis situations in the future. Moreover, the study points to opportunities for collaboration in the educational sector and in cross-cultural interdisciplinary settings.

The IW is a social innovation program that brings together people from all over the world and from a variety of backgrounds. Innovators and entrepreneurs, who are willing to give their time to coach and mentor others, are joined by university scientists and lecturers, who want to bring their innovation skills to the next level and pass them on to their students. The IW teaches and applies Design Thinking (DT). As an innovation approach made popular at Stanford University and increasingly used and demanded in modern and dynamic corporate settings, DT offers a mindset, skillset and toolset to effectively master and surmount the challenges of complex and changing environments (Brown and Wyatt 2010; Liedtka 2018). As such, DT is a cutting-edge change and innovation process whose people-centric nature is aptly suited to develop world stage-experienced human services practitioners and corporate innovation and change professionals (cf. Kotter 2012; Kolb 2014; Kotter and Cohen 2012).

Expressed in simple language, IW participants learn (1) how to understand and analyse real world problems; (2) how to observe and conduct qualitative research; (3) how to extract meaning from qualitative data (synthesis); (4) how to generate ideas to solve existing problems; (5) how to make ideas tangible and ready for testing (prototyping); (6) how to work together in diverse teams, use different perspectives and make decisions; (7) how to pitch their ideas in front of a jury. Furthermore, IW coaches additionally learn how to facilitate the DT process.

With this background and rationale, this study builds on the past experience and success of the award-winning IW, which was recognised by the Government of the Federal Republic of Germany as one of the 100 most innovative social concepts in 2018[1] for connecting worlds and strengthening cohesion. Furthermore, a research paper jointly authored by two members of this author group (Unger and Luetz 2019) was recognised at the World Symposium on Social Responsibility and Sustainability in Edinburgh, Scotland, 27–29 June 2018, with the Best Paper Award.[2] It digests the unique people-centric contribution of the IW within the format of a multi case study

[1] https://land-der-ideen.de/en/project/impact-week-workshops-for-thinkers-in-developing-countr ies-3859.

[2] https://chc.edu.au/news/best-paper-award-for-chc-researcher-at-world-symposium-in-scotland/.

(Unger and Luetz 2019). Finally, a short video presents the conceptual DT approach advanced by the IW.[3]

It is against this professional, practical, historical and scholarly background that this study expands upon the progressive evolution of the IW over the years in the current context of the global COVID-19 pandemic. The IW itself started with a one-week on-site project in Nairobi, Kenya in 2015 and was initiated by a number of entrepreneurs and corporate innovation experts[4] (Unger and Luetz 2019). From inception, the goal has been to foster innovation and entrepreneurship in a sustainable way, thereby putting human needs at the center of developing solutions using the DT approach (Kolko 2015; Grots and Creuznacher 2016). At the core of the program has been the encouragement and utilization of diversity and intercultural exchange as a source of inspiration and innovation, and to strengthen participants' intercultural competence, an ability that has been described as the 'literacy of the future' (Deardorff 2011; UNESCO 2013; OECD 2018; Nelson et al. 2019). In a 2–3 day train-the-trainer, local university professors, lecturers, and representatives of international corporations learned DT methods, tools and mindset. After completing the train-the-trainer they formed tandems of one local and one international participant each to coach and support local teams of students to develop ideas and business models in a four-day event.

After the second IW in Kenya in 2016 a growing movement has been created from within the community bringing the IW to 9 countries on 3 continents, with 23 events in total until the beginning of 2020. Of the 23 IWs, 8 events have been organized independently—planned and run by local universities or community leaders. Much of the success of the IW growth and sustainability was due to a strong feeling of fellowship, shared values and collaboration on eye level across countries and cultures. As one local organizer put it: "In Impact Week we are all equal, all loved, and all valued, one people, different color, fun and feedback is our way of life, and solving challenges is what we love."[5]

While the event itself has also been viewed as a success based on feedback from all stakeholders involved (student participants, corporate participants, professional volunteers, partners and sponsors cf. Unger and Luetz 2019), there have been at least two major challenges in the past. The first challenge concerns stakeholders' hope that ideas developed during the IW will be implemented following each event. This issue has been discussed by Unger and Luetz (2019) and is by and large a matter of intended scope of the program. The IW format has been developed with a primary focus on fostering innovation, entrepreneurial skills and mindset and to encourage global collaboration on eye level rather than focusing on implementing ideas. Implementation of ideas will depend on local incubators and support systems available to IW participants. The second challenge concerns the effort and investments required to conduct an IW abroad. Typically, centrally organized IWs have been planned starting 10 months before the actual date. Time spent at the event

[3]https://youtu.be/AYnON5aODjM.

[4]With Michael Huebel as co-founder and lead organizer in 2015 and 2016.

[5]https://www.facebook.com/photo?fbid=10214706159789617&set=a.10205472383390978.

location was typically between 10 and 14 days. Costs ranged between US$40.000 and US$80.000 including materials, accommodation and food plus air travel for at least 25 people. With the onset of the coronavirus crisis the future perspective of the IW changed drastically. Travelling, of course, was not possible any more. The concept, that heavily relied on teamwork, physical collaboration including visualizations, building prototypes and a cultural experience, was not feasible anymore due to health risks. Primary program sponsors and partners suspended or canceled their financial commitments. At the same time, partners in different countries reached out to the IW organization with an immediate expressed need to connect with others and to find solutions to successfully managing the challenges brought about by the crisis. This prompted the development of a virtual event, the Country-vs-COVID-19 program which we describe and analyze in our study. We wanted to find out: Would it be possible to carry out the IW online? Is it possible to foster innovation and intercultural exchange in a virtual setting?

1.1 Methodological Considerations: A Case Study Approach

In respect of its research methodology, this study may be most appropriately situated within the case study methodological design. Given the prevalence of complex multidisciplinary and multicausal interrelationships, the case study approach offered a fitting research design for holistic analysis (Bryman 2016; Johnson and Christensen 2017; Punch 2014). More specifically, Punch (2014) has described the analytical benefits of the case study method as follows:

> All data relevant to the case are gathered, and all available data are organized in terms of the case. The case study method gives a unitary character to the data being studied by interrelating a variety of facts to a single case. It also provides an opportunity for the intensive analysis of many specific details that are often overlooked with other methods. (p. 121)

Relatedly, the case study design is aptly suited to manage complex research contexts and therefore often surpasses the capacity of other research designs (Punch 2014), wherefore case studies are well-known as effective methodological tools for conceptualising future research:

> Discovering the important features, developing an understanding of them and conceptualising them for further study, is often best achieved through the case study strategy. Following this line of argument, it may be that too much research has tried to go straight to measurement and quantitative mapping without a fuller understanding of the phenomena and processes involved that are best achieved by case studies. (Punch 2014, p. 124)

Given that "some of the best-known studies in sociology are based on this kind of design" (Bryman 2016, p. 60), it seemed most appropriate to feature the analysis and synthesis presented by this paper within a case study methodological frame of reference. Given the need for both speed and innovation the IW organisers needed to move promptly, wherefore this case study may be appropriately described as a 'pilot' study that has the potential to be longitudinally continued as future learnings emerge from program replication in other locations.

More specifically, in this pilot study we report data collected between April 2020 and June 2020. We use multiple sources and integrate perspectives from multiple stakeholders including organizers, local hosts and community leaders, coaches, and participants. Data were collected on the final day of the event using a brief live online survey with all stakeholders and a short post event feedback survey with participants. Moreover, data were extracted from two separate event retrospectives—one with participants focusing on qualitative feedback and one with the responsible coaches and mentors focusing on lessons learned from the event. Finally, we integrate insights and observations taken together by the event organizing team.

In the next section we describe initiation, framework, execution, and outcomes of the virtual DT challenge using the example of Nepal-vs-COVID-19. We then synthesize key learnings followed by a presentation of how the initial event spread to other countries. In the final section , we discuss opportunities for future research, present applications for practice and highlight implications for sustainability.

2 Virtual Impact Week to Fight COVID-19 in Nepal

2.1 *Initiation of Nepal-Vs-COVID-19*

Impact Week Kathmandu, Nepal, was among the historical IWs that took place in 2019. The event was organized by employees of the Lufthansa Group in collaboration with King's College, Kathmandu. It brought together 28 international professionals from all over the world and alongside academic professionals from Kings College, more than 100 students went through a DT challenge to tackle social and environmental challenges in the country. The event received a lot of attention in the country as the first of its kind in Nepal. It also saw involvement and support from ministerial leadership, diplomatic authorities, governmental organizations and leading private entities. Following the success of the event, the IW team was scheduled to organize the second edition of the IW 2020 in Nepal, targeting to bring together young students and teachers from all over the nation. Regrettably, due to the COVID-19 outbreak and global borders shutting down, the program was canceled in March 2020.

On the week when the Impact Week Nepal 2020 was canceled, the German Federal Government, with support from leading incubators and companies, successfully organized the largest 'hackathon'[6] in the world, which trended under the hashtag #WirVsVirus. It brought together 28,000+ participants who developed more than 1.500 solutions over a period of two days. This triggered a discussion between Prof. Dr. Bettina Maisch, Dr. Jens Unger and the lead organizer of the Impact Week Nepal,

[6] A so-called 'hackathon' (also known as 'hackfest' or 'codefest') is an event in which computer programmers and software designers collaborate on projects with the aim of speedily developing functional software and/or hardware by the end of the event.

Aman Bhattarai, on how to bring the physical IW edition into a completely virtual environment, bringing global minds from all over the world together to tackle the challenges caused by COVID-19 in Nepal. Within a span of a few hours, the trio wrote a concept paper and came up with a concept of Nepal-vs-COVID-19 as an initiative of the Impact Week.

To the authors' knowledge, Nepal-vs-COVID-19 was among the first virtual DT challenges in the world aiming to bring together Nepalese diaspora and international DT facilitators from all over the globe to develop user-centric solutions to tackle challenges caused by COVID-19 in Nepal. The naming was inspired by the #WirVsVirus 'hackathon' and was curated as a Country-vs-COVID-19 challenge. The virtual DT challenge was the first of its kind worldwide, bringing together innovation experts and participants from diverse countries, continents and professional backgrounds. Hence, it was a prototype aimed to test if intercultural collaboration and innovation could be done virtually using state-of-art digital collaboration platforms. The idea was primarily to 'think global and act local' from the very beginning. Upon the potential success, Nepal-vs-COVID-19 could expand as Country-vs-COVID-19 movement.

2.2 Framework of the Virtual Impact Week 'Country-Vs-COVID-19

The core team was convinced that it was both possible and that there was a sense of urgency to do it quickly in order to address the needs during a time of crises. Through a written call for participation within the existing network interested people were then able to sign up for the project. An initial ZOOM call set up by the main organizer on 10th of April 2020 was used to communicate the purpose and the idea of a virtual IW for Nepal, later termed Nepal-vs-COVID-19, as well as to get the planning process underway.

2.2.1 Team Structure, Roles and Responsibilities

The team structure, roles and responsibilities in the virtual IW were similar to the physical one with the exception that an additional tech team had to be built. The participants within the core team took over the following roles, responsibilities and tasks:

- *General Organizing Team*
  A three-member organizing team was responsible for overall alignment between sub-teams and ensuring the overall execution in terms of time in quality from planning, conducting to post-event work as well as taking care of the overall collaboration culture and living the values of the IW.
- *Content Team*

Four DT experts from Germany were part of the content team who ideated on the possible collaboration platforms and translated the DT process and method content to a much shorter and virtual format. The team created information and guiding material for coaches and team members as well as acting as 'head coach' to address questions.

- *Marketing & Communications Team*
  A cross national marketing and communication team consisted of marketing and communications experts from Nepal and Germany. This team was responsible for overall organizing, executing and monitoring all external and internal communications throughout the event. External communication involved attracting the right participants as well as possible partners from all over the world by spreading the word in the media. Internal communication involved designing and executing actions to create positive spirit also for the virtual team collaboration which is always a unique feature of the IW.
- *Design Thinking (DT) Coaches/Team Mentors*
  The main tasks of the coaches/mentors were as follows: guiding the team in their application of DT along their challenge, giving examples and recommendations how to execute the steps of the human-centered approach on their selected challenge, moderating the conversation, helping to validate the final solution and improving the quality of the final pitch presentation.
- *Tech Team*
  Given the digital execution of Nepal-vs-COVID-19, technical challenges were anticipated. The three-member tech team ensured that all technical questions or problems of participants could be addressed immediately and professionally.

2.2.2 Technology, Communication and Collaboration Platforms

Due to the fact that this IW had to be conducted virtually, the appropriate selection of digital communication and collaboration tools was crucial. The responsible tech team looked for digital platforms that provide easy open access globally, allowing real time collaboration with a good user experience as well as the required reliability and stability. The biggest challenge was to find a tool that is able to transfer DT working behavior, collaboratively working with data shared on post-its and enfolding insights through clustering. This visual and interactive element is a crucial success factor for the kind of creative and collaborative working approach. In the end, the team selected a variety of tools to best address the different working demands. Figure 1 displays the different phases of the project, tasks to be achieved in each phase as well as an overview of tools that were used. The real-time conversation and working sessions were conducted via the telco solution, ZOOM (Fig. 2).

Through a sponsorship from MURAL, a digital white board solution, we were able to provide a DT working style in the virtual world. The content team created visual templates on MURAL to support each team in the DT process. Additionally, Google G-Suite was used as a platform to collaboratively develop and share content (Fig. 3), including all general information to coaches and participants, agenda with timeline

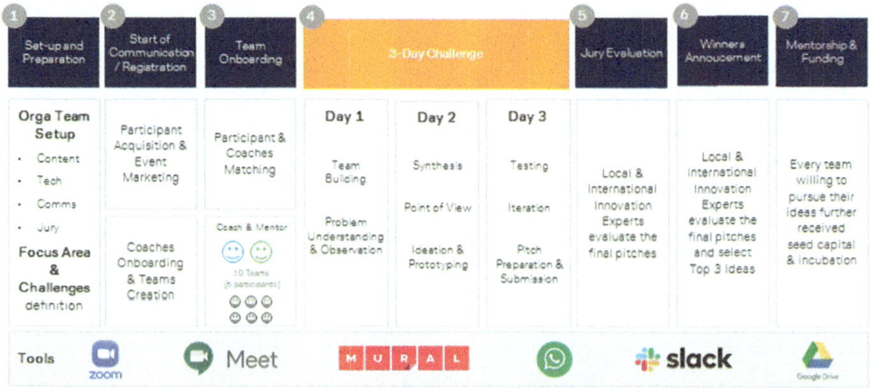

Fig. 1 Different phases of virtual DT challenge Nepal-vs-COVID-19 from preparation to implementation of 3-day challenge to supporting teams with most promising ideas

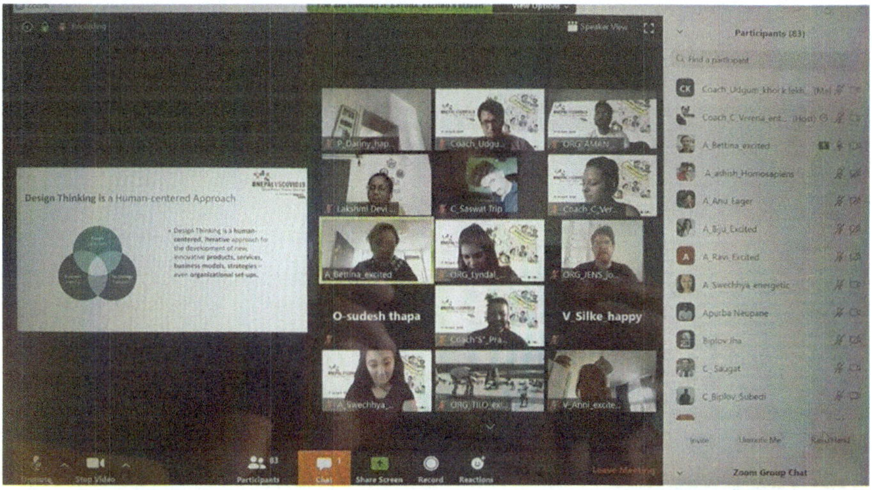

Fig. 2 Screenshot of a ZOOM based "All-Hands" session of the participants during the challenge

and milestones, a playbook that answered all possible questions as well as separated team folders for their project artefacts. Additionally, we used SLACK for asynchronous communication, for bringing all participants together and for addressing the various needs of all involved parties (Fig. 4).

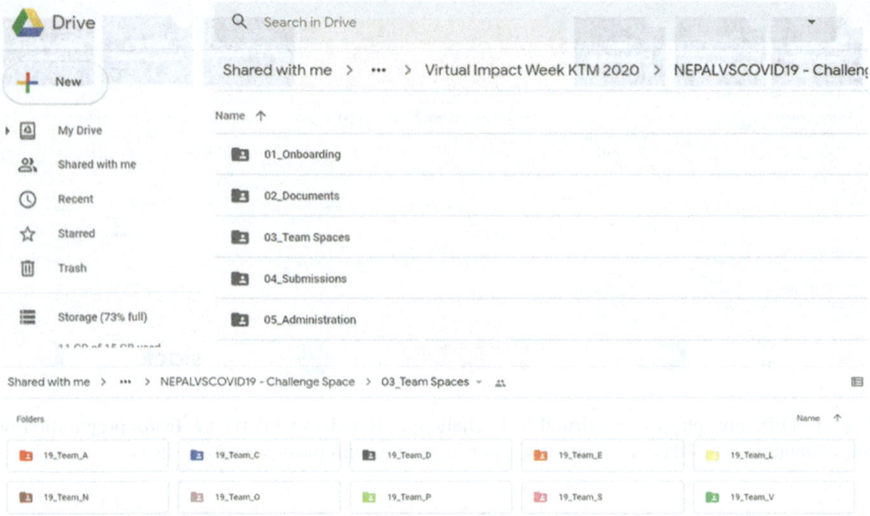

Fig. 3 Screenshot of challenge space on Google Drive used for documents sharing within teams

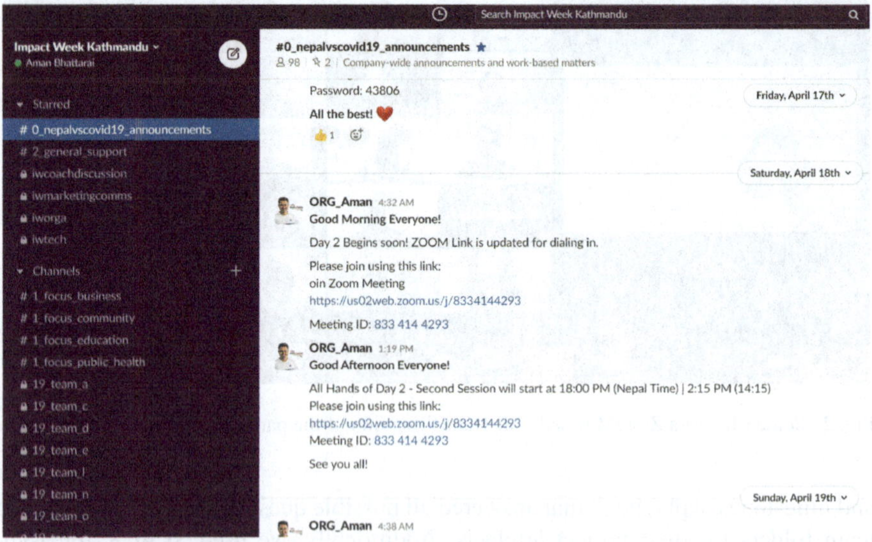

Fig. 4 Screenshot of SLACK platform with multiple channels created for team collaboration

2.3 Design Thinking Method Application

An integral part of the IW is the application of Design Thinking (DT) as the fundamental philosophical and methodological human-centered approach for problem solving (Brown 2008; Brown and Wyatt 2010; Melles et al. 2012). For the physical IW, the teams have 4–5 days to work on their projects, starting with the so-called "Problem Space". This involves a deep dive into the topic and getting a first-hand understanding of the users involved and their real needs through observation and interviews. Based on their insights the teams decide which particular challenge they would like to address and generate ideas for the second phase, the "Solution Space" (Council 2015). Due to the urgent matter in the context of COVID-19 the content team decided to open the stringent process and provide method support that allowed coaches and their teams to start with the challenge as a Research to Design approach or with an idea as a Design to Research approach (Downton 2003). The goal of the virtual IW for Nepal was to generate impact in the region. Therefore, also the role of the coaches changed from a mere DT facilitator to an expert mentor who would work more closely with the teams and build on existing solutions around the globe.

2.4 Execution and Outcomes of Nepal-Vs-COVID-19

Virtual Impact Week Nepal or Nepal-vs-COVID-19 had a lead time from planning to execution of 15 days. A concept paper was written with thorough detail on the format and expectations of the challenge. Through social media, academic networks and word of mouth, information about the event was disseminated across the Nepalese diaspora around the world. While Impact Week Kathmandu 2019 already was a big success and had a strong brand recognition, it led the virtual edition of the challenge to gather larger interest and participants from across the world applied to be part of it.

Out of 300 applications, 54 participants were selected based on diversity in academia, gender and location. Participants were allowed to apply as a pre-set team or as an individual. Individuals with relevant background and interest were paired into a team by the organizers. Those 54 participants formed 10 strong teams who were mentored by 2–3 coaches per team.

The coaches consisted of DT experts from Europe and Nepal. The majority of the teams consisted of a local Nepali expert paired with an international DT expert. This helped bring about a cultural and knowledge diversity within the team. For 10 teams, 21 coaches were selected by the organizers who came from 6 different countries.

10 days before the event, all coaches had a virtual onboarding. They were introduced to their coaching pair as well as briefed by the content team about technological tools, methodology and approach for executing the challenge. Each coaching pair were given a set of guidelines leaving specific decisions on format and detailed methods to be applied during the DT process to the coaches. Following the selection of participants, coaches were briefed about their participants and were asked to get in touch and onboard their team members. Participant onboarding comprised the following elements: tech onboarding and introduction to DT and framework of the event. Each coaching pair carried out the onboarding with their teams separately.

One day before the kick-off, all coaches and the organizing team came together for a mock session to update the larger group on their status quo and clarify any issues arising prior to the challenge. The challenge started on 17th April at 12:00 Noon (GMT +5:45). All participating teams and coaches came together in a so-called “All-Hands” ZOOM call hosted by the organizing team. Following the general introduction and keynote speeches, each team went to their team spaces i.e. separate ZOOM calls and started the DT process using MURAL & SLACK. Figures 5 and 6 show the digital space of one of the teams, the use of visualization to get participants engaged as well as digital post-its for brainstorm and clustering of ideas. The next two days started early at 09:00 AM (GMT −5:45) and always ended with an “All-Hands” in the evening at 5:30 PM (GMT −5:45) to bring the larger teams and their associated excitement together.

Due to strict lockdown in the country, participants were not allowed to go out for understanding the problem or testing their ideas on the users. Thus, all the interviews were carried out virtually through ZOOM, WhatsApp or telephone calls. To support the teams with getting appropriate experts for verifying the feasibility and relevance of their ideas, the organizers created an “Expert Pool” document with contact details of Nepalese and international experts in the field of medicine, education, business or social services. Participants had the possibility to contact experts on the list who were relevant to their domain of solution and gather feedback to further improve their prototype idea.

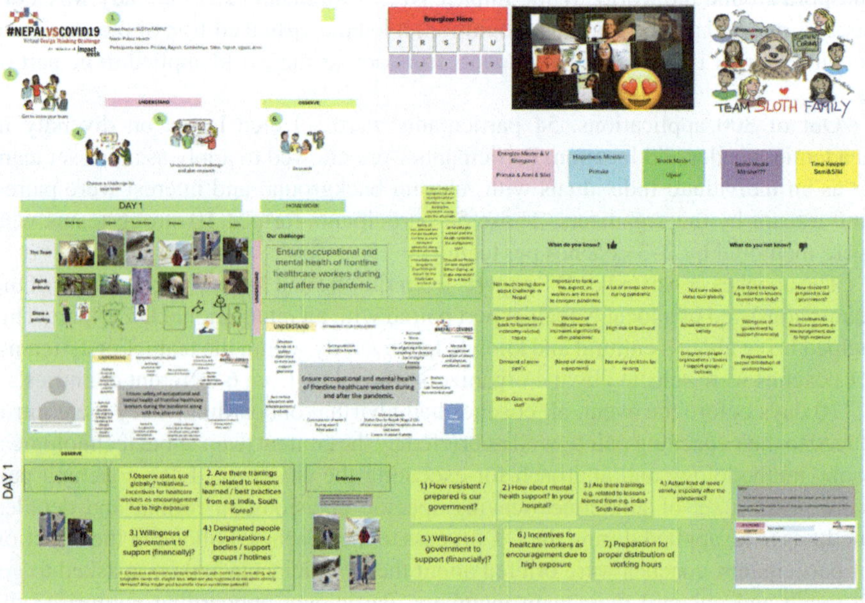

Fig. 5 MURAL screenshot illustrating virtual workspace for understanding phase during the challenge of a participating team

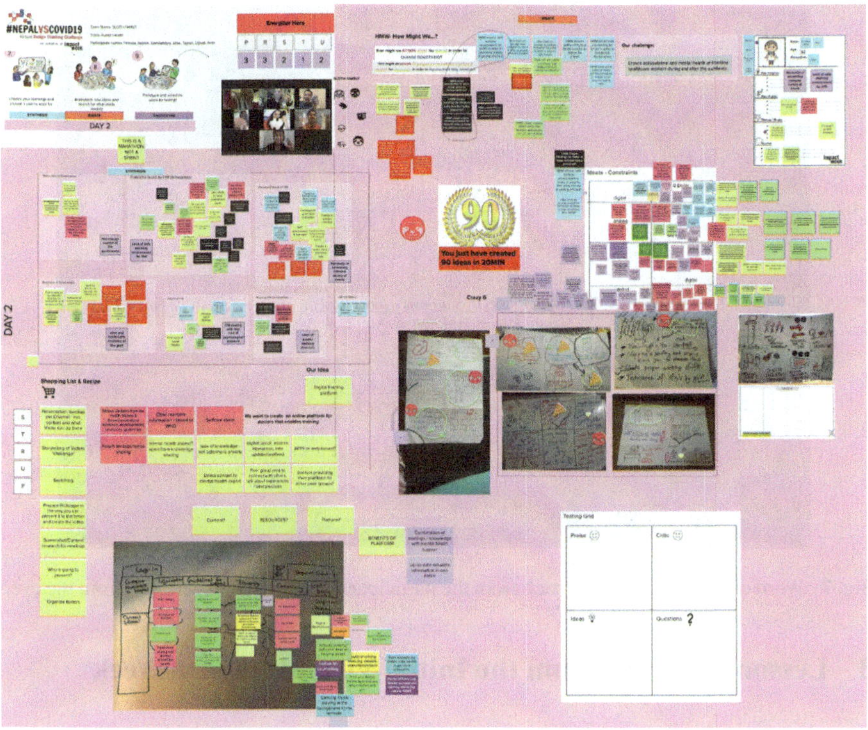

Fig. 6 MURAL screenshot for ideation phase during the challenge of a participating team

Each team had to provide an official submission at 5 PM on Sunday afternoon, 19th April, which consisted of a one-page executive summary of their solution idea, as well as a video pitch presentation of 3 min. The jury members comprised 3 local Nepalese entrepreneurs and 3 international innovation experts who evaluated the solutions and wrote feedback to individual teams. Three teams were selected as the "Top 3 Ideas" and a winning team was announced who would later also receive a seed capital of $500 to take next steps towards implementation. When asked to describe Nepal-vs-COVID-19 at the end of the event, participants used attributes ranging from "amazing", "inspiring", "creative" to describing the technical challenges like "there was power outage" (see Fig. 7). On average, the participants rated the overall experience as 8.9 on a scale from 1 to 10 (1: poor; 10: amazing) (N = 50). When asked to rate their overall learning from the event in a post-event feedback report 3 months after the event the score was 4.5 on a 5-point Likert scale (1: poor; 5: amazing) (N = 20).

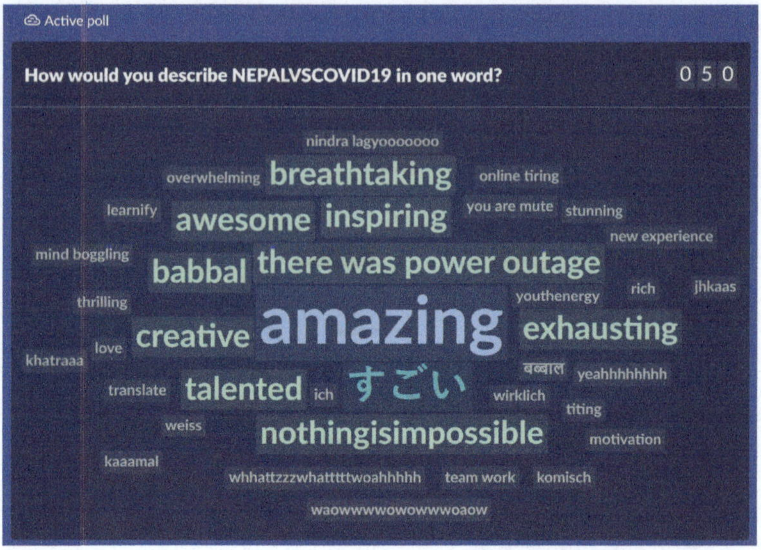

Fig. 7 Word clouding of live feedback from the participants about their experience from the event

3 Lessons Learned from the Initial Virtual Impact Week

The event was conceptualized and pursued as a prototype to see whether it was possible to bring diverse minds from around the world in a virtual space to do user-centric innovation. The biggest learning was simply: yes, it is possible! To the surprise of the organization team, the virtual edition of IW was engaging, enlightening and impactful and created a similar energy as the physical ones despite the limitations associated with the virtual technological set-up. Sitting in front of a computer for 10 h a day for 3 consecutive days evidently drained some energy, but the satisfaction of learning, innovating and contributing back to the society was nevertheless vividly seen across participants. Participants were determined and stayed together as a larger virtual community. According to an experienced jury, the ideas and prototypes that were developed under intense time pressure were judged to be equally good (or even better) when compared to those previously developed during physical IW events. Figure 8 shows the solution prototype from one of the teams: an online offer for students to prepare for their final exam during lockdown.

Participants in their post-event retrospective highlighted their gratitude and feeling of meaningfulness to have contributed to society in some form during the crisis. It was a new learning opportunity for them—both on using digital collaboration platforms as well as experiencing intercultural collaboration. This is exemplified by the following qualitative feedback, offered by a participant and coach:

> Nepal-vs-COVID-19 was not just about coming up with innovative ideas to help deal with the challenges posed by Covid-19 in Nepal but more of turning these ideas into tangible concepts making sure that those are doable right today. It felt amazing to be a part of this

Fig. 8 Prototype of online platform "Ejalo" from one of the participating teams

> diverse team with such amazing team spirit along with the support and guidance from the coach and mentor. It was a great learning experience and definitely looking forward to it next year. (Bijana Pantha, participant)
>
> Coaching the virtual IW for Nepal showed me that it is indeed possible to create powerful ideas, think outside the box and build solutions across virtual international borders. I was amazed by the spirit, motivation and eagerness to create an impact by the team - if YOU want anything is possible, even without touching and being with each other for real. (Anna-Lena Drewitz, coach)

A defining element of the regular global IWs is the unique spirit between participating students, coaches and the rest of the team. Everybody is working together shoulder to shoulder all day long for almost a week where you see and hear everybody. There are energetic discussions and even louder laughs, cheering within the team and informal chatters between the teams in the breaks and joint lunches. There are also celebrations in the cultural context of the country within which the IW is taking place. And there are expressions of joy of working together on something meaningful and being a part of a change for good. The organizers and the coaches struggled to bring this unique spirit to the virtual world. Likewise, technology was an enabler as well as a bottleneck. This manifested on some occasions when collaboration platforms went into sudden maintenance or their server crashed, thus requiring teams to find alternative platforms at the last minute to visualize their thoughts. Sudden power cuts in Kathmandu due to weather changes also created internet connectivity problems and PC access issues. As expected, time zone difference was also a major issue, but at the same time also manifested as an exciting experience for coaches joining from different parts of the world. Additional learning for any future virtual IWs will be predicated on local mentors bringing local perspective as well as supporting the facilitation of international coaches.

Upon reflecting on these experiences and lessons learned, several core factors seemed to be critical for the overall success of the event. As a pioneering event, things were expected to change at the last minute. Despite careful planning, virtual coordination was sometimes difficult, and things did not always work out as planned. Having an improvisation mindset from the beginning among both organizers and coaches helped to overcome last-minute obstacles throughout the event. Likewise, the selection of intrinsically motivated participants and coaches was a major success driver. During every "All-Hands" session, the organizers used icebreakers and humorous activities to engage the crowd and maintain a feeling of togetherness. For instance, on the last day the organizers organized a so-called "Pink It or Sing It! Challenge" where everyone was asked to wear pink clothes and those without would have to sing to the entire crowd virtually. This mini activity enriched the group dynamics and helped people to engage together as a group. Organizers had the skill to entertain and encourage the participants throughout the challenge, which became a major factor for overall success of the program.

4 Impact of Nepal-Vs-COVID-19 to Other Countries

The Nepal-vs-COVID-19 had sustainable success factors that led to spillover effects in other countries. Figure 9 shows the timelines for the virtual DT challenges following the Nepal-vs-COVID-19. These are Kenya-vs-COVID-19, Nigeria-vs-COVID-19, GreenImpactWeek in Germany, India-vs-COVID-19 and South Africa-vs-COVID-19. Many more events are in the pipeline without concrete dates yet but with initiators who have already taken the lead.

The Nepal-vs-COVID-19 provided the foundational structure and format which has become a blueprint for other DT challenge events. These factors include matters related to content, tools and coaches. This formed a major and easy starting point for organisers of any future events of the same kind. As a starting point, handover calls were adopted as a means to enlighten organisers of subsequent events. For instance, the handover call to the Kenya-vs-COVID-19 organising team clearly shed light on organisation strategies that worked well and those that didn't work well for the Nepal-vs-COVID-19. Challenges faced during the Nepal-vs-COVID-19 were also brought out as a way to take precautions. This included MURAL maintenance downtime on the first day of the event that affected team startup. The same happened for Nigeria-vs-COVID-19, GreenImpactWeek, India-vs-COVID-19 and SouthAfrica-vs-COVID-19.

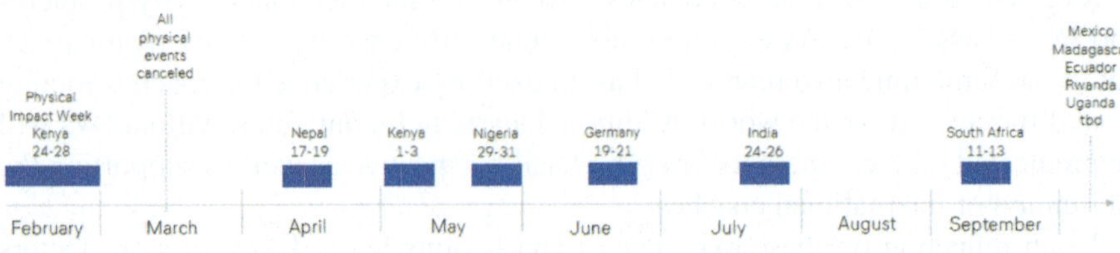

Fig. 9 Timeline for the virtual DT challenges following cancelation of physical events

Fig. 10 Team 'The Kibera Young Innovators' from Kenya-vs-COVID-19 during team building phase using a digital whiteboard

To give a snapshot of how Kenya-vs-COVID-19 took over from Nepal-vs-COVID-19, the organising team for the Kenya-vs-COVID-19 adopted a big percentage of the Nepal-vs-COVID-19 structure and format. It is however worth noting that the setting for Kenya-vs-COVID-19 was purely at the Kibera slum as opposed to Nepal-vs-COVID-19 which targeted University students. Such a slum setting meant that participants had no access to home internet, no guarantee of access to electronic devices and little knowledge on technological tools for collaboration. Despite the adoption of the structure and format established by Nepal-vs-COVID-19, the slum setting prompted consideration of new approaches including: having Google JamBoard as a lightweight option and backup for MURAL; participants being given data bundles to enable them to have access to technology; for better internet connectivity, participants were encouraged to join the sessions using Zoom on their phones while running MURAL on the laptops; and limiting team pitch submissions to only a one-page summary and recorded pitch, to ease the workload for coaches and participants. As a way to enrich the ideas that emerged from the DT challenge, announcement of results included jury providing feedback to all teams before announcement of the winning team. This was a completely new approach that was initiated by the Kenya-vs-COVID-19 organising team that has since been adopted by subsequent event organizers. Figure 10 shows "The Kibera Young Innovators" from Kenya-vs-COVID-19 during team building phase using a digital whiteboard and how they put their idea of a clean and healthier Kibera into practice by running the first waste management system in Kibera supported by funds from online crowdfunding (see Fig. 11). A number of participants from the team expressed in their own words, what they learned from the event and what impact the IW had on them.

> I would like to tell other youngsters that it is time for us to bring change with the wildest ideas that we have in mind. (Hillary Omuga, participant from team 'The Kibera Young Innovators')

> With Impact Week we learn how to apply the Design Thinking process in coming up with business ideas. It makes the complex process of developing an idea very easy, understandable and most important: fun! The coaches were very helpful to us, while we had the idea in place, it wouldn't be this great without their guidance on making the business plan and following the process. (James Mugambi, participant from team 'The Kibera Young Innovators')

> It's only through trial and error that we get to learn and each day we ought to be willing to be students. (Loghan Otieno, participant from team 'The Kibera Young Innovators')

A major success factor that made the dissemination of the event possible in other countries was the existing IW community. So far the selection of coaches has not been a major challenge for the organisers. The majority of the coaches had at some point been senior or junior coaches in past physical IWs. As a key boost to Design Challenges following Nepal-vs-COVID-19, many of the coaches opted to join Kenya-vs-COVID-19. This made it easier for the organising team to do coach onboarding. The

Fig. 11 Participants from 'The Kibera Young Innovators' cleaning a sewage in Kibera (Klinbera Facebook Page: https://www.facebook.com/klinberaservices/, Funding for Klinbera: https://www.gofundme.com/f/klinbera, Video Klinbera: https://www.youtube.com/watch?v=Fd6Ca2cJarc)

same happened to Nigeria-vs-COVID-19, the GreenImpactWeek, India-vs-COVID-19 and the upcoming SouthAfrica-vs-COVID-19. During India-vs-COVID-19 for example there was big support from the African IW community with coaches from Kenya, Nigeria, Ghana, Rwanda. For the final online session in India, a Kenyan coach even organized a live dance performance given by children from Kibera as a sign of connectedness across cultures within the IW community which was much appreciated by the Indian participants. The cross-country and intercultural collaboration of organisers and coaches may in fact add to international understanding and has the potential to decrease the risk of the rise of nationalism. We take the continuing cross-country collaboration also as an indicator of longevity and sustainability of the virtual DT challenge concept.

5 Opportunities for Future Research and Applications

In our case study, we presented a virtual program meant to foster innovation and intercultural exchange during the COVID-19 pandemic. As with all new programs and despite a strong sense of overall success, there have been limitations and remaining challenges. In the following we discuss these challenges and open questions, suggest avenues for future research and practice, and present implications for sustainability.

5.1 Limitations and Avenues for Future Research

The main goal of the virtual DT challenge was to offer a platform to connect people around the world during this crisis, to foster a "can-do" mindset and enable teams to solve local challenges related to the crisis and to offer quick and feasible solutions that may have an immediate positive effect during the pandemic. While the event indeed offered a functional space for a diverse group of people to innovate and collaborate, actually implementing the ideas that were developed during the three-day program proved to be difficult—these challenges had been similarly noted in physical IW events. Most teams, with very few exceptions, were highly motivated to continue to work on their ideas. Since the end of the Nepal-vs-COVID-19 event, a small number of teams have stayed together and are still working on their solutions after three and a half months (to our knowledge there are at least two Nepal-vs-COVID-19 teams who fall in this category). Teams indicated they were lacking knowledge and mentorship primarily to get necessary funding. Others also had other activities (studies or work) that did not allow them to invest enough time for implementing their ideas or even starting a business based on their prototype. Importantly, teams were more likely to pursue their ideas if they registered as a team and if they joined the program with an already existing idea. Increasing the likelihood and speed of implementing ideas would therefore require respective measures during the selection process (focus on teams, pre-existing ideas, participants with sufficient time and motivated to start a business) and a functional support system for teams after the event (e.g. incubators, support through local policy makers and institutions).

Moreover, when solutions are already obvious (like making masks available on a larger scale), the DT approach, which typically involves phases of research and understanding, may also not be the method of choice. An improved program may therefore require even more flexibility or preselected challenges that are best tackled by DT. Concerning the desired immediate solutions and effects of the virtual challenge it is worth noting, however, that the platform itself created an opportunity for everyone to actually be social again, to meet new people, and to even collaborate and do something meaningful. This was articulated by many participants after the event. In this way, the program also had an immediate positive social effect on all people who were involved.

The case study also has limitations. The organizers' focus has clearly been on delivering the program and much less on carrying out rigorous research. Organizers of future Country-vs-COVID-19 events should aim to properly evaluate the event qualitatively and quantitatively and in terms of different outcomes and on different levels (Kirkpatrick 1994; Kirkpatrick and Kayser Kirkpatrick 2016; Chen and Luetz 2020) as well as over time. Collected and analysed data across events could also allow for cross-country and cross-cultural comparisons yielding insights into global versus differential and contingent effects of the program.

5.2 *Opportunities for Future Practice*

Variations of the virtual program presented in this chapter could be used in a number of other contexts as well as similar crisis situations. For companies interested in enhancing their social corporate responsibility, collaborating in a virtual environment may offer additional if not completely new possibilities to invest. Projects may be started without or with less time abroad, saving costs and time plus reducing emissions associated with air travel. Digitally based initiatives would be a lot faster to initiate and would more easily allow to include people from all over the world with the benefit of using a much broader spectrum of knowledge and backgrounds for any challenge at hand. Some companies that had employees volunteering to take part in the new virtual format during the crisis seem to have already discovered the advantage of virtual Corporate Social Responsibility (CSR) projects and communicated on social media how while at home in lockdown, their employees were able to help other countries in need. As an additional advantage for companies, employees working in virtual settings inevitably boost their digital competencies—skills that we know are much needed at work with many companies struggling to sufficiently qualify their workforce.

The domain of education is another area with great potential and vested interest in virtual formats. This could also be seen in the engagement of King's College Kathmandu in Nepal-vs-COVID-19. In total 5 lecturers and 10 participants signed up for the event. Many of the lecturers have continued to teach virtually and to further develop their virtual teaching skills. The crisis clearly showed the possibilities and interest in online learning and teaching in general as well as an obvious need during pandemics like COVID-19 (also demonstrated by Team "Ejalo" offering online preparation courses for students in their final year). In the future, virtual IWs could even be offered to universities to foster online collaboration, innovation, and intercultural exchange. Some interest has already been expressed by universities in Australia.

Finally, the virtual format proved to be an effective and efficient tool during a global crisis. It enabled collaboration across countries, cultures, and time zones. It was fast to implement, resource friendly and inexpensive. The virtual DT challenge in particular proved to be a suitable intervention when confronted with new situations and unknown challenges like in the coronavirus crisis. It would become even more effective in terms of implementing solutions when officially commissioned by respective governments or investors.

5.3 Implications for Sustainability

In summary, the analysis of the virtual innovation program shows several implications for paving the way for a more sustainable world. As probably the most obvious sustainable outcome, the digital format developed for Nepal subsequently sparked events in a number of other countries, including Kenya, Nigeria, Germany, India, and South Africa, where the concept was adopted by local stakeholders. A global community of DT experts is educating students to understand the principles of this human-centered approach and apply it to real world problems in their communities. Creative thinking, problem-solving capabilities and teamwork (Dorst 2011) are crucial capabilities and not just increase the chance of sustainable employability but also set the base for a possible entrepreneurial pathway. Three aspects stood out that made the dissemination of the program and its educational efforts possible. First, organizers could recruit from an existing and active community with many team members taking part in previous virtual events and bringing in their experience. Second, in most cases local partners from previous years signaled their interest and subsequently helped promote the event in the respective countries. Third, organizers could build on a set of shared values like curiosity, respect and diversity that pre-existed in the IW community and made collaboration across cultures possible.

The study further adds to our understanding of sustainability by presenting a format that is flexible, scalable, and cost efficient. It can easily be applied to the management of future crises or other contexts such as education and business. At the same time, it is environmentally friendly without relying on international air travel and—as an increase in virtual collaboration in general—can be expected to contribute to widespread reduction of greenhouse gas emissions. Next, the virtual format supports the grassroots level development of solutions to local problems in a developing country context. In line with the IW philosophy, solutions in the new format are developed by local participants themselves and with feedback from local experts and mentors. International coaches merely facilitate the process. Furthermore, the study provides a very concrete example how cross-cultural, interdisciplinary collaboration and knowledge exchange beyond differences and national boundaries may look like in a global crisis, thus nurturing hope that similar approaches may increasingly be adopted for successful management of future global challenges on the whole. Finally, the study points to opportunities for bridging the divide between majority world countries and the rest of the world through encouraging investments in digital infrastructure and skills that in turn will provide more equal opportunities for people around the globe.

6 Conclusion

Is it possible to transfer a DT challenge into the virtual world during a global pandemic? Yes it is! Our motivation for contributing to this book was to share concept, outcomes and lessons learned with a broader target group interested in hands-on good practice for a more sustainable world. The virtual version of the IW was a success in a number of ways. First, it offered a platform for people from around the globe to connect during a global crisis, to create a sense of community and to work on specific challenges related to the crisis. Second, solutions developed by the teams were of high quality overall, with some solutions being even more advanced than in previous physical events. Third, the platform created an opportunity for both participants as well as coaches to boost their digital literacy by working in a fully virtual setting and using various virtual collaboration tools. Fourth, the event proved to be sustainable as it was subsequently adopted by other countries during the pandemic by local initiators and other members of the IW community. Fifth, the digital format presents a sustainable concept relevant for applications in business, education, and future crisis. Finally, and most importantly, the format connects people across the world in a meaningful way that promises to foster desirable outcomes far beyond innovation: mutual understanding, collaboration across borders, and the confidence that we together can shape the future of our planet.

Acknowledgements In the end, it is always about people and we would like to thank all of the people who helped to make the Impact Week (IW) happen—the past, current and upcoming one. With their passion and professionalism, as well as their positive and constructive attitude, almost everything is possible. We also want to thank all the contact partners and corporate sponsors who were involved in supporting the IW. Your contribution is important and very much appreciated, thank you!

If you want to become a part of the global IW community or host an event please get in contact with us at https://www.impactweek.net.

References

Brown T, Wyatt J (2010) Design thinking for social innovation. Stanf Soc Innov Rev Winter:30–35
Brown T (2008) Design thinking. Harv Bus Rev 86(6):84–92, 141
Bryman A (2016) Social research methods, 5th edn. Oxford University Press, Oxford, UK
Carmichael F, Darko C (2020) The global health pandemic and education in developing countries. University of Birmingham, Retrieved from https://www.birmingham.ac.uk/research/wirc/news/global-health-pandemic-education-developing-countries.aspx. Accessed 27 May 2020
Chen J-M, Luetz JM (2020). Mono-/Inter-/Multi-/Trans-/Anti-disciplinarity in Research. In: Leal Filho W, Marisa Azul A, Brandli L, Gökcin Özuyar P, Wall T (eds) Quality education: Encyclopedia of the UN sustainable development goals. Springer Nature, Switzerland, pp 1–17. https://doi.org/10.1007/978-3-319-69902-8_33-1
Council D (2015) What is the framework for innovation? Design Council's evolved Double Diamond. The Design Council. Retrieved from https://www.designcouncil.org.uk/news-opinion/what-framework-innovation-design-councils-evolved-double-diamond. Accessed 21 May 2020
Deardorff D (2011) Assessing intercultural competence. In: Penn J (ed) Assessing complex general education student learning outcomes. Wiley Periodical Inc., Raleigh CA
Dorst K (2011) The core of 'design thinking' and its application. Des Stud 32(6):521–532
Downton P (2003) Design research. RMIT Press, Melbourne
Goldin I (2020) Coronavirus is the biggest disaster for developing nations in our lifetime. The Guardian. Retrieved from https://www.theguardian.com/commentisfree/2020/apr/21/coronavirus-disaster-developing-nations-global-marshall-plan. Accessed 21 April 2020
Goodman PS, Politi D, Rai S, Chutel L, Dahir AL (2020) In world's most vulnerable countries, the pandemic rivals the 2008 crisis. New York Time. Retrieved from https://www.nytimes.com/2020/03/24/business/coronavirus-per-country-pandemic.html. Accessed 24 April 2020
Grots A, Creuznacher I (2016) Design thinking: process or culture? In: Brenner W, Uebernickel F (eds) Design thinking for innovation. Springer, Cham. https://doi.org/10.1007/978-3-319-26100-3_13
Johnson RB, Christensen L (2017) Educational research: quantitative, qualitative and mixed methods approaches, 6th edn. Sage, Thousand Oaks, CA
Kirkpatrick DL (1994) Evaluating training programs, 2nd edn. Berrett-Koehler, San Francisco, CA
Kirkpatrick JD, Kayser Kirkpatrick W (2016) Kirkpatrick's four levels of training evaluation. ATD Press, Alexandria
Kolb DA (2014) Experiential learning: experience as the source of learning and development, 2nd edn. Pearson Education, Upper Saddle River, NJ
Kolko J (2015) Design thinking comes of age: organizational culture. Harv Bus Rev. Retrieved from https://hbr.org/2015/09/design-thinking-comes-of-age. Accessed 28 April 2020
Kotter JP (2012) Leading change. Harvard Business Review Press, Cambridge, MA
Kotter JP, Cohen DS (2012) The heart of change: real-life stories of how people change their organizations. Harvard Business Review Press, Cambridge, MA
Leal Filho W, Luetz JM, Sattler DN, Nunn PD (2020) Coronavirus: COVID-19 transmission in Pacific Small Island developing states. Int J Environ Res Public Health 17:5409. https://doi.org/10.3390/ijerph17155409
Liedtka J (2018) Why design thinking works. Harvard Business Review, September–October Issue, Retrieved from https://hbr.org/2018/09/why-design-thinking-works. Accessed 14 April 2020
Melles G, Howard Z, Thompson-Whiteside S (2012) Teaching design thinking: expanding horizons in design education. Procedia: Soc Behav Sci 31:162–166
Nelson W, Fowler CF, Luetz JM (2019) Intercultural education for intercultural competence: a new kind of literacy for sustainable development. In: Leal Filho W, Marisa Azul A, Brandli L, Gökcin Özuyar P, Wall T (eds) Quality education: Encyclopedia of the UN sustainable development goals. Springer Nature, Switzerland, pp 1–16. https://doi.org/10.1007/978-3-319-69902-8_28-1

OECD (2018) PISA global competence framework: preparing our Youth for an inclusive and sustainable world. 33pp. Retrieved from https://www.oecd.org/education/Global-competency-for-an-inclusive-world.pdf. Accessed 5 April 2020

Punch KF (2014) Introduction to social research: quantitative and qualitative approaches, 3rd edn. Sage, Thousand Oaks, CA

Unger JM, Luetz JM (2019) Engaging employees in corporate social responsibility projects—a case study from the Lufthansa Group showcasing experiences and lessons gathered in Kenya, Rwanda, Nigeria and Columbia. In: Leal Filho W (ed) Social responsibility and sustainability—how businesses and organizations can operate in a sustainable and socially responsible way, World Sustainability Series. Cham, Springer Nature, pp 241–275. https://doi.org/10.1007/978-3-030-03562-4_13

United Nations Educational Scientific and Cultural Organization UNESCO (2013) Intercultural competences—conceptual and operational frameworks. New York, Retrieved from http://unesdoc.unesco.org/images/0021/002197/219768e.pdf Accessed 18 April 2020

Walker A (2020) Developing world economies hit hard by coronavirus. BBC News. Retrieved from https://www.bbc.com/news/business-52352395. Accessed 23 April 2020

Dr. Jens M. Unger is a Senior Manager in the area of Cultural Transformation at Lufthansa Group where he is responsible for fostering innovation and future competencies in top management and employees. He has a Diploma in Work and Organizational Psychology from Goethe University, Frankfurt and a Ph.D. focusing on success factors in entrepreneurship from the University of Giessen, Germany. Dr. Unger is managing director and global organizer of the Impact Week and has introduced the program to the learning portfolio of big corporations like the Lufthansa Group combining personal development, business needs and corporate responsibility.

Aman Bhattarai is an Analytics Consultant & Design Thinking Evangelist at zeroG—A Lufthansa Group company. Aman holds a M.Sc. in Supply Chain Engineering & Management and a B.Sc. in Industrial Engineering & Management from Jacobs University, Bremen, Germany. At his current role, he advises airlines and aviation executives on driving digital transformation programs and develops analytical products to derive insights from big data. As a visiting lecturer, he also teaches MBA class on data-driven decision making at leading B-schools in Nepal. Prior to zeroG, he spent several years in various analytical roles at Amazon in Munich, ArcelorMittal in Bremen & Lufthansa Cargo in Frankfurt. As an avid Design Thinking practitioner and supporter, he led first Impact Week Nepal in 2019 and later conceptualized Country-vs-COVID-19 movement in 2020. He led the Nepal-vs-COVID-19 and post its major success, he drove Nigeria-vs-COVID-19 and has been supporting other virtual Design Thinking challenges across the world. In 2015, German logistics magazine "DVZ" had featured him as "Next Generation of Managers".

Dr. Bettina Maisch is Professor for Entrepreneurship at the University of Applied Sciences Munich and Head of Education and Qualification at the Strascheg Center for Entrepreneurship (SCE). Dr. Maisch has two diplomas from the University of the Arts Berlin—in public and business communication as well as in electronic business. After a position as a marketing manager for the eGovernment Lab at the Fraunhofer Institute of Open Communication Systems she started her Ph.D. at the University St. Gallen and worked at the Media and Communication Institute at the University. During her Ph.D. she spent time at the Center for Design Research at Stanford University, Chair Prof. Larry Leifer, supported with a scholarship by the Swiss National Science Foundation as well as the Hasso Plattner Design Thinking Research Program. Before joining the current position, Bettina was Senior Key Expert Industrial Design Thinking at Siemens Corporate Technology (CT) in Beijing and Munich from 2012 to 2020, responsible for promoting customer-oriented development within the company.

Dr. Johannes M. Luetz is Research Chair at Christian Heritage College (CHC) in Brisbane, Australia. He has consulted for World Vision on research projects raising awareness of the growing effects of demographic, environmental and climate change on vulnerable communities in Asia, Africa and Latin America. He was raised in Sierra Leone and Switzerland, cares deeply about poverty reduction, holistic education, justice and equity, and has researched and published conceptually and empirically in these areas. He is a big picture thinker and has interdisciplinary research interests at the science-policy interface. His research has attracted grants, prizes, scholarships, awards and distinctions, including best thesis prize (2006), UNSW ASPIRE Award (2012) and best paper awards (2017, 2018). Dr. Luetz is an Adjunct Senior Lecturer at Sydney's University of New South Wales (UNSW), Adjunct Fellow at the University of the Sunshine Coast (USC), Deputy Editor at Emerald journal IJCCSM and Editorial Board member at Springer Nature journal Discover Sustainability.

Dr. James Obuhuma is a Computer Science faculty and the current Chairman, Computer and Information Technology Department, Africa Nazarene University, Kenya. James completed his Ph.D. in Computer Science at Maseno University, Kenya in March 2020. He holds an M.Sc. in Computer Science from the University of Nairobi and a B.Sc. in Computer Science and Technology from Maseno University. His M.Sc. thesis focused on Road Traffic Analysis using GPS Technology which opened his interest in Intelligent Systems. This heavily influenced his Ph.D. research topic which was in the area of Intelligent Transportation Systems (ITS) with a focus on Vehicle Driver Behaviour Modeling. His research interest is in the application of Machine Learning in different domains, including, ITS, Cybersecurity and IoT. James got passionate about Design Thinking in 2015 since the inception of the Impact Week in Kenya. He has since then participated in a number of Design Thinking workshops in Kenya, Uganda and Rwanda in different capacities, including, coaching, workshop overall moderation and even as an organiser.

Labor Market Sustainability and Corporate Governance—COVID-19 a Game-Changer

Diana-Mihaela Ţîrcă, Laura Mariana Cismaş, and Cornelia Dumitru

Abstract Sustainability cannot exist without the actual engine driving it: the active people, involved in the socio-economic sectors and who decide, finally, through various means whether economic policies addressing pressing concerns are feasible and acceptable, or not. The issue of the relationship between sustainability, labor market, and corporate governance needs to be addressed from a complex, holistic perspective combining both heterodox and orthodox economic perspectives considering the reality of COVID-19 as game-changer. The current pandemic has raised the issue of the necessity of 'renegotiating' the institutional-economic status-quo on labor markets showing the need of closer links between economic and social perspectives more than ever, especially with emphasis on the vulnerable groups. The chapter intends to analyze how operational terms at strategic institutional and economic policy level might support or hinder initiatives of preserving jobs and ensuring labor market resilience, while accomplishing the overall objectives of sustainable development. The methodology relies on desk research, based on empirical data collected in various international and European databases as well as on articles, papers, and working papers published by national and international academic and business journals, as well as in daily newspapers and will use to equal extent findings according to orthodox economy, combining them with findings resulting from the use of institutional indicators' measurement, thus providing an innovative perspective for the EU-level and national labor market, with respect to the labor market, and to the internal labor market, by considering developments in corporate governance.

Keywords Labor market sustainability · Corporate governance · COVID-19

D.-M. Ţîrcă (✉)
"Constantin Brancusi" University of Targu-Jiu, Faculty of Economics, Tineretului Street, No. 4, Targu-Jiu, Gorj, Romania

L. M. Cismaş
West University of Timisoara, Faculty of Economy and Business Administration, J.H. Pestalozzi Street, nr. 16, Timisoara, Romania
e-mail: laura.cismas@e-uvt.ro

C. Dumitru
Institute of National Economy, Romanian Academy, Calea 13 Septembrie nr. 13, sector 5, Bucharest, Romania

W. Leal Filho (ed.), *COVID-19: Paving the Way for a More Sustainable World*,
World Sustainability Series, https://doi.org/10.1007/978-3-030-69284-1_11

1 Introduction

The current period is described increasingly more as a period requiring a reset of the economic, social, cultural and political frameworks. All these frameworks operate in a complex environment, consisting of several stakeholders that assume different roles in various instances: they might be entrepreneurs, CEOs, public officials, employers and employees and also voices of the society at large in other circumstances. At the same time, the imperative of sustainability has been assumed by most countries of the world, but the Sustainable Development Goals (SDGs), while addressing issues that must be solved, somehow omit the core issue: actual deliverables depend on labor market and its sustainability in a context fraught with threats at all levels more than ever due to the COVID-19 pandemic. The same threats could be converted into opportunities if adequate policies, measures, instruments and tools are developed. In this context, the core reuniting peoples and their roles is the labor market and all the other corollaries significant for it from households to the way in which economic and social policies find agreement with large segments of the population and hence support ensuring overall welfare.

Already before the COVID-19 pandemic outbreak, economy and society were at crossroads caught between the indicators describing economic growth and wealth in terms of GDP and incomes' distribution, and the need for new sets of indicators, for instance the Green GDP (GGDP) (Stiglitz et al. 2009), and the ones indicating the level of distributed welfare inside and among countries grouped in various supranational economic organizations. In this context, COVID-19 has on one hand accelerated processes of technological progress and pressure, putting at risk not economic growth per se, but the workforce involved in national, regional and international value creation, and while attempts are made to identify the best ways to involve the active labor force and, at the same time, transform the risks into opportunities on which employers' various organizations and trade unions agree, currently the pandemic continues to pose more questions than answers.

There are some key words associated with the current pandemic, and which indicate a shift from the traditional economic approaches to ones that have been of concern more for the representatives of institutional, evolutionary and behavioral economics: "reset" or the "great reset" calling to action global stakeholders to act on the direct effects of COVID-19 It attempts to present the pandemic as a "window of opportunity" and indicates clearly the role of socio-economic institutions as important indicators about the capacity and ability of governments to generate the favorable environment for the three components, respectively fairer market outcomes—a market viewed in general terms of trade, incomes, taxation, subsidies and rules governing a core social-economic institution, the property rights with direct emphasis on intellectual property rights, and competition, viewed as fair, and even competitive solidarity. The second goal is even more related to the aforementioned concepts as it mentions investments as the basis for achieving the targets related to equality and sustainability, and the third component, the innovation agenda which should be more applied to goals dealing with public goods that contribute in safeguarding health and ensuring the premises for better facing social challenges that technology at unprecedented pace brings with it, underpinning the pivotal role for resilience (Schwab et al. 2020). Institutional indicators and mechanisms that are relevant and mentioned in the context, supporting the argument of (re)considering the relevance of socio-economic institutions for governments and business alike are the coordination capacity of governments, implying their ability to create medium- and long-term visions and to build sound public-private partnerships.

Other terms that need be more included are 'uncertainty', 'economic sentiment', transparency in both communication and cooperation, especially between public and private stakeholders that by their policy decisions decide the future of work.

The paper intends to emphasize the relevance of considering jointly the traditional macroeconomic indicators and the less used institutional indicators in analyzing the capacity of countries to achieve SDGs, as in the immediate future GDP growth will not necessarily mean also improvement of the indicators related to inequality, poverty and fair access to education and health (Millard 2020; Eurofound 2020).

The methodology relies on desk research, based on empirical data collected from various international and European databases (OECD, EUROSTAT, ILO, WGI and the Institutional Profile Database (CEPII)), along with surveying the literature dedicated to the social and economic impact of the coronavirus disseminated by academic and business journals, but also in the newspapers, in attempting to identify not only the widely accepted key terms of the necessary interventions for resuming economic growth and achieving set targets, but also to underpin how relevant the socio-economic institutions are and will be for all the economic initiatives and interventions dedicated to shaping a new socio-economic environment more in agreement with the SDG goals. The paper makes use to equal extent of findings according to orthodox economy, combining them with findings resulting from institutional indicators' measurements, thus providing an innovative perspective for the EU-level and national labor market, and to the internal labor market, by considering developments in corporate governance. However, a caveat is necessary: due to the COVID-19 pandemic, the economic data, irrespective of the socio-economic sector has still a high level of volatility, and the data and forecast models used before 2020 for time horizons like 2025 or 2030, by considering developments based on regressions, over the period 2010–2018/2019 has lost some of the significance, as the pandemic contradicts all economic growth models due to the necessity of taking into account mortality rates, number and duration of lockdowns, shutdowns and quarantines at national, sub-national/regional and local level, and how these measures were coordinated or not in the same geographic region. Lately, mathematical-statistical attempts have focused mostly on non-linear predictive models based on estimating the combined effects of infection outspread, and type and duration of national/sub-national quarantine measures (Vasiljeva et al. 2020) or econometric augmented models (Chudik et al. 2020) involving threshold effects based on a multi-country analysis and considering factors like global volatility, spillovers, interconnections of countries and markets. Yet, other estimation exercises developed possible macroeconomic development scenarios that resulted, in all instances in significant impact for the global economy of the pandemic (McKibbin et al. 2020). Other researches attempted to use indicators based on the infection rate at national and sub-national levels, and the type and duration of quarantine measures related to factors such as the relevance of certain services' sectors, dependency on remittances from other EU countries, oil prices, and/or energy consumption and mobility which is assumed to be mostly commuters from home to work, or students travelling across borders for studying, to registered unemployment, and measures taken for those whose labor contracts, (Coibion et al. 2020; Michaillat et al. 2020; Adams-Prassi et al. 2020) to mention but a few. Overall, it was difficult to capture in time the impact of COVID-19, save for the most vulnerable sectors where the effects were visible. One such sector is the tourism and travel sector which has dropped dramatically, the latest World Tourism Barometer from October 2020 (UNWTO) indicating that tourist arrivals dropped over the first eight months of the year by 70%, translated into 700 million fewer international tourist arrivals. All traditional tourist regions were hit, worst the Asia and Pacific region where the pandemic hit first, with a drop by 79%, and Europe, with a decline by 68%. The effects are severe, as in EU in main and partial tourism related activities account for more than 6% employment at EU-level, and for some countries for instance, Greece and Cyprus, tourism and tourism-related activities account for 16.4%, respectively 13.5% of total direct employment (Santos et al. 2020).

All these estimates are faced with essential key factors, that are more indicative about how institutional changes might occur, and other emerge for ensuring sustainability, starting from the socio-economic institutions and institutional arrangements in place: 'uncertainty', 'volatility', 'economic sentiment' that need be correlated with performances of institutional variables and indicators like public and economic stakeholders trust in the economic statements of governments, and of the business sector, etc.

Hence, turning the current difficult threats into opportunities for achieving SDGs at world level, and implementing successfully the "Green Deal" Agenda at European Union (EU) level requires addressing first the multiple economic uncertainties heightened by the pandemic (Baker et al. 2020).

We enumerate some of the known uncertainties, the first being the one induced by the technological pressure/progress which sped up over the last two decades and implied huge shifts in the way businesses and enterprises operate, and just as many challenges for the workforce present on the labor market. Large corporations to small businesses tend currently to be evaluated on how quickly they adapt and absorb new 'greener', profit generating technologies in their production chains, on how innovative they are or could become by making use of the technological-based identified advantages. All these changes are reflected in terms of productivity, employment, skills, and how incomes are distributed, Incomes especially seem to be the traditional quantifiable measure of how well adjusted is the economy of any country to these changes. One fact noticed already before the pandemic was that incomes become increasingly more polarized between high-earners and low earners, with a disappearing middle-class that used to be the backbone of free markets all over the world for more than half of the twentieth century. The issue of narrowing the income gaps among and even inside socio-economic sectors is complicated, as several issues need be taken into account from relevance of the work performed, required skill level to socio-demographic factors like the types of families, their origins, their level of and access to education, beliefs and values etc. While it might be considered that technology-driven globalization is at the root of these challenges, some are the outcome of how policies performed and managed to adapt to the requirements of the globalized world and its accelerated pace as of the nineties. The division is not only between rich, developed countries, but also exists also at the level of each country, and some patterns could be identified (OECD 2011) from the way incomes are distributed, to how support is granted for the unemployed, to the weight held by the informal sector in the economy, to work arrangements (short-, and part-time work) and contracts (full, for determined or undetermined time) etc. Moreover, other related challenges are related to spatial-geographic characteristics, ethnic or race considerations, access to education and health services, and the types of incentives used to boost employment at a reasonable pace with technological changes and improvements. These were all issues of economic concern before the pandemic, with high emphasis on the income inequality (OECD 2017). Consequently, COVID-19 has impacted on world and European economies when the economies were already pressured by increasing inequality, and wider gaps among and inside countries regarding access to good jobs, education, health care while attempting to manage an innovative Agenda which included tangible goals related to environmental sustainability.

The other huge uncertainty is how long the pandemic will last, and how current efforts of developing vaccines and initiating campaigns of vaccination will prove effective in allowing economies and societies to open up.

COVID-19 might be regarded in this context as a litmus paper for businesses and governments alike. The range of issues it has brought to light, especially as regards labor and incomes is wide, and some key terms might be identified: uncertainty on increase (Baker et al. 2020) for economies and labor markets, but also for society at large related to the confusing and oft times hasty measures that needed to be implemented for hindering the dramatic increase of persons infected and of deaths, uncertainty in the relationship between government and experts in the medical field, one example in point being that frequently Medical Pharmaceutical Interventions (MPI) and Non-Pharmaceutical Interventions (NPI) were more or less uncorrelated, and the outcomes of the latter being widely debated. A simple internet search has delivered more than 51.000.000 million results related to NPIs and their effects. Most emphasize the beneficial effects of reducing transmission by imposed lockdowns, shutdowns and social distancing (Flaxman et al. 2020), while some are more concerned in assessing not only efficiency, but also the outcomes of such lockdowns and shutdowns for all society minus the representatives of the health care workers (Seale et al. 2020). It shows that while some measures have been proved more effective in protecting the health of the general population, closing schools, for instance results in putting at risk a rather large group of children whose nutritional health was threatened (as their one sure meal per day was provided during the schooling hours), along with the higher risk of missed education opportunities as not all children have access to computers and/or internet (Masonbrink et al. 2020; UNICEF, FAO, and

WFP 2020). While the situation was commonplace for world and European countries, the responses of the society have brought to the fore new issues related to needed behavioural change as to encourage increased solidarity within the communities, but also behavioural change of main stakeholders from governmental to business level, one focal point being the ability to provide innovative solutions to ensuring access to education during the pandemic. The wide range of initiatives identified at EU-level shows that while adjusted to the local epidemiological context some issues were identified—and answers are yet to be found—as regards the psycho-social and nutritional wellbeing of the students the most worrying was the highlighting of the still fragile access to education and ensured nutritional health (King et al. 2020). All the issues enumerated, highlight the reasons why socio-economic institutions and institutional arrangements need to be addressed, and why sustainability could attempt to create an own socio-economic institutional agenda which highlights explicitly the existing socio-economic institutions relevant for sustainability, while progressively initiating the complex process of developing own institutions seen as sets of rules, regulations, standards, incentives and sanctions, and working towards identifying valid and measurable institutional variables in this respect.

The argument is that sustainability cannot exist without the actual engine driving it: the active people, involved in the socio-economic sectors and who decide, finally, through various means whether economic policies addressing pressing concerns are feasible and acceptable, or not.

Therefore, the issue of the relationship between sustainability, labor market, and corporate governance needs to be addressed from a complex, holistic perspective combining both heterodox and orthodox economic perspectives considering the reality of COVID-19 as game-changer. The current pandemic has raised the issue of the necessity of 'renegotiating' the institutional-economic status-quo on labor markets showing the need of closer links between economic and social perspectives more than ever, especially with emphasis on the vulnerable groups. The chapter intends to analyze how operational terms at strategic institutional and economic policy level might support or hinder initiatives of preserving jobs and ensuring labor market resilience, while accomplishing the overall objectives of sustainable development.

2 Economic and Social Institutions—Possible Framework for Developing the Socio-Economic Institutions of Sustainability

Over the past thirty years, there were several attempts to draw attention to the need of building and strengthening own socio-economic, cultural and political institutions for sustainable development, a phrase that seeks to identify and underpin the shared interests and concerns that cover the entire spectrum of socio-economic active individuals from the profit-oriented and pursuing entrepreneurs and industrialists' to those working in subsistence farming and seeking to minimize their risks, and to the social workers pleading for more social justice, and finally the environmentally concerned individuals advocating for wildlife protection and pollution elimination (Lele, 1991). Very often, these socio-economic stakeholders have goals and objectives that are at odds, and require sound, transparent, clear, predictable frameworks. Such frameworks are those based on functional and acknowledged socio-economic institutions. Therefore, in the absence of clear and specific institutions for sustainable development (O'Riordan and Voisey, 1997) a relevant framework of reference could be best designed at the labor market level, the place where government and business meet, and where institutional changes have occurred at rapid speed due to technological progress and pressure, and some of these changes have been even accelerated due to the current health crisis.

In this context, the cooperation and transparency in the communication and decision-making processes between government, corporations/companies could contribute top-down to providing a new basis for general consensus about how the very often vehiculated concepts of this period could ensure a sounder and fairer restart of economies at European and international level.

First, the new policies, actions and measures imposed by the changed circumstances, especially on the labor market and its specific socio-economic sectors, will require a deep and consistent analysis of the current socio-economic institutions, arrangements and architectures for operating improvements. As actual example, that still lacks more comprehensive analyses, and concrete measures dealing with the status of the workers involved in such socio-economic activities, from their rights nowadays to their rights in older age, we would like to mention the new work patterns of the gig economy and of platform workers which might register further considerable increases in the aftermath of the current difficult period. It is known that the financial-economic crisis has triggered the emergence of more workers in the 'gig economy' and of 'platform workers' both categories covering a wide range of occupations, from consultants, freelancers and experts' in various fields to unskilled workers—in the 'gig economy' and the platform workers for whom, the current nonpharmaceutical interventions (NPI) during the pandemic had, theoretically, little impact on the way they perform their work, as their activities in services' were already conducted digitally, and they were already location-independent, and web-based. While the definition is based on their earning 50% or more of their income via platforms, and by the working time of more than 20 h per week, their departure from the working arrangements on the traditional labor markets is obvious, while they still face the risk of being less experienced in a traditional labor market, than the workers present on the typical labor markets. Therefore, their status remains unclear, despite the fact that about 68.1% of them claim to be employees, while other 7.6% claim to be self-employed, and this results in a lot of risks regarding their rights of property over the work they deliver, or even over some intellectual property rights (Pesole et al. 2018).

To these two atypical workers' categories were added now, workers in both public and private sectors, who maintained their jobs by working from home (WFH) due to the conditions generated by COVID-19, as essential measures were NPI in nature, the most severe being represented by lockdowns and other various forms of confinement for the entire population of the world. Both types of 'new' work patterns pose similar policy issues, like solving the ambivalence of their status, which on one hand shows the potential of labor market innovations that are favorable to achieving full-employment and decent work goals, but on the other hand, might have undesirable socio-economic impacts that might undermine basic principles of social protection, equal opportunities and equity.

Technological change perceived as a factor for progress, but also as factor of pressure by some of those active in the labor market and lacking digital skills to higher and lesser extent, but also by those who benefit by putting to good use their up to date advanced digital/technological skills is the signal—long ignored—that it is time for policy intervention, respectively changes in policies and regulations that deal with the above-mentioned ambivalence but also with the issues that are sure to emerge, in the aftermath of the pandemic for large segments of the current economically active population.

The starting point, for governments would be to analyze where they are positioned with respect to the overall confidence of the public in the economic information they deliver, in particular at the national/regional and local level. Taking account of these issue, we opted for an analysis performed for selected countries of the EU based on the Institutional Profile Database as it shows some interesting results, regarding countries belonging to the Old Member-States, to the Former Member-States of Convergence and Cohesion and for the New Member-States. For purposes of more visibility, the selection does not include all member-states, but the ones which were considered most relevant for emphasizing the relevance of the institutional setting in the post-COVID-19 period. The selection includes four Old Member-States, four Former Member-States of Cohesion and Convergence, and four New Member-States based on criteria of vicinity and shared history, respectively Romania, Bulgaria, Hungary and Poland.

The premise for a well-functioning economy, implicitly labor market consists of several elements, and as most relevant for the current context when "reset" is the most used as regards European and world economies, we consider some of the institutional indicators proposed by the Institutional Profile Database managed by CEPII. The considered institutional variables are in our opinion indicative for how policy interventions, aiming for and ensuring sustainability, should be developed over the following period, based on monitoring the evolution of these variables that are critical in ensuring overall support for the set goals.

Moreover, these also could contribute, by improving the quality of the economic information, and the transparency in developing the economic policies—an attribute of the national governments in accordance with the agreements regarding the various socio-economic sectors achieved at EU-27/28 level—to achieving wide agreements between various industrial stakeholders, trade unions and the communities at large over a green, sustainable restart of the European economy by reindustrializing, developing new and safer products, services, and widening the customers' base all the time ensuring the safety of own supply-, distribution-, and value chains. The pandemic has proven that relying on the production of goods and services in other parts of the world, while a natural outcome of the interconnected and globalized world, is not satisfying and enough for ensuring safety, health, and welfare at the level of the European continent.

The stand-still of some of the economic sectors, in particular in the services' sector, and more specifically in HORECA and related industries (for instances) shows that while confinement measures were disposed by the governments', same governments have up to date failed in outlining safety nets for workers that might be exposed to risks, not only in conditions of health hazards, but in all other circumstances that bear the characteristics of a catastrophe. For instance, closing down the tourism facilities, restaurants, and other leisure and entertainment locations has brought with it not only risks for many of the most vulnerable workers employed in hotels, restaurants, etc. but also antagonism from the stakeholders in the branch, from employers, and led to claims of 'unfairness' as comparisons were made with other sectors dealing directly with customers. Moreover, European and world economies were faced with the reality of a lacking labor market intervention framework for possible instances that would imply closing down, partially closing down, or restricting some of the socio-economic activities that represented the source of income for many workers in vulnerable categories, and even in sectors that are not faced with vulnerability regarding their workers.

While before the pandemic labor market segmentation and polarization was apparently clarified enough, and the search for proper interventions on the right path, the pandemic has signaled the potential for a new type of segmentation, that will need to be taken into account, considering the current uncertainty which will continue after the pandemic: jobs that are fit for remote working and those that are not. Even more, inside this rough classification, other issues will need further inquiry, like the level of acceptability of corporations and public sector regarding WFH after the pandemic, the implications of such work arrangements for other segments of the workforce, and impact on generated incomes and revenues for sectors that are not fitted for such working arrangements.

Therefore, in order to ensure a good "reset" of economies, the beginning should be represented by meaningful top-down interventions on labor market that would have to deal with the most pressing policy issues, that could build the confidence of the employers and employees alike, and boost reforms intended to slow-down and reverse segmentation and polarization of the labor market that will become even harsher in the absence of meaningful dialogue of all stakeholders.

The building blocks that are necessary for developing a long-term vision, and disposing of stronger capacity for structural reforms, at sectoral level, including here how willing and able will be the national multi-level governments to cooperate with the private sector for creating long-term sustainability on the labor market, considering several vulnerabilities that have been highlighted by the SARS-COV19 outbreak in the EU-27/28 and at world level (Table 1).

We notice over the period 2012–2016 that all countries have undergone some changes as regards completeness, credibility and performance with respect to the economic information delivered, and in some instances, also with respect to the transparency of economic policies, with agrees with the view that in some cases the

Table 1 Building blocks of general socio-economic trust in selected European Union Member States in 2012 and 2016 (Authors' processing after the Institutional Profile Database, CEPII, 2012 and 2016)

Countries	Year			
	2012		2016	
	Economic information reliability	Economic Policy Transparency	Economic information reliability	Economic Policy Transparency
Germany	4.00	4.00	3.33	4.00
France	3.83	4.00	3.50	4.00
Italy	4.00	3.50	–	4.00
Netherlands	4.00	4.00	3.50	4.00
Austria	4.00	4.00	3.33	3.00
Greece	3.17	3.00	2.50	3.00
Portugal	3.50	4.00	2.83	3.00
Spain	3.50	3.50	2.67	2.50
Bulgaria	3.17	3.00	2.67	3.50
Hungary	3.00	3.50	2.67	3.50
Poland	4.00	2.50	2.67	3.50
Romania	3.17	3.00	2.50	3.00

influence and power of trade unions representative at branch, and sector level has continued to weaken.

These changes, as reflected in these two institutional indicators' components show that attention should be paid as to how institutional changes occur, and to the capacity for institutional responsiveness.

The macroeconomic policies promoted at world and European level have been focused for more than half of a century only on the estimates, assessments and measurements performed in the framework of mainstream economics, and less on the economic and social institutions on which they were founded. However, the increasing more frequent economic crises, triggering unavoidably social crises have brought again into attention institutional economics as possible means of explaining why certain economies and nations thrive, while others fail (North 1990; Hall and Jones 1999; Acemoglu et al. 2001, 2004, 2012; Rodrik et al. 2002), providing a center role to the Neo-institutional Economics as bridge between mainstream macroeconomics and institutionalism in economics. Based on the new perspectives of North, who defined institutions as the rules of the game, an entire literature has since been continuously enriched on the role of institutions for economic growth and development. Most scholars in the field regard institutions as per North's definition as 'rules of the game' and support the argument that they are the basic ingredient ensuring the good or bad functioning of the economies, while providing information over the attitudes and reactions of the main stakeholders according to their respective operational framework.

For the purposes of the present chapter, we intend to pursue the Northian definition, while arguing that sustainability though formalized institutionally by the shared agreement at world level about the SDGs, with few exceptions, still needs to create and strengthen own institutions and institutional arrangements, for fulfilling its purpose, and that these institutions should be grounded in the institutional arrangements of the labor market, as it reflects the most important goals that are pursued, and is the main 'supplier' of the necessary ingredients for achieving the SDGs. Hence, sustainability of labor market and the outcomes it delivers are decisive for the generations active in the labor market, and for the ones either preparing to enter or exit it by retirement, and for promoting the SDG agenda based on supporting economic growth in agreement with the objectives of social justice.

For supporting this argument, we provide yet two other indicators, that are relevant for both SDG and labor markets if the aimed at improvement in the quality of using resources and achieving quality and equity in distribution while maintaining the ability of the future generations to enjoy benefits and welfare of their own (see Brundtland Report 1987) (Table 2).

Overall, it seems that countries that have a strong dialogue between the public and private sectors with shared interests and good cooperation in decisions about the direction of the economic policies, and associated reforms, including the necessary reforms on the labor market have more consistency and constancy for instance,

Table 2 'Ingredients' for ensuring improved labor market sustainability, 2012 and 2016 (Authors' processing after the Institutional Profile Database, CEPII, 2012 and 2016)

Countries	Year			
	2012		2016	
	Capacity for sectorial reform	Public-private cooperation	Capacity for sectorial reform	Public-private cooperation
Germany	3.80	3.33	3.80	3.33
France	2.60	3.00	3.40	2.67
Italy	2.80	2.33	2.60	3.00
Netherlands	3.00	4.00	3.20	4.00
Austria	3.00	2.33	2.20	2.67
Greece	2.20	3.00	1.40	2.00
Portugal	3.40	4.00	3.60	3.67
Spain	3.50	3.50	2.60	2.67
Bulgaria	2.60	2.00	2.60	2.67
Hungary	3.00	1.33	2.67	3.50
Poland	3.40	1.00	2.80	2.67
Romania	2.40	2.33	1.80	1.67

Germany with no changes in this respect, whereas countries that have been previously more affected by the financial-economic crisis have fewer perspectives and suffer even a worsening of these institutional indicators that could be regarded as warning signals with respect to the economic growth perspectives, but also for the perspectives as regards smart, inclusive and sustainable growth, as the economic growth will be strongly conditioned, in particular in the post-pandemic period by the ability to cooperate between the public and private stakeholders if sustainability is to be achieved on labor markets, but also in the other socio-economic sectors. Worrying for the future, considering the impact of the health crisis that will have according to most estimates deeper effects that the financial-economic crisis of 2008–2011, is that countries in the two groups of former member-states of convergence and cohesion, and the New Member States are more threatened as there seems to be a decrease in their capacities of sectoral reform and in ensuring the private-public cooperation.

One last issue that should be included into the aims of the SDGs, and directly related to the labor market is the property rights' regime and the distributional and allocation mechanisms incorporated in them are relevant as regards sustainability for both natural capital, and labor, as a property right for both employer and employee. The two facets of labor right—rather restrictive if we consider the position of the employer, and aiming to improved and fairer work conditions and security on the side of the workers—has particular relevance for labor market sustainability in the context of the current accelerated changes. If labor right of the employer might limit, or influence negatively the industrial relations which have been weakened lately by the financial-economic crisis and perhaps it will be even more weakened by the current pandemic, on workers' side, the same consideration of labor right as property right might ensure for past, present and future employees the guarantee of fairer work conditions, improved contractual terms, better retirement perspectives as regards pensions, and even warranties for new entrants on labor market (Novitz 2012).

3 Reset and the 'New Labor Market'

The current and ongoing SARS-COV2 pandemic triggered the intensive, almost fashionable use of the key-word "reset" as regards economies and societies together, as it highlighted all vulnerabilities of the world economies in all sectors and at multiple levels. In agriculture, production and the services' sector multiple distribution and value chains were disrupted, and the frail recovery of jobs was endangered, especially in sectors that are not teleworkable and rely on multiple interactions between suppliers and consumers. One such high-risk sector is, for instance, the tourism and travel industry, which is faced with high losses, the estimates circling around a drop by 60–80% for the entire year 2020. At the other end of the spectrum we encounter essential jobs that need constant interaction like those of ensuring public order, medical assistance, essential transports etc. The common denominator for all jobs is that most threatened continue to be the low-paid not teleworkable ones, and the teleworkable jobs which might be replaced in the immediate future by automated and even AI coordinated means.

However, if we consider the wider framework on medium- and long-term, it can be seen that the knowledge-based and digitalized economy and society needed a reset already that was merely accelerated by the pandemic. Therefore, any approaches to sustainability should begin by focusing on the core issue of labor market as key-factor in achieving the SDGs for the time-horizon 2030 and beyond.

In this context, some caveats are necessary: the traditional definitions of labor market while maintaining their value and relevance need be improved for meeting the complex demands of an economy increasingly more digitalized, automatized, and implementing at accelerated speed AI in all fields of activity. Moreover, a knowledge-society which changes knowledge into competitive advantage and improved value added, should reflect on the institutions and institutional arrangements and frameworks governing the activities from company level to top governmental level for ensuring sustainability of labor markets. In this respect, some steps have been taken by researchers and academia involved in analyzing the possible future of institutions representative for the labor markets (Freeman et al. 2005).

Sustainability of labor markets is crucial, as they are a reflection of the people involved in the various socio-economic sectors and they are ultimately the most valuable resource in achieving the SDG goals by 2030. Hence, the second caveat is that sustainability as overall goal, and for all its 17 goals requires stronger rooting in institutions and institutional arrangements of the labor market along with developing own institutions/institutional arrangements that would be measurable and inform better about results, progress and intervention needs.

The third and last caveat is that we define institutions in the Northian understanding as rules of the game, respectively as the set of pre-existing norms, conditions, provisions, incentives and sanctions according to which operate economic agents and societies alike be it in economic, social or cultural environments (North 1990).

The 'reset' paradigm that seems to dominate the current discussions from the governmental and academic level to the one of corporations and even inside the societies needs to be addressed, therefore from a holistic and complex perspective, seeking to define the "reset" as all-encompassing concept, and form there, the definition about what is and how the "new labor market" of the twenty-first century will function, both from the perspective of policies, and of the ways these will be implemented considering the challenges posed by the increased polarization.

The currently recognized polarization, is worsened by the new type of segmentation that emerged over the period in which the COVID-19 outspread worsened, respectively essential socio-economic sectors and non-essential, associated to the ability of those present in the respective sector to work or not remotely. A new segmentation emerged and became measurable on the labor market: teleworkable and non-teleworkable essential and non-essential economic activities. Tackling the amounting issues resulting from this new type of segmentation is essential, as it creates a new threat to the SDGs that deal with inequality, decent work, and inclusiveness and even with the first SDG priority of no poverty.

Next, one overall goal should be involving decision-factors, non-governmental and corporate stakeholders in defining the characteristic features of a 'new' sustainable labor market which begins to take shape even during the pandemic, and will be key in resuming sustainable economic growth thereafter, the skills and competences' needs, the main sectors' and activities that could contribute towards achieving the SDGs, the level of involvement required and necessary for achieving short- and medium-targets, and the way these targets could support achieving long-term targets and setting new goals beyond 2030.

"The reset" is a triggering concept for complex initiatives encompassing changes determined and accelerated by the current pandemic, which needs to take into account as building blocks several shifts and transitions that are ongoing and will be finalized in the foreseeable future, or which have been finalized and are consolidated due to the pandemic. (1) It means shifting to a more integrated and interactive relationship between economic and social considerations, while taking holistic approaches as regards the interests of the stakeholders and shareholders by shifting the emphasis from profit/rents seeking and achieving company/corporation targets and focusing not only on the own shareholders, to identifying opportunities and shared/complementary goals with other interconnected companies/corporations dealing with comparable issues for creating the premises for sound and sustainable recovery of the economy and society in the aftermath of the pandemic. (2) It implies focusing on outcomes that meet the agreement and expectations of the society and support a more sustainable, environmentally friendly and resilient rebuild of the activity/industry/sector while acting as driver for resuming growth not only in their sphere of interest, but also in the other complementary, related and non-related spheres of socio-economic activity. (3) It will require a rethinking of the actual measurements about what counts for wellbeing and welfare, a fact already emphasized by the Sen-Stiglitz-Fitoussi report (2009) which underpinned that GDP as measurement gives but a narrow and restrictive information. In fact, by focusing only on outcomes regarding material production it eludes considering the most relevant facts that are representative for a society based on knowledge and aiming to inclusiveness, eliminating inequalities, poverty and all types of discrimination. (4) Governments, companies and non-governmental organizations will need to improve and strengthen further socio-economic institutions and mechanisms that could create a holistic environment at both strategic and operational level that could function as platform for identifying new measurements of economic growth that are more focused on essential future priorities and are circumscribed to the imperatives of the knowledge-society: education, health, wellbeing and environment.

"The new labor market" should be defined as the repository of all economic and societal gains, as the backbone for ensuring inclusiveness, eliminating inequity, discrimination and for generating the outcomes providing for personal wellbeing and general welfare of the societies involved in the process of shifting to a new type of economy that combines the interests of stakeholders in various fields with the ones of the other members of the society. It is defined by all the characteristics it has developed over time, where increased con old behaviors of which some will disappear, while others need permanent adjustment, and emerging new behaviors with respect to work, work ethics, and productivity. At the same time employment and unemployment change their features due to opportunities and threats generated by the policy solutions identified for improving the former and combating the latter. The new perspective and definition of labor market is no longer only as the place where demand and supply for labor force meets, but as a fluid conglomerate of demand and supply for labor, knowledge, skills and competences that become complex and need new types of institutions and institutional arrangements to arbitrate over the changed working environment created by the increasing cooperation and coordination between human activity involvement and use of digitalized-automated-and AI driven processes in all socio economic sectors.

The question about labor market sustainability becomes pertinent precisely in this context, as there are several questions and issues that need to be addressed relying on the relevance of corporate governance which will need to assume new roles for meeting not only the assumed goals and targets of the SDG 2030 Agenda, but also the goals and targets that emerge ongoing as result of the effects triggered by COVID-19.

One pressing concern is the future of labor market, as 'teleworkable' is the effect of changing work-rules due to a health crisis. However, recent analyses pursuing to identify how the labor market has responded to the challenges of the pandemic have shown changes that might turn into new work practices for the future, even after the pandemic is past.

The division of the sectors into essential and fully active, active but conditioned on the capacity to work remotely (active via telework), followed by essential and partly active sectors but that cannot operate remotely, and non-essential and inactive lacking the capacity of telework, and closed sectors was realized initially based on how restrictive were the confinement measures.

The results were two-digit drops of the GDP for the entire European Union, and in the world, and the results were shown in the increased and forecasted to increase further unemployment rates (Table 3).

This evolution of real GDP and unemployment could also be defined in terms of the adjustability of the economies to operating based on their digital and automation/robotization and IT-capabilities. It shows, that the GDP increase does not necessarily imply increases in employment, as automation, robotization and digitalization are preponderant.

In this context we mention that GDP is not an accurate measure for indicating economic growth associated to welfare at national, European and world level, but a convention employed by economists and accountants for delivering an image about the actual development of the respective economy. The suggested measurement of the 'green GDP' (GGDP) which attempts to reconcile objectives of economic growth and of environmental sustainability would need to be associated to new types of socio-economic measurements (Costanza et al. 2009; Stiglitz et al. 2009), for instance a green employment rate (GER) that would reflect the progress registered in creating new jobs or improving current jobs as to meet the demands of the green economy which is the shared aim at European and global level.

Hence, the imperative of taking account of measures to resume the activity on labor markets, should be related to active green labor market policies. These green

Table 3 Estimated real GDP and unemployment rates for 2020 and 2021 (*Source* European Commission, Autumn Economic Forecast, November 2020)

Countries	Year			
	2020		2021	
	Real GDP	Unemployment rate	Real GDP	Unemployment rate
Germany	−5.6	4.0	3.5	4.0
France	−9.4	8.5	5.8	10.7
Italy	−9.9	9.9	4.1	11.6
Netherlands	−5.3	4.4	2.2	6.4
Austria	−7.1	5.5	4.1	5.1
Greece	−9.0	18.0	5.0	17.5
Portugal	−9.3	8.0	5.4	7.7
Spain	−12.4	16.7	5.4	17.9
Bulgaria	−5.1	5.8	2.6	5.6
Hungary	−6.4	4.4	4.0	4.4
Poland	−3.6	4.0	3.3	5.3
Romania	−5.2	5.9	3.3	6.2

active labor market policies should focus first on solving the issues related to lost working hours, labor force exits during the pandemic, but also in the post-pandemic period dedicated to rebuilding all markets. Especially as regards those that left the traditional labor force by opting for the gig economy, or platform work and who are most represented in the high-skilled category, should be encouraged by means of taxation policies, and other incentives to become more innovative and greener in their economic activities, and platform workers should have a status comparable to those in the gig economy, by creating better frameworks regarding their insurance opportunities and access. These actions, in turn would contribute in creating new social funds for assisting those at the opposite pole, and who are included in the 'vulnerable group' based on their low—or lacking skills. Therefore, new and institutionalized economic provisions should have their emphasis also on the social side by providing various incentive schemes that by their components would support inter- and intra-generational equity.

4 Corporate Governance and COVID-19 Triggered Game Changes

The resilience of corporations has been tested sharply, and in this context, they will have to reconsider the relationships built and to be built in the next future not only with their own stakeholders, suppliers, consumers and the public, but to analyze the possible options for a 'stakeholder' capitalism model (Hunt et al. 2020) for being able to negotiate with their shareholders, but also with their workforce and even the society as trust will need to be restored at the level of the entire society about the shared interest in a common sustainable future.

The general feeling, according to a recent survey, even before the corona-crisis showed that from the persons interviewed 92% considered that companies should be more active on issues related to training, to the effects of automation on jobs, and to the ones related to migration (72% of the interviewed). These are but a few issues that need tackled with emphasis on training and education, and ensuring a healthier work environment to which, despite the horrendous negative effects and deaths, COVID-19 has opened the path. For instance, with proper top-down leadership and direction, teleworking might become the future for many jobs both in the public and private sector in administration, health (telemedicine is turning into reality in many countries), education, etc.

Corporate governance will need in this instance to prove its involvement by strengthening its social involvement, changing the model centered exclusively on own shareholders to a model centered on the resilience of the company, and to consider avoiding immediate profit maximizing behaviors in favor of more involved long-term benefit warranting behaviors, that generate long-lasting values.

The corporate governance models will have to shift emphasis of the corporate purpose to a more equity ensuring one to all partners: shareholders, employees, suppliers, and communities. This will imply developing new institutionalized informal networks for constant and consistent dialogue with all the groups it enters into constant relationship: shareholders, employees, consumers, communities, etc.

Another target, which corporations will need to strive to achieve is the one of developing clear, transparent and cooperative tools for transparent and collaborative dialogue with partners from inter-related sectors, in order to prevent a major crisis, irrespective of its nature (health crisis or as result of a natural disaster) to destabilize the economy and the labor market as the current pandemic has done.

Moreover, all corporations will need to acknowledge that by their activities they impact either positively or negatively on the systemic issues that generated the inequities and disparities on the labor market and will have to contribute to improving conditions as regards labor market polarization, the increasing inequities in incomes and wealth, the environmental degradation, and provide for a climate that contributes to improving the deteriorating environment in some public institutions.

Following the Six-sigma principles companies will have to follow an improved organizational culture that will provide for long-term competitive advantage. This means pursuing the DMAIC model more consistently, respectively respect the steps in defining the issues hindering performance, and identify the requirements for improvement, measuring the distance towards the aimed level of performance, analyzing the hypothesis for the lacking performance and certify their validity, than is able to intervene rapidly for improvement based on tested solutions, and exert control for maintaining the required and aimed at level of performance.

This would also mean strengthening in the post-COVID-19 future the capacity of concluding partnerships and alliances that will be centered more on the 'collective self-interest' of the environment in which they develop the business, and use their digital and technological capabilities also for the benefit of the community. These partnership and alliances will need to rely increasingly more on sectors of public interest like education and research-development, if the rate of innovation and work disruption is to be tempered so as to avoid increased unemployment.

This means, corporations will need to be involved more in closing the skills gaps by concerted actions of improving the capabilities of their present workforce and preparing in collaboration with the educational system the workforce of the future. The reasoning is that by now, already 84% of all employers in the world have already began the rapid digitalization of their working processes associating them with remote working. The reverse is, that in absence of decisive measures, this development will increase even more the inequality as it will grow under the impact of the two forces transforming currently the world and European economy: technology and pandemic.

5 Conclusions

The reset envisaged at world and European level requires a renewal of the "social and economic contract" in developed, developing and even less-developed countries (LDCs) alike. In order to achieve this, we believe that sustainability needs to develop own institutions and institutional arrangements both formal and informal, that could provide a better basis for delivering the intended outcomes, by addressing first priorities related to labor market, as its institutions and institutional arrangements could be a tentative 'roadmap' for creating own institutions. Labor market can provide support for sustainability as it encompasses all stakeholders, from traditional ones represented by employers' organizations, federations to the ones of trade unions in both public and private sector, and more important the emerging new types of gig economy and platform workers—who, by right incentives could act as vectors of change towards a green economy, by being incentivized and involved in processes that turn greener the economies at their countries' level.

The key institutions and institutional arrangements that could serve these purposes are based on the credibility and transparency in the communication about the reliability of economic information and transparency in developing economic policies that target employment, but also in social policies that are directly affected by them.

These institutions contributing to improving the understanding and action capacity of peoples and corporations regarding the overall economic environment, could be used also in terms of sustainability, where the goals should be accompanied by comprehensive, transparent and clear analyses about short-, and medium- risks and opportunities on the road to achieving the targets.

For instance, the paper has analyzed institutional indicators that could deliver a good basis for sustainability goals: respectively, the ones related to reform capacity at sectorial level, could be translated for use in specific sectors that are still intensive-polluters (energy, various manufacturing and processing enterprises, transportation, tourism etc.). Evidence over the need of stronger sustainability institutions is the fact that an agreement has not been achieved over a single definition by all interested parties and stakeholders. Institutions and own formal and informal arrangements built inside the "sustainability' concept would contribute better to arriving at a unitary definition, and provide for a sound framework in pursuing the goals aimed for the time-horizon 2030 and beyond.

The renewal of the social and economic contract is based on the reality that after the crisis a 'new labor market' will emerge, and ensuring its sustainability implies closer links between the stakeholders of public and private governance that will need to act based on the reality that the pandemic has accelerated the emergence of new patterns of work, but which will be for the foreseeable future disruptive for the workforce, in absence of developing new educational and training models.

The four starting pillars for 'reconciling' public and private sphere with respect to the sustainability of the labor market in the context of changed rules (remote work, remote education etc. changed work processes and value chains), are: economic information credibility as provided at supranational (EU), national and sub-national levels, associated to transparency of the economic policies that should be debated openly, not only with stakeholders of the targeted sector, but also with stakeholders from complementary and related sectors, with representatives of the employees in the sector, and the communities. These two pillars will require also, for EU member-states improving their capacities of sectoral reforms under the combined pressure of technology and skills gap, together with their capability for concluding sustainable, transparent and reliable public-partnerships. These partnerships could provide sound basis for reskilling needs and labor force demands, ensuring that the current disruptions that are fraught with uncertainties could be changed into competitive advantages based on sustained dialogue, transparency and cooperation for achieving labor market sustainability.

The combination of these factors could contribute beyond the digital capability building which takes place at an accelerated rate during the pandemic to actual behavioral changes for both corporations that will lead to increased business resilience of their operations, but also to labor force upskilling and increased resilience in the market.

References

Acemoglu D, Johnson S, Robinson JA (2001) The Colonial origins of comparative development: an empirical investigation. Am Econ Rev 91(5):1369–1401

Acemoglu D, Johnson S, Robinson JA (2004) Institutions as the fundamental cause of long-run growth. National Bureau of Economic Research, Working Paper, No. 10481

Acemoglu D et al (2012) Why nations fail: The origins of power, prosperity and poverty. Asean Economic Bulletin 29(2):168

Adams-Prassi A et al (2020) The large and unequal impact of COVID-19 on workers. https://voxeu.org/article/large-and-unequal-impact-covid-19-workers

Baker S et al (2020) COVID-induced economic uncertainty and its consequences. https://voxeu.org/article/covid-induced-economic-uncertainty-and-its-consequences

Brundtland G (1987) Report of the World commission on environment and development: Our common future. United Nations General Assembly document A/42/427

Chudik A et al (2020) Economic consequences of Covid-19: a counterfactual multi-country analysis. https://voxeu.org/article/economic-consequences-covid-19-multi-country-analysis
Coibion O et al (2020) Labor markets during the Covid 19 crisis: a preliminary view. Working Paper. (April 28, 2020). University of Chicago, Becker Friedman Institute for Economics Working Paper No. 2020-41, Chicago Booth Research Paper No. 20-06, Fama-Miller Working Paper. http://dx.doi.org/10.2139/ssrn.3574736
Costanza R et al (2009) Beyond GDP: the need for new measures of progress. Technical Report 4, The Frederick S. Pardee Center for the Study of the Longer-Range Future, Boston University, Boston, MA
European Commission, Directorate-General for Economic and Financial Affairs, European Economic Forecast, Autumn 2020, Institutional Paper 136, November 2020, Luxembourg Publications Office of the European Union 2020
Eurofound (2020) Living, working, and COVID-19, COVID-19 series, Publications Office of the European Union, Luxembourg. https://www.eurofound.europa.eu/publications/report/2020/living-working-and-covid-19
Flaxman S et al (2020) Estimating the effects of non-pharmaceutical interventions on COVID-19 in Europe. Nature 584:257–261. https://doi.org/10.1038/s41586-020-2405-7
Freeman RB, Hersch J, Mishel LR (eds) (2005) Emerging labor market institutions for the twenty-first century. University of Chicago Press
Hall RE, Jones CI (1999) Why do some countries produce so much more output per worker than others? Quarterly J Econ 144(1):83–116
Hunt V et al (2020) The case for stakeholder capitalism, McKinsey&Company, November. https://www.mckinsey.com/business-functions/strategy-and-corporate-finance/our-insights/the-case-for-stakeholder-capitalism
King K et al (2020) What strategies and approaches are countries implementing within schools both in response to COVID-19 and the localized outbreaks? Sept 15 2020. https://analysis.covid19healthsystem.org/index.php/2020/09/15/what-strategies-and-approaches-are-countries-implementing-within-schools-both-in-response-to-covid-19-and-to-localized-outbreaks/
Lele S (1991) Sustainable development: a critical review. World Development, 19(6):607:621
Masonbrink AR et al (2020) Advocating for children during the COVID-19 school closures. Pediatrics. https://doi.org/10.1542/peds.2020-1440
McKibbin WJ et al (2020) The global macroeconomic impacts of COVID-19: seven scenarios (March 2, 2020), CAMA Working Paper No. 19/2020. http://dx.doi.org/10.2139/ssrn.3547729
Michaillat E et al (2020) Keeping track of the unemployment gap, https://voxeu.org/article/keeping-track-unemployment-gap
Millard J (2020) Impacts of COVID-19 on social development and implications for the just transition to sustainable development, prepared for UNDESA Virtual Expert Group Meeting on the "Socially just transition towards sustainable development. The role of digital technologies on social development and well-being of all". https://www.un.org/development/desa/dspd/wp-content/uploads/sites/22/2020/08/Impacts-of-COVID-19-on-social-development-and-implications-for-the-just-transition-to-sustainable-development-4-8-20.pdf
North DC (1990) Institutions, institutional change and economic performance. Cambridge University Press, Cambridge
Novitz T (2012) Labour rights and property rights: implications for (and beyond) redundancy payments and pensions? Ind Law J 41:136–165. https://doi.org/10.1093/indlaw/dws015
OECD (2017) The next production revolution: implications for governments and business. OECD Publishing, Paris, http://dx.doi.org/10.1787/9789264271036-en
OECD (2011) Divided we stand: why inequality keeps rising. OECD Publishing. http://dx.doi.org/10.1787/9789264119536-en
O'Riordan T, Voisey H (1997) The political economy of sustainable development. Environ Polit 6(1):1:23. https://doi.org/10.1080/09644019708414309
Pesole A et al (2018) Platform workers in Europe, EUR 29275 EN. Publications Office of the European Union, Luxembourg, 2018. https://doi.org/10.2760/742789

Rodrik D, Subramanian A, Trebbi F (2002) Institutions rule: the primacy of institutions over integration and geography in economic development, International Monetary Found, Working Paper, no. 189, 131 https://www.imf.org/external/pubs/ft/wp/2002/wp02189.pdf
Santos M et al., Science for Policy report, Joint Research Centre, European Commission (2020) Behavioural changes in tourism in times of COVID-19 employment scenarios and policy options. Luxembourg Publications Office of the European Union. https://doi.org/10.2760/00411
Schwab K et al (2020) World Economic Forum, Now is the time for a 'great reset', https://www.weforum.org/agenda/2020/06/now-is-the-time-for-a-great-reset/
Seale H, Dyer CEF, Abdi I et al (2020) Improving the impact of non-pharmaceutical interventions during COVID-19: examining the factors that influence engagement and the impact on individuals. BMC Infect Dis 20:607. https://doi.org/10.1186/s12879-020-05340-9
Stiglitz J, Sen A, Fitoussi J (2009) Report of the commission on the measurement of economic performance and social progress (CMEPSP)
UNICEF, FAO and WFP (2020) Mitigating the effects of Covid-19 Pandemic on Food and the Nutrition of School Children, Interim Guidance Note, March 2020. https://www.unicef.org/media/68291/file/Mitigating-the-Effects-of-the-COVID-19-Pandemic-on-Food-and-Nutrition-of-school-children.pdf
Vasiljeva M et al (2020) A predictive model for assessing the impact of the COVID-19 pandemic on the economies of some Eastern European countries. J Open Innov, Technol, Mark Complex, 6:92. www.mdpi.com/journal/joitmc, https://doi.org/10.3390/joitmc6030092

Diana-Mihaela ŢAÎRCĂ, is currently University Professor with a PhD in Economics at the "Constantin Brancusi" University of Targu-Jiu and Faculty of Economics, Department of Management and Business Administration. She also develops intense research activity published in numerous economic studies and participation to conferences and in research projects. Her research interests include sustainable development, labour market, green jobs, rural economy, green economy, education. She is Director of the Center for Fundamental and Applied Economic Studies (CSEFA). She was Visiting Professor at the University of Szczecin, Poland, and had research internships at the University of A Coruna, Spain. She is Guest Editor of Special Issue on Sustainability Energy Use and Management (Progress in Industrial Ecology, An International Journal, 2015), Special Issue—Innovation for Sustainable Development (Journal for Cleaner Production, 2018), Special Issue—"Information Society and Sustainable Development—Selected Papers from the 5th International Conference ISSD 2018" (Sustainability, 2019), and the Special Issue "Information Society and Sustainable Development—Selected Papers from the 5th International Conference ISSD 2019" (Sustainability, 2020), Special Issue—Impact of Sustainable Financial, and Economic Development on Greenhouse Gas Emission (Energies, 2021).

Laura Mariana Cismaş, is currently University Professor with a PhD in Economics, at the Department of Economics and Economic Modelling, at the Faculty of Economics and Business Administration, from West University of Timisoara (WUT). Her teaching experience with the West University of Timisoara and in the field of scientific research regarding Macroeconomics, Microeconomics, Regional Economics, European Macroeconomics, European Institutional Economics, and European Economic Policies goes back more than 29 years. In 2007–2008, she was Visiting Professor at Università degli Studi di Padova (Italy), teaching the master's course: "Costruzioni dello Spazio Pubblico Europeo", specialization in European Union Policies (Politiche dell 'Unione Europea), master developed within the joint project between: WUT and Università degli Studi di Padova. Next year, in 2008–2009, she held the same position for the master's course: "Politica monetari dell'UE". In 2010, Laura Mariana Cismas was Visiting Professor at the Academy of Economic Studies of Moldova (ASEM), in Chisinau (Moldova), supporting the master's course: "Economic Communication and International Business Negotiation," in the European project between WUT, ASEM, and Universite d'Auvergne (France). She has coordinated, and participated as a member of more than 12 research and grant projects, in both national and international teams. She has elaborated more than 120 scientific papers, has published 12 books and university courses in Romanian, English, and Italian languages, and has participated in more than 100 national and international conferences. She is Member of the Scientific Board and Member of the Editor Board of more than 12 Romanian and international journals. As of 2008, she has guided 15 doctoral students and 5 postdoctoral students. Moreover, she is Member of several professional associations: The Regional Science Association International (RSAI), the European Regional Science Association (ERSA), and Romanian Association of Regional Science (ARSR). Also, she is Member of the network of the countries from the Vysegrad Group (Reseaux des Pays du Groupe de Vysegrad-PGV-ERECO, Grenoble, France) and Member of the East-European Center of Research in Economics and Business (ECREB) from WUT, etc.

Cornelia Dumitru, is currently a PhD in Economics and scientific researcher at the Institute of National Economy of the Romanian Academy. She has been involved in several projects regarding the structural restructuring of the Romanian economy over the first years of transition, financed by the Dutch Matra programmes. Her main interests are in institutional economics with emphasis on the economic institutions of the labour market, sustainable development, the Green Deal, and the implications for the economic components of the social contract.

Public Financing and Management for a Sustainable Healthcare Sector: Some Lessons from the Covid-19 Pandemic

Stella Tsani, Elena Riza, Panagiota Tsiamagka, and Margarita Nassi

Abstract As the Covid-19 pandemic evolves, healthcare system sustainability and resilience worldwide has been in the spotlight. Observations indicate the significant human and financial capital requirements to address the immense pressure. While the pandemic constitutes an unprecedented shock, the coordinated mobilization of public capital and intervention in the healthcare systems worldwide provide a prime opportunity for sustainable change and transformation. This can materialize if efforts and funding focus on socio-economic and healthcare system innovation, infrastructure and green investment support. This chapter discusses the public financing responses worldwide towards the support of healthcare systems in view of the Covid-19 pandemic. It also discusses the organizational challenges and changes that should be put in place to make the healthcare system resilient to unforeseen shocks. These regard innovation support, healthcare personnel planning, establishing new models of care delivery, and other interventions that have the potential of securing uninterrupted, sustainable and efficient healthcare systems both at the personal and at the population level.

Keywords Public financing · Sustainability · Innovation · Public healthcare policies · Fiscal policy · Covid-19

1 Introduction

The Covid-19 pandemic constitutes an unexpected shock to the societies and the economies at a global level. The pandemic is not the first that humanity has experienced in the recent past neither the most dangerous for human existence. Past experience includes for instance the 2002–2004 SARS outbreak and HIV/AIDS outbreak

S. Tsani (✉) · P. Tsiamagka
Department of Economics, University of Ioannina, Ioannina, Greece
e-mail: stellatsani@uoi.gr

E. Riza · M. Nassi
Department of Hygiene, Epidemiology and Medical Statistics, Medical School, National and Kapodistrian University of Athens, Athens, Greece

W. Leal Filho (ed.), *COVID-19: Paving the Way for a More Sustainable World*, World Sustainability Series, https://doi.org/10.1007/978-3-030-69284-1_12

in the 1980s. Nevertheless, is seems to be one of the most challenging pandemics in recent years for the healthcare systems worldwide, if not the one that brings the national and global systems on the verge of the collapse (Ortega and Orsini 2020; Zheng 2020). Realizing the urgency of the matter, the speed of spread, the absence of scientific knowledge on how to efficiently prevent the spread of the virus (i.e. vaccine) and the imminent threat for the healthcare systems, governments and health authorities around the globe activated immediate social measures so as to limit the spread and the high infection rates. First response included self-isolation recommendations for vulnerable groups (e.g. elderly, people with underlying health conditions), or for the general population (these are enforced and usually referred to as "lockdown"). While these measures temporarily reduce the transmission rates of Covid-19, they also subject the national, regional and global economies and societies to a significant stress.

According to the World Bank Global Economic Prospects, the baseline scenario for 2020 shows a 5.2% contraction in global GDP in 2020. This constitutes the deepest global recession in decades (World Bank 2020a, b). The pandemic is expected to plunge most economies in recession in 2020, with per capita income contracting in the largest fraction of countries globally since 1870. Given the gloom realizations and estimations on the economic impact of the pandemic, significant fiscal effort has been put worldwide to support the economies and limit the socio-economic impact of the pandemic. Countries have mobilized resources and public spending and have implemented specific measures including tax reliefs, social security contribution deferrals, healthcare spending, and sector-specific support (e.g. transport, tourism, exporting sectors).

The pandemic evolves in a time of debate and action on sustainability transition, global and regional initiatives, such as the Paris Agreement, the UN Agenda to 2030 and the European Green Deal, to address climate change and environmental degradation. This has triggered a fresh debate in the literature on the impact of the pandemic on sustainable development of the health care sector (Leite et al. 2020; Kruse and Jeurissen 2020; Almeida 2020) and of the economies (Barbier and Burgess 2020; Arora and Mishra 2020; Schaltegger 2020; Amankwah-Amoah 2020), and on how the health crisis, if properly managed, can constitute an window of opportunity for sustainability transition (Bodenheimer and Leidenberger 2020; Ioannides and Gyimóthy 2020; Markard and Rosenbloom 2020). While Covid-19 constitutes an unprecedented shock, the mobilization of significant public capital and the coordination of fiscal policies worldwide indicate that this shock constitutes a prime chance for sustainable change and transformation, primarily in the healthcare sector, but in overall production and consumption structures as well. This can materialize if efforts and funding focus, among other, on healthcare and socio-economic system innovation (Lambert et al. 2020; Zeitoun et al. 2020; Li-Ying and Nell 2020), infrastructure and green investment support (Engström et al. 2020; Naidoo and Fisher 2020), cross-border cooperation (Mallinson 2020), action against poverty and access to healthcare for different socio-economic groups and regions (Patel et al. 2020; Tsani et al.2021; Abedi et al. 2020).

This chapter reviews fiscal policy responses in selected countries worldwide, with the aim to identify healthcare system-specific interventions, infrastructure and innovation support that can ensure resilience in the current shock, sustainable change and transition. The review reveals that little Covid-19-related and/or triggered steps have been taken globally towards this end. Fiscal responses have been limited in scope, neglecting in large the need for sustainability transition and innovation support in the healthcare system and beyond. The review is complemented with a brief discussion of potential interventions discussed in the recent literature that could support sustainability in the healthcare system. The remainder of the chapter develops as follows: Sect. 2 reviews the fiscal response in selected countries. Section 3 discusses micro-level healthcare sector-specific interventions that could foster sustainability and sector resilience. Last section concludes.

2 Review of Selected Fiscal Responses to the Covid-19 Pandemic in 2020

The policy response to Covid-19 pandemic includes the application of measures in terms of fiscal, macroeconomic and monetary policy. Fiscal policy responses have been first and most in use (IMF 2020). This has brought to light the importance of regulatory measures that allow for the emergency use of budget funds (Barroy et al. 2020). For the purposes of this chapter, fiscal measures adopted worldwide have been reviewed. For this analysis, information has been retrieved from the IMF Policy tracker of responses to Covid-19 (IMF 2020). The policy tracker provides details on the fiscal, macroeconomic and monetary policies that have been implemented in each country. As the focus of this chapter rests with fiscal measures, only fiscal policies are presented and discussed here. Fiscal responses include packages for the health care system, long-term care, short-term work, compensation packages for economic groups (e.g. self-employed, small businesses) for loss of earnings related to forced closure or sickness, deferral tax payments (e.g. VAT, personal or income tax), social security contributions, compensation of companies or labor for forced leave, tax relief measures for the sectors mostly hit by the pandemic (like hospitality, tourism, transport, agriculture), hardship funding for businesses, unemployment assistance and several other variations of fiscal support.

Table 1 summarizes the fiscal measures adopted in selected economies. To facilitate the reader fiscal measures are presented in bullet points with reference to the specific fiscal tool/approach used and targeted group in receipt of support. Readers interested in the details of the fiscal policies used, and on the amounts allocated to each measure (where such information is available), are advised to refer to the detailed presentation of the IMF Policy tracker of responses to Covid-19 (IMF 2020). Selected economies include the 10 richest and the 10 poorest as classified and measured by the average of Gross Domestic Product, in purchasing power parity terms (DGP, PPP constant 2017 international $) in the period 2009–2019. Gross Domestic Product, in

Table 1 Fiscal response to Covid-19 in selected economies worldwide

United States	• Unemployment benefits • Student loan payment relief • Deferring collections of employee social security payroll taxes • Options for avoidance of evictions and foreclosures • Small Business Administration loans and guarantees to retain workers • Support to hospitals and virus testing • One-time tax rebates to individuals • Provision of food and safety net for the most vulnerable • Prevention of bankruptcy • Hospital support • International assistance • Virus testing • Medicaid funding • Vaccine development, therapeutics, diagnostics, support for disease control and prevention • Support for paid sick or emergency leave
European Union	• Financing of health-related spending • Creation of a temporary loan-based instrument (SURE) to protect workers and jobs • Coronavirus Response Investment Initiative (CRII) and the Coronavirus Response Investment Initiative Plus (CRII +) to support public investment for hospitals • SMEs support • Labor markets support • Stressed regions support • European Investment Fund to incentivize banks to provide liquidity to SMEs and midcaps • Credit holidays to crisis-affected debtors • Activation of the general escape clause in the EU fiscal rules • Allowance for flexible interpretation of EU State Aid rules

(continued)

Table 1 (continued)

China	• Increased spending on epidemic prevention and control • Production of medical equipment • Unemployment insurance and extension • Tax relief • Waived social security contributions • Additional public investment
India	• Deferred public revenues • Expedited spending for primarily social protection and healthcare • Business credit provisioning • In-kind (food, cooking gas) and cash transfers to lower-income households • Wage support and employment provision to low-wage workers • Insurance coverage for workers in the healthcare sector and provision of healthcare infrastructure • Additional public investment • Fertilizer subsidy allocation and infrastructure support in the agriculture sector • Support for urban housing construction • Ease of tax compliance burden across a range of sectors • Credit support to businesses, poor households, migrants and farmers • Credit support to distressed electricity distribution companies
Japan	• Preventive measures against the spread of infection • Strengthening of treatment capacity • Protection of employment and businesses • Enhancement of future response capacity • Handouts to affected individuals and firms • Deferral of tax payments and social security contributions • Concessional loans from public and private financial institutions • Expansion of the work subsidies • International aid: contribution to the IMF's Catastrophe Containment and Relief Trust, Poverty Reduction and Growth Trust and the COVID-19 Crisis Development Initiative

(continued)

Table 1 (continued)

Germany	• Spending on healthcare equipment, hospital capacity and R&D (vaccine) • Expanded access to short-term work • Expanded childcare benefits for low-income parents • Easier access to basic income support for the self-employed • Grants to affected small business owners and self-employed persons • Interest-free tax deferrals until the end of 2020 • Venture capital funding for start-ups • Temporarily expanded duration of unemployment insurance • Parental leave benefits • Temporary VAT reduction • Income support for families • Grants for hart-hit SME's • Financial support for local governments • Expanded credit guarantees for exporters and export-financing banks • Subsidies/investment in green energy and digitalization • Extension of maximum duration of short-term work benefits from 12 to 24 months • Guarantees, access to public guarantees for firms and facilities for public equity injection into firms with strategic importance through the newly created economic stabilization fund (WSF) and the public development bank KfW • Additional federal government's fiscal package won measures through direct support and state-level loan guarantees • Revenue compensation (of up to 75%), as well as, public loan guarantees and basic income provision for affected businesses

(continued)

Table 1 (continued)

Russian Federation	• Increased compensation for frontline medical staff as well as health and safety inspectors • Sick leave benefits and sick leave pay • Lumpsum benefit transfers to children of age 1–16 • Lumpsum benefit transfers to families with children in case of employment occupation loss • Interest rate subsidies for SMEs and systemically important enterprises • Tax deferrals for most affected companies on most taxes • Deferrals on social contributions for SMEs • Permanent reduction of SMEs social contributions on wages in excess of the minimum wage • Social contributions and CIT permanently reduced for IT firms • Tax holiday on all taxes (excluding VAT) and social contributions for Q2 for SMEs, sole proprietors, and NGOs providing social services • Tax refund (whole or partial) on registered self-employed for 2019 and 2020 • Eligibility age to register as self-employed lowered from 18 to 16 • Sole proprietors partial refund on their social contributions • Deferrals on rent payments to all levels of government until the end of the year • Zero rent to the federal government for three months for SMEs in affected sectors • Budget grants for SMEs in affected industries to cover salaries at the rate of one minimum salary per employee for two months plus subsidized and forgivable loans for all enterprises in affected industries to pay minimum wages for 6 months • Zero import duties for pharmaceuticals and medical supplies and equipment • Guaranteed loans to SMEs and affected industries • Subsidies to airlines, airports, automakers, and others • Expanded eligibility for subsidized mortgage lending

(continued)

Table 1 (continued)

Brazil	• Lifting of government's obligation to comply with the primary balance target in 2020 and of the constitutional expenditure ceiling • Temporary income support to vulnerable households • Cash transfers to informal and low-income workers • Advance payments of salary bonuses to low income workers • Partial compensation to suspended or reduced employment • Temporary tax breaks • Lower taxes and import levies on essential medical supplies • Federal transfers to state governments to support higher health spending • Financial assistance to municipalities - temporary stay of debt payments • Expansion of credit lines of public banks for businesses and households • Government back up to SMEs and micro-businesses to cover payroll costs, working capital and investment
France	• Public guarantees for bank loans and credit reinsurance schemes • Support to health insurance for the sick or their caregivers • Increased spending on health supplies • Postponing of social security and tax payments for companies • Accelerated refund of tax credits • Wager support for workers under the reduced-hour scheme • Financial support for affected microenterprises, liberal professions, independent workers and low-income households • Postponement of rent and utility payments for affected microenterprises and SMEs • Additional allocation for equity investments or nationalizations of companies in difficulty • Facilitating granting of exceptional bonuses exempt from social security contributions • Extension of expiring unemployment benefits until the end of the lockdown • Preservation of rights and benefits under the disability and active solidarity income schemes • Support to the hardest-hit sectors • Incentives to purchase greener vehicles, green investment support for the auto and aerospace sectors

(continued)

Table 1 (continued)

Indonesia	• Support to the health care sector to boost testing and treatment capacity • Increased benefits and broader coverage of existing social assistance schemes to low-income households (food aid, conditional cash transfers, electricity subsidies) • Expanded unemployment benefits, including the informal sector • Tax reliefs and lowering of corporate income tax • Capital injections into state-owned enterprises • Interest subsidies, credit guarantees, and loan restructuring funds for micro, small, and medium enterprises • State funds in selected commercial banks to increase leverage and capital loans for labour-intensive corporations
Dominica	• Extensions to filing personal and corporate income tax returns • Extension for payment of corporate income tax • Reduction in the corporate income tax rate to companies which commit to continue to employ at least 80% of their staffing • Reduction to 0% in the import duty and the value-added tax charged on disinfectants, cleaning supplies, protective gears and face masks • Increased budgetary funding to the Ministries of Health and Agriculture • Cash grants to individual crop farmers • Infrastructure projects and construction investments • Payments to small contractors and merchants utilizing the resources approved by the IMF under the Rapid Credit Facility • Income support for the unemployed

(continued)

Table 1 (continued)

Vanuatu	• State budget envelope and help from Australia, China, New Zealand, UNICEF, WHO, other NGOs/CSOs and some local businesses for the expansion of health facilities, restocking personal protective equipment and supplies, community education and awareness • Test analysis with assistance from Australia, France and New Zealand in New Caledonia • Vanuatu National Provident Fund Hardship Loans (interest-free withdrawal for 6 months, after which the member choses a repayment plan or permanently withdrawal penalty) • Deferred and cancelled taxes, license fees and charges for businesses in 2020 • Cash reimbursement of business license fees • Commodity Support Grant provided to producers of copra, kava, cocoa and coffee • Shipping Support Grant to facilitate farmers' access to major market centres • Suspension of secondary school tuition fees for 2020 • Reprioritization of expenditures, debt, and development partner assistance
Sao Tome and Principe	• Health contingency plan prepared with the WHO and increased health spending (on medicine, equipment, staffing, and treatment centers) • Expansion of social assistance to the most vulnerable, including expansion of the WB-supported cash-transfer program, increased support to the disadvantaged (the elderly, disabled and abandoned children • Salary contributions support to small businesses • Financial assistance to suspended employment in the formal and in the informal sector • Provision of procure seeds, feedstock, and other essential inputs • Solidarity tax on workers and public servants
Tonga	• Support of the health sector, tourism, transport, agriculture, education and security • 3-month moratorium on Government Development Loans & TC Gita Recovery Loan Fund • Deferral of retirement contributions and hardship allowances for laid-off employees • Needs-based financial assistance, tax and duty relief • Public utility bills payment assistance

(continued)

Table 1 (continued)

Micronesia, Fed. Sts.	• Develop quarantine and isolation facilities • Mandatory infection control training • Increase testing capacity and ventilators for each island state • Pandemic Unemployment Assistance Program supported by the U.S. Department of Labor • Wage subsidies • Debt relief • Social security tax and other tax rebates
Palau	• Hospital Trust Fund to the to help with prevention and preparation • Unemployment benefit scheme • Temporary subsidies for utility bills • Job creation scheme for public works • Private sector lending scheme
Kiribati	• Donor-funded support • Unemployment support and benefit via partial income substitution • Private business and state-owned enterprises stimulus • Employer cost sharing for offshore observers, sea farers, and fruit packers • Reduction in social security contributions for both employers and employees • Loan support through government-owned financial intermediaries
Marshall Islands	• National preparedness plan including infrastructure, medical supplies and equipment, and surge support • Increased expenditure for major infrastructure projects such as new isolation and quarantine buildings, building of hand-washing stations • Economic relief payouts • Response Plan for the Neighboring Islands/Outer Islands (food baskets, fishing gears and farming tools)
Nauru	• Containment and expenditures on keeping the national airline and other state-owned enterprises afloat • Increase health expenditures and isolation costs • Budget support to SOEs in managing inventory from limited freight and cargo services • Cash buffers and general reserves to support the fiscal measures, including necessary medical expenditures • Approvement of the Asian Development Bank US$5 million policy-based grant to support public investment management, fiscal sustainability, management of public expenditure and national infrastructure, as well as the governance of state-owned enterprises

(continued)

Table 1 (continued)

Tuvalu	• Procurement of personal protection equipment, ventilators, testing equipment and other essential specialize equipment • Repatriation of Tuvalu students studying abroad • Maintenance of relevant infrastructures for quarantine purposes • Improvements in broadband for internet connectivity • Additional police personnel • Grant assistance to the private sector by the Development Bank of Tuvalu

purchasing power parity terms data for 2009–2019 are extracted from the World Bank World Development Indicators (World Bank 2020a, b). Series have been used for the estimation of the average of Gross Domestic Product, in the period 2009–2019. Based on this calculation, the 10 richest economies, the fiscal policy responses of which are summarized and discussed therein, are (in descending order): United States, European Union, China, India, Japan, Germany, Russian Federation, Brazil, France and Indonesia while the poorest economies (in descending order) are: Dominica, Vanuatu, Sao Tome and Principe, Tonga, Micronesia, Palau, Kiribati, Marshall Islands, Nauru and Tuvalu.

Selection is based on the wealth of countries with the aim to obtain a snapshot of different responses worldwide from both poor and rich countries. The selection does not consider the exposure of the economies to the Covid-19 (i.e. number of cases recorded in the respective countries/economies, mortality, morbidity, infection rate etc.). The severity of the pandemic in each economy might indeed impact on the extent of the fiscal policy response. Nevertheless as the economies reviewed operate in a globalised market and form part of the larger global economy, it is expected that they will respond to the pandemic acknowledging that they are exposed to an imminent global threat, irrespective of whether the pandemic fully evolves in the respective territories in the specific period of time or not.

All economies reviewed therein have channeled in 2020 public funds towards the support of the healthcare system. This is made mainly through increased funding for virus testing and medical funding (e.g. China, EU, USA, Indonesia), therapeutics, diagnostics, support for disease control and prevention. USA, Germany and Russia have explicitly allocated funds for vaccine development. India and Russia make explicit reference to funds for increased compensation and support of healthcare workers. Funds for virus testing are provided for in some countries (USA, Tuvalu, Micronesia, Vanuatu and Indonesia). Almost all fiscal measures reviewed provide for healthcare related infrastructure (hospitals, testing units, isolation units). Some of the high-end countries make provision for international support and WHO guidance (USA, Japan) while in the lower end Vanuatu, Marshal Islands, Micronesia and Sao Tome and Principe receive international aid and support to respond to the Covid-19 crisis.

Beyond the healthcare sector, both rich and poor countries chose to support employment and business operations through unemployment benefits, incentives to retain workers, provision of low-cost inputs and reduced work time schemes. Also, countries chose to support sectors most in stress or sectors that are traditionally essential to the economic activity like tourism, agriculture or exporting sectors. Some countries chose to support the most vulnerable with in kind contributions (like food provision in Indonesia, or seeds and feedstock in Sao Tome and Principe). Provisions are also made with regards to securing housing and food supplies (e.g. eviction protection in the US, food provision in India). Only Russia and Germany make explicit reference to funds for child protection and parental support. USA and Vanuatu make explicit reference to education support through students' loans support and suspension of education fees. Germany and France are the only countries in the sample reviewed and in the group of the rich countries that make explicit reference to support of green investments and innovation. Infrastructure spending is included in the measures in India, Dominica, Marshall Islands, Nauru and Tuvalu. Little information is provided on whether this infrastructure upgrade regards advanced technology infrastructure.

The measures reviewed make use of the available funds for the direct response to the pandemic. Nevertheless, they make limited provisions for capital mobilization and direction towards socio-economic and healthcare system innovation, infrastructure and green investment support, cross-border support, poverty and access to healthcare for different socio-economic groups and regions. Most countries chose to support existing patterns of production and consumption and provide little evidence of perceiving the socio-economic shock caused by the Covid-19 pandemic as an incentive provider for sustainability transformation, and increase in its speed. The recorded response may be sensible in the case of low-end income countries. These countries struggle to raise capital for the much-needed transition while at the same time are projected to be most hit by climate change and environmental degradation. Despite the understanding of the limitations present for low-income countries, recognition of this imminent threat should have been included in their fiscal policy response packages indicating a first recognition of the future ahead if no structural shift and first mobilization effort (possibly under a shock-therapy approach) is implemented.

In the same manner rich countries ought to mobilize Covid-19 related funds and actions in a way that also supports a faster sustainability transition. Rich countries, except for Germany and France, make no provision for innovation support or structural shifts in production and consumption. This could be supported though funds provided for technological upgrade of the producing sectors so as to help them cope with Covid-19 emergency (e.g. remote working, improved IT access and capacity, digitalization of work and public services, improved rural IT infrastructure so as to support move away from urban centers and easing the burden in urban environments), support of green investments so as to speed up greening of the economies and of the production and employment generation in the emerging sectors. Additionally, infrastructure investments as a response to the pandemic economic effects could potentially have several positive outcomes. These are related to employment and capital income generation, but also to the long-term productivity impact of infrastructure availability and upgrade.

Most countries opt for tax reliefs and income support of the affected labor force. It would be a valid case to support the most impacted and vulnerable groups income through transfers coupled with education and training in new technologies and skills rather than just through income and unemployment support with no skills upgrade. It is not yet clear how the next day after the pandemic will be for the global markets, for labor and production. Neither it is well understood how long the impact of the pandemic will be on the existing production and consumption structure and if this "traditional" structure will continue to exit after the end of the pandemic. Thus, training and education of the most vulnerable (non-skilled, low income labor) and the increase in capacity to rapid adjustments to uncertain requirements emerges as a low risk alternative for the future.

3 Micro-level Sustainability and Innovation Considerations in the Health-Care System

The pandemic calls for a response at the fiscal and macro-economic level, but it also indicates the need to reconsider the micro-environment of the healthcare systems into which interventions can support resilience and speed up sustainability transition at sectoral level. The Covid-19 pandemic is caused by a highly contagious pathogen that disproportionally affects individuals with vulnerabilities expressed as comorbidities e.g. chronic diseases such as diabetes or cardiovascular disease, or social determinants of health e.g. race, ethnicity, social status, lack of social insurance, precarious employment (Lacobucci 2020; Sze et al. 2020; CDC 2020). At the same time Covid-19 has put immense pressure on healthcare personnel calling for a reconsideration of training and support to the human capital in healthcare provision (WHO 2020). These considerations have been addressed in the recent literature. The review of relevant studies, performed for the purposes of the present chapter, identifies interventions that can be made in the healthcare sector which can support sustainability transition and sector resilience. These are discussed in brief next.

A first realisation at the outbreak of the pandemic was the importance of the reduction of operational obstacles in healthcare access. Insufficient medical supplies, sources and tools in emergency settings such as transportation services, availability of isolation beds, and hospital room capacity, protective equipment for staff and patients, lack of medical and administrational workforce with expertise in emergency management are common obstacles to healthcare delivery in times of high demand (Dekkers and Hertroijs 2018). This has been realised in the Covid-19 pandemic and the fiscal response has attempted to address these obstacles through emergency funding and action. However, this is rather a short-term short-planned response rather than a long lasting, forward looking response, if no appropriate structural and system changes are implemented.

An important characteristic in the management of the Covid-19 pandemic was the limited use of Primary Health Care services which should normally be the first point of entry into the healthcare system. The Primary Health Care sector can manage the mild cases of Covid-19, absorb the non-Covid-19 related incidents and help reducing the overload of the tertiary care sector. It should be noted however, that in order to avoid an overwhelming effect during health crisis response, primary and community health care services must be strengthened, especially in deprived and densely populated areas as well as in the geographically remote ones, to secure an optimum healthcare system performance. Since the Covid-19 outbreak, healthcare personnel worldwide is carrying an excess work load. Combined with personnel shortages, increased average workforce age and reduced supply of personal protective equipment (PPE) is exposing healthcare labour force to additional occupational risk factors potentially jeopardizing care delivery.

The pandemic calls for a closer look at technological innovations that can change the way healthcare services are delivered. The use of telemedicine can contribute to the successful delivery of care for chronic disease patients or those in isolation and to decrease the risk of exposure to infection from direct contact with the hospital environment (Monaghesh and Hajizadeh 2020; Salisburry et al. 2015). A more generalized use of remote consultation services can alleviate the overloading of hospitals with patients with minor health concerns, can support regular mental health service provision and can cover the needs of vulnerable population groups such as people with special needs, the ones living in remote areas or in isolation (Iancu et al. 2020). An additional tool to the remote provision of healthcare services is electronic prescription of medicines (e-prescription) especially in chronic disease patients to ensure uninterrupted medication intake.

A significant number of health care workers worldwide have been directly affected by the Covid-19 pandemic, as a result of higher exposure risk through the contact with infected patients (Bielicki et al. 2020). Physical and mental support (The Lancet 2020), monitoring a potential infection, easy access and self-use of diagnostic tools to health care workforce along with proper use, adequate supply of personal protective equipment and training may be the most effective strategies to strengthen a well-organized working environment, to improve the delivery of care (Chou et al. 2020) and to effectively control the rate of infection among health care professionals (Chowdhury et al. 2020). Organization and allocation of human resources during pandemics can address the increased needs within the existing and/or new facilities. Despite the fact that during health crises, a part of health care services' personnel is unable to work due to illness or burnout, there are numerous occupational specialties, which could shift posts to contribute to the emergency response, as well as the option to use external health care personnel, such as volunteers, retired medical practitioners, senior medical students etc.

Personnel's resilient training with focus on personal safety and continuous education on the management of cases of new emerging diseases within the healthcare settings can help to minimize the risk of infection and to reduce workforce shortages especially in the frontline (Adams and Walls 2020). Moreover, regular personnel briefings on new conditions and treatment protocols can significantly strengthen workforce's performance and can facilitate the adaptation in health sector operations, especially during pandemic emergency response. Such educational support, in the context of Covid-19, includes training on personal safety rules and guidance, proper use of mask and PPE, safe disposal practices, patient care with ventilators, mental support and encouragement of patients' families, co-workers and colleagues.

Further strategies to improve an advanced and sustainable health care system ready to respond to health crises, include the use of assessment and evaluation tools of the system and its services. A health care qualitative evaluation on indicators linked to services provision, personnel's self-assessment, capacity measurement in meeting the population's needs, working environment and its economic aspects can help to improve the performance of the healthcare sector and its readiness to respond to public health threats.

Healthcare professionals are put under high levels of stress during the disaster emergency response phase, which challenges their mental health status. Diagnosis of Post-Traumatic Stress Disorder (PTSD) is common in the post-response phase (Lee et al. 2018). Factors such as gender, family status, working position, economic aspect, environment, available and adequate protective equipment, beliefs or stigmatization and post experience, education and training to perform under these circumstances partly determine the extent of mental health symptoms (Devnani 2012; Badrfam et al. 2020). Stress, depression, anxiety disorders and insomnia are commonly reported by medical and nursing health care staff working in emergency response hospitals and services (Badahdah et al. 2020). In addition, non-medical staff in healthcare settings is also susceptible to mental disorders due to the increased demand for longer and higher working performance. Securing a stable and safe working environment for the healthcare personnel has a beneficial impact on their mental status and works towards a sustainable healthcare provision. Training in crisis management within the health care units has proven effective in maintaining performance during the management of health emergencies (Tan et al. 2020). The provision of psychological support services for the health care professionals and their families is also important as it helps to also deal with other sources of anxiety such as the risk of infecting family members and the disrupted family life (Shanafelt et al. 2020, Walton et al. 2020).

On the other hand, psychological support is necessary for professionals who have recovered from the viral infection. These individuals having already experienced the severity of self-isolation and quarantine, often feel reluctant returning to work for fear of stigma or possible psychological war within their working environment as potential sources of contagion (Koh et al. 2005).

The establishment of a national crisis management centre operating under a central public authority is essential, especially in countries frequently affected by disasters. A systematic training programme supervised by the national centre can be organized and delivered to health professionals to provide hands on emergency response and disaster management practice using appropriate educational tools and experienced trainers. This can decrease the gaps in knowledge in disaster response of the health workforce, can build their confidence and willingness to work during an emergency or disaster event, while at the same time can ensure a high productivity and functionality of the healthcare system (Sultan et al. 2020). In addition, training in global and public health strategies needs to be incorporated in the centre's line of operation, as this is a vital element in the management and evaluation of the crisis response. The systematic collection of data and the use of epidemiological surveillance tools can help formulate and amend plans for the effective management of public health emergency incidents.

The efficiency of the health system's response in the management of pandemics relies on preparedness and on the ability of the system to cater for unexpected demands. The operation of a single-authority institution for the prevention, shock preparedness and emergency response, the planning and coordinating post-urgency mitigation and management in close cooperation with various government and/or regional agencies can increase resilience in the affected areas and achieve community competence towards crises.

Last, the Covid-19 pandemic saw an explosion of information through all types of media, which resulted in various misconceptions about health, prevention strategies or extreme ways of infection from infectious diseases (Sahni and Sharma 2020) and use of unproven treatments (Radwan and Radwan 2020). It becomes evident that health education interventions to raise public awareness by trained health professionals and specialists are required considering the characteristics of each target group and, particularly, their health literacy level. Such interventions are considered an important investment on social well-being, which indirectly interacts with the health system, especially during a pandemic.

4 Concluding Remarks

The Covid-19 pandemic is ongoing at a time where countries have regionally (e.g. EU Green Deal) and globally (United Nations' Agenda to 2030 and the 17 Sustainable Development Goals-SDGs) committed to sustainable green development. Implementation of the SDGs envisages a brighter future in social, economic and environmental terms. At the same time future developments may hold environmental degradation, biodiversity loss and climate change to a point of no return if no timely and adequate actions are implemented at a global scale. Acknowledgment of these trends indicates that the net effect and future outcome will depend on the speed and the robustness of sustainability transition. If this transition will not take place in a fast and robust manner, humanity and economies face the prospects of a grim future. Covid-19 creates a prime opportunity for a globally coordinated response, capital mobilization and fiscal action that can have a significant impact on the speed of sustainability transition. This transition regards primarily the healthcare system, but also other economic sectors and the societies.

Recent scientific analysis, as discussed in this chapter, on the emergency response funding to the healthcare system calls for a careful management of the additional funds and of the human capital. This leads to recommendations that call for the sustainable technological upgrade of the systems (e.g. rapid tests development, remote medical services, training of personnel, etc.), appropriate and adequate support to frontline medical workforce and continuous monitoring and evaluation of the system so as to support the sustainable transformation of the sector.

Beyond the healthcare system, the analysis in this chapter draws the attention to the use of the public funds in the fight against Covid-19. The mobilization of public funds to address the Covid-19 pandemic can be used either for the maintenance of the current consumption and production patterns or for incentive provision to change the existing consumption and production patterns towards more sustainable models, through the financing of innovation, green investments, education and technological upgrade. Recommendations into this direction urge for a more careful look on the structural shifts in the existing production and consumption patterns as a response to the Covid-19 pandemic that can have far-reaching implications for the sustainability transition. Policy making should consider the importance of digitalization, IT and green infrastructure so as to speed up greening of the economies and of the production and employment generation in the emerging sectors. Infrastructure investments may pay off in terms of employment, and long-term productivity of the inputs to production. Last, fiscal measures should look beyond the direct financial support to the most vulnerable and should actively seek for a combination of financial, training and education support package that can ensure resilience of the most vulnerable under the current conditions and readiness for future uncertainties and skill requirements.

Overall developments with the Covid-19 pandemic in 2020 show that sustainability implementation requires a rethinking of the healthcare systems in terms of design, funding and delivery of services. At the same time, they indicate the importance of proper management of the coordinated fiscal response at national, regional and global level. In this regard future works should systematically investigate the implications of fiscal responses on the speed and the direction of sustainability transition, at sectoral, e.g. the healthcare sector, and economy-wide level.

References

Abedi V, Olulana O, Avula V et al (2020) Racial, economic, and health inequality and COVID-19 Infection in the United States. J Racial Ethnic Health Dispar (2020). https://doi.org/10.1007/s40615-020-00833-4

Adams JG, Walls RM (2020) Supporting the health care workforce during the COVID-19 global epidemic. JAMA 323(15):1439–1440. https://doi.org/10.1001/jama.2020.3972

Almeida F (2020). Exploring the impact of COVID-19 on the sustainability of health critical care systems in South America. Int J Health Policy Manag. https://doi.org/10.34172/ijhpm.2020.116

Amankwah-Amoah J (2020) Stepping up and stepping out of COVID-19: new challenges for environmental sustainability policies in the global airline industry. J Clean Prod 271:123000. https://doi.org/10.1016/j.jclepro.2020.123000. ISSN 0959-6526

Arora NK, Mishra J (2020) COVID-19 and importance of environmental sustainability. Environ Sustain 3:117–119. https://doi.org/10.1007/s42398-020-00107-z

Badahdah A, Khamis F, Al Mahyijari N, Al Balushi M, Al Hatmi H, Al Salmi I, et al. (2020) The mental health of health care workers in Oman during the COVID-19 pandemic. International Journal of Social Psychiatry. 002076402093959.

Badrfam R, Zandifar A, Arbabi M (2020) Mental health of medical workers in COVID-19 pandemic: restrictions and barriers. J Res Health Sci 20(2):e00481, Published online 2020 Jun 18. https://doi.org/10.34172/jrhs.2020.16

Barbier EB, Burgess JC (2020) Sustainability and development after COVID-19. World Dev 135:105082. https://doi.org/10.1016/j.worlddev.2020.105082. ISSN 0305-750X

Barroy H, Wang D, Pescetto C, Kutzin Z (2020) How to budget for COVID-19 response? A rapid scan of budgetary mechanisms in highly affected countries. World Health Organization. (Posted 27 Mar 2020, Originally published 25 Mar 2020). Available online at: https://www.who.int/publications/m/item/how-to-budget-for-covid-19-response. Accessed 30 March 2020

Bielicki JA, Duval X, Gobat N, Goossens H, Koopmans M, Tacconelli E et al (2020) Monitoring approaches for health-care workers during the COVID-19 pandemic. Lancet Infect Dis 20(10):261–267. https://doi.org/10.1016/S1473-3099(20)30458-8

Bodenheimer M, Leidenberger J (2020) COVID-19 as a window of opportunity for sustainability transitions? Narratives and communication strategies beyond the pandemic. Sustain: Sci, Pract Policy 16(1):61–66. https://doi.org/10.1080/15487733.2020.1766318

Centers for Disease Control (2020) People with certain medical conditions. https://www.cdc.gov/coronavirus/2019-ncov/need-extra-precautions/people-with-medical-conditions.html-accessed. 17 Dec 2020

Chou R, Dana T, Buckley DI, Selph S, Fu R, Totten AM (2020) Epidemiology of and risk factors for coronavirus infection in health care workers. A living rapid review. Ann Internal Med 173(2):120–136. https://doi.org/10.7326/m20-1632

Chowdhury R, Luhar S, Khan N, Choudhury SR, Matin I, Franco OH (2020) Long-term strategies to control COVID-19 in low and middle-income countries: An options overview of community-based, non-pharmacological interventions. European Journal of Epidemiology 13 35(8):743–8.
Dekkers T, Hertroijs D (2018) Tailored healthcare: two perspectives on the development and use of patient profiles. Adv Therapy 35(9):1453–1459. https://doi.org/10.1007/s12325-018-0765-2
Devnani M (2012) Factors associated with the willingness of health care personnel to work during an influenza public health emergency: an integrative review. Prehospital Disaster Med 27(6):551–566. https://doi.org/10.1017/S1049023X12001331
Engström G, Gars J, Jaakkola N et al (2020) What policies address both the coronavirus crisis and the climate crisis? Environ Resource Econ 76:789–810. https://doi.org/10.1007/s10640-020-00451-y
Iancu AM, Kemp MT, Alam HB (2020) Unmuting medical students' education: utilizing telemedicine during the COVID-19 pandemic and beyond. J Med Internet Res 22(7). https://doi.org/10.2196/19667
IMF (2020) Policy tracker. Policy responses to Covid-19. https://www.imf.org/en/Topics/imf-and-covid19/Policy-Responses-to-COVID-19. Last accessed 30 Nov 2020
Ioannides D, Gyimóthy S (2020) The COVID-19 crisis as an opportunity for escaping the unsustainable global tourism path. Tour Geogr 22(3):624–632. https://doi.org/10.1080/14616688.2020.1763445
Koh D, Lim MK, Chia SE, Ko SM, Qian F, Ng V et al (2005) Risk perception and impact of severe acute respiratory syndrome (SARS) on work and personal lives of healthcare workers in Singapore: what can we learn? Med Care 43(7):676–682. https://doi.org/10.1097/01.mlr.0000167181.36730.cc
Kruse FM, Jeurissen PPT (2020) For-profit hospitals out of business? Financial sustainability during the COVID-19 epidemic emergency response [published online ahead of print, 2020 May 4]. Int J Health Policy Manag 9(10):423–428. https://doi.org/10.34172/ijhpm.2020.67
Lacobucci G (2020) Covid-19: increased risk among ethnic minorities is largely due to poverty and social disparities, review finds. BMJ 2020(371):m4099. https://doi.org/10.1136/bmj.m4099
Lambert H, Gupte J, Fletcher H, Hammond L, Lowe N, Pelling M et al (2020) COVID-19 as a global challenge: towards an inclusive and sustainable future. Lancet Planet Health 4(8):E312–E314. https://doi.org/10.1016/52542-5196(20)30166-6
Lee SM, Kang WS, Cho A-R, Kim T, Park JK (2018) Psychological impact of the 2015 MERS outbreak on hospital workers and quarantined hemodialysis patients. Compr Psychiatry 87:123–127. https://doi.org/10.1016/j.comppsych.2018.10.003
Leite H, Lindsay C, Kumar M (2020) COVID-19 outbreak: implications on healthcare operations, The TQM J, vol. ahead-of-print, no. ahead-of-print. https://doi.org/10.1108/TQM-05-2020-0111
Li-Ying J, Nell P (2020) Navigating opportunities for innovation and entrepreneurship under COVID-19. Calif Manag Rev 2020. https://cmr.berkeley.edu/2020/06/innovation-entrepreneurship/
Mallinson DJ (2020) Cooperation and conflict in state and local innovation during COVID-19. Am Rev Public Adm 50(6–7):543–550. https://doi.org/10.1177/0275074020941699
Markard J, Rosenbloom D (2020) A tale of two crises: COVID-19 and climate. Sustain: Sci, Pract Policy 16(1):53–60. https://doi.org/10.1080/15487733.2020.1765679
Monaghesh E, Hajizadeh A (2020) The role of telehealth during COVID-19 outbreak: a systematic review based on current evidence. BMC Public Health 20:1193. https://doi.org/10.1186/s12889-020-09301-4
Naidoo R, Fisher B (2020) Reset sustainable development goals for a pandemic world. Nature 583:198–201. https://doi.org/10.1038/d41586-020-01999-x
Ortega F, Orsini M (2020) Governing COVID-19 without government in Brazil: ignorance, neoliberal authoritarianism, and the collapse of public health leadership. Glob Public Health 15(9):1257–1277. https://doi.org/10.1080/17441692.2020.1795223

Patel JA, Nielsen FBH, Badiani AA, Assi S, Unadkat VA, Patel B, Ravindrane R, Wardle H (2020) Poverty, inequality and COVID-19: The forgotten vulnerable. Public Health. 2020 Jun; 183:110–111. https://doi.org/10.1016/j.puhe.2020.05.006.

Radwan E, Radwan A (2020) The Spread of the Pandemic of Social Media Panic during the COVID-19 Outbreak. Eur J Environ Public Health 4(2):em0044. https://doi.org/10.29333/ejeph/8277

Sahni H, Sharma H (2020) Role of social media during the COVID-19 pandemic: beneficial, destructive, or reconstructive? Int J Acad Med 6(2):70–75. https://doi.org/10.4103/IJAM.IJAM_50_20

Salisbury C, Thomas C, O'Cathain A, et al (2015) TElehealth in CHronic disease: mixed-methods study to develop the TECH conceptual model for intervention design and evaluation. BMJ Open. 5(2):e006448. https://doi.org/10.1136/bmjopen-2014-006448.

Schaltegger S (2020) Sustainability learnings from the COVID-19 crisis. Opportunities for resilient industry and business development. Sustain Account, Manag Policy J, vol. ahead-of-print no. ahead-of-print. https://doi.org/10.1108/SAMPJ-08-2020-0296

Shanafelt T, Ripp J, Trockel M (2020) Understanding and addressing sources of anxiety among health care professionals during the COVID-19 pandemic. JAMA 323(21):2133–2134. https://doi.org/10.1001/jama.2020.5893

Sultan M, Løwe Sørensen J, Carlström E, Mortelmans L, Khorram-Manesh A (2020) Emergency healthcare providers' perceptions of preparedness and willingness to work during disasters and public health emergencies. Healthcare (Basel) 8(4):442. https://doi.org/10.3390/healthcare8040442

Sze S, Pan D, Clareece N, Gray L, Martin C, Narareth Joshua N et al (2020) Ethnicity and clinical outcomes in COVID-19 a systematic review and meta-analysis. EClinicalMedicine 29:100630. https://doi.org/10.1016/eclinm.2020.100630

Tan BYQ, Chew NWS, Lee GKH, Jing M, Goh Y, Yeo LLL et al (2020) Psychological impact of the COVID-19 pandemic on health care workers in Singapore. Ann Intern Med 173(4):317–320. https://doi.org/10.7326/M20-1083

Tsani S, Riza E, Tsiamagka P, Nassi M (2021) Public policies, one-health and global Inequalities under the Covid-19 lens In Leal Filho W, Azul A, Brandli L, Özuyar P, Wall T. (eds) Reduced Inequalities. Encyclopedia of the UN Sustainable Development Goals. Springer, Cham. ISBN: 978-3-319-71060-0.

The Lancet (2020) COVID-19: protecting health-care workers. The Lancet. 2020 395(10228):922. https://doi.org/10.1016/s0140-6736(20)30644-9

Walton M, Murray E, Christian MD (2020) Mental health care for medical staff and affiliated healthcare workers during the COVID-19 pandemic. Eur Heart J: Acute Cardiovasc Care 9(3):241–247

World Bank (2020) World bank databank. Data on GDP, PPP (constant 2017 international $) from database: World Development Indicators. https://databank.worldbank.org/source/world-development-indicators. Last accessed 30 Nov 2020

World Bank (2020) Global economic prospects, June 2020. Washington, DC: World Bank. © World Bank. https://openknowledge.worldbank.org/handle/10986/33748 License: CC BY 3.0 IGO

World Health Organisation (2020) Weekly epidemiological update 15 December 2020 https://www.who.int/publications/m/item/weekly-epidemiological-update—15-december-2020. Accessed 17 Dec 2020

Zeitoun JD, de TSerclaes P, Lefevre JH (2020) Making healthcare cleaner and more sustainable after covid-19. https://blogs.bmj.com/bmj/2020/10/07/making-healthcare-cleaner-and-more-sustainable-after-covid-19/. Accessed 17 Dec 2020

Zheng J (2020) SARS-CoV-2: an emerging coronavirus that causes a global threat. Int J Biol Sci 16(10):1678–1685. https://doi.org/10.7150/ijbs.45053

Stella Tsani is Assistant Professor at the Department of Economics, University of Ioannina, Greece. She holds a PhD in Economics and Business from the University of Reading, UK. She has worked with multi-sectoral networks across the knowledge triangle and has participated in more than 30 research projects funded by national and international bodies (European Commission FP7, Horizon 2020, EU Directorates, National Bank of Greece, World Bank, Ministry of the Environment of Cyprus, etc.). Her research focuses on public policy assessment, natural resources with focus on renewables, hydrocarbons and water, the waterenergy-food-ecosystems nexus, sustainable development, institutions, labour markets, technology innovation, economy energy and environmental modelling and scenario analysis. She has consulted nationally and internationally on sustainable development, energy, environment, and public policy issues. She has published in leading economics peer reviewed journals (Economics letters, Environmental Science and Policy, Energy Economics, Economic Systems, Science of the Total Environment, Resources Policy, etc.) in edited volumes and in the media.

Elena Riza is an epidemiologist, trained in public health medicine, a Research and Teaching Fellow at the Department of Hygiene, Epidemiology & Medical Statistics, Medical School National and Kapodistrian University of Athens. She holds a PhD in chronic disease epidemiology (London School of Hygiene and Tropical Medicine, University of London). Her main research interests are in the areas of heath crises, disease management and response, gyneacological cancer screening, occupational and environmental risk factors, the effects of climate change on health, health needs assessment of vulnerable and hard-to-reach populations with special focus on women and minority groups. She is a coordinating member and author of the WHO Strategic Plan for Public Health in Greece-2017–2021 and of the WHO first national report on Migrant Health in Greece-2019. She participates in humanitarian projects offering voluntary work in collaboration with NGOs in Greece for vulnerable and disadvantaged population groups. She has numerous publications in scientific peer-reviewed journals and books.

Panagiota Tsiamagka holds a Master's degree in economic analysis and policy from the University of Ioannina. Her research interest includes sustainable development, environment, green financing and public policy issues.

Margarita Nassi holds a Master's degree in Global Health - Disaster Medicine from the department of Medicine, National and Kapodistrian University of Athens Medical School. She is a research assistant at the Department of Hygiene, Epidemiology & Medical Statistics of National and Kapodistrian University of Athens Medical School.

COVID-19 and Sustainability Learning

Innovative and Sustainable Research-Based Learning & Community Services During Lockdown by COVID-19

Ika Dewi Ana, Cahyono Agus, Hatma Suryatmojo, Wirastuti Widyatmanti, Irwan Endrayanto Aluicius, Sri Suning Kusumawardani, Mustofa, Mirwan Ushada, Ririn Tri Nurhayati, Adhy Kurniawan, Irfan Dwidya Prijambada, and Ambar Kusumandari

Abstract The COVID-19 pandemic has caused severe human suffering and substantial damage to life throughout the world. The lockdown caused by COVID-19 also significantly impacted three main elements in UGM education processes, including teaching, research, and community service, where most of their activities involving a

I. D. Ana
Faculty of Dentistry, UGM, Yogyakarta 55281, Indonesia

Research and Community Services, UGM, Yogyakarta 55281, Indonesia

I. D. Ana
e-mail: ikadewiana@ugm.ac.id

C. Agus (✉) · A. Kusumandari
Faculty of Forestry, UGM, Yogyakarta 55281, Indonesia
e-mail: cahyonoagus@gadjahmada.edu; acahyono@ugm.ac.id

A. Kusumandari
e-mail: ambar_kusumandari@ugm.ac.id

C. Agus
PKBTS Tamansiswa Alumni Board, Yogyakarta 55151, Indonesia

H. Suryatmojo
Department of Forest Resource Conservation, Faculty of Forestry, UGM, Yogyakarta 55281, Indonesia
e-mail: hsuryatmojo@ugm.ac.id

H. Suryatmojo · W. Widyatmanti · I. E. Aluicius · S. S. Kusumawardani · A. Kusumandari
Center for Academic Innovation and Studies, UGM, Yogyakarta 55281, Indonesia
e-mail: endrayanto@ugm.ac.id

S. S. Kusumawardani
e-mail: suning@ugm.ac.id

W. Widyatmanti
Department of Geographic Information Science, Faculty of Geography, UGM, Yogyakarta 55281, Indonesia
e-mail: wwidyatmanti@ugm.ac.id

I. E. Aluicius
Department of Mathematics, Faculty of Mathematics and Natural Sciences, UGM, Yogyakarta 55281, Indonesia

W. Leal Filho (ed.), *COVID-19: Paving the Way for a More Sustainable World*, World Sustainability Series, https://doi.org/10.1007/978-3-030-69284-1_13

large group of people, field, and outdoor programs. Innovation in learning, research, and community service without mobility from home is needed to overcome the impact of Covid-19. Methods of internet-based learning, research, and community service must be established during the pandemic to generate a reliable yet efficient blended program that can accommodate their outcomes. This paper will generally look at the initiative and strategy taken by UGM in planning and developing a sustainable research-based learning and community services. It is based on integrated research-based learning and community service, which integrates former individual education, research, and community service programs. The policy of considering COVID-19 as an emergency at UGM has eliminated teaching and learning activities on campus and replaced them with online learning activities from home. The application of the Three-Centra Education concept, which consists of families, schools, and communities, is truly relevant for developing research-based learning and service programs. Besides, academic and non-academic atmospheres to support integrated education, research, and community service processes in universities in the current era must be developed under millennial students' style. The millennial approach's adjustment is possible by developing tutoring systems through student-centered learning with

S. S. Kusumawardani
Department of Electrical Engineering and Information Technology, Faculty of Engineering, UGM, Yogyakarta 55281, Indonesia

Mustofa
Department of Pharmacology and Therapy, Faculty of Medicine, Public Health and Nursing, UGM, Yogyakarta 55281, Indonesia
e-mail: mustofafk@ugm.ac.id

Mustofa · M. Ushada · R. T. Nurhayati · A. Kurniawan
Directorate of Research, UGM, Yogyakarta 55281, Indonesia
e-mail: mirwan_ushada@ugm.ac.id

R. T. Nurhayati
e-mail: ririn_hi@ugm.ac.id

A. Kurniawan
e-mail: adhy_kurniawan@ugm.ac.id

M. Ushada
Department of Agro-Industrial Technology, Faculty of Agricultural Technology, UGM, Yogyakarta 55281, Indonesia

R. T. Nurhayati
Department of International Relations, Faculty of Social Political Science, UGM, Yogyakarta 55281, Indonesia

A. Kurniawan
Department of Civil Engineering, Vocational School, UGM, Yogyakarta 55281, Indonesia

I. D. Prijambada
Faculty of Agriculture, UGM, Yogyakarta 55281, Indonesia
e-mail: Irfan_prijambada@mail.ugm.ac.id

Directorate of Community Services, UGM, Yogyakarta 55281, Indonesia

the edutainment process that triggers the learning system to be more enjoyable, but the quality is still maintained. This quality of online learning during an emergency in UGM has been improved based on conducted surveys indicating that the advantages of online lectures, according to students, are its flexibility and relaxation; no preparation needed to visit campus; more courageous to ask questions during lecture; most materials are well documented for rereading; time and cost-efficient. Meanwhile, online learning disadvantages covered the fact that the lectures confused with online assignments that caused an excessive workload; the lecturers need to be more interactive and communicative; the schedule should not be frequently changed until later evening, and decrease concentration ability among students. University leaders, the academic community, including alumni, have been building solidarity, compassion, and empathy through food assistance, credit, academic fee reduction, non-bureaucratic procedures for students affected by COVID-19. Co-learning system and work from home in the cyber campus 4.0 for the millennial student in the era of lockdown and destructive innovation with the support of sophisticated information technology and significant data access seems to be the most appropriate learning media. The empowerment online learning system must implement the principles of win–win solution, co-creation, co-finance, flexibility, and sustainability of the proposed programs to strengthen education's real meaning during this pandemic.

1 Introduction

The Coronavirus disease-19 (COVID-19) pandemic outbreak has resulted in a widespread and tragic humanitarian disaster and various life aspects. As of August 27th, 2020, 24,335,741 patients have spread across almost all countries globally. It has resulted in more than 829,676 people died, while more than 16,874,717 people have been able to heal (Worldometers 2020), and more than half of the world's population has been affected by the lockdown. The COVID-19 caused by severe acute respiratory syndrome coronavirus 2 (SARS-CoV-2) was firstly reported in Wuhan, Hubei Province, China, on December 31st, 2019. This virus has infected dozens of people in Wuhan and spreads quickly to other provinces in China. World Health Organization (WHO) declared COVID-19 as a global pandemic and call on all parties to fight together to prevent and handle the virus (WHO 2020).

UNESCO reports that national closure in 193 countries affects more than 91.4% of the world's student population (UNESCO 2020). Several other countries have implemented local closures, which have an impact on millions of students. Coronavirus's disastrous impact also has dramatically and radically changed the model of teaching and learning activities. All university leaders in the affected areas of the COVID-19 pandemic were requested to defer the academic activities temporarily, especially face to face lectures, by issuing policies related to the mechanism of online learning from home for students and its supporting infrastructure. Research also needs different approaches since the access to the laboratories and fieldwork were minimized.

Indonesian government stated the COVID-19 pandemic as a national disaster and prevented the spread of the virus and avoided large crowds, so calm-down, soft-down, and lockdown are carried out. The escalation of this spreading virus pushes the government to implement several pandemic mitigation and management policies in many sectors. To minimize the transmission chain of this virus, instead of implementing a nationwide lockdown, the government had approved large-scale social restrictions for some regencies and cities, which restricts the social mobility of its citizens, reduces markets, factories, offices, and school's operational hour resulted in declining economic and social activities. Business activities, education, worship, performances, arts, travel, sports, and tours are suspended and must be carried out from home (Work from Home/WFH). The virtual life of society 5.0 is based on modern remote online information technology, forced to be applied today. Including the School from Home (SfH) program must also be implemented by all educational personnel from various regions.

The concept of technology-based e-learning has excellent potential to become an alternative and development of education (Valverde-Berrocoso et al. 2020), research, and community service during the COVID-19 pandemic and the last few decades. Online learning is incredibly attractive to many students because it offers flexibility in participation, effortless accessibility, and practical convenience. Online learning has emerged as an alternative or an excellent complement to traditional learning models (Ni 2013; Angelino and Natvig 2010; Mansour and Mupinga 2007). It is also expected to be the backbone of tertiary education expansion that accomplish lifelong learning necessities. A conducive learning environment has an essential role in encouraging students' decision to conclude online learning. Nevertheless, La Velle et al. (2020) reported that working online from home has resulted in new stresses on the workloads of tutors, trainees, teachers, and household circumstances.

The paradigm of three pillars of mandatory tasks have shifted at Universitas Gadjah Mada (UGM), from former: education, research, and community service individually, became more integrated into research-based community service and education. The term innovation is identic with continuous improvement in the design and development of innovative and sustainable research-based learning and community services. COVID-19 pandemic could be generalized as a complex, unpredictable behavior problem, especially in uncertain conditions. A similar pandemic could be occurred in the future world, with lower or higher severity, probability, and detectability. COVID-19 could be viewed as multi-loop or iteration, which could be occurred anytime.

This chapter described the impact, strategy, and the bright side of the COVID-19 pandemic, as an initiative to develop an innovative and sustainable research-based learning and community services system in Indonesia. It includes determining the appropriate justification for policy changes in response to the COVID-19 disaster emergency response period, focusing on a radical change and disruption from traditional methods to full online research-based learning and community services methods. The academic community's high resilience factors are also identified, including redefining resiliency, based on its real meaning from education perspectives and challenges during the pandemic.

2 Material and Methods

This paper was written using primary and secondary research data. Primary data were collected from observations, monitoring, and evaluation of program, survey, and field interviews, whereas secondary data were obtained from reports, books, journals, and the internet, and included theories, precedents, and standards used in the field. Qualitative and quantitative analyses of the collected site data are ongoing to identify solutions for similar problems.

3 Tri-Centra Universal Education

Indonesia's education index is ranked the lowest by several international survey institutions. Finland is ranked first in the Global Education Rank, although in the 80s, it was no better than Indonesia (Agus et al. 2020). Nowadays, there is a lot of mental breakdown and evil disputes in various fields and professions. Even educated and religious people have lost their power of reason, objectivity, empathy, family values, and humanity.

Indonesia launched five packages of revolutionary education policies of "Freedom of Learning." The first is a package of primary and secondary education policies; the second is a policy of Independent Campus; the third is the Student Operational Assistance Policy (BOS); the fourth is the Movement Organization Program (POP), and the fifth is the Motivating Teacher (Kemendikbud 2020). The understanding of the word of freedom itself is still multi-perceptual and does not have essential similarities.

Freedom education seeks to develop the talents of students as independent as possible. The methods developed by international systems tend to give absolute freedom. Meanwhile, the education of an independent soul by Ki Hadjar Dewantara (KHD) to galvanize the nation's mentality is also limited by responsible self-discipline, without disturbing an orderly life of peace, greetings, and happiness (Agus

2017) and developed through the *Among* system, to be independent, able to determine one's destiny, independent of orders, on one's strength and competent in an orderly manner.

Ki Hadjar Dewantara develops national education based on its own local culture with a peaceful acculturation process from around the world by developing the school like a garden without walls. With the Tri-Central Education system covering families, schools, and communities, formal, informal, and non-formal education in a synergistic and integrated manner, it is not the school's sole responsibility (Agus 2017). Like outbound activities, Freedom Park also aims to form team cohesiveness, cooperation, information communication, leadership, focus and mind concentration, new creative creativity, strategic management, analytical power, and increased self-confidence (Cahyanti et al. 2019). The Tri-Center Education concept was also developed at *Muhammadiyah* by KH Ahmad Dahlan and *Nahdlatul Ulama* by KH Wahid Hasyim.

Excessive independence will cause chaos for all parties. Absolute freedom in the name of human rights must also be followed by obligations and under the community's human rights and the universe. Every child has his specific talents, strengths, and advantages, different between children and relatively not the same. KHD's teachings expect responsibly, independent soul education is reflected in the concept of "IKIGAI" from Japan and the Javanese philosophy of becoming '*Jalma Kang Utama*' (Main Human Being). Those who can sharpen the potential, aura, hobbies, spirit, passion, and talents of students, and according to nature's nature, the soul's will are independent and responsible (Agus et al. 2020). Furthermore, they can have a career and contribute significantly to society and the universe under their potential and education to be useful and blessed.

Independent soul education requires guidance and an academic atmosphere based on the principles of kinship, kindness, empathy, love, and respect for others. Students are encouraged to have an independent personality, physically & mentally healthy, intelligent, become useful members of society, and are responsible for their happiness and welfare. Tri Education centers educate human beings to have a unique character, ethics, noble morals, and individual responsibility and respect others' rights, nature, and diversity (Agus 2017, 2018). Have smart, broad, deep, and futuristic insights to contribute significantly to complete development in the present and future.

Indonesian government regulates learning adjustments that do not burden teachers and students but are full of character strengthening values and the Covid-19 emergency status (Kemendikbud 2020). Educational people are encouraged not to complete all material in the curriculum, but to be still involved in relevant learning such as life skills, health, and empathy. Also, have the courage to develop the concept of "Out of the Box, within the System." Education policy reforms need to be strengthened, but still provide room for diversity to innovate and be creative in realizing the main path of real human (Agus et al. 2020).

The limitations and diversity of facilities, infrastructure, technology superstructure, and support networks in implementing SfH in Indonesia are still causing concern. In big cities, it is still not right, especially in remote areas that are entirely unreached. The SfH program is tricky to implement thoroughly and thoughtfully. The level of absorption, independence, responsibility, culture, understanding, and character of each student and teacher varies according to mental development and nature (Agus 2017). Thus, the implementation of online learning at each level of education must also be adjusted.

Educational reform needs to be carried out gradually to adapt to changes in the new life order, not necessarily frontal and spontaneous. Education policies during the Covid-19 pandemic cannot be uninformed because the diversity of education elements is very high. It is necessary to adjust educational operations according to each education unit's potential, readiness, and characteristics, to facilitate, create, and innovate responsibly independently. It is estimated that the impact of the Covid-19 pandemic outbreak will last a long time, so in the long term, of course, the curriculum must be adjusted. However, in the short term, the learning outcomes and delivery methods are more comfortable to adjust to the current situation. Increasing students' literacy and numerical power are no longer done through dogmatic learning and rote memorization, but through deep meaning to contribute significantly to solving our common problems.

The education 4.0 concept revolution must also match the 4.0 industrial revolution based on advanced information technology (Amiron et al. 2019; 1.Shahroom and Hussin 2020). Including the use of artificial intelligence, integrated data, remote, mobile games, which seems appropriate for the millennial, golden, and futuristic generations. However, it turns out that it makes human beings tend to be impersonal, ego-centric, like a factory, manufacturer, business, less interactive, less cultured, loses empathy, loses social and human values. Currently, millennial co-working spaces and cafes are always crowded with millennial students. Co-learning system 4.0 for millennial children in the era of destructive innovation with advanced information technology support and big data access seems to be the most suitable learning media.

4 Online Learning System in Higher Education

A shifting paradigm in higher education has migrated the education system from a traditional classroom to distance education programs through online platforms and has increased students' number in the learning process. The advances in information and communication technologies have elicited universities to begin to offer a significant number of online distance learning ranging from offering specific/single topics (courses) to educational degree packages (Allen and Seaman 2013; Rovai and Jordan 2004; Rovai and Downey 2010; Ferre and Wiest 2012).

Indonesia has initiated a higher education information system and network development program, the Indonesia Higher Education Network (INHERENT), since 2006. This initiation has connected many universities in Indonesia, utilizing some development programs of application systems that support teaching and learning activities based on the internet, content development for e-learning needs, and information service system development. In 2014, Indonesia developed the Indonesian Open and Integrated Online Learning program (PDITT), later referred to as the Online Learning System (SPADA Indonesia). A learning management system (LMS) was developed to improve equitable access to quality learning and Higher Education Raw Admission Rate that is targeted to reach 50% in 2024 (UGM 2020). The development of LMS has several benefits, including accessibility expansion to learning content; guarantee and balance of the content quality; college performance improvement; opportunities increase for cooperation between universities; overcoming the limitations of human resources; facilities and technology in partner universities; development of credit transfer and course transfer processes; and media of knowledge dissemination to the broader community. For students, participating in online learning will help expand access to better education, build networks between students, increase self-confidence, and expand digital experiences and digital horizons.

Universitas Gadjah Mada (UGM) has developed the eLisa Learning Management System (elisa.ugm.ac.id) since 2004 and has more than 4,500 learning communities developed by 1,300 lecturers. In 2017, UGM then developed eLOK (elok.ugm.ac.id), based on Moodle, which is more flexible, adaptive and supports learning models blended learning and entirely online based on Massive Open Online Courses (MOOCs), to anticipate technological developments, the need for rapid system development, integration connections and easiness of various systems between universities. UGM has established policies to encourage blended learning by up to 40% in one semester. This policy has encouraged more than 100 courses at UGM, carried out in a mix, and collaboration with university partners in Indonesia since 2018 (UGM 2020). This online learning system establishment in UGM became a role model for other universities and colleges in Indonesia, proven by routine visitors and comparison studies conducted by other large to newly established universities in Indonesia. Latest world university ranking organization (QS) that positions UGM as the best university in Indonesia is another benchmark that portrays its education quality, including the online learning system.

UGM responded to the development of the Covid-19 disastrous impact by increasing awareness and take steps on health and self-protection measures. UGM is replacing offline teaching and learning activities on campus and classroom with online and/or other teaching and learning methods to ensure the continuation of the learning processes and quality. The online lecturing manual for the COVID-19 emergency response period was launched with the emphasis that the implementation of online lectures must combine lecture methods asynchronously through various LMS where lecturers and students are familiar with, or using lecture method synchronously through various social media applications and/or online interactive face-to-face applications that are the lecturers' preferences. The flexibility of utilizing different learning methods and platforms is particularly crucial during this pandemic to situate lecturers and students in the condition that is as convenient as they could when undergoing the learning process. This setting will contribute to the development of high resilience among themselves and the environment surrounding them.

UGM fully comprehends that evaluation is needed to identify the problems or obstacles during the implementation. 52.3% of students used an internet connection from mobile gadgets, and 46.3% used a Wi-Fi connection from their home, dormitory, boarding house, or rented house. Meanwhile, 67.6% of students chose to use laptops for online lectures, and 31% used mobile phones (UGM 2020). The portion of laptops use to attend online lectures is consistent with the quality of Wi-Fi connections and mobile internet users as the sources of student internet access. The students (50.5%) mentioned that the online lectures during the two weeks of the COVID-19 emergency response period were like the quality of face-to-face learning. Meanwhile, 38% of respondents experienced different teaching quality than in-person classrooms, whereas 11.5% stated dissimilar. Furthermore, 66.9% of students can adequately understand the material in online lectures (very good, fair, and moderate).

The 85% of students perceive that the lecturers have an excellent ability to deliver material online using asynchronous and synchronous methods, and 83% stated the quality of online learning materials presentation is good. This result indicates that using various internal and external learning resources through the right social media and audio-visual-based content was adequate and well accepted by students. It also supports the fulfillment of learning outcomes. Most students have an assessment to the satisfaction and excellence of the online lectures, presented below: (i) more flexible and relaxed in its implementation, (ii) no preparation needed to go to campus, (iii) increasing the courage to ask questions, and discuss with lecturers, (iv) becoming new experience and challenge, (v) material can be well documented in the system and can be re-studied, (vi) more efficient in terms of time and cost when compared to offline learning on campus (UGM 2020). Meanwhile, some aspects which were assessed by students as the weakness online lectures were as follows: (i) increasing learning time because some lecturers consider online lectures can be replaced by giving an assignment, (ii) excessive workloads thereby reducing students' stamina, (iii) concentration decreases sometimes because of students is in a relaxed situation, (iv) interaction and good communication from lecturers need improvement, (v) online lecture schedules frequently change, even at night, (vi) the internet network is sometimes unstable, so it interferes with the learning process (UGM 2020).

The 79.6% of UGM lecturers already have the provision to conduct online lectures through internal training from UGM (61%), independent study (38%), and training from the government (1%). Besides, about 67% of UGM lecturers have held online lectures before the COVID-19 emergency response period online and in full blended learning to implement online lectures up to 40% (UGM 2020). At the same time, there are 33% of UGM lecturers who still always teach face to face class/offline.

Infrastructure preference is an essential factor in ensuring the convenience of lecturers in conducting lectures online. Around 87.4% of lecturers chose to use laptops to support online lectures, and only 10.1% chose to use mobile phones or tablets. The majority reasons for choosing a laptop include the need to use a widescreen monitor, flexibility, and ease of control of menus and applications. In the implementation of online learning with face-to-face interaction, the internet connection's quality is a significant factor that needs attention. About 64.5% of lecturers chose to use a Wi-Fi connection from a home internet network provider, 30.8% chose to do internet tethering from a mobile phone, and the rest used another connection line, such as a LAN cable on a desktop computer (UGM 2020). There are many aspects and perspectives of the activities carried out in implementing synchronous online learning using various online interactive applications.

Table 1 shows the implementation of online learning at UGM during the COVID-19 disaster emergency response period, which was relatively well implemented. The experiences that lecturers and students have had in the previous period have made it easier to implement online learning with the mixing method since 2018. The

Table 1 Activities and perspectives of synchronous online learning implementation by lecturers

Activities	Perspectives				
The activity of material delivery	64.5% (synchronous with direct interaction)	49.7% (asynchronous using LMS)	32.7% (sync audio/text-based on social media)	32.7% (asynchronous with video recording)	
Discussion interaction activities	61.9% (direct interaction)	54.7% (text-based interaction with social media)	30.5% (interaction via the LMS discussion forum)		
Assessment and evaluation activities	61.6% (using LMS)	45% (using email)	28.9% (using Google form)		
Online college implementation perspective	78.2% (ease of use of technology)	74.8% (visual display quality)	72.3% (effectiveness supports CPMK)	70.5% (internet network quality)	70.1% (quality of lecturer and student interaction)
What students need to do for online lectures	80.8% (read the material first)	72% (prepare the tools needed)	60.7% (pay attention to well & actively)	43.1% (know online college etiquette)	
Advantages of online courses for lecturers	82.4% (flexibility)	48.1% (efficiency)	48.1% (convenience)	70.5% (comfort)	
Disadvantages of online courses for lecturers	58.8% (lack of interaction with students)	50.3% (internet costs surge)	50% (need more preparation)	48.4% (an internet connection is not adequate)	38.4% (difficulty of assessment and evaluation)

Source UGM (2020)

implementation of online learning requires quite a long preparation and uses many supporting applications so that the process of interaction and delivery of material lecturers to students can be appropriately conveyed and when carrying out offline learning. This need for more business is considered by most lecturers to be difficult, especially for those who have never had the experience of conducting online learning.

Most respondents provide good feedback (3.5, range 1–5) on the mid-term examination process, and it is improved on the final examination (3.7). This improvement is a significant achievement among lecturers and students in this emergency learning process. Its obstacles faced at an earlier stage of a pandemic was able to be solved in less than three months, indicating that the resilience of human resources dealt with online learning processes in UGM is fast, sturdy, and secure (UGM 2020).

In terms of the competency accomplishment, students were asked to measure their ability to achieve the course competency when completing their exam, based on the type of assessment or questions delivered by the lecturers. It included multiple choices, short answer, essay, or description answer, writing assignment, and creating video/vlog. The most assessment types were considered moderate to good ways (3.53–3.84) to help them accomplish the course competency (UGM 2020). In contrast, the video/vlog type of assessment was not good enough to support competency accomplishment.

Ever since UGM provides flexibility for students and lecturers to use any platform to conduct online learning, it is also applied for a midterm or final examination. UGM LMS (eLISA, eLOK, and Simaster), email, online storage, and social media were the leading platforms used by lecturers to deliver their course assessment (UGM 2020). The result portrays that students prefer UGM LMS compared to other platforms, with social/media being the least preferred platform. In term of exam result submission method, understanding level to exam instruction, the system usage effectivity, and the easiness to receive exam result, the perspective of the application effectiveness for midterm and final exam were considered as good, and these scores during the final exam are improved compared to the one from mid-term exam.

During the midterm exam, four main online assessment elements needed improvement, covering: time-length (78.9%), exam instruction (51.4%), exam types (43.3%), and the platform used during the exam (42.4%) (UGM 2020). The time-length concern appeared since some technical difficulties often occurred while completing the exam, related to internet stability, the smooth-running system's level, supporting device condition, the complexity of exam types, and the course size. The dissemination of midterm survey results to the lecturers resulted in significant improvement in implementing an online final examination. Percentage decrease until more than 30% for previously mentioned elements that needed improvement indicates that the ability and willingness of lecturers and students in UGM are remarkable to increase the quality and level of achievement in their learning and general education framework. Intense training courses and close supervision by the university's academic task force after the midterm exam justify the improvement in the UGM online learning process as part of resilience development among students and lecturers.

5 The Impacts of COVID-19 Towards Research Activities

The COVID-19 pandemic has impacted the university includes research and publication. Limiting citizens' activities also limits the researchers' mobility in conducting research activities in gathering data, limiting access to laboratories and other supporting facilities. The confirmation of national disaster status has pushed the government to reduce its national budget from multiple sectors and ministries and divert it to fulfill the needs to fight the COVID-19 pandemic (UGM 2020). This policy implicates budget rationalization for research activities, which is managed by the Indonesian government.

Indonesia's government facilized the research topics refocusing towards the research related to the mitigation and management of COVID-19 (UGM 2020). Through establishing COVID-19 Research and Innovation Consortium, which gathers several ministries, research institutions, universities, hospitals, and industries. Various parties incorporated in this consortium collaborate to produce innovation in products and assessments to answer the citizens' real needs to fight COVID-19. Besides that, it also focuses on COVID-19 mitigation and management topics and initiating innovative and strategic ideas to prepare citizens to face the new normal.

The regional government's policy to mitigate the broadening spread of COVID-19 has limited research activities. This regulation has also impacted the limiting access to research facilities, closing research objects. The limitation also impacts the changes research design and methodologies which initially plans through participatory research methods and field observations to gather primary data, should postpone the collecting of that data or change their methods, wherein some cases, the difference of the methods would also influence the nature and quality of the gathered data. Social distancing also gives impacts towards the conduct of research initially designed with interview techniques, focus group discussion, and workshop inviting stakeholders gathering and discussing the targeted research topics should experience adjustment by using an online survey method or online meeting.

The research activities had significantly impacted the campus's temporary closure, budget cuts, and adjustments to administrative and communication procedures (UGM 2020). UGM also applies for the home policy work so that all lecture, research, and community service activities cannot be carried out on campus, and adjustments need to be made. Temporary closure of the campus directly results in the closure of researchers' access to laboratories, research centers, libraries, and other research facilities. The adjustment is also enacted towards research and community service activities being held inside or outside the campus. In enacting the policies, UGM provides supports for students such as phone credits and free internet service to ease the online-based learning activities. UGM also provides free essential foods for students from outside cities during this pandemic.

The central government's policy related to the rationalization of research budgets also impacts the implementation of research activities at UGM by cutting research budgets received by UGM and delaying some research schemes to the following year (UGM 2020). The limitation of inter-city and international mobility also forced researchers to postpone or cancel travel to other cities for data collection purposes or visit research objects in other cities or provinces. Some researchers then alter their research design by making more use of various online platforms to facilitate communication, meetings, and discussions with their research stakeholders. The temporary suspension of public transportation has also resulted in slow shipping materials and research equipment imported from other cities or countries. The research activities should be managed online.

It was reported that among 605 proposals funded by Indonesia Government, 298 (49%) researchers continue their research, 218 (36%) reschedule their research in 2020, 80 (13%) suspend their research in 2021, and only 9 (2%) cancel their research in 2020 (UGM 2020). Furthermore, ten proposals conducting in 2020 were refocused on COVID-19. It is indicated that almost all UGM researchers want to conduct and continue their research, even the research funding reduced due to the COVID-19 pandemic.

Many researchers of UGM are working intensively to combat COVID-19 and its negative impacts mitigated (Table 2). UGM has also supported any research concerning COVID-19 from various disciplines, including health, engineering, agriculture, humanity, and social sciences. UGM committed to allocate internal funding for COVID-19 research. Furthermore, research collaborations with the governments, research institutions, and private sectors were also conducted. The majority of research activities are focused on producing medical device prototypes needed to protect COVID-19 transmission during the pandemic, such as maskers, hand sanitizer, face shield, and personal protective equipment (PPE), swab chamber, ventilator, and rapid diagnostic test. The research and innovation activities related to Covid-19 are directly applied in the community.

The commitment of UGM in the solidarity and collaboration in research is demonstrated by the involvement of the UGM researchers on the research activities funded by national and international agencies. These researches focusing on COVID-19 involved multidisciplinary researcher teams in six topics include (1) Prevention; (2) Screening and Diagnostics; (3) Medical Devices and their Components; (4) Humanity and Social Sciences; (5) Therapy; (6) Multicenter Clinical Trial. Various schemes of Competitive, Fundamental, and Postgraduate Research Gants from various research topics include food safety and security, health and medicine, information and technology, art, and culture refocused on COVID-19 topics (UGM 2020).

Among 57 Indonesia's innovation research products fight COVID-19 launched by the Indonesia Government, 32 products involved the scientists or researchers or innovators from UGM (Table 1). These research activities are supported by a strong partnership between UGM and industry partners to accelerate the process of scale-up and mass production. UGM also contributed to the innovation research activities in the humanity and social sciences resulting in scientific publications, book chapters, and policy recommendations to the government (UGM 2020).

Although the COVID-19 pandemic had negative impacts, however, it also provided a positive impact on research activities. UGM is actively involved in the solidarity and collaboration in various research at national and international levels,

Table 2 Products of creativity and innovation fight to COVID-19 involving researchers of UGM

No.	Name	Description
1	Rapid Diagnostic Test RI-GHA	Detection of IgM/IgG against COVID-19
2	Rapid Diagnostic Test Microchip	Detection of COVID-19 antigen
3	Viral Transport Medium	Swab media for PCR assay
4	Venindo V01	High-performance ventilator for ICU
5	Venindo R03	Valve bag ventilator conversion
6	Sequence protein S SARS CoV-2	Indonesian SARS CoV-2 isolate
7	OST D	Vitamin D preparation
8	Teh Jahe	Ginger herbal drink
9	Teh Sereh	Citratus herbal drink
10	Immunogama	Herbal imunomodulator
11	Wedang Uwuh Celup	Mix herbal drink
12	Lowkol	Probiotic milk
13	Herbal Hand Sanitizer	Hand sanitizer containing herbal
14	Imboost Flu Herbal	Herbal preparation for influenza
15	Smart Biosafety Swab Chamber (BCL-UGM)	Swab sterile chamber
16	GAMA Swab Sampling Chamber	The positive pressure sampling chamber
17	Insert Masker	Masker with membrane ultrafiltration
18	Air Cleaner	Air cleaner for COVID-19 cleaning
19	Portable Air Purifier	Portable air purifier contaminants
20	COV Watch	Health monitor for COVID-19 patients
21	Touchless Mobile Hand Washer	Automatic mobile hand washer
22	Platform Digital (Cared+)	Self-screening COVID-19
23	Radiography X-ray Digital	Digital X-ray radiography
24	Box Sterilizers Masker N95	Sterilizer for masker N95
25	Aero Dental Suction Unit	Droplet suckers from dental patient
26	HR COMED	Heat-Radiation Compartment
27	Win-MTA	Smart telemedicine robot
28	Robot servant	Servant robot for COVID-19 patient
29	IMUNOCOV-1	Herbals combination supplement
30	Machine PCR Isothermal Lamp 2	Isothermal PCR machine for LAMP-2
31	E-Nose	Diagnosis of infection through exhalation
32	Clinical test: remdisivir	Solidarity trial of remdisivir

Sources Kemenristek/BRIN (2020), UGM (2020)

involving multidisciplinary researchers from various faculties. In UGM, all civitas academics developed the multidisciplinary applied research to educate communities against COVID-19, create policy recommendations to the government, and evaluate the government policy during the COVID-19 pandemic. Significant research activities show collective intelligence in the Triple-Helix (university-industry-government) synergy model (Etzkowitz and Leydesdorff 2000).

The Indonesia COVID-19 Response Acceleration Task Force projects need around 29.9 thousand unit ventilators (UGM 2020). UGM and Dr. Sardjito General Hospital, Yogyakarta, collaborated with some local manufacturers and encouraged by the government policy to produce a local ventilator that was more economical and good. Two prototypes of ventilators named Ventilator R-03 and Ventilator V-01 have been successfully produced, and now they are in an ongoing clinical trial.

UGM coordinated a research collaboration to produce a local rapid diagnostic test (RDT) for COVID-19 called RI-GHA. The diagnostic test developing based on antibodies to detect IgM and IgG produced by the body against COVID-19 is produced to respond to this product's great need during the pandemic (UGM 2020). It will support the standard diagnostic test using real-time reverse-transcriptase-polymerase chain reaction (rRT-PCR), which is time-consuming, high technology, and expensive. Currently, the validation test of the RI-GHA is being conducted in some hospitals in Indonesia. Hopefully, soon, the rapid, comfortable, practical, high sensitivity and very specific and cheap RTD will be available.

Real-time reverse transcription-PCR (RT-PCR) assay remains the test of choice for the diagnostic of COVID-19. UGM has contributed to setting up two laboratories among 89 the laboratories for the COVID-19 diagnostics test throughout Indonesia. UGM/Dr. Sardjito General Hospital researchers have also actively participated in a joint Solidarity Trial led by the WHO to obtain drugs against COVID-19. The four different drugs, i.e., (1) remdesivir, (2) a combination of two drugs, lopinavir and ritonavir, (3) the two drugs plus interferon-beta, and (4) chloroquine, are tested (UGM 2020). The multi-country clinical trial is specifically designed to shorten the time needed to produce substantial proofs of the drugs without neglecting the good clinical practice (WHO 2020).

The COVID-19 pandemic has posed significant challenges in research activities in UGM; however, at the same time, the pandemic is a strong driver of creativity and innovation. Re-formulation of the 2020 state budget has been allocated to address COVID-19 and its impacts. As much as 677.2 trillion IDR (US$ 48 billion) of the State Budget was allocated to fight COVID-19, led to a reduction of the State Budget for other sectors, including research sectors (UGM 2020). The research budget of UGM allocated by the government reduced by 20% in 2020, which caused research projects should be suspended, refocused on COVID-19, and rescheduled, even canceled.

UGM should manage 1100 ongoing research programs and 110 new research programs concerning COVID-19 in 2020 (UGM 2020). The research collaborations with industry or government remain to continue; moreover, new research collaboration focusing COVID-19 is being initiated. Surprisingly, registration of the intellectual property (IP) submitted by researchers of UGM increase during the pandemic. It is indicated that work from home has been efficiently used to prepared IP drafts.

The work from home is a new culture for the directorate staff who previously commonly work from the office. In order to boost the productivity of their staff, the Directorate of Research UGM conducts some things as follow (1) preparing the role of work from home; (2) supporting and facilitating online regular meeting and communication; (3) relaxing of regulation such as the implementation of paperless office; (4) preparing work instruction that meets the health protocol for COVID-19; (5) preparing facilities and work protocol for work from office face up to new normal (UGM 2020).

6 Student Community Services During the Covid-19 Pandemic

Student mobilization in the village community became a formal program as an initial student community service (SCS) in UGM practiced since the 1970s. The empowerment paradigm of the Research-Based Community Empowerment Learning Program (SCS-CEL) through a multidisciplinary approach had been applied since 2007. Every year, around 7,000 students are deployed to the community for two months, equivalent to the 3-semester credit unit for undergraduate students (UGM 2020). This empowerment paradigm must follow the principles of win–win solution, co-creation, co-finance, flexibility, and sustainability. The program theme's results and impacts must be measured to ensure the sustainable improvement of the theme. The program must be implemented through network collaboration between UGM and the regional government, industry, community, professional associations, etc. SCL-CEL students are actively involved in three main pillars: personality, institutional, and community empowerment. The program is implemented through a continuous learning and empowerment process based on the priority of the local community's basic needs, profession, and superior potential.

The global COVID-19 pandemic has shifted the implementation of SCS to an online program. UGM is still obliged to organize SCS for the following reasons: (1) SCS is a compulsory subject, so it must continue to be implemented. (2) The learning process must be continued; otherwise, it will impact the student graduation period. (3) If the SCS is postponed, the student's study period will increase, and a higher tuition fee is required. (4) If the study period increases, it will affect the accreditation and performance of the university. (5) If SCS is not carried out according to the schedule, graduates will lose their opportunity to find work (UGM 2020).

The development of SCS CEL UGM during the Covid-19 Pandemic consist of 4 periods. The SCS CEL UGM period two was launched by the Minister of Education and Culture of the Republic of Indonesia as a model for SCS online in Indonesia. UGM placed 4,504 students from 19 faculties, accompanied by Field Supervisors, in 178 locations in 27 provinces, 77 districts/cities, 143 sub-districts, and 263 villages (UGM 2020). Students were still able to work together with the community during online community services to result in regional mapping, including strategic planning for heritage and culture identification.

The implementation of the 2nd period of online SCS is pervasive; dozens of village master plan documents, tourism master plans, village profiles, and pocketbooks have been produced. Likewise, hundreds of video tutorials for information and education materials for the community and thousands of infographics have been completed. The community also were greatly helped and benefited from the existence of SCS even though they were online. Even so, they prefer that students can stay directly in the field to live with the community in carrying out their programs. Students developed Vontripo as a platform to accommodate volunteers to experience tourism and contribute directly to local tourism development. Vontripo combines the objectives of volunteers and tourism in one concept with the tourism area community's social value and welfare. Vontripo tries to explore the positive side of tourism that can create more value for providers and participants alike. This platform is developed during the students' community services through an online approach.

The next SCS theme of Cares for Education is motivated by the community's difficulties, which must do a study from home. These constraints include the lack of strength of the existing internet network, the lack of laptop and mobile phone facilities owned by the community, and inadequate learning assistance in personnel and competence. Their program consists of helping the teacher teach and prepare teaching materials such as power points, posters, videos, and accompanying them during their study time using the online platform.

7 Conclusions

The governance model, institutional management, the learning process policy and practice, and scientific meetings are transformed into an online model, even though it causes much turmoil due to this disruption. The COVID-19 pandemic status finally changed all education systems in Indonesia from the offline classroom face-to-face model to full online face-to-face, without going through transition and adaptation.

Multidisciplinary applied research to educate communities against COVID-19, policy recommendations to the government, and evaluate the government policy during the COVID-19 pandemic should be developed to challenge any uncertainty, any scenario, and circumstances by COVID-19 and in the future. Students were still able to work together with the community during online community services to result in regional mapping, including strategic planning for heritage and culture identification.

The 4.0 Collective Education Park for millennial children in the pandemic era and the future seems to be an appropriate medium to learn, work and worship together at the integrated Tri-Centra Universal Education at home, school, and community. The development of innovative learning, research, and community service systems in response to the emergency period of the COVID-19 disaster and the future is critical by focusing on radical changes and disruption from traditional methods to full online research-based learning and community service methods.

References

Agus C (2017) Revitalisasi Ajaran Luhur Ki Hadjar Dewantara untuk Pendidikan Karakter Bagi Generasi Emas Sebagai Cucuk Lampah Kebangkitan Nasional II Indonesia. Jurnal ABAD 1(1):51–56 (In Indonesian)

Agus C (2018) Development of blue revolution through integrated bio-cycles system on tropical natural resources management. In: Leal Filho W, Pociovălişteanu D, Borges de Brito P, Borges de Lima I (eds) World sustainability series: towards a sustainable bioeconomy: principles, challenges and perspectives. Springer, Cham, pp 155–172

Agus C, Cahyanti PAB, Widodo B, Yulia Y, Rochmiyati S (2020) Cultural-based education of Tamansiswa as a locomotive of Indonesian education system. In: Leal Filho W et al (eds) Universities as living labs for sustainable development. World Sustainability Series. Springer, Cham, pp 471–486

Allen IE, Seaman J (2013) Changing course: ten years of tracking online education in the United States. Babson Survey Research Group and Quahog Research Group, Babson Park, MA. Retrieved from https://www.onlinelearningsurvey.com/reports/changingcourse.pdf. Access 27 Aug 2020

Amiron E, Latib AA, Subari K (2019) Industry revolution 4.0 skills and enablers in technical and vocational education and training curriculum. IJRTE 8(1C2): 484–490

Angelino LM, Natvig D (2009) A conceptual model for engagement of the online learner. JEO 6(1):1–19

Cahyanti PAB, Widiastuti K, Agus C, Noviyani P, Kurniawan KR (2019) Development of an edutainment shaft garden for integrated waste management in the UGM Green Campus. IOP Conf Ser Earth Environ Sci 398(2019):012001

Etzkowitz H, Leydesdorff L (2000) The dynamics of innovation: from national systems and 'mode 2' to a triple helix of university-industry-government relations. Res Policy 29(2):109–123

Ferre HGC, Wiest LR (2012) Effective online instruction in higher education. Q Rev Dist Educ 13(1):11–14

Kemendikbud (2020) KBBI Daring. https://kbbi.kemdikbud.go.id/entri/disrupsi. Access 27 Aug 2020

Kemenristek BRIN (2020) Tanggap Hadapi Covid-19 Katalog Inovasi Karya Peneliti Dan Perekayasa Kemenristek/Brin Untuk Mengatasi Pandemi. Kemenristek/BRIN, Jakarta

la Velle L, Newman S, Montgomery C, Hyatt D (2020) Initial teacher education in England and the Covid-19 pandemic: challenges and opportunities. J Educ Teach. https://doi.org/10.1080/02607476.2020.1803051

Mansour B, Mupinga DM (2007) Students' positive and negative experiences in hybrid and online classes. College Student J. pdfs.semanticscholar.org. Access 27 Aug 2020

Ni AY (2013) Comparing the effectiveness of classroom and online learning: teaching research methods. J Publ Affairs Educ 19(2):199–215. https://doi.org/10.1080/15236803.2013.12001730

Rovai AP, Jordan H (2004) blended learning and sense of community: a comparative analysis with traditional and fully online graduate courses. Int Rev Res Open Distrib Learn 5(2). https://doi.org/10.19173/irrodl.v5i2.192

Rovai AP, Downey JR (2010) Why some distance education programs fail while others succeed in a global environment. Internet High Educ 13(3):141–147. https://doi.org/10.1016/j.iheduc.2009.07.001

Shahroom AA, Hussin N (2018) Industrial revolution 4.0 and education. IJARBSS 8(9):314–319

UGM (2020) Monitoring and Evaluation of 2020 Program Report. Yogyakarta

UNESCO (2020) Covid 19 educational disruption and response. https://en.unesco.org/covid19/educationresponse
Valverde-Berrocoso J, Garrido-Arroyo MC, Burgos-Videla C, Morales-Cevallos MB (2020) Trends in educational research about e-learning: a systematic literature review (2009–2018). Sustainability 12:5153. https://doi.org/10.3390/su12125153
WHO-World Health Organization (2020) Solidarity clinical trial for COVID-19 treatments. Access: 20 June 2020. Available from: https://www.who.int/emergencies/diseases/novel-coronavirus-2019/global-research-on-novel-coronavirus-2019-ncov/solidarity-clinical-trial-for-covid-19-treatments
Worldometers (2020) COVID-19 Coronavirus Pandemic. https://www.worldometers.info/coronavirus/. Access 27 Aug 2020

Ika Dewi Ana is currently an associate professor at the Department of Dental Biomedical Sciences, Faculty of Dentistry, Universitas Gadjah Mada (UGM). Her area of interest is bioceramics, tissue engineering, and regenerative medical therapy. There are 2 research works of her (CHA Bone Graft and CHA-based Hemostatic Sponge) have been already registered and translated into the market in Indonesia by a university holding company and state-owned pharmaceutical company in Indonesia. She received national and international awards from tissue engineering and biomaterials societies related to her research. She serves as an Industrial Committee Member of TERMIS-AP (Tissue Engineering and Regenerative Medicine Society – Asia Pacific) since 2015 and TERMIS-AP Council since 2019. She is now a Vice Rector for Research and Community Services of UGM.

Cahyono Agus is a Professor at Universitas Gadjah Mada Yogyakarta Indonesia, was born in Yogyakarta, March 10, 1965. The Doctorate was obtained from Tokyo University of Agriculture & Technology, Tokyo, Japan, in 2003. He was head of UGM University Farm 2008–2015. He currently serves as Chairman of the Central Board of Tamansiswa Alumni (PP PKBTS) 2016–2021, a Majelis Luhur Persatuan Tamansiswa (MLPTS) member 2016–2021, and member of Education Board Daerah Istimewa Indonesia. Active as a reviewer in research, community development, scientific publications, and institutional development in Higher Education, Indonesia. He published many scientific works in international seminars and journals and has several awards and copyrights from various agencies.

Hatma Suryatmojo born in Yogyakarta, November 21, 1974. Lecturer at Faculty of Forestry, Universitas Gadjah Mada, Yogyakarta, Indonesia. His researcher interests are forest hydrology, erosion control and watershed conservation. He obtained Doctorate degree from Kyoto University in 2013 on watershed hydrology in the Indonesian tropical rainforest. Active in research and community development at Center for Disaster Studies, UGM. He managed the International Journal of Sustainable Future for Human Security as chief editor. He is Head of Center for Academic Innovation and Studies UGM since 2015 to manage the learning innovation and academic development.

Wirastuti Widyatmanti is a lecturer and researcher in Laboratory Remote Sensing, Department of Geographic Information System, Faculty of Geography, UGM, majoring in Soil Landscape Mapping and Atmospheric/Climate Change. She finished her Ph.D. from University of New South Wales, Sydney, Australia. Currently, she is also active at the Center for Innovation and Academic Studies (PIKA) UGM. She has become a reviewer of various national and international journals, such as Acta Geophysica (SpringNature), and Journal of Geochemical Exploration (Elsevier), as well as a member of the Technical Committee for the Formulation of Indonesian National Standards 07-01 Geographic Informations/Geomatics.

Irwan Endrayanto Aluicius is a lecturer at the Department of Mathematics, Faculty of Mathematics and Natural Sciences, UGM. He obtained his Doctorate degree from Universiteit Twente, Enschede, Netherlands. His current research topic is the application of queuing theory in various applications in hospitals. Since December 2016, he has been appointed as the Head of Academic Studies, in Center for Innovation and Academic Studies (PIKA) UGM.

Sri Suning Kusumawardani, is a lecturer in Department of Electrical Engineering and Information Technology, Faculty of Engineering UGM. She holds the position of Secretary at the UGM Center for Innovation and Academic Studies (PIKA). Her research fields are related to Educational Data Mining (EDM), Elearning / MOOCs (Semantic Technology / Ontology, Predictive Analysis and Learning Analytics, Machine Learning and Artificial Intelligence).

Mustofa is a professor in the Department of Pharmacology and Therapy, Faculty of Medicine, Public Health and Nursing, UGM. He is currently appointed as Director for Research, UGM. He obtained his Master from Postgraduate Program, Faculty of Medicine, UGM and his PhD from Faculty of Pharmacy, University of Montpellier I, France. His research activities concerned drug discovery and development both from natural and synthetic compounds. He is authors/co-authors in 70 international publications concerning antiplasmodial, anticancer, antibacterial, antibiofilm, antidiabetic activity of natural and synthetic compounds.

Mirwan Ushada is a lecturer at the Department of Agroindustrial Technology, Faculty of Agricultural Technology, UGM. He obtained his Master and Doctorate degree from Osaka Prefecture University, Japan under support of Hitachi Scholarship Foundation. His research interest are Ergonomic system, Product design & development, Kansei (Affective) engineering and Agroindustry 4.0. The current research topic is modeling collective-affective human parameters in agroindustry using artificial intelligence. He was a grantee of 2006 Armand Blanc-CIGR International Commission of Agricultural and Biosystems Engineering, and 2019 President Award of Indonesian Ergonomics Association (PEI). Since October 2017, he has been appointed as Deputy Director for Research UGM.

Ririn Tri Nurhayati is a lecturer at the Department of International Relations, Faculty of Social and Political Sciences, UGM. Her areas of expertise are international politics, human rights and international humanitarian law. She finished her doctoral study at the School of Political Science and International Studies, the University of Queensland, Australia in 2019 and worked on research on Sovereignty and Responsibility in Indonesia. Currently, she is the Head of Research Program at the Directorate of Research UGM.

Adhy Kurniawan is a lecturer at the Department of Civil Engineering, Vocational School and at the Environmental Science Study Program, Graduate School of UGM, in the field of civil engineering for hydro-environment and water resources. He obtained his Doctorate degree in Fluvial Hydraulics from Ecole Polytechnique Federale de Lausanne (EPFL Switzerland) in 2003. Then he got a Post Doctor scholarship from The Hitachi Scholarship Foundation at the Disaster Prevention Research Institute (DPRI) Kyoto University, Japan.

Irfan Dwidya Prijambada is a Professor of Soil and Environmental Microbiology and Biotechnology, Director of Community Services, Universitas Gadjah Mada, Indonesia. He got a Philosophical Doctor of Engineering, Dept of Fermentation Technology, Fac. of Engineering, Osaka University, Osaka, Japan. In the last five years, He published ten articles in the field of environment in reputable international periodicals. He also has been a guest speaker at various scientific meetings related to community empowerment.

Ambar Kusumandari has a degree in Agriculture at Universitas Gadjah Mada (1986), master in Environmental Studies at the University of Waterloo, Canada (1994), and Doctor in Forestry at Universitas Gadjah Mada (2012). She is a senior lecturer at the Faculty of Forestry, Universitas Gadjah Mada since 1990. From 2018-nowadays she is the Head of Sub Directorate for Student Community Services-UGM and member of LPPM Forum in Yogyakarta, Indonesia. She has experience in Watershed management and conservation with emphasis on Soil and water conservation and Tropical forest hydrology. She published several scientific works in international seminars, workshops and International Scopus indexed journals, such as Malaysian Applied Biology (MAB), Agriculture and Natural Resources (ANRES), Journal of Tropical Forest Management (JMHT), and International Journal of Geography (IJG).

Challenging the Plague of Indifference: COVID-19 and Posthumanistic Education for Sustainability

Alison Sammel and Peter Blaze Corcoran

Abstract Reflective of this COVID-19 era, Maxine Greene's analysis of Albert Camus' great novel, *The Plague*, reminds us "that the plague can be understood as a metaphor for people's indifference or distancing or thoughtlessness" where we "organize people into sanitary squads to fight the plague…because everyone carries the microbe for the plague of the body, the potential for the plague of indifference". As our species and planet respond to current unprecedented times, opportunities emerge that encourage educational shifts toward ecological sustainability. We argue that education must not be accepting of the plague of indifference because that is to be complicitous with it. As such, this paper explores opportunities to destabilize the enduring assumptions of indifference towards ecological sustainability within humanistic education, while building capacities to see beyond these assumptions. Since human subjectivity is shaped by educational agendas, we advocate for a provocative posthumanistic discourse where caring and kindness shape the future of teaching and learning within this deeply interconnected, beautiful world.

Keywords Education for sustainability · Posthumanism · Humanism · Caring and kindness · Discourse and subjectivity

1 Introduction

> While intensifying their desire to be set free, the terrible months they had lived through had taught them prudence, and they had come to count less and less on a speedy end of the epidemic…One of the signs that a return to the golden age of health was secretly awaited

A. Sammel (✉)
School of Education and Professional Studies, Griffith University, G30 3.37, Gold Coast campus, Southport, QLD 4222, Australia
e-mail: a.sammel@griffith.edu.au

P. B. Corcoran
Environmental and Sustainability Education, University of the Sunshine Coast, Maroochydore, QLD, Australia

c/o Post Office Box 171, South Freeport, ME 04078, USA

W. Leal Filho (ed.), *COVID-19: Paving the Way for a More Sustainable World*, World Sustainability Series, https://doi.org/10.1007/978-3-030-69284-1_14

> was that our fellow citizens…now began to talk…of the new order of life that would set in after the plague.
>
> Albert Camus (1948), *The Plague*, p. 269

Returning in a time of COVID-19 to Albert Camus' great novel, *The Plague*, we were reminded of philosopher Maxine Greene's brilliant analysis of the text; she writes, "Tarrou recognizes that the plague can be understood as a metaphor for people's indifference or distancing or thoughtlessness. He finds the imagination to organize people into sanitary squads to fight the plague and, critically, make it the moral concern of all, because everyone carries the microbe for the plague of the body, the potential for the plague of indifference" (Personal communication, n.d.). We believe it is this discourse of indifference that educators must organize against. The time of COVID-19 dramatically shows our indifference to inequality, our indifference to an ongoing unsustainable economy, our indifference to the vulnerability of those who live in extreme poverty. Indeed, we see our indifference to the agony of all the social world. We see our indifference to massive species extinction, our indifference to the dramatic impacts of climate change, and our indifference to the increasing of zoonotic viruses as burgeoning populations push into the wild. And, indeed, we see our indifference to the agony of all the natural world.

Who can say we weren't warned? Rachel Carson (1962) in *Silent Spring*, perhaps the most important book of the twentieth century, cautioned us as to the impact of modernity upon the very systems that support life as we know it. She wrote, "future generations are unlikely to condone our lack of prudent concern for the integrity of the natural world that supports all life" (p. 13). For the book's epigraph, she chose the bleak words of Albert Schweitzer, "Man has lost the capacity… to foresee and to forestall. He will end by destroying the Earth—from which he and other living creatures draw their food. Poor bees, poor birds, poor men" (1962, p. IV).

It behooves us to prove Dr. Schweitzer wrong! But what to do? We already know most of what needs to be done—forbid wet markets where viruses jump to humans, create and protect large biodiversity preserves, create a sustainable economy that functions within the Earth's limits, advance a scientific literacy that promotes ecological sustainability, cut consumption dramatically, transition from the use of fossil fuels to renewable sources of energy. But to know is not to do.

What will motivate us to action? What values and attitudes and ethics will lead us to change our society's behavior? One of the answers to this fundamental question which emerges from the research in this book is the prioritizing of an alternative education discourse for the post-COVID-19 world. The project of this book is to understand the implications of COVID-19 on sustainable development, on the United Nations Sustainable Development Goals, and on teaching and research. In this chapter, we advocate for a provocative posthumanistic discourse to be embedded in education where the relational entanglements of all species with whom we share our daily lives shape teaching and learning within this deeply interconnected world.

Greene uses a metaphor for the reification that must be achieved. In *The Plague*, she writes, "Dr. Rieux fights the plague for the most abstract of reasons at first, because that is his job. Only later, when the unspeakable tragedies he witnesses make him think about what he is doing, does he reconceive his practice and realize that the most important thing he can do is not to be accepting of the pestilence because that is to be complicitous with it" (Personal communication, n.d.). If we accept the destructive impacts of the current political, economic, and educational systems on the ecological systems that support life on Earth, we are complicitous. If we accept the devastating impacts on the poor and on future generations, we are complicitous.

To overcome this complicity, we valorize the deep ethical discourse underpinning the Earth Charter. Although not written explicitly for education but rather as an ethical framework for an interconnected, ecologically sustainable future, the Earth Charter (2000) reminds us:

> fundamental changes are needed in our values, institutions and ways of living. We must realize when basic needs have been met, human development is primarily about being more, not having more" (Earth Charter Commission, Preamble, paragraph four).

Pope Francis (2015) in *Laudato Si'* states, "the Earth Charter asks us to leave behind a period of self-destruction and make a new start. But we have not as yet developed a universal awareness needed to achieve this" (Paragraph 207, p.137).

One of the great insights of the Earth Charter is the concept of a broadening sense of identity and caring. In Western cultures we care for ourselves first, as the capitalist culture places great emphasis on the individual. Beyond ourselves, we care for those close to us such as our families. Beyond that, we care for the communities of which we are a part. The Earth Charter calls upon us to expand our sense of identity and moral responsibility to include all living things, indeed, to include the larger universe. By articulating common concerns and common values, the Earth Charter provides a rich vision and gives us hope that this vision is a viable path to take. It is not the only path and surely it is not the last word on a global ethical framework, but it is an invitation to reflect on the significance of globalization for life on Earth and an invitation to promote another more caring way of being. It is also a call to consider the responsibility of the academy to raise issues related to the direction in which globalization has gone. If our way of life is to be an ecological sustainable one, if globalization can induce caring, we in the educational field, need to engage deeply in such ethical challenges.

2 Reflecting on Caring and Kindness

Perhaps COVID-19 will help us to develop such an awareness as we see how mutually dependent and interconnected, we truly are. Leonardo Boff (2008), in a brilliant essay on the Ethic of Care, writes,

> humanity and Earth stand together facing the future. This future is not guaranteed by the forces leading the universe. We have to want it. Hence the Earth Charter goes on to say realistically 'we must decide to live with a sense of universal responsibility.' Accordingly, the principle of self-destruction must be counteracted with the principle of care and of universal responsibility…This is the context in which the ethic of care proposed by the Earth

Charter gains relevance as one of the axes around which the sustainable way of life revolves. It will either be oriented by care or it will not be sustainable" (p. 131).

Schweitzer, despite his fatalism for human prospects, never stopped caring in his own life's work. The comforting sign at his hospital in Gabon, Central Africa, was reputed to have read "At whatever hour you come, you will find light, and help, and human kindness" (Elliott n.d.).

But what is human kindness and are our societies effective in expressing it? For the authors of this chapter, caring and kindness are *actions*, rather than cognitive experiences. They occur when thoughts, feelings and intentions of love, empathy, compassion and altruism are deliberately acted upon. Unfortunately, this is not perceived as common in Western societies, although the growing cries for human kindness to be practiced, are. It seems that social infrastructures—politics, healthcare, education, and wider assumptions and social contracts that citizens share—are increasingly understood as not caring or supporting citizens or the planet (Sammel et al. 2020). In seeking to foster caring or kindness, we must resolve to overcome the rising of isolationism, tribalism, racism and authoritarianism underpinned by discourses of indifference and complicity. Dedication to the ethos of caring and kindness is needed to rethink our social systems. This ethos could be understood as showing care and kindness for self, other humans, animals, plants, nature, ecosystems, as well as ideas and collegiality. These thoughts are not new to disciplines that were created to care for children. Marginalized educational philosophies such as education for sustainability, critical education, feminist education, multicultural education, and peace education, at their core, have all organized around an ethos of care and kindness. This agenda hasn't been fully incorporated within educational systems. Noddings (2003) advises that hegemonic education focuses on 'caring' for a limited selection of ideas and excludes other realms of caring. This lack of care in education is significant: as what is taught and learned in schools largely determines what students believe is important or unimportant, possible or not possible, good or bad, normal or abnormal (Eisler and Fry 2019). The assumptions, beliefs and ideas seeded through schooling not only influences how a child's brain develops, but also how people learn to perceive and interact with the world, and how they learn to perceive themselves.

Globally, in response to the unprecedented global challenge all humans are facing with the recent outbreak of COVID-19, many are asking how we arrived at this lack of caring and the social plague of indifference. How did our social and education systems become so devoid of kindness and so complicit towards the continuing environmental crises? How did people come to view every aspect of our environment as a resource to be consumed or utilized, rather than to be cared for and viewed as inherently connected to our well-being?

3 Reflecting on Western Assumptions

Since the industrial revolution, Western civilizations have been founded on the assumption that the natural world is inert or mechanical. In this way, nature operated like a grand machine, and by investigating the specific functions and functioning of each part of the machine, it was believed that science could come to understand how the whole machine worked (Sammel 2020). This reductionist understanding (reducing the machine down to smaller parts) assumed that the whole could be understood by isolating and investigating parts. To various degrees this assumption still frames Western education where concepts are chunked into distinct sections and taught for the mastery of each section. It is assumed that students will inherently connect learned sections together to gain a fuller picture of how the natural world works. In this discourse, learning is understood as an outcome. This process to teach the workings of nature by separating knowledge into distinct categories is reflective of the Enlightenment period more broadly.

In the Enlightenment era, alongside this reductionist science, many dominant faith traditions continued to espouse that humans were the only species with intelligent souls and divinely connected. Constructed as being less significant than that of humans, the rest of nature was ultimately viewed in service to humanity. Philosophy was also built on this understanding, with Aristotle advocating only humans possessed a 'rational soul' that provided access to the divine. Descartes proposed 'the Great Chain of Being' that provided a hierarchy of living things, dichotomizing unthinking, mechanical objects (the human body, non-human animals, plants etc.) and the thinking human mind, which was understood as being reflective of the divine. With humans perceived as the top of this Earthly hierarchy, Western civilizations progressed with the belief that by virtue of their intellect, humans were exceptional in regard to other species, and above or separate from things classified as "nature". This dualistic thinking of humans as distinct from nature still justifies the exploitation of those classified as objects, and the dismissal of cultures who were perceived as being closer to animals and nature. This perception of separateness comes from the story of humanism and is unfortunately endorsed in many people's everyday experiences. For example, large populations in the Western world live in cities where natural systems may be rendered invisible and the infinite ways humans are supported by what science classifies as 'nature' can be ignored or discredited. Legacies of these beliefs still circulate in the collective consciousness, policies and infrastructures of contemporary colonial cultures. These foundational assumptions need to be challenged as they underpin the Anthropocene and the Chthulucene.

4 Reflecting on the Anthropocene Epoch

The Anthropocene is the name given to this historical era or geological epoch in which human activities have changed planetary ecosystems, generated mass species extinction, and altered the composition of air, water, and land globally. In this era, humans have become the dominant driver of accelerating changes in the Earth's climate. However, the Anthropocene era is not just about weather patterns, or the disappearance of flora or fauna, or even the social or political unrest that follows these kinds of changes. This era is about the unpredictable effects unsustainable human ideologies and practices are having on deeply integrated social and natural systems. The deeply integrated, complex relationship humans have with all aspects of the planet is not easily understood by Science or Western cultures: indeed, the English language does not have a word that fully captures this embeddedness. Even the concept of 'humanity' defines our separateness from nature.

Far from being separate and superior to nature, the emerging consequences of Western ideologies and actions illustrates how complex our relationship with nature is. Morton (2010) suggests humans share an "unbearable intimacy" (p. 50) with all of nature and the ecological emergencies we are now facing illustrate how deeply our actions influence this complexity and how wrong Western ideology is that distances humans and their beliefs and actions from nature. However, Haraway (2016) suggests that it is not really 'humans' per say who are to blame for this environmental crisis, but the era of Capitalocene or hegemonic world systems. She suggests that the narrative of Anthropocene points to a destructive ending and proposes a reframing of the way we think about this era through her use of the term Chthulucene. The Chthulecene explores how multiple species connect and interact in this multispecies world. She believes that the world is made up of ongoing multispecies stories and practices that highlight how we are all beings from, with and of the Earth (Haraway 2016). Haraway believes that *all earthlings are kin* but making and recognizing kin is the hardest and most urgent challenge for humans. Interestingly, Western science has always shown that humans have coevolved with and are connected to ecosystems, yet within this understanding lurked a disconnection from nature. This disconnect was and is underpinned by the ideologies of humanism, reductionism and the belief of humanity's unique and privileged position above animals, plants and all other objects classified as natural.

5 Reflecting on Humanism, Discourses and Subjectivity

Humanism is the story that frames Western societies, including formal education. Foucault (1984) describes humanism as a set of themes that, though varied in their content, have reappeared over time in European societies. Forming an almost invisible thread that ties together our conscious and unconscious; individual and social thoughts; understandings, actions and infrastructures, humanism has shaped the power relations and value judgments that support certain ways of knowing and understanding the world. St. Pierre (2000) says humanism is "the air we breathe, the language we speak, the shape of the homes we live in, the relations we are able to have with others, the politics we practice, the map that locates us on the earth, the futures we imagine, the limits of our pleasures" (p. 478). Humanism underpins the assumption that nature can be known, represented and predicted by science. This paradigm provides a structured logic and credible, verifiable practices that perpetuate human exceptionalism (Plumwood 2002) which promotes and endorses the human/nature divide. It mandates that knowledge of the world is centered around humans, and humans are the judge and jury for what counts as intelligence. The impact of this discourse on Western culture cannot be understated.

Human exceptionalism should not be perceived as an idea or a belief, as it is much more than this: it is a key player in the discourse of humanism. Bove (1990) argues that the concept of discourse "aims to describe surface linkages between power, knowledge, institutions, intellectuals, the control of populations, and the modern state as these intersect in the functions of systems of thought" (p. 35). Foucault (1984) suggests that discourses highlight how language gathers itself together according to socially constructed rules and regularities, allowing certain statements to be made and not others. Discourses organize certain ways of thinking within a culture and create and maintain the ways of acting in the world that over time become normal or natural. These historically constructed ways of knowing and acting limit thinking, understanding, being and questioning to only that inside the boundaries of the language and concepts prioritized within that discourse (St. Pierre and Pillow 2000). The fixed meanings or closed systems of conceptualizing words, constructed within the discursive, produce accepted notions of truth and common-sense ways of knowing (Weedon 1987). These truths are believed to be absolute rather than belonging to specific cultural narratives. Discourses, therefore, are social practices through which 'reality' is made intelligible. In other words, dominant Western discourses shape how individuals come to understand themselves and their place in the world. It speaks to power relations and value judgments that support certain ways of knowing that are viewed as natural. Therefore, as much as science shows we are part of nature, it seems *normal* to perceive humans as distinct or above nature. A critique of this normalized perception makes us question if we, the authors, do believe this? What do we think as individuals and why do we hold these views? To explore these questions, we need to understand the concept of the subject versus the individual.

The concept of subject illuminates the role discourse plays in our lives, the depths to which language constitutes the very nature of our conscious and unconscious understandings of self (St. Pierre 2000). Weedon (1987) maintains "neither the body nor thoughts and feelings have meaning outside their discursive articulation, but the way in which discourse constitutes the minds and bodies of individuals is always part of a wider network of power relations, often with institutional bases" (p. 108). Therefore, language is not merely a tool for describing, and a discourse is not just a narrative in which an individual locates herself or himself (St. Pierre and Pillow 2000). They both form the basis through which persons understand themselves as knowable individuals with a collective reality. Thus, the term *subject* acknowledges the constitutive force of discourse. A person inevitably understands the world from within a repertoire and from the vantage point of the particular images, storylines, metaphors and concepts of their discursive practices (Davies 2000). In this way a person *is* the product of the discourses they are exposed to, or engage in, rather than an autonomous individual who can choose to be independent of all discourses.

The individual is continually produced by and situated within similar or competing discourses, and this understanding is fundamentally different from the humanist perspective. The humanistic perspective assumes that a person is an individual agent, a relatively fixed product whose sense of *self* or *identity* is established before birth. This notion of identity is linked to talk of an authentic self (Davies 2000). It is assumed there is a stable and coherent self that can look objectively and reliably on knowledge gathered through the correct use of reason and obedience to scientific laws (St. Pierre and Pillow 2000). The humanistic notion of self and individual agency is key in education. Learning is viewed as an end point, rather than a process of becoming.

When we move away from seeing teachers and students as having a fixed or stable identity, to understanding how their subjectivity is produced by the dominant political, economic and social educational discourse, we can develop more clarity in analyzing their struggles. As subjects who have been produced by the dominant educational discourse, their language and the meanings hidden within that language are reflective of this humanistic discourse (Martusewicz 2001). Hence, invisible understandings about the role of education becomes that of *securing future economic wealth* and the assumptions and methods of achieving that directly link to isolationism, tribalism, racism and authoritarianism underpinned by indifference, complicity, and lack of kindness or care for self, other humans, animals, plants, nature, or ecosystems.

6 Reflecting on Educational Discourses

So, who is to blame? Are the teachers? The students? If this dominant educational discourse represented the *only* discourse teachers and students had been exposed to, then these assumptions would be the only way these teachers or students *could* understand what it means to teach and learn (Davies 2000). Educational discourses are structured and restructured to suit particular political agendas which most often perpetuate human exceptionalism and the nature/human perception as being *normal*. To this end, dominant Western educational curricula, pedagogies, teacher identities, and administrative practices inherently promote ecological unsustainability rather than sustainability by controlling what counts as knowledge, truth, normalizing the educational experience while silencing minority perspectives and voices (Sammel et al. 2020). Pedagogies are examples of this, as they are structured around people rather than nature: either teacher centered, or more progressively, student centered. Teachers are normally portrayed as lesson-dispensers, controlling and being controlled by the dominant narratives that perpetuates turning the 'empty vessels' sitting in front of them into workforce ready producers and consumers. Giroux (1988) speaks to this when he says that powerful invisible narratives that continuously promote and maintain dominant power structures are evident in most school curricula. Curricula explicitly and dramatically influences teachers' work, supporting and confirming the political agendas through the daily rhetoric of schools. All players within educational systems have been historically produced to understand what is socially acceptable to teach and learn, and what is not (Davies 1990). As such, our educational discourses are complicitous with the discourse of pestilence.

Much has been written for five decades on this problem. However, Pacini-Ketchabaw et al. (2016) suggest it is easier to theorize about challenging this reality then it is to practice it, as these are powerful and core beliefs within educational systems. But there is always hope! Discourses that shape our world have a situated character, shaped through and by social understandings and practices, and are continually open to contestation and negotiation. This continually emerging work that challenges discourses associated with human exceptionalism and human/nature dualisms, can be referred to as posthumanism. When teachers and students are exposed to other educational discourses, such as posthumanism, they are offered other ways of understanding what it means to teach and learn. They may be able to shift among available discourses. However, rarely does exposing someone to new discourses lead to causal or linear change relationships. Power dynamics always influence how people position ourselves within the similar and competing discourses they are exposed to. In our experiences, teachers struggle with power dynamics that play out between dominant and marginalized educational discourses which can lead to a disconnection between what they believe they should be teaching and what they want to teach (Sammel 2020). Recognizing and resisting dominant educational discourses operating in Western societies is a significant challenge for anyone who wants to generate change.

7 Reflecting on a Posthumanistic Education for Sustainability

Posthumanism invites us to appreciate the constitutive potential that underlies any discourse and the impressive power of the way discourses are overtly taken up but rarely recognized. It makes visible the productive power of educational discourses that explains how education systems, teachers and students have been produced within humanism's grids of regularity and normalcy. By decentering the individual (politician, administrator, teacher, students etc.) our focus turns to how subjects are produced through the humanistic discourse rather than standing outside of this discourse. It encourages us to grasp the difference and possibility of another discourse of teaching and learning, one that explores shifting educational conversations towards a more relational framing of how we can understand ourselves and the world around us. Posthumanistic education for sustainability directly challenges the idea of human alienation from nature and rejects the notion that humans are the crown of evolutionary processes. This discourse focuses on making visible the deep interconnections and relationships that link everything on the planet. It highlights that humans are just one part of the biological fabric of life on Earth and acknowledges humans are incapable of an independent existence. At its core, posthumanistic education for sustainability promotes care and kindness towards all aspects of the planet by illustrating how we are all deeply interconnected, and by harming one aspect, we are harming all aspects.

Posthumanistic education for sustainability offers the chance to view learning as a relational process rather than product or outcome. Learning requires the fusing of horizons, or the melding of discourses. When Gadamer (1997) writes of horizons, he is referring to conscious and subconscious perceptions, beliefs, and biases that are brought into any discussion. These horizons are twofold: a historical horizon—defined by the past and the traditions that have resulted from it—and a present horizon—that encompasses all that is believed and understood by a person at this moment in their current situation. They are interconnected as the historical horizon influences the present horizon and as such and must be acknowledged and examined so that the present horizon can be better understood. The explorations of these horizons result in what Gadamer (1997) refers to as their fusing.

In this fusion, the historical horizon remains fixed, while the present horizon is continually in the process of being formed because we are continually having to test all our prejudices. It is through dialogue between the two horizons that understanding can grow. Gadamer (1997) proposes that developing a rich understanding is not something that can be achieved individually, but through dialogue when people lay open their experiences and horizons, and entertain the possibility of change and growth. Posthumanistic education for sustainability seeks to enter this type of Gadamerian dialogue where historic and present horizons fuse in an attempt to challenge dualistic ontologies about the human/nature divide within schools. It disrupts this dominant discourse by embedding diverse ways of knowing the world, and self, into daily teaching and learning practices. Posthuman must therefore be based in dialogue, where there is respect for diverse ways to understand the world that does not privilege this human/nature dualism.

Practically, all citizens live in worlds that include non-human life, environmental forces and entities (Pacini-Ketchabaw et al. 2016). Latour (2005) speaks to the idea of 'common worlds' as the intersection where all life and ecosystems meets. A posthumanistic pedagogy advocates that students explore how their lives are intertwined and embedded with, and mutually dependent on, this common world. This way students can begin to understand and acknowledge that all life shares common vulnerabilities, and engages in common daily actions, and all have a responsibility to protect, and show care and kindness for our common world. By decentering the human and directing students to make sense of their relationship with this common world, posthumanistic education *for* sustainability encourages pedagogies that explore the entirety of what constitutes local environments. Students are encouraged to see connections between all species and to recognize we are all interconnected and deserving of the right to dignity, the right to food, the right to security, to shelter and to kindness and compassion. If *coming-to-know* is inseparable from *coming-to-being* (Higgins 2016), then it is imperative that citizens are exposed to alternative discourses that nurture capacities for caring and build students capacities and neural networks to *come to be* more compassionate citizens. By infusing small acts of kindness into daily educational practices, formal education may not directly save the planet, but hopefully it will create new modes of attention to the entangled relationships that thread our common world together (Haraway 2010; Tsing 2013).

Citizens of today and the future need to be exposed to alternative discourses promoting attention to this common world and ecological sustainability, rather than those associated with human growth and entitlement. If not, the plague of indifference will continue to bring us new challenges we are ill-equipped to resolve. Posthumanistic education *for* sustainability emerges as one discourse that focuses on nurturing the relational agenda of *human as nature*, in order to promote an ecological sustainable way of citizens *coming to be*. It encourages another way of using socially available repertoires of concepts, philosophies, words, and gestures to promote ecologically sustainable perspectives, actions and societies.

8 Conclusion

In this chapter, we advocate for systemic, long-term educational change—change that encourages a clearer understanding of our ethical responsibility within this COVID-19 era. By analyzing the historical development of the dominant hegemonic discourse of humanism, we seek to challenge how Western cultures have come to understand their connection to the Earth. We explain how the humanistic discourse perpetuates the belief that it is normal to support human exceptionalism and endorse the human/nature divide. We maintain that if humanism is the only educational discourse teachers and students are exposed to, then education will inherently promote ecological unsustainability rather than sustainability.

We call for teachers and students to be offered other educational discourses that promote human potential, ecological consciousness, and acknowledge our responsibility to the future of all species - and to Earth itself. Instead of identifying humanity as above nature, education needs discourses that highlight humanity's deep interconnections with non-human animals, plants, and ecosystems. To find a way through the plague of indifference and the challenges the Anthropocene is bringing, we need a different way of thinking than that of the humanistic discourse which caused it. We believe that posthumanistic education for sustainability is one way to generate citizens who can develop compassion and kindness toward their local ecosystems. It provides opportunities to engage with learning as a *way of becoming* rather than just a *way of knowing*, embedded within local ecologies. It advocates for ecological sustainable agendas that promote what it means to live in relationship with all aspects of the biosphere.

Looking ahead, we believe that COVID-19 provides a transformational opportunity. In this period of social disruption and crisis, it is vital not simply to get the educational cart out of the ditch and back on the same road. Rather, we must find new paths that create revolutionary ways of learning and teaching that move students and teachers beyond indifference and complicity and toward caring and kindness.

References

Boff L (2008) The ethic of care. In Corcoran PB, Wohlpart AJ (eds) A voice for Earth: American writers respond to the Earth Charter. University of Georgia Press, Athens, Georgia

Bove P (1990) Discourse. In: Lentricchia F, McKaughlin T (eds) Critical terms for literary study. University of Chicago Press, Chicago, IL, pp 50–65

Camus A (1948) The plague. Vintage International, New York, NY

Carson R (1962) Silent spring. Houghton Mifflin, Boston, MA

Davies B (1990) The problem of desire. Social Prob 37(4):501–516

Davies B (2000) Eclipsing the constitutive power of discourse: the writing of Janette Turner Hospital. In St. Pierre E, Pillow WS (eds) Working the ruins: feminist poststructural theory and methods in education. Routledge, New York, NY

Earth Charter Commission (2000) The Earth Charter. Earth Charter International, San Jose, Costa Rica

Eisler R, Fry D (2019) Nurturing our humanity: how domination and partnership shape our brains, lives, and future. Oxford University Press, Oxford, England. Retrieved from https://www-oxfordscholarship-com.libraryproxy.griffith.edu.au/view/10.1093/oso/9780190935726.001.0001/oso-9780190935726

Elliott GP (n.d.) Kindness spoken here. Retrieved from https://www.livinglifefully.com/flo/flokindnessspokenhere.htm

Foucault M (1984) What is an author? In Harari PR (ed) The Foucault reader. Pantheon, New York, NY

Francis (2015) Laudato Si': On care for our common home. Retrieved from https://w2.vatican.va/content/francesco/en/encyclicals/documents/papa-francesco_20150524_enciclica-laudato-si.html

Gadamer HG (1997) Truth and method, 2nd edn. Continuum, New York, NY

Giroux HA (1988) Teachers as intellectuals: Toward a critical pedagogy of learning. Bergin & Garvey, Granby, MA

Haraway DJ (2010) When species meet: staying with the trouble. Environ Plan D Soc Space 28(1):53–55. https://doi.org/10.1068/d2706wsh

Haraway DJ (2016) Staying with the trouble: Making kin in the Chthulucene. Duke University Press, Durham, NC. Retrieved from https://ebookcentral-proquest-com.libraryproxy.griffith.edu.au/lib/griffith/detail.action?docID=4649739

Higgins M (2016) Decolonizing school science: pedagogically enacting agential literacy and ecologies of relationships. In Taylor CA, Hughes C (eds) Posthuman research practices in education. Palgrave Macmillan, London, England

Latour B (2005) Reassembling the social: an introduction to Actor Network Theory. Oxford University Press, Oxford, England

Martusewicz RA (2001) Seeking passage: post-structuralism, pedagogy, ethics. Teachers College Press, New York, NY

Morton T (2010) The ecological thought. Harvard University Press, Cambridge, MA

Noddings N (2003) Happiness and education. Cambridge University Press, Cambridge, England

Pacini-Ketchabaw V, Taylor A, Blaise M (2016) Decentering the human in multispecies ethnographies. In Taylor CA, Hughes C (eds) Posthuman research practices in education. Palgrave Macmillan, London, England

Plumbwood V (2002) Environmental culture: the ecological crisis of reason. Routledge, Abingdon, England

Sammel A (2020) How embedding indigenous knowledge systems will help the teaching and learning of western science to evolve. In Sammel A, Whatman AS, Levon B (eds) Indigenizing education: discussions and case studies from Australia and Canada. Springer Publishing, New York, NY

Sammel A, Whatman S, Levon B (2020) Indigenizing education: discussions and case studies from Australia and Canada. Springer Publishing, New York, NY

St. Pierre EA (2000) Poststructural feminism in education: an overview. Qual Stud Educ 13:477–515
St. Pierre EA, Pillow WS (2000) Working the ruins: feminist poststructural theory and methods in education. Routledge, New York, NY
Tsing A (2013) More-than-human sociality: a call for critical description. In: Hastrup K (ed) Anthropology and nature. Routledge, New York, NY
Weedon C (1987) Feminist practice and poststructuralist theory. Blackwell, Oxford, England

Dr Alison Sammel specialises in the teaching, learning and communication of Science, ecological sustainability, climate change and fresh water. Alison teaches Science from a critical posthumanistic perspective. She promotes teaching and learning of ecologically sustainability, climate change, ecopsychology, the Sustainable Development Goals and authentically Indigenising [Science] education.

The Impact of COVID-19 on the Progress towards Sustainable Development Goal 4 in the Early Years: A Rapid Review

Jane Spiteri

Abstract As a result of the COVID-19 pandemic, many events have been interrupted, but young child development has not. This rapid review of the literature explores some of the potential impacts of the pandemic on young children, early childhood education and care (ECEC) and Sustainable Development Goal (SDG) 4 (United Nations 2015). In times of crises, ECEC can play a critical role in keeping children, healthy and safe, while promoting the foundations for lifelong learning and wellbeing. Yet, with the closure of schools as one of the measures to control the spread of the virus, many opportunities for learning and development in these formative years will be missed. This will cause major issues especially for vulnerable children. The pandemic also has long-lasting impacts on child development, lifelong learning opportunities, as well as economic and human capital. It also has implications on the progress towards SDG 4 (United Nations 2015), which had already fallen behind before the crisis. While it is impossible to say at this point how long the crisis will last, if school closure and the COVID-19 pandemic were to last into the next academic year, there will be enormous losses for young children, with implications for human and economic capital.

Keywords Early childhood education · COVID-19 · Pandemic · Young children · Sustainable development goal 4 · Rapid review

1 Introduction

The World Health Organization (WHO) declared the novel coronavirus, or the virus SARSCoV-2 which causes COVID-19 pneumonia, a pandemic on March 11th, 2020 (WHO 2020a). In a short timeframe since the outbreak of the virus in Wuhan, China, in December 2019, COVID-19 has spead rapidly and sustainability from person-to-person in 210 countries, claiming thousands of victims, mostly the elderly and

---

J. Spiteri (✉)
University of Malta, Msida, Malta
e-mail: jane.spiteri@um.edu.mt

W. Leal Filho (ed.), *COVID-19: Paving the Way for a More Sustainable World*,
World Sustainability Series, https://doi.org/10.1007/978-3-030-69284-1_15

people with underlying health conditions (WHO 2020b). As the pandemic continues to unfold, disrupting many lives, it is presenting a series of challenges to the world.

In the absence of an effective vaccine for COVID-19, many countries opted for partial or total lockdown to enforce physical distancing measures required to control the outbreak. During lockdown, people are urged to stay home and practise physical distancing, where they go out only to meet urgent needs, such as going to work or to buy food and medicine (Reimers and Schleicher 2020). In this context, physical distancing means that people get further apart in terms of social interactions and contact but not in terms of social connection, the most critical dimension of child development. In these unprecedented times, while life has been brought to a halt because of lockdowns and physical distancing measures, child development has not.

2 Early Childhood Education and Care

The first five years of a child's life is a highly sensitive and critical period for human development (Organisation for the Economic Co-operation and Development [OECD] 2019; Schleicher 2019; Spiteri 2020; United Nations Children's Fund [UNICEF] 2020a). During early childhood the brian matures faster than any other time and children learn at a faster pace than any other period in their lives too (OECD 2019). Relationships with adults and peers in safe, predictable and nurturing environments together with stimulation learning experiences assit young children in developing cognitive, social and emotional skills fundamental to development and wellbeing throughout their lifetime (OECD 2019; Schleicher 2019). Such learning experiences are the hallmark of early childhood education and care (ECEC), which is the education of children from birth till age eight (Spiteri 2020).

Strong evidence suggests that high-quality ECEC has numerous benefits to children, society and for the achievement of a sustainable future (OECD 2019; Pramling Samuelson and Kaga 2008; Schleicher 2019; Spiteri 2020; United Nations 2015; United Nations Educational, Scientific and Cultural Organization [UNESCO] 2017). While there is no concensus on a universally-accepted definition of high-quality in ECEC (La Paro et al. 2012), this chapter draws on OECD's (2019) definition of high-quality ECEC as including meanginful interations and learning experiences between educators and children via a variety of developmental and educational activities based on children's interests and peer interactions, which in turn encourage learning, wellbeing and holistic development. The importance of high-quality ECEC is also reflected in United Nations Sustainable Development Goal (SDG) 4, which focuses on the provision of quality education for all children around the world. In particular, Target 4.2 aims to "Ensure that all children have access to quality early childhood education and care so that they are ready for primary education" (United Nations 2015). Target 4.2 highlights the importance of school settings and systems that support the learning and development of young children (United Nations 2015). Most importantly, high-quality ECEC has the potential to improve life chances of children, especially for those hailing from disadvantaged backgrouds

(Schleicher 2019). The converse is also true. In fact, research indicates that poor-quality ECEC can either have no effect on child development or it can negatively impact the child's socio-emotional development, therefore impacting wellbeing from an early age, which has serious implications especially for highly-vulnerable children (OECD 2019; Schleicher 2019).

High-quality ECEC is even more important in times of crises, such as during the COVID-19 pandemic, when its proper implementation and the monitoring of its progress become even more problematic due to several restrictions imposed upon the public to reduce the spread of the virus, including school closure. Additionally, the benefits of ECEC across a lifetime are also evident in that this sector is transversal to all SDGs, including SDG 1 (reducing poverty), SDG 3 (health), SDG 4 (education), SDG 8 (employment and economic growth), SDG 10 (reduction of inequalities) and SDG 16 (achieving peace and justice and reducing violence) (Ponguta et al. 2018), making ECEC of utmost importance for the achievement of the 17 SDGs.

Building on the SDG 4 framework (United Nations 2015), and particularly its recognition of the importance of ECEC, and the current COVID-19 crisis, this chapter attempts to answer the question:

What is the impact of the COVID-19 pandemic on

(a) ECEC, the progress towards SDG4 and young children?

This chapter begins by examining the COVID-19 crisis, and how it impacts the ECEC sector, young children, families, educators, teaching and learning in the early years and the implimentation of SDG 4. Then, it moves on the unpack the challenges therein, especially the issues faced by vulnerable young children. Finally, it discusses implications of these for the aftermath of the pandemic, identifies some research gaps, and concludes with some broad recommendations for the way forward.

3 Methods

To produce a succint, timely, but in-depth sythesis of the current evidence on the impact of COVID-19 on ECEC and the implementation of SDG 4, a rapid review of the literature was conducted (Khangura et al. 2012). Rapid reviews use streamlined traditional methods for systematic reviews to help synthesise and communicate evidence within a short period of time (Khangura et al. 2012). In this chapter, the stages suggested by Khangura et al.'s (2012) were followed. These included:

- *Needs assessment*—A call for proposals from Springer for chapters related to COVID-19 and the SDGs was issued.
- *Question development and refinement*—Research questions were formulated based on the needs' assessment.
- *Proposal development and approval*—A proposal was sent to the editors for peer-review, and was approved.

- *Literature search*—With the research questions in mind, a rapid review was conducted.
- *Screening and selection of studies.* A decision was made to draw the information exclusively from evidence reported recently related to the topic under study to allow the review to be undertaken in a timely manner, reflecting the nature and purpose of a rapid review and the needs of this study. This review was restricted to articles written in English, even if this choice may have excluded pertinent material. Additionally, the search has also been partially limited by the specific terminology used in the context of COVID-19, ECEC and SDG 4, and time constraints. The search criteria were conducted through EBSCO Host via University of Malta library, focusing on 13 databases: GreenFILE, British Educational Index, Academic Search Complete, Education Source, Humanities International Complete, Education Resources Information Centre (ERIC), ProQuest, JSTOR, SCOPUS, ERIC, Web of Science, PubMed and PsychINFO.

Boolean searches were conducted using a combination of keywords with AND/OR operators to produce more relevant results. The keywords used were determined following careful consideration of a number of factors; such as awareness of scientific terminology, commonly used terms and the relationship between the AND/OR operators. Further research parameters were also included to reduce the volume of records returned. The full inclusion criteria are detailed in Table 1.

The searches were conducted from 2nd May 2020 up until 10th June 2020. A total of 43 searches were returned and nine of which were eliminated because they were not written in English or they did not fit the parameters of the review as described above. This screening reduced the records to 34. There may be more publications within the range, however the university database only revealed these titles. Taking that sample, journals/websites which were deemed relevant were considered more closely. A thematic approach to the data analysis was adopted.

Table 1 Inclusion criteria

Topic	Inclusion criteria
Keywords	AND—Using the Boolean Operator AND will narrow your search results. In this case, using AND will retrieve search results containing all keywords, in this case 'COVID-19', 'early childhood education', 'educational outcomes', 'school', 'family', 'Sustainable Development Goal 4' OR—Using the Boolean Operator OR will broaden your search results. In this case, using OR will retrieve search results containing either of the keywords, in this case 'attainment', 'early childhood', 'pandemic', 'technology', 'wellbeing'
Date range	2nd May 2020—10th June 2020
Publication/Document type	Scholarly peer-reviewed publications, journals, websites, research reports, government reports, peridicals
Language	English
Education level	Early childhood education for children aged 0 to 8 years

- *Narrative synthesis of included studies*—No sysnonyms were used in order to exclude any unrelated results, which is prohibitive in a rapid review. It should be noted that there are some key articles that pre-date the period under review and, where appropriate, these are cited in the narrative. There is also additional literature relating to the methodology, general conceptual issues and analysis, and these are also cited where appropriate.
- *Report production*—Themes from the datasets were synthesised into a coherent narrative to address the research question.
- *Ongoing follow-up and dialogue with knowledge users.* This step was followed through via the peer-review process of this chapter.

4 Results and Discussion

Reading the articles generated by the searches led to the development of some overall impressions of the impacts of the COVID-19 pandemic on young children's learning in ECEC contexts and on the implementation of SDG 4. These impressions were grouped into four themes: school closure; education and technology; impact on SDG 4; and, direct and indirect impacts on young children. These themes are supported by direct references to specific research and are described in more detail in the sections below. It is important to note the absence of research, or perhaps the limitation, that was related to the COVID-19 pandemic but not situated in relation to SDG 4 and the ECEC context. Hence, this chapter aims to fill this gap.

4.1 School Closure

Education is key to ensure good health, economic growth, and protection of the planet (UNESCO 2020e; United Nations 2015). Education can play a critical role in keeping children healthy and safe, while ensuring learning and wellbeing. For many children, school is often much more than a place for learning. In a context for human interaction, school is a safetynet, a place for health and nutrition, and the provider of essential health services which save lives (UNICEF 2020b), particularly in the early years. Yet, like other sectors of the economy, ECEC programmes are being heavily impacted by the COVID-19 outbreak.

In repsonse to the coronavirus pandemic, school closure has been implemented in 187 countries worldwide, currently affecting 1.5 billion children (UNESCO 2020g; United Nations 2020) and nearly 743 million girls (Giannini 2020). This unprecedented action was taken in an attempt to control the spread of the virus, affecting almost 91% of the world's student population (Armitage and Nellums 2020; OECD 2020). The evidence for the effectiveness of school closure as part of a physical distancing measure during an outbreak is based on evidence and assumptions from past influenza outbreaks. This action is based on the assumption that transmission of the virus tends to be driven by children, and it is believed that school closure may help reduce social contacts between children, thus also reducing the rapid transmission of the virus. In fact, studies show compelling evidence that early school closure during an influenza outbreak reduced transmission of the virus and indicated a reduction in the outbreak peak (Viner et al. 2020). With that said, lessons learnt from past epidemics of coronavirus outbreaks, such as SARS in China in 2002, Ebola in West Africa in 2014 and MERS in 2015, while they provide limited information about the effectiveness of school closure and physical distancing measures, and transmission of the virus (Viner et al. 2020), can help shed light on the current crisis. Taking the Ebola outbreak as an example, which affected 5 million children by school closure, evidence suggests that poverty increased significantly, the youngest were the most vulnerable during the crisis, and children were more prone to experience violence, abuse and neglect as families struggled to cope with stress, anxiety and unease (Giannini 2020). Overall, past epidemics also show the impact on any already scarce resources and services (Devercelli 2020). Therefore, during the COVID-19 crisis, promoting education for holistic child development is even more important.

While the effectiveness of school closure in preventing the spread of COVID-19 is somewhat unclear as transmission dynamics appear to be different than influenza, one thing is clear—the economic costs and the potential harm of school closure cannot be underestimated (Viner et al. 2020). When schools get disrupted, they carry high social and economic costs for all, but vulnerable and marginalised children and children living in low-income countries are the worst hit (UNESCO 2020c). School closure impedes learning and deepen inequities, disproportionately affecting disadvantaged children the most (Armitage and Nellums 2020; UNESCO 2020a, f). In such circumstances, the potential loss of human capital are hard to imagine, let alone comprehend.

During school closure, the world's most vulnerable children are missing out on more than just education. Many are missing out on social contact too, which is so essential for their learning and development (UNESCO 2020a), particularly in the early years. Schools provide safeguarding and supervision of vulnerable children, and school closure exacerbates the economic burden of working parents and older, and vulnerable, relatives who care for young children (Armitage and Nellums 2020). This situation presents many challenges for parents who cannot go to work to care for their children. In this context, the most vulnerable, marginalised and fragile populations are being left behind even during the pandemic, especially in areas where high conflict increases the already existing educational exclusion and inequalities, and the deepening of new inequalities and exclusions which threaten social cohesion (UNESCO 2020e).

Furthermore, school closure could lead to an increase in drop-out rates, child labour, violence against children, teen pregnancies, and exacerbate already existing socio-economic disparities, which will most likely negatively impact girls more than boys, especially those living in economically-disadvantaged areas, thus increasing gender disparities in education (Armitage and Nellums 2020; Giannini 2020; UNESCO 2020c). The current pandemic, and the unexpected closure of schools for an unexpected duration, have created lots of interruptions in opportunities for growth, development and learning, especially for underprivilidged learners who tend to have less opportunities for learning beyond the school context (UNESCO 2020a). Together, these issues will further hinder the implementation of SDG 4, pushing it further out of reach for the most vulernable children (UNESCO 2020c).

The negative impact of COVID-19 is likely to be felt for years to come (UNESCO 2020b). However, the world seems to be paying more attention to the health of the global population than children's education, leading to loss of learning time, which could in turn impact children's learning, financial and career potentials in the future (UNESCO 2020c). This situation calls for governments worldwide to make education a priority, starting in ECEC, in an attempt to help mitage the impact of COVID-19 on children's educational achievements during the necessary physical distancing periods, particularly by having clear plans to preprioritise curricular goals and define what should be learned during this periods of physical distancing (Reimers and Schleicher 2020). The wellbeing of children and educators, and the need for meaningful relationships between educators and children to contribute towards learning need to be prioritised during these unprecedented times (Reimers and Schleicher 2020). Additionally, support for teachers during the crisis is essential to ensure adequate political commitment and inverstment in education during the recovery phase (Giannini 2020; UNESCO 2020g). To this end, education must prioritise inclusion and equity; a transformation for quality, relevance and impact across the SDGs; better data and intermediate benchmarks to inform action; increased funding for equitable and resilient public education systems; and, strengthened partnership between stakeholders (UNESCO 2020e).

4.2 Education and Technology

As the world grapples with the pandemic, the education sector is experiencing a global social experiment with the new ways of teaching and learning for children, even in the early years. Led by the constraints imposed on the delivery of education programmes by the lockdown, physical distancing and school closure, the use of online teaching and learning has certainly accelerated the process of uptake of online education technologies by educational instituations worldwide. Indeed, in response to this situation, several educators have also discovered how to efficiently use new technologies in the early years. Currently, distance learning systems are being used by many education providers to ensure the continuation of education programmes and to limit the loss of learning opportunities in most countries (Martin et al. 2020; UNESCO 2020c, f). This has dramatically altered the educational landscape over the course of a few weeks for many children, educators and parents. With this new wave of online teaching being implemented, access to digital technology is crucial to ensure that young children have equal opportunities to learn despite the current dire circumstance.

Transition to distance teaching and learning may be easy for some but challenging and frustrating for others (UNESCO 2020a). It certainly has increased inequalitities among children (Armitage and Nellums 2020) because not all children start out digitally equal. For example, while children living in most Western countries and those hailing from high socio-economic backgrounds, who have many resources available to them, are taking the current emergency in their stride, other less well-resourced peers, are not. For some children, online learning is possible because they have easy access to the internet and electronic devices at home. For others, living in disadvantaged communities however, it is not because they do not have internet access at home or some socio-economic factors, such as family poverty and deliterious living conditions, limit access to technology needed for learning. Additionally, children living in large families, sharing limited space and limited digital divices may get left behind too. Coupled with already existing stress within families, like for example families facing domestic violence or sudden unemployment and anxiety, successful educational outcomes for vulnerable children are slim to none. Overall, it seems that the pandemic has also exposed additional socio-economic inequality in terms of accessibility to online education, thus increasing the digital divide and the attainment gap from an early age.

This unprecedented situation poses difficulties in the acquisition of early literacy too. Considering that many young children experience immersion in digital technologies from a young age, educators and parents need to help them build their digital 'funds of knowledge' (Moll et al. 1992), and provide them with opportunities to engage with the technological, social and cultural demands of the knowledge economy (Nutbrown 2018), away from the classroom setting. Perhaps what is most worrying is the fact that while the use of remote learning may benefit children in acquiring numeracy and literacy skills from home during times of crises, the issue of building close and supportive relationships via human interaction and play, the building blocks of early child development, remains unresolved. As a result, young children still miss out on building meaningful relationships with people outside the family context. Therefore, young children are disproportionately at risk and their needs and development must be priortised because severe lifelong impacts resulting from deprevations in the early years can hinder their development. Most importantly, education authorities need to act quickly to lessen the impacts caused by the sudden onset of the pandemic and the unplanned introduction of online teaching and learning. Certainly, underinvesting in ECEC now, will have lifelong concequences for children and long-lasting negative impacts on the world's economic and human capital.

Parental involvement in young children's education enhances academic success and wellbeing later on in life (OECD 2011). In the context of school closure and the subsequent take on of remote teaching and learning, parents are key sources of education provisions and motivators for children's engagement, as well as facilitators of children's learning (OECD 2020). However, many parents are either unprepared for homeschooling or distance learning (UNESCO 2020a), and/or do not afford to stay home to teach and/or support their children. This is especially true for parents with limited education and financial resources, illiterate parents, or parents who cannot afford to stay home despite the current emergency. In the absences of alternative options, working parents struggle to find care for their children. Working parents are more likely to miss work in order to care for their children, which negatively impacts the economic outcomes of families and the productivity of countries too. Professionals, like health-care workers with children, struggle with finding appropriate childcare at times when they are needed the most. Such high economic pressures tend to encourage abuse. Consequently, increased exposure to violence is likely to ensue and young children are the ones who tend to suffer the most in such circumstances (UNESCO 2020a).

The pandemic has certainly necessitated the accelerated development of skills sets that were previously underused and potentially undervalued in an ECEC setting. Nevertheless, this situation has also created enromous technical and human challenges for many, the outcome of which is pressure and uncertainty for children, parents and educators alike (UNESCO 2020a) because many adults, including educators, lack effective training in using these new platforms. The pandemic exposed the lack of preparedness by many educational institutions to rise up to the challenge, in terms of both infrastructure and human capital. While many educators have taken on the challenge of online teaching, others are still uncertain of what is expected of them in times of crises. Others might find it difficult to learn how to use technology overnight and adapt content and delivery to an online environment. Even in the best circumstances, young children, parents and educators are struggling to cope with the situation and the new demands imposed upon them unexpectedly. Even worse, the OECD (2020) estimates that in the absence of a vaccine, it might be increasingly difficult for schools to go back to normal any time soon. In this context, remote teaching and learning is likely to remain in the future and online teaching seems to be the way forward, especially if the pandemic goes on for longer than expected. Consequently, an investment in remote teaching platform infrastructure and teacher training are a must. Educators need to develop their expertise in using ICT for teaching and learning, and make use of digital training opportunities and resources, as well as collaborative (possibly virtual) professional learning opportunities (OECD 2020). In this regard, there is urgent need for equity measures, supporting educators, children and parents during this time since learning inequalities are more likely to deepen the education crisis, particularly in the ECEC context.

Assuming children will not go back to school anytime soon, special interventions are required to help many catch up with their missed opportunities for learning. To mitigate these educational inequalities and the negative effects, ECEC programmes must provide resouces to target young children who are especially hard hit. One way of doing this could be to mail material to parents, relax curriculum content and allow for the introduction of alternative educational activities, such as outdoor free play, physical activity or music, to mention but a few activities which could possibly be organised by parents in collaboration with educators. Another possibility is for resources to be shared among a much larger number of children, and the flexibility of being able to follow classes at any time even if without the possibility of real-time interaction afforded by real-life classroom situations.

4.3 Impact on SDG4

In July 2019, the High-level Political Forum (HLPF) reviewed the progress towards SDG 4 and concluded that it was already behind in being achieved by 2030 (Sharma et al. 2019). Later, in December 2019, UNESCO (2020d, p. 11) confirmed that many of the targets of SDG 4 are hard to reach by 2030 and indeed 'we are far from reaching our promise to "leave no one behind."' With children not able to go to school, SDG 4 is falling behind more than ever. Indeed, lockdowns and school closure have disrupted the progress towards SDG 4, and have added more strain to its achievement, making it even harder to reach by 2030 (UNESCO 2020d; United Nations 2020).

The intersection of SDG 4 and ECEC has never been clearer. However, the pandemic will certainly have a major impact on the global economy resulting in a drop in government revenues. The resulting economic slowdown will have implications on the finances and resources of global education systems in the future (UNESCO 2020c). This economic downturn will make it more difficult to achieve the progress towards SDG 4 and any progress made before the pandemic may even be reversed. Now, more than ever, progress towards SDG 4 requires international cooperation. Consquently, Governments all over the world must take action to intensify efforts to accelerate access to ECEC for all. Since it is unclear for how long school closure will last, there is also an urgent need to identify a safe return to school for children and a safe return to work for parents. In turn, this calls for education authorities to anticipate challenges and ways to address them for the benefit of all involved (OECD 2020).

Nothing is certain right now and the aftermath of the pandemic could accentuate the serious risks affecting the development trajectories of millions of young children today, while majorly hindering the progress towards achieving the targets of SDG 4 even more. Since so much is still unknown about COVID-19, it is hard to predict the future. What is certain though is that there is trauma during the pandemic. When the pandemic ends, trauma will probably kick in even more. One consequence of trauma is toxic stress, or the body's physiological response to stress, which happens when individuals, including children, feel they cannot manage the stress, or feel unsafe and out of control. To help young children deal with toxic stress and enhance their wellbeing, adults must provide a sense of safety in children by caring for them and develop interventions to improve their optimal cognitive and socio-emotional development. To better understand the effects of the pandemic and the multiple environmental risk factors on children's cognitive and socio-emotional development, a holistic, multidisciplinary, and multilevel approach that encompasses the complex interactions between biological, physical, and psychosocial factors impacting children's developmental outcomes is needed. Such an understanding will make room for more effective educational interventions with special focus on relationships, mental health, wellbeing and socio-emotional learning. While the COVID-19 crisis threatens to stall, or reverse, progress towards SDG 4, it also offers opportunities to teach us how to behave responsibly and protect everyone else. With the appropriate support, children will come out of the pandemic quite strong, not only in terms of learning but also in terms of relationships, a move in the right direction for the achievement of SDG 4, if used wisely. Surely, in this regard, the pandemic provides an opportunity to redesign ECEC programs towards safeguarding the planet, more than human species and humanity.

4.4 Direct and Indirect Impacts on Young Children

While children seem to have been mildly affected by the direct health effects of the virus during this pandemic, their right to a voice and their wellbeing have both been negatively impacted (United Nations 2020), seriously undermining the United Nations Convention for the Rights of the Child (United Nations 1989). Undoubtedly, children all over the world will not be impacted equally by the pandemic (United Nations 2020). As COVID-19 unleashes its wrath across countries, killing thousands of people, the voices of other casualties of this disease also often go unheard—young children, particularly the vulnerable and disadvantaged. Disadvantaged and vulnerable children are more likely to be negatively affected in several ways by this crisis. Indeed, the pandemic has a myriad of direct, indirect and long-lasting health, social, and educational consequences for young children and families living in poor socio-economic neighbourhoods and facing an insecure future due to the huge economic implications (Rosenthal et al. 2020). Other issues, such as the infection with the virus itself, and the socio-economic outcomes during and post-COVID-19, will also have long-lasting impacts on children (United Nations 2020). Therefore, the coronavirus global outbreak is not only a global health crisis and a human crisis but also a children's rights issue. This is especially so since young children's voices seem to have been absent from discourse related to the COVID-19 outbreak research so far.

The significance of the physical environment on children's cognitive and socio-emotional development across the lifespan, from the prenatal period through adulthood, has been well-documented (Ferguson et al. 2013). Since young children depend on their parents or caregivers for emotional and physiological regulation, who act as a protection from adverse exposures and stress potential (Black and Merseth 2018), nurturing environments, which include proper nutrition, health, safety and responsive caring adults, are essential to ensure optimal and holistic child development (OECD 2019; Ponguta et al. 2018). In the absence of nurturing and supportive environments, for example during the current pandemic, young children tend to have more self-regulation and anxiety issues, putting them at risk of not reaching their full developmental potential (Black and Merseth 2018; Ferguson et al. 2013; Rosenthal et al. 2020).

As a result of the negative impact of the pandemic on the global economy, parents or caregivers may lose their jobs, making children more prone experience violence, stress and abuse (Armitage and Nellums 2020). As parents struggle with unemployment and the economic downturn, child poverty is likely to increase (Martin et al. 2020) and parents are more likely to struggle with difficult child behaviour. Poverty is frequently associated with multiple environmental risks (Black and Merseth 2018; Ferguson et al. 2013; Rosenthal et al 2020). Poverty can also lead to homelessness, both of which prevent many children from reaching their developmental potential, and in this regard COVID-19 has added another barrier to child development. Homelessness and poverty make self-isolation and physical distancing during the pandemic more challenging, thus exacerbating inequalities and in access to education, health care and other essential services.

Since families spend more time together at home due to restrictions imposed by physical distancing measures, the economic uncertainty, parental conflicts and domestic abuse are likely to increase. Unfortunately, research indicates a correlation between the number of exposures to adversities during early childhood and the rates of lifelong adverse consequences (Black and Merseth 2018; Reimers and Schleicher 2020; Shonkoff et al. 2012). Furthermore, during the COVID-19 crisis, many support services will experience significant disruptions and will be delivered remotely most of the time. Given the fewer availability of essential services, vulenrable children who spend a lot of time in abusive households are more likely to experience abuse and neglect. Consquently, the developmental gaps are likely to widen especially for economically-disadvantaged children. With these caveats in mind, the evidence to date documents adverse impacts of individual environmental risk factors on children's cognitive development (Black and Merseth 2018; Ferguson et al. 2013), which are likely to be amplified for vulerable children during the pandemic, thus having implications for their development and educational outcomes.

5 Conclusion

While the full repercussions of the pandemic are yet to be seen, much remains unknown at this stage in terms of the duration of the crisis, the challenges countries will face in the aftermath of COVID-19 (UNESCO 2020c). In this context however, it is difficult to predict what the ECEC and the SDG 4 landscape will look like in the aftermath of the pandemic, in part because of the possibility of a second wave (or more) in the future, which might prolong the need for physical distancing measures, thus prolonging children's time away from school. Given the unexpected and unprecedented obstacles and challenges caused by the pandemic, it is important to understand how best to support young children as they take their place in the world as confident and competent learners during times of global crises.

The impact of emergency remote learning on children cannot be discounted, especially since it hit vulnerable children the hardest. The rapid move towards the digitisation of education, while essential and inevitable, is influenced by costs of infrastructure, which could be a huge burden on some countries and families, and the level of preparedness of educators and parents in assiting young children to access these resouces. Possibly, this will increase the learning crisis, which will be further amplified because of inequalities in access to education and technology (UNESCO 2020c), thus increasing inequalities. Unless offered the right support, vulnerable children will suffer the most and this will have long-term implications for their future.

Nevertheless, the COVID-19 crisis has the potential to provide lessons that if taken advantage of can be turned into new opportunities for learning in ECEC and possibly enable the successful implimentation of SDG 4 by improving the educational outcomes of many young children worldwide. For this to happen, many ECEC pedagogies will have to be rethought, redesigned and repositioned in new and creative ways. Clearly, this requires a rethink children's engagement and motivation, possibly by including more focus on children's wellbeing.

The pandemic has accelerated new opportunities for research. Young children's lives have changed because of the COVID-19 outbreak, yet their voices about the pandemic remain unheard. In light of this new reality, research needs to capture young children's voices through the pandemic. To do this, researchers need to engage in collaborative consultation with young children about key issues during the pandemic affecting them and how growing up in a pandemic affects them. More than ever, policymakers need to recognise children's voices when redesigning new ways of teaching and learning in ECEC that contribute, and are resposive, to young children's development and wellbeing. Otherwise, ECEC will continue to increase disparities in educational access and outcomes, and limit the progress towards SDG 4, often to the detriment of those who are most vulnerable in times of crises.

Finally, it remains to be seen whether the ECEC sector keeps up with the sustained efforts to build on the strengths and opportunities presented by the challenges of this pandemic. Certainly, this is an opportunity not to be missed.

References

Armitage R, Nellums LB (2020) Considering inequalities in the school closure response to COVID-19. Lancet Glob Health 2020(8):e644. https://doi.org/10.1016/S2214-109X(20)30116-9

Black MM, Merseth KA (2018) First 1000 days and beyond: Strategies to achieve the sustainable dvelopment goals. In: Suman V, Petersen AC (eds) Developmental science and sustainable development goals for children and youth. Springer Nature, Cham, Switzerland, pp 97–112

Devercelli A (2020). Supporting the youngest learners and their families in the COVID-19 (Coronavirus) respones. The World Bank Blogs. Retrieved from: https://blogs.worldbank.org/education/supporting-youngest-learners-and-their-families-covid-19-coronavirus-response

Ferguson KT, Cassells RC, MacAllister JW, Evans GW (2013) The physical environment and child development: an international review. Int J Psychol 48(4):437–468. https://doi.org/10.1080/00207594.2013.804190

Giannini S (2020) COVID-19 school closures around the world his girls hardest. UNESCO, Paris. Retrieved from: https://plan-international.org/blog/2020/03/covid-19-school-closures-hit-girls-hardest.

Khangura S, Konnyu K, Cushman R, Grimshaw J, Moher D (2012) Evidence summaries: the evolution of a rapid review approach. Syst Rev 1(10):1–10. https://pubmed.ncbi.nlm.nih.gov/22587960/

La Paro KM, Williamson Payton AC, Lower JK, Kintner-Duffy VL, Cassidy DJ (2012) Examining the definition and measurement of quality in early childhood education: a review of studies using

the ECERS-R from 2003 to 2010. Early Childhood Res. Pract. **14**(1), online. https://files.eric.ed.gov/fulltext/EJ975649.pdf
Martin J, McBride T, Masterman T, Pote I, Mokhtar N, Oprea E, Sorgenfrei M (2020) Covid-19 and early intervention: Evidence, challenges and risks relating to vitual and digital delivery. Early Intervention Foundation, London
Moll LC, Amanti C, Neff D, Gonzalez N (1992) Funds of knowledge for teaching: using qualitative approach to connect homes and classrooms. Theory into Practice 31(2):132–141
Nutbrown C (2018) Early childhood educational research: international perspectives. Sage Publications Ltd., London
OECD (2011) Starting strong III: a quality toolbox for early childhood education and care, starting strong. OECD Publishing, Paris. Retrieved from:https://doi.org/10.1787/9789264123564-en
OECD (2019) TALIS Starting Strong 2018 Technical Report. OECD Publishing, Paris. Retrieved from: https://doi.org/10.1787/301005d-en
OECD (2020) Teaching in focus #32. How prepared are teachers and schools to face the challenges to learning caused by the coronavirus pandemic? OECD Publishing, Paris. Retrieved from: https://www.oecd-ilibrary.org/docserver/2fe27ad7-en.pdf?expires=1591790477&id=id&accname=guest&checksum=24B9F5B223ABFD0884EBA56EAFC2732C
Ponguta LA, Donaldson C, Affolter F, Connolly P, Dunne L, Miller S, Britto P, Salah R, Leckman J (2018) Early childhood development programs, peacebuilding, and the sustainable development goals: Opportuniites for interdisciplinary research and multisectoral partnerships. In: Suman V, Petersen AC (eds) Developmental science and sustainable development goals for children and youth. Springer Nature, Cham, Switzerland, pp 77–95
Pramling Samuelsson I, Kaga Y (eds) (2008) The contribution of early childhood education to a sustainable society. UNESCO, Paris. Retrieved from: https://unesdoc.unesco.org/images/0015/001593/159355e.pdf
Reimers FM, Schleicher A (2020) A framework to guide an education response to the COVID-19 pandemic of 2020. OECD, Paris. Retrieved from: https://read.oecd-ilibrary.org/view/?ref=126_126988-t63lxosohs&title=A-framework-to-guide-an-education-response-to-the-Covid-19-Pandemic-of-2020
Rosenthal DM, Ucci M, Heys M, Hayward A, Lakhanpaul M (2020) Impacts of COVID-19 on vulnerable children in temporary accomodation in the UK. The Lancet Public Health 5(5):e241–e242. https://doi.org/10.1016/S2468-2667(20)30080-3
Schleicher A (2019) Helping our youngest to learn and grow: policies for early learning. International summit on the teaching profession. OECD Publishing, Paris. Retrieved from: https://www.oecd-ilibrary.org/docserver/9789264313873-en.pdf?expires=1592067117&id=id&accname=guest&checksum=9E3133273FA1AE1CCE5F1875E9586EEA
Sharma A, Bhandary RR, Lebada A, Nyigi DW (2019) Summary of the 2019 meeting of the high-level political forum on sustainable development: 9–19 July 2019. Earth Negotiations Bull Reporting Serv Environ Dev Negotiations 33(5):1–21. https://enb.iisd.org/download/pdf/enb3355e.pdf
Shonkoff JP, Richter L, van der Gaag J, Bhutta ZA (2012) An integrated scientific framework for child survival and early childhood development. Pediatrics 129(2):e460–e472
Spiteri J (2020) Early childhood education for sustainability. In Leal Filho W, Azul A, Brandli L, Ozuyar P, Wall T (eds) Quality education. Encyclopedia of the UN Sustainable Development Goals. Springer, Cham
UNESCO (2020a) Adverse consequences of school closures: More on UNESCO's COVID-19 education response. UNESCO, Paris. Retrieved from: https://en.unesco.org/covid19/educationresponse/consequences
UNESCO (2020b) Back to school: Preparing and managing reopening of schools—COVID-19 education webinar # 6. UNESCO, Paris. Retrieved from: https://en.unesco.org/events/back-school-preparing-and-managing-reopening-schools-covid-19-education-webinar-6
UNESCO (2020c) What price will education pay for COVID-19? UNESCO, Paris. Retrieved from: https://www.iiep.unesco.org/en/what-price-will-education-pay-covid-19-13366

UNESCO (2020d) Educating for an inclusive and sustainable future. UNESCO, Paris. Retrieved from: https://unesdoc.unesco.org/ark:/48223/pf0000372607
UNESCO (2020e) Accelarated action and transformative pathways: Prioritize education to achieve sustainable development. SDG-Education 2030 Steering Committee, Paris: UNESCO. Retrieved from: https://sdg4education2030.org/sites/default/files/2020-03/2020-03-19%20SC%20HLPF%20submission%20short%20final.pdf
UNESCO (2020f) SDG-Education 2030 steering Committee urges protection of education, now and post-crisis. SDG-Education 2030 Steering Committee, Paris: UNESCO. Retrieved from: https://www.sdg4education2030.org/index.php/sdg-education-2030-steering-committee-urges-protection-education-now-and-post-crisis
UNESCO (2020g) The SDG-Education 2030 steering committee recommendations for COVID-19 education response. SDG-Education 2030 Steering Committee, Paris: UNESCO. Retrieved from: https://www.sdg4education2030.org/sites/default/files/2020-04/SDG-Education%202030%20SC%20recommendations%20-%20COVID-19%20education%20response.pdf
UNICEF (2020a) Early moments matter. UNICEF, New York. Retrieved from: https://www.unicef.org/early-moments
UNICEF (2020b) Futures of 370 million children in jeopardy as school closures deprive them of school meals—UNICEF and WFP. UNICEF, New York. Retrieved from: https://www.unicef.org/press-releases/futures-370-million-children-jeopardy-school-closures-deprive-them-school-meals
United Nations (1989) Convention on the rights of the child. Adopted and opened for signature, ratification and accession by General Assembly resolution 44/25 of 20 November 1989 entry into force 2 September 1990, in accordance with article 49. Geneva: United Nations High Commissioner for Human Rights. Geneva: Office of the United Nations High Commissioner for Human Rights (OHCHR). Retrieved from: https://www.ohchr.org/en/professionalinterest/pages/crc.aspx
United Nations (2015) Transforming our world: The 2030 Agenda for sustainable development. A/RES/70/1. United Nations, Geneva. Retrieved from: https://sustainabledevelopment.un.org/content/documents/21252030%20Agenda%20for%20Sustainable%20Development%20web.pdf
United Nations (2020) Policy brief: The impact of COVID-19 on children. United Nations, Geneva. Retrieved from: https://www.un.org/sites/un2.un.org/files/policy_brief_on_covid_impact_on_children_16_april_2020.pdf
Viner RM, Russell SJ, Croker H, Packer J, Ward J, Stansfield C, Mytton O, Bonell C, Booy R (2020) School closure and management practices during coronavirus outbreaks including COVID-19: a rapid systematic review. Lancet Child Adolesc Health 2020(4):397–404. https://doi.org/10.1016/S2352-4642(20)30095-X
WHO (2020a) WHO timeline—COVID-19. WHO, Geneva. Retrieved from: https://www.who.int/news-room/detail/27-04-2020-who-timeline---covid-19
WHO (2020b) Coronavirus disease (COVID-19) Pandemic. WHO, Geneva. Retrieved from: https://www.who.int/emergencies/diseases/novel-coronavirus-2019

Jane Spiteri holds a PhD in early childhood education for sustainability from the University of Edinburgh. She is a full-time lecturer at the Department of Early Childhood and Primary Education, within the Faculty of Education, at the University of Malta. Her work focuses on early childhood education for sustainability, and gifted and talented education.

COVID 19: Causal Loop Diagramming (CLD) of Social-Ecological Interactions for Teaching Sustainable Development

Gavin Melles, Stefan Lodewyckx, and Hariharan Thangatur Sukumar

Abstract Sustainable development requires integrated systems thinking about environmental, economic, and social drivers to achieve long term solutions. Systems Thinking and Modelling (STM) is now taught widely in disciplines such as management, business, biology, and sustainability education. Recent Emerging Infectious Diseases (EID) such as COVID-19 in 'hotspot' rural locations highlight systemic socio-economic and health interactions in social-ecological systems (SES). While popular susceptible-infected-recovered (SIR) models of EID focus on transmission factors, such as contact rates, SES drivers focus on the broader economic, ecological, and social factors that can enable EID. While STM has been applied to disease propagation aspects of EID, no current models attempt to model the multi-scalar interactions between health systems, economics, and the environment in rural settings where virus hosts and humans come into contact. Systems Dynamics Modelling (SDM) tools, such as causal loop diagrams (CLD) and Archetype diagrams can visualise the complex and multi-scalar 'panarchic' interactions influencing outbreaks of EID. In this chapter, we show that such multi-scalar CLD models can articulate the economic, environmental, social and health interactions that perpetuate unsustainable development and contribute to potential EID. Such models help visualise the complex dynamics of sustainable development in resource-constrained rural settings in ways that textbook descriptions do not always capture.

Keywords Systems thinking · Causal loop diagrams · Emergent infectious diseases · Archetype · Social-ecological system

G. Melles (✉)
Centre for Design Innovation, Social and Environmental Sustainability in Organisations (SESIO) Research Group, Swinburne University, Melbourne, VIC 3122, Australia
e-mail: gmelles@swin.edu.au

S. Lodewyckx
Faculty of Business and Law and SESIO Research Group, Swinburne University, Melbourne, Australia
e-mail: slodewyckx@swin.edu.au

H. T. Sukumar
Department of Management Studies, IIT Madras, Chennai, India

W. Leal Filho (ed.), *COVID-19: Paving the Way for a More Sustainable World*,
World Sustainability Series, https://doi.org/10.1007/978-3-030-69284-1_16

1 From an Epidemiological to a Social-Ecological Systems Focus on EID

There are many systemic links between sustainable development, health, and the environment (Redding et al. 2019), however, 'little attention has been paid to the interactions between environmental change and infectious disease emergence, despite growing evidence causally linking these two phenomena' (Di Marco et al. 2020, p. 3889). From a systems perspective, beyond standard susceptible-infected-recovered (SIR) epidemiological models, a focus on broader social and environmental drivers in developing world rural 'hotspots' is recommended. The 'reactive' epidemiological approach to viruses and airborne disease transmission is not coping well with preparing for local and global threats (Graham and Sullivan 2018). Despite the explanatory and predictive value of these environmental-disease and systems-based approaches, they remain under-utilised in public health, education and public discussion (Luke and Stamatakis 2012), except in the global One Health framework (e.g. Deem et al. 2019; Rabinowitz et al. 2018).

It is broadly accepted that wildlife exploitation and anthropogenic activities have increased the likelihood of human-animal (zoonotic) EID (Johnson et al. 2020). Thus, EID in 'hotspots' where wildlife diversity is significant 'correlates strongly with human population density, supporting the hypothesis that disease emergence is driven by largely anthropogenic changes, such as the expansion of agriculture, travel routes and trade, and changes in land-use' (Morse et al. 2012, p. 1957). Although, a systems view is essential to understand challenges in interlinked social and ecological (SES) systems (Meadows 2008), Singh observes that 'little is known about the socio-ecological history of interaction, the implications of changing human behaviour, or the socio-ecological drivers of significant or mundane human–animal contact' (Singh 2013, p. 79), However, current discussions of and approaches to EID have not given way to social-ecological (SES) system-focused approaches (Waltner-Toews 2017), which is a view that is widely shared among sustainability experts (Wood et al. 2012).

A systemic to this 'wicked problem' shows that there are 'many drivers that are likely to influence the emergence, persistence and/or prevalence of zoonotic diseases, which are interconnected, multi-level, and multi-scalar' (Ahmed et al. 2019, p. 450). Epidemiological studies suggest that natural selection pressures on wildlife virus hosts, such as bats, resulted in zoonosis of COVID-19 through direct contact with humans or transmission via intermediate mammal hosts, e.g. pangolins and livestock (Andersen et al. 2020). It is further known that certain species are viral reservoirs, e.g. bats and rodents (Mollentze and Streicker 2020), and viruses can spread to intermediate hosts, e.g. livestock, or directly to humans in hotspots that correlate with disease emergence (Johnson et al. 2020). The increasing prevalence of activities in hotspots is putting pressure on biodiversity and wildlife and promoting animal–animal and ultimately animal–human transmission. Studies suggest that bushmeat hunting, therefore, should be the target of integrated policy interventions (Friant et al. 2015). However, the argument of a blanket ban for bushmeat hunting tends to ignore the multiple socio-economic and cultural factors promoting bushmeat hunting and consumption. In fact, no measure, e.g. alternative livelihood promotion, has been successful thus far in addressing such transmission factors (van Vliet 2018).

The macro-level interactions alluded to above are illustrated by the Ebola outbreak, where the scope and severity of recent outbreaks may be related to deforestation, 'which is in turn connected to local food insecurity and other uses of forests by expanding human and livestock populations. These are themselves related to cultural beliefs and practices, governance challenges, and economic pressures' (Rabinowitz et al. 2018, p. 3). The diverse nature and time scales associated with these variables require more than a model of days and months but rather years of slow environmental degradation and related consequences.

In contrast to typical short-term symptomatic treatments of proximal causes, e.g. forest fire management, system models have enabled a better appreciation of the need for the preventative and strategic management of threats to society and environment (Collins et al. 2013). There are multiple modelling approaches possible, and such disease models include social roles and relations, unsafe practices, mobility and movement, temporal changes, and long-term dynamics (Scoones et al. 2017). As Sterman (2000) notes, multiple limitations to human reasoning persist in the face of dynamic systems, including bounded rationality, information gaps, defensive routines, flawed cognitive maps, and other factors. These limitations lead not only to flawed decisions at an individual and community level, but also policy resistance that exacerbates the original problems (see Moxnes 2000). It appears that even highly educated and experienced individuals fail to comprehend basic interactions between stocks and flows in systems (Cronin et al. 2009), although experiments also show that learning about system interactions and better decision making is possible (Moxnes and Saysel 2009).

As a result, there is real learning potential for the tools and methods of modelling 'to gain useful insight into situations of dynamic complexity and policy resistance' (Sterman 2000, p. 39). In this chapter, we aim to illustrate how that can be achieved.

2 Social-Ecological Systems, Panarchy and Resilience

The concept of SES is core to sustainability (e.g. Ostrom 2009), and suggests that we 'all live and operate in social systems that are inextricably linked with the ecological systems in which they are embedded' (Walker and Salt 2006, p. 31). An SES view sees systems as complex adaptive systems (CAS) with degrees of dynamic stability and resilience to perturbations. Such systems self-organise with the help of feedbacks and respond to environmental drivers or shocks either by retaining their structure or tipping over thresholds (Levin 1998). As Dewitte et al. (2017) note in their study of Black Death, 'acute shocks to systems occur against a background of existing conditions, which are crucial determinants of the eventual public health outcomes of those shocks, and in the context of complex dependencies among and between ecological and societal elements' (Dewitte et al. 2017, p. 254). The current COVID-19 pandemic and its consequences for society is a typical case of epidemic emergence and spread challenging the resilience of regional and global CAS (Walker et al. 2005), which a systems perspective can elucidate (e.g. Bishai et al. 2014).

Thus, not only do diseases emerge as an outcome of intertwined SES factors but also the relevant systems, e.g. viral pools, poverty traps and forest ecologies, are self-organising CAS in degrees of equilibrium and vulnerability to external tipping points and threshold crossing (Rockström and Steffen 2009). Threshold crossing and system breakdown/transformations occur, such as virus outbreaks, when critical variables are overwhelmed by the combined feedbacks of relevant variables. Thus, EID emergence is 'the result of complex multifactor interactions, requiring pathogens to overcome numerous ecological and evolutionary barriers to switch hosts and establish themselves in human populations' (Fornace et al. 2014). In resilience thinking and systems literature, interlocking multi-level CAS drivers and processes are described as panarchies (Allen et al. 2014). Higher-level systems, such as the existing SES, will operate more slowly towards consolidation or change than, for example, the immediate transmission framework, which can have faster time units of days and months.

Anthropogenic activities can perturb EID existing in stable equilibrium within reservoirs: Humans can exert sufficient pressure on this stability through economic and other drivers to trigger exponential growth, which must be eventually stabilised by precautionary management and other measures. Villages or even wildlife–host incursions may be manageable until multiple incursions lead to threshold-crossing transmission. This indicates a lack of resilience in the relevant locations to these forces. Higher-level environmental, social, and other interlinked drivers such as agro-industry, economic or political forces, typically work at a slower overall rate of change and can likewise trigger instability and breakdown.

The SES perspective, which is also linked to resilience thinking, is seen as a new paradigm for modelling and understanding EID (Everard et al. 2020; Lewis 2005; Parkes et al. 2005; Singh 2013). The SES concept of panarchy, meanwhile aims to link the multi-scalar systems that influence the response of downstream systems to perturbations (Holling et al. 2002). This concept is relevant in understanding the dynamics of pandemics and zoonoses (Berkes and Ross 2016) by combining levels of SES systems with levels of system change cycles below and above.

As illustrated below in Fig. 1, any SES environment has a stability landscape which may be transitioning to instability, threshold crossing, and regime change. A local 'hot spot' rural community that is being driven towards high input intensive agriculture and resource extraction (see Fig. 7) can reach a state of rigidity with respect to its capacity to adapt to change or shocks. Thus, precarious livelihoods can be overwhelmed by external 'pushes', soil quality deterioration or monoculture crop failures and lead to breakdown. For virus spread, the village-level (small-scale) breakdown of the natural equilibrium between population, environment and local virus breaks down due to multiple incursions. This lower-level 'revolt' triggers-higher level community collapse. The slower moving socio-economic system dominated by exploitative growth also provides context for this failure by slowly eroding the capacity of the population to weather such storms.

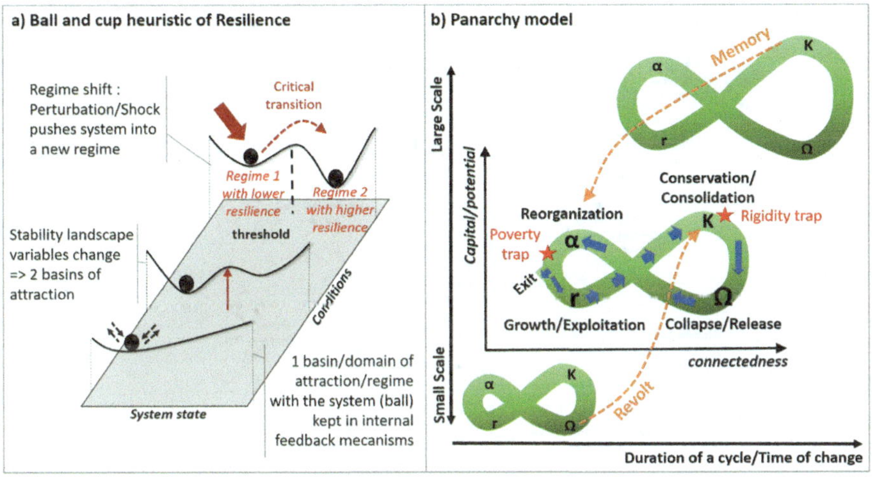

Fig. 1 Models of resilience thinking and panarchies (from Ollivier et al. 2018)

Panarchy reminds us of the multi-scalar SES relationships across geographies and time that create either community resilience or vulnerability to EID (see Berkes and Ross 2016). We propose that integration with thinking around multi-scalar SES, CAS, and resilience provides a better appreciation for EID vulnerabilities in communities and extends the current conventional systems perspective.

3 Causal Loop Diagrams and Archetypes

One of the obstacles to dissemination of SDMs is the complexity inherent in models. Hence, there are increasing calls for so-called low order (simplified) models, capable of demonstrating non-linear outcomes, that are more appropriate for communication between and with relevant stakeholders (Newell and Siri 2016). One key challenge is 'a conceptualisation of these systems that support visualisation of the interfaces of animal and human health' (Coker et al. 2011, p. 327). Causal Loop Diagrams (CLD) and Archetype structures model feedbacks and variables without quantifying parameters. Such diagrams (see below) map cause and effect relationships between variables in systems and sub-systems (see below under Methodology). Recently, Bradley et al. (2020) published an example CLD system diagram integrating public perception and governance sub-systems linkages that promote COVID-19 transmission. The model in Fig. 2 exemplifies reinforcing and balancing loops, quantitative and qualitative variables, time delays, and other properties of CLDs. Note that there are several scale/level variables, rates/flows, and time delays.

Such approaches recognise that the 'relationship between human-driven environmental change and human health can be non-linear that are affected by delays,

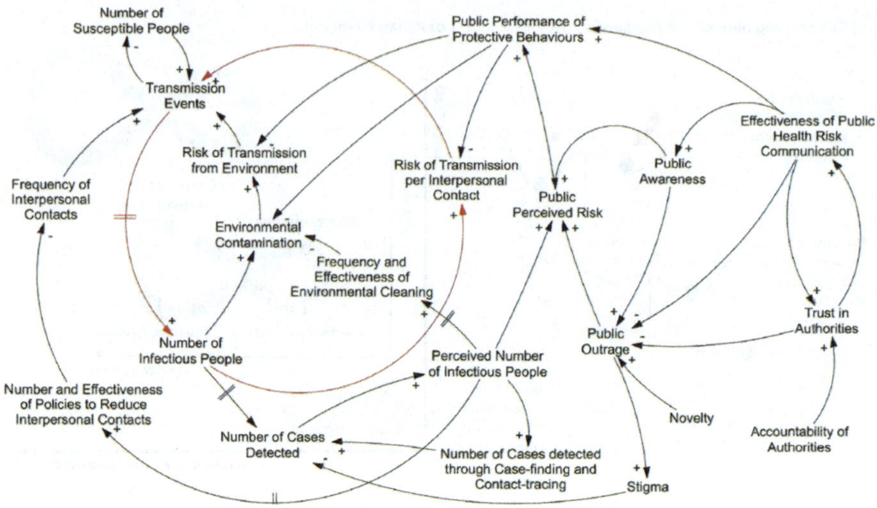

Fig. 2 CLD integrating EID transmission and perception issues (from Bradley et al. 2020)

and dominated by feedback loops across social, economic, and environmental dimensions' (Pongsiri et al. 2017, p. e257).

Not only should diagrams highlight social and ecological interactions, but they can also include reference to so-called archetypal feedback processes that generalise common structures observed across system models (Kim and Anderson 1998). We identified several potentially relevant system archetypes (see below in results) including: fixes that fail, shifting the burden, and limits to growth. The latter archetype explains the exponential growth of infections, which eventually flatlines as the susceptible pool is decreased by herd immunity or vaccines. In the fixes-that-fail archetype, a problem is addressed symptomatically while the fundamentals, e.g. unsustainable environmental degradation and pressure, remain unanswered. In our results section below, we identity several relevant Archetypes.

From the SES perspective above, the symptomatic solution is at the level of disease transmission and involves immediate health prevention such as masks, contact tracing, and social distancing. This strategy may resolve the symptomatic problem, e.g. COVID-19 spread, while leaving the fundamental issue of environmental degradation and poverty unaddressed. Thirdly, in the shifting-the-burden archetype, a response to a problem may make the apparent problem disappear while generating side effects that exacerbate other fundamental problems. The fundamental problems can then generate more symptomatic problems, which are likewise addressed, and so the cycle continues. This archetype seems typical of the current treatment of EID as discrete outbreaks.

4 Existing SES Systems Modelling Examples of Zoonosis

To identify the relevant variables and feedbacks, we reviewed the latest literature using Google Scholar looking for articles with an overlapping focus on viruses, zoonosis, health, and environment. In addition to the general literature referred to above, we selected articles with these focal areas that included systems-perspectives and modelling; we excluded those with only a narrow epidemiological transmission focus. As a result, we identified articles mentioned throughout this chapter. The ability to conduct a rapid review and develop a subsequent model is a conventional modelling step, yet crucial in new situations, e.g. COVID19. We emphasise again that we see this particular virus as an example of a broader category of environmentally and socially driven EID.

For bat-derived EID including COVID-19, an intermediate species, e.g. pig, horse, chicken, is essential in many cases. Bat ecology and populations are influenced by *land-use*, *wildlife management*, and *conservation*. Human–bat interactions are also influenced by livelihood and ritual practices. Public health impacts and detection of these interactions vary with local circumstances. Finally, political, cultural and policy framings of disease—and how these are communicated and treated in communities—have a key influence (Wood et al. 2012, p. 2885). Morse et al. (2012) developed a three-stage model, which includes the environmental enabling conditions. At stage one, disturbances to disease reservoirs due to land-use changes and other pressures lead to increasing intra-species interactions and transmission to wildlife or livestock hosts.

Stage two consists of localised emergence in humans in 'hotspots' through exposure to wildlife and infected livestock. However, a threshold needs to be crossed via multiple incursions to overwhelm the natural resistance of the human population and to create spillover dynamics (Wood et al. 2012, p. 2884). Subsequent human–human spread through other channels, e.g. international travel, then follows (Morse et al. 2012, p. 1958). Hence the reduction of human exposure to pathogens in such hotspots is seen as key to controlling emergence and spread (Graham and Sullivan 2018, p. 21). A recent study of coronaviruses visualises some such pathways, as seen in Fig. 3 (Shereen et al. 2020).

Particular practices, especially bushmeat hunting and consumption, are embedded in complex socio-cultural norms and behaviours (van Vliet 2018). They are targets for interventions not only because hunting and consumption threaten wildlife biodiversity and conservation targets, but because targeting them has implications for reducing the risk of zoonosis in hotspot areas (Ripple et al. 2016). Such practices also include the supply of bushmeat for urban consumption. Nonetheless, blanket bans are not likely (Vliet and Mbazza 2011) considering that alternative livelihood programs 'specifically targeting hunters and aimed at providing alternative protein sources that would satisfy local taste preferences' (Friant et al. 2015, p. 13) have not succeeded to date. Instead, hunting practices and perceptions of risky wildlife hunters are key factors that need to be included in any analysis and subsequent interventions (Friant et al. 2015).

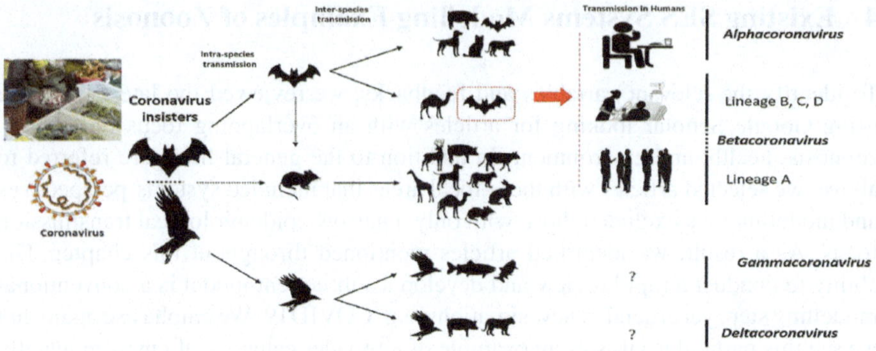

Fig. 3 Zoonotic pathways implicated in coronavirus EID (from Shereen et al. 2020, p. 93)

Wilderness encroachment and movement of people and products 'are rapidly changing environmental conditions and also generating favourable grounds for the (re)emergence of infectious disease vectors with increasing epidemiological complexities' (Ahmed et al. 2019, p. 444). Urbanisation drivers emerge from 'rapid urban growth and increased density of human occupation, increased movement of people and animals, increased complexity in the value chains around animal-sourced products, rural-to-urban migration, intra-city inequalities and land-use changes' (Ahmed et al. 2019, p. 455). For the urban poor, these consequences compound into further health, poverty and environmental consequences.

Given the systemic multi-scalar (panarchic) nature of the links between environment, health, and disease, there have been attempts to develop broad qualitative system models at various levels. These have produced specific parameters and variables that also aggregate into higher-level sub-system elements. A relevant recent example employing an SES and systems perspective is by Finucane et al. (2014), and models the spread and testing of the HN51 influenza virus in Fig. 4.

They identify 12 variables grouped into three main sub-system level elements leading to a virus outbreak and its transition across the virus–wildlife threshold: urbanisation, habitat alteration, and agricultural change. Allen et al. (2017) identify other regression correlations, which they group into three broad variables—human, animal and environmental—that enhance the likelihood of EID emergence (see Table 1); these are then mapped to hotspots of emergence.

More recently, Everard et al. (2020) employed the STEEP framework in Fig. 5 as a way of capturing the interacting influences across systems. Their high-level identification of drivers helps identify five broad interacting sub-system factors: social, environmental, political/governance, economic and technological. The details within each dimension, e.g. ecosystem degradation, reliance on wild food, distributional equity, provide sufficient detail for modelling.

Through our review, CLD modelling and analysis, we identify economic, environmental, agriculture, technological, health and governance issues as relevant for rural hotspot areas. We detail these below.

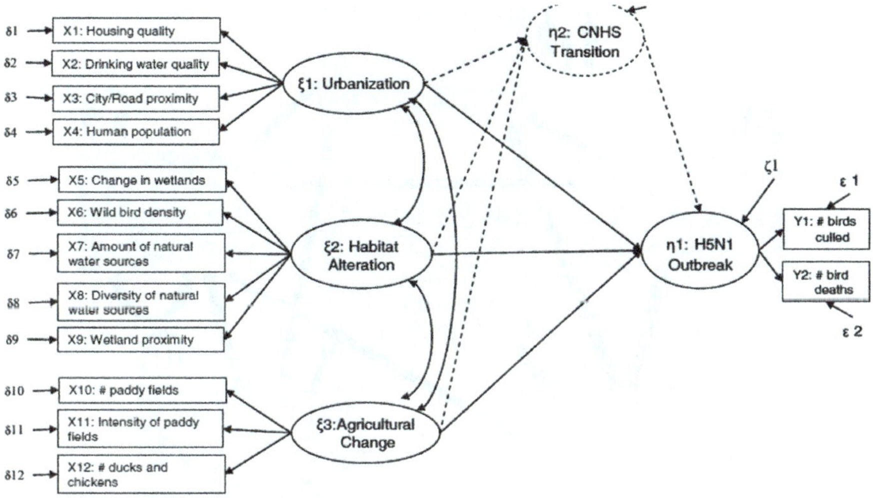

Fig. 4 Factors affecting H5N1 Outbreak (from Finucane et al. 2014, p. 103)

Table 1 Anthropogenic factors influencing virus emergence

Human Activity	Animals	Environment
Population and Change in Population	Mammal / Biodiversity	Nature of Forest (Evergreen, Deciduous, Mixed)
Cultivates/Managed Vegetation	Livestock and Mammal Headcount	Shrubs and Herbs
Nature of Grazing of Pasture	Poultry	Regularly Flooded Vegetation
Cropland and Cropland Change		
Change in Urban/Built-Up Area		

5 Methodology: Causal Loop Diagram

Opinions differ in describing the modelling process, but general agreement exists on setting boundaries and gradual specification of the problem and its factors before detailed parameter setting. We have already detailed above our process of variable discovery through rapid review, whereupon we set our boundary at the rural community hotspot level. Subsequent modelling followed a group-building protocol between the three authors as detailed below. All approaches agree on identifying endogenous, exogenous and excluded variables, where endogenous variables are embedded in feedback loops and interact within the system to produce non-linear outcomes. Excluded variables are those with an influence on outcomes and processes but not included within the model boundary (Sterman 2000, pp. 83–105). As noted,

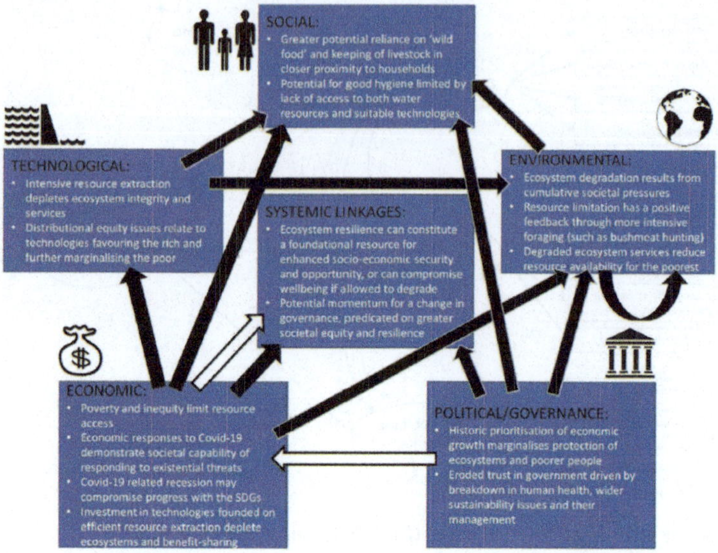

Illustrative representation of systemic relationships across the STEEP system of current impacts leading to a degrading cycle in the socio-ecological system, including increasing vulnerability to zoonoses. Solid arrows represent negative influences; hollow arrows represent potentially positive influences.

Fig. 5 A representation of the STEEP framework in terms of dominant sub-system factors (from Everard et al. 2020, p. 12)

however, what is excluded in a typical CLD or system diagram will be included from a multi-scalar SES perspective as part of an interlinked panarchy.

CLD is a qualitative graphical representation of the structure of influences characterising a system as perceived by the modeller(s). In CLD, variables are causally connected by arrows (Sterman 2000). The variable at the arrow's tail is referred to as 'cause', and the variable at the head as 'effect' (Lane et al. 2016) Each arrow carries a sign in its head, which takes (' + ' or 's') or ('–' or 'o'). A ' + ' or 's' signifies that the change in the variable in the arrowhead is in the same direction as that of the change of the variable in the tail. A '–' or 'o' signifies a change in the opposite direction. Two types of feedback loop can be represented using CLDs, viz., (a) reinforcing loops, and (b) balancing loops.

- Reinforcing loop—a type of feedback loop in which the effect of variation of a variable transmits through the loop to amplify variations in the same direction. Reinforcing loops seek to achieve momentum (exponential)
- Balancing loop—a type of feedback loop in which the effect of variation of a variable transmits through the loop to oppose variations, thus tending to nullify the change. Balancing loops seek to achieve equilibrium.

These feedback loops become the basis for a qualitative understanding of systems (Wolstenholme and Coyle 1983). In this study, we have attempted to represent the

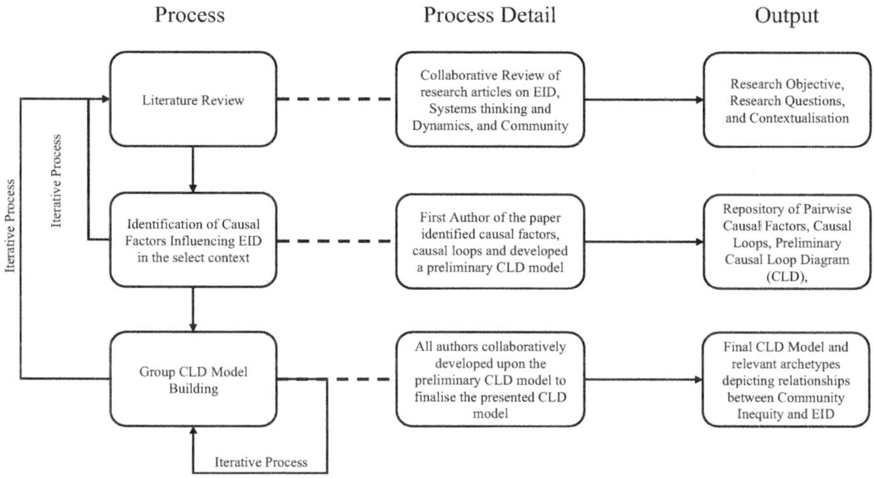

Fig. 6 Process for developing CLD detailing linkages between rural communities and EID

underlying factors their interactions that cause EIDs. CLD[1] can be developed through identification of factors using combinations of methods such as interviews and the Delphi technique. In this study, the three authors have used a Group Model Building (GMB) technique to model the causal factors. GMB was first proposed by Vennix (1999) and has been used since then in system dynamics literature.

The CAS approach we take here focuses on multi-scalar panarchy system interactions. In the current case, we propose what others describe as higher-level environmental 'drivers', rather than causes. Building on the SES approach and the concept of multi-scalar CAS, we agree with recent systems approaches which see key processes in 'dynamic interactions between bats, viruses, intermediate livestock hosts and people in a local system, influenced by wider environmental, social and politico-economic drivers' (Wood et al. 2012, p. 2883). We used recent sources addressing zoonosis, environmental drivers and systems thinking to identify drivers of the emergence of diseases such as COVID-19. We underscore that much of this literature and these recommendations appeared before the emergence of COVID-19.

In sum, there are multiple environmental, social and economic variables and drivers influencing zoonosis in hotspot areas and beyond. We limit our focus in this paper on the first and second stage contexts of growth in the virus–host pool due to multiple pressures. These include agriculture, habit destruction, and the second stage expansion of the human population in relevant hotspots through a variety of channels, such as hunting, wildlife consumption, and livestock transmission. We do not consider the widely developed epidemiology of disease spread and global transmission through air travel etc. in this chapter. Based on this background review, we identify a number of sub-system elements and key variables. Figure 6 depicts

[1]CLD is drawn using Vensim Software by Venetana Systems, which is available free of cost for academic purposes. Their generosity is much appreciated.

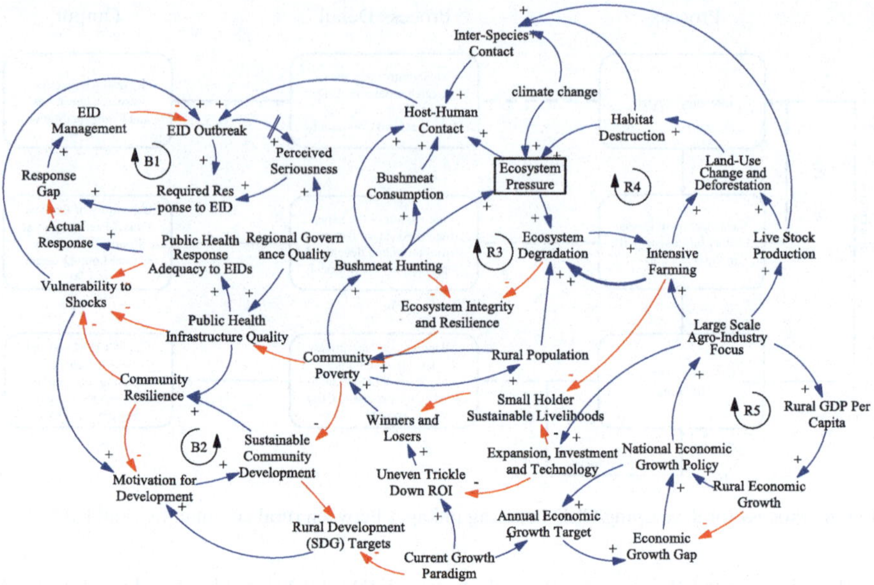

Fig. 7 Overall system CLD detailing the linkages between rural 'hot spot' populations and EID

the process involved in developmenting the CLD model and identifying relevant archetypes. The authors of the paper initially reviewed relevant and topical research articles on EID, System Thinking and Dynamics, and other anthropogenic factors influencing EID. We excluded articles about epidemiological propagation models such as SIR. The first author then identified the pairwise relationship between anthropogenic causal factors that influence EID, from which a preliminary CLD model was developed. Following which, the authors of this article collaborated iteratively for approximately 20 h as a group through teleconferencing to finalise the CLD model as presented in Fig. 7.

6 Results

We present results under three main headings: balancing and reinforcing archetype loops, the overall CLD, and the accompanying narrative.

6.1 Balancing and Reinforcing Loops and Archetypes

As a complement to our overall system diagram below, we provide some isolated examples of archetype reinforcing EID loops to show the value and application of these system tools also (Tables 2 and 3).

6.2 Overall System CLD

Finally, in this section we provide an overall diagram of system variable interactions both balancing and reinforcing; below a narrative aims to explain these connections in more detail. We present this as one potential interpretation of the environmental, social, health and economic drivers of EID emergence and persistence.

6.3 System Narrative: Balancing Parsimony and Detail

As already mentioned, we considered the boundary of our analysis to encompass poorer rural communities owing to zoonotic disease emergence being observed in such conditions on numerous occasions. We are aware of the effect of mobility, such as rural–urban migration, wildlife transport to cities, high population density, and poor hygiene and health conditions in nearby urban centres as promoters of virus spread and ultimately pandemic potential. However, in this model, we aimed to focus on rural communities, whose development and vulnerability is driven by higher (panarchic) level economic drivers such as a growth preference versus development focus. The developed systemic picture treats ecological threats to EID outbreaks as driven by growth versus development policies and their effects. Here, factors pertaining to development, local ecosystem health, agriculture expansion, and public health infrastructure are conceptualised as variables that can ultimately affect the conditions conducive to EID outbreaks and the efficacy of their subsequent management.

Choices were made regarding the level of aggregation in each variable such that it contributed to a parsimonious output. In a more detailed diagram, *Inter-species Contact* driven by *Habitat Destruction* and *Live Stock Production* could be further unpacked into separate contact interactions between virus-host and species to cover events such as bat host–wildlife intermediate transmission and other wildlife–livestock channels. However, this level of detailing would have overcomplicated the diagram. This need for balance between aggregation and detail, for purposes of clear communication, is a constant in system modelling for public purposes (Newell and Siri 2016). In this way, choices and priorities in line with an overall narrative of plausible connections have to be made; our diagram serves as an invitation to discussion and debate rather than a complete articulation of all potential factors.

Table 2 Balancing and reinforcing Loops related to community development and EID

Loop	Loop Description
Symptomatic Response Loop—B1 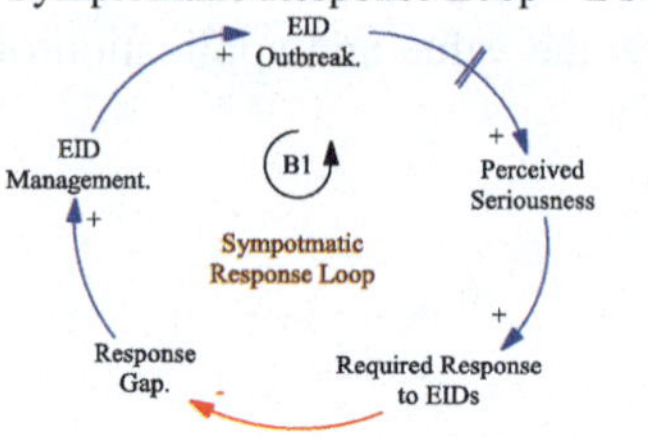 	The Symptomatic Response Loop explains the behaviour of public health providers and experts to an *EID Outbreak* event. The gap between the *Required response to EID* and the actual response is the '*Response Gap*'. The *Response Gap* prompts *EID Management* that has a balancing effect on the *EID Outbreak*. The rate of response and the rate of growth of the *EID outbreak* determines the level of *EID Management*. However, an *EID Outbreak* may not be considered serious by health care providers, and response delayed. A shorter response delay will lessen the stress the public health system will be under
Loop of Lethargy—B2 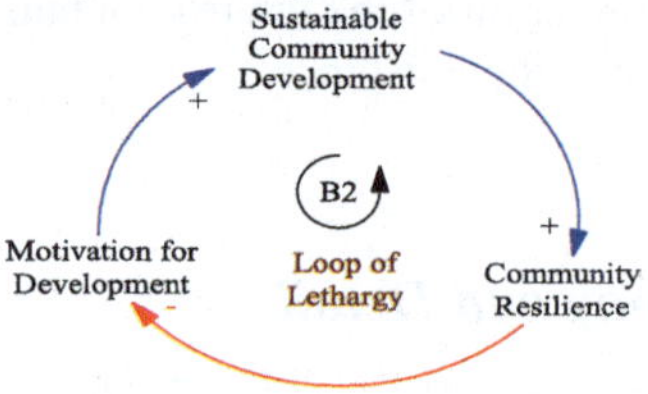 	*Sustainable Community Development* results in *Community Resilience* to disturbances such as pandemics. However, the more resilient a community, the less the further *Motivation For Community Development*. This decreased motivation risks the community against a potentially stronger disturbance. Hence, policymakers must evolve to always focus on community development
Poverty Perpetuity Loop—R3 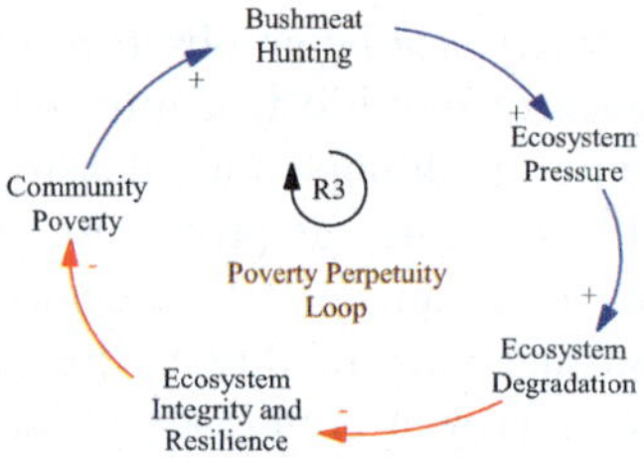 	While poverty is a multi-dimensional construct, we have contextualised it around food and nutrition here. *Community Poverty* is the strongest determining factor of *Bushmeat Hunting* in the rural context. Excessive *Bushmeat Hunting* affects the ecosystem in complicated ways. *Bushmeat Hunting* adds to *Ecosystem Pressure*, which leads to *Ecosystem Degradation*. Hunting not only results in the *Bushmeat Consumption*, but also result in illegal wet and exotic markets (not scoped inside this study). Ultimately, *Ecosystem Degradation* results in reduced *Ecosystem Integrity and Resilience*, contributing to poverty within communities who are reliant on the ecosystem
Resource Depletion Loop—R4 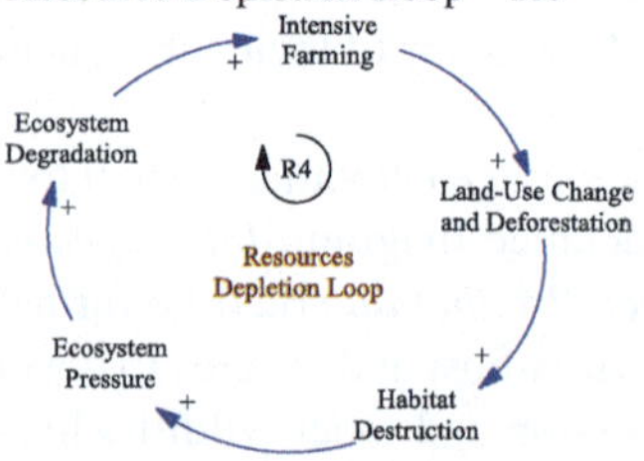 	A *Large Scale Agro-Industry Focus* results in *Intensive Farming* practices to meet growth and profit requirements of industry players. As such, *Intensive Farming* promotes excessive use of fertilisers and other agricultural resources to bolster yields. Further, intense and landscape-scale monoculture practices entailed in *Intensive Farming* promote *Land-Use Change and Deforestation*. Large scale deforestation over a long period results in *Habitat Destruction*, which adds to *Ecosystem Pressure* that ultimately results in *Ecosystem Degradation*. Greater *Ecosystem Degradation* prompts agro-industry to intensify Farming activities further to maintain previous yield levels

(continued)

Table 2 (continued)

Loop	Loop Description
Engine of Growth Loop—R5 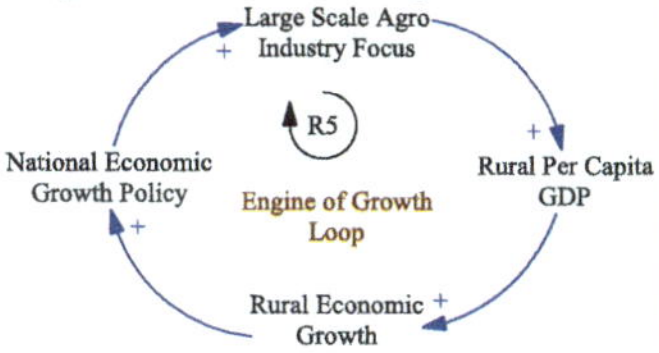	Economic Growth in terms of GDP or GNI per capita is the conventional measure of national progress. This economic growth cycle assumes that prioritising economic growth problem will naturally solve a range of other community maladies. While a *Large Scale Agro-Industry Focus* improves *Rural Per Capita GDP* and thereby *Rural Economic Growth*, the effect of increased GDP is spread unevenly. Since the policymakers' (government) targets are measured predominantly based on a *National Economic Growth Policy*, this loop is perpetuated

6.4 Stocks, Rates and Auxiliaries and Parameters

This diagram, like all CLDs, builds around stocks/levels, rates, auxiliaries and other parameters. There is also a mix of so-called informational and material variables. Stock variables, such as *Sustainable Community Development*, *Community Poverty* and *Ecosystem Degradation* tend to have respective inflows, outflows and relevant units of measure in fuller modelling, although they are not specified here. Deforestation and Land-use Change could likewise be quantified and cumulatively measured with inflow rates of change, further specified by rate constants per time unit. A fuller system model containing these additional levels of integration coupled with differential equations is planned for future development.

As already noted, once an EID crosses the relevant threshold for outbreak transmission, existing SIR or similar models can be coupled to handle infection spread from an epidemiological perspective. The narrative developed and reported in this chapter attempts to unpack some of the 'black box' variables alluded to above. As such, we wish to reiterate our aim of a plausible account of interconnections between several sub-systems—social, economic, environmental, and health—as proposed in the literature to date.

We argue that mainstream growth, rather than a more balanced development policy implementation, drives ecosystem pressure and degradation through intensive farming and other channels. Thus, we see small-holder farmers and similar community members not as deep causes of ecosystem degradation but as channels for agro-industry focus that lead to community poverty. There is admittedly an aspirational driver in such communities for increasing income and capital through more intensive farming that can contribute to further ecosystem loss. The kinds of capital required to engage in more intensive scale farming tends to favour those with existing resources and power and can, therefore, contribute to increased inequity in communities.

Consequentially, the widening gap (and outflows of income) exacerbates community poverty and detracts from (real) community development that include equitable distribution of capitals thatconstitute the basis for sustainable livelihoods (human,

Table 3 Archetype model

Standardised Archetype (Adopted From Kim and Anderson 1998)	Adapted Archetype	Archetype Explanation
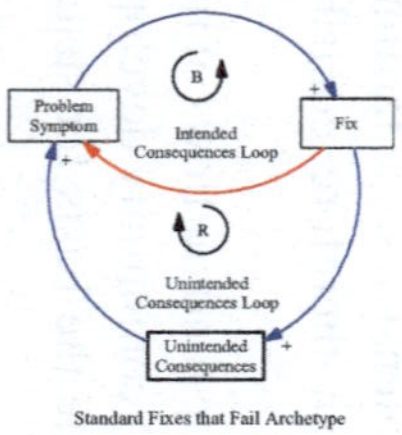	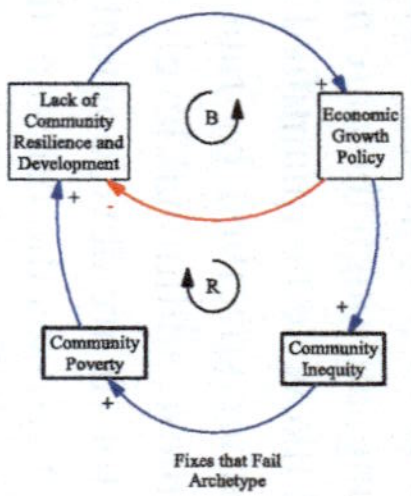	This Archetype explains that crude fixes may lead to unintended consequences that eventually undermine the very purpose for which they were intended A lack of community development and resilience is a common problem across many developing countries: the solution policy makers in these countries adopt is typically an economic growth policy. Many socio-economic systems experts have debunked the promise of economic growth relieving all community issues. In most cases, economic growth alone results in uneven wealth distribution that increases community poverty, which undermines community development and resilience If the unintended consequences loop is stronger than the intended consequences loop, the Fixes that Fail Archetype will result in what is called 'Rebound Backfire'

financial, socio-political, natural, etc.). The key role played by public health infrastructure and institutions in such areas depends on regional and rural development policy and implementation, which is often deficient. Hence an adequate response and subsequent management of EID is typically lacking both in terms of speed and depth of response. In addition, 'hot spots' are the location of recurrent virus EIDs. Thus, EID management both here and in other settings continues to treat symptoms rather than deep causes of disease.

7 Conclusion

Hopeful voices have suggested that current experience with COVID-19 might promote progress towards sustainable development, particularly through reduced consumption, although experience has shown this is unlikely (Cohen 2020). One way in which the current pandemic might promote permanent change is through learning about the system-level processes and behaviours which drive the emergence of disease. Consistent with the multi-dimensional nature of sustainable development and studies to date on SES within which EID emerge, systems thinking and CLD in particular offer a way of visualising the interactions between economics, society, environment, and health. There is a real need to emphasise these linkages, according to many scholars (McKinnon et al. 2016).

Through rapid review, CLD model process, and group building we have presented in this chapter a model of the interacting ecological, socio-economic, health, and governance issues related to the emergence and spread of zoonotic EID. The model process and outcome is illustrative of what is possible in a teaching and learning setting to engender discussion and understanding. We have presented one plausible account in this chapter, which in essence aims to capture many of the characteristics of development issues globally. Inviting students to convert their textbook accounts of health and virus interactions with other development aspects can provide a particularly lively discussion point. We encourage others to employ similar methods to engage students, faculty, practitioners and the public in the debate about the consequences of unsustainable development.

References

Ahmed S, Dávila JD, Allen A, Haklay M (MUKI), Tacoli C, Fèvre EM (2019) Does surbanisation make emergence of zoonosis more likely? Evidence, myths and gaps. Environ Urbanisation 31(2):443–460. https://doi.org/10.1177/0956247819866124

Allen CR, Angeler DG, Garmestani AS, Gunderson LH, Holling CS (2014) Panarchy: theory and application. Ecosystems 17(4):578–589. https://doi.org/10.1007/s10021-013-9744-2

Allen T, Murray KA, Zambrana-Torrelio C, Morse SS, Rondinini C, Di Marco M, Breit N, Olival KJ, Daszak P (2017) Global hotspots and correlates of emerging zoonotic diseases. Nat Commun 8(1):1–10. https://doi.org/10.1038/s41467-017-00923-8

Andersen KG, Rambaut A, Ian Lipkin W, Holmes EC, Garry RF, Lipkin WI, Holmes EC, Garry RF (2020) The proximal origin of SARS-CoV-2. Nat Med 26(4):450–452. https://doi.org/10.1038/s41591-020-0820-9

Berkes F, Ross H (2016) Panarchy and community resilience: sustainability science and policy implications. Environ Sci Policy 61:185–193. https://doi.org/10.1016/j.envsci.2016.04.004

Bishai D, Paina L, Li Q, Peters DH, Hyder AA (2014) Advancing the application of systems thinking in health: why cure crowds out prevention. Health Res Policy Syst 12(1):1–12. https://doi.org/10.1186/1478-4505-12-28

Bradley DT, Mansouri MA, Kee F, Garcia LMT (2020) A systems approach to preventing and responding to COVID-19. EClinicalMedicine 21:100325. https://doi.org/10.1016/j.eclinm.2020.100325

Cohen MJ (2020) Does the COVID-19 outbreak mark the onset of a sustainable consumption transition? Sustainability Sci Pract Policy 16(1):1–3. https://doi.org/10.1080/15487733.2020.1740472

Coker R, Rushton J, Mounier-Jack S, Karimuribo E, Lutumba P, Kambarage D, Pfeiffer DU, Stärk K, Rweyemamu M (2011) Towards a conceptual framework to support one-health research for policy on emerging zoonoses. Lancet Infect Dis 11(4):326–331. https://doi.org/10.1016/S1473-3099(10)70312-1

Collins RD, de Neufville R, Claro J, Oliveira T, Pacheco AP (2013) Forest fire management to avoid unintended consequences: a case study of Portugal using system dynamics. J Environ Manage 130:1–9. https://doi.org/10.1016/j.jenvman.2013.08.033

Cronin MA, Gonzalez C, Sterman JD (2009) Why don't well-educated adults understand accumulation? A challenge to researchers, educators, and citizens. Organisational Behav Human Decis Processes 108(1):116–130. https://doi.org/10.1016/j.obhdp.2008.03.003

Deem SL, Lane-deGraaf KE, Rayhel EA (2019) Why one health? Introduction to one health: an interdisciplinary approach to planetary health. Wiley Blackwell, Hoboken (NJ), pp 3–14

Dewitte SN, Kurth MH, Allen CR, Linkov I (2017) Disease epidemics: Lessons for resilience in an increasingly connected world. J Public Health 39(2):254–257. https://doi.org/10.1093/pubmed/fdw044

Di Marco M, Baker ML, Daszak P, de Barro P, Eskew EA, Godde CM, Harwood TD, Herrero M, Hoskins AJ, Johnson E, Karesh WB, Machalaba C, Garcia JN, Paini D, Pirzl R, Smith MS, Zambrana-Torrelio C, Ferrier S (2020) Sustainable development must account for pandemic risk. Proc Natl Acad Sci USA 117(8):3888–3892. https://doi.org/10.1073/pnas.2001655117

Everard M, Johnston P, Santillo D, Staddon C (2020) The role of ecosystems in mitigation and management of COVID-19 and other zoonoses. Environ Sci Policy 111:7–17. https://doi.org/10.1016/j.envsci.2020.05.017

Finucane ML, Fox J, Saksena S, Spencer JH (2014) A conceptual framework for analysing social-ecological models of emerging infectious diseases. In: Understanding society and natural resources (pp. 93–109). Springer, Netherlands. https://doi.org/10.1007/978-94-017-8959-2_5

Fornace K, College RV, Rushton J, College RV, Coker R (2014) Agricultural practices on the emergence and reemergence of human viral diseases. In: Singh SK (ed) Viral infections and global change. Wiley Blackwell, Hoboken (NJ), pp 133–150

Friant S, Paige SB, Goldberg TL (2015) Drivers of bushmeat hunting and perceptions of zoonoses in nigerian hunting communities. PLoS Negl Trop Dis 9(5):e0003792. https://doi.org/10.1371/journal.pntd.0003792

Graham BS, Sullivan NJ (2018) Emerging viral diseases from a vaccinology perspective: preparing for the next pandemic. Nat Immunol 19(1):20–28. https://doi.org/10.1038/s41590-017-0007-9

Holling CS, Gunderson LH, Peterson GD (2002) Sustainability and panarchies. In: Panarchy: understanding transformations in human and natural systems. Island Press, Washington, pp. 63–102

Johnson CK, Hitchens PL, Pandit PS, Rushmore J, Evans TS, Young CCW, Doyle MM (2020) Global shifts in mammalian population trends reveal key predictors of virus spillover risk. Proc R Soc B Biol Sci 287(1924):20192736. https://doi.org/10.1098/rspb.2019.2736

Kim DH, Anderson V (1998) Systems archetype basics: from story to structure. In the systems thinker. Pegasus Communications, Cambridge (MA)
Lane DC, Munro E, Husemann E (2016) Blending systems thinking approaches for organisational analysis: Reviewing child protection in England. Eur J Oper Res 251(2):613–623. https://doi.org/10.1016/j.ejor.2015.10.041
Levin SA (1998) Ecosystems and the biosphere as complex adaptive systems. Ecosystems 1(5):431–436. https://doi.org/10.1007/s100219900037
Lewis ND (2005) Is the social-ecological framework useful in understanding infectious diseases? The case of HIV/AIDS. EcoHealth 2(4):343–348. https://doi.org/10.1007/s10393-005-8477-x
Luke DA, Stamatakis KA (2012) Systems science methods in public health: dynamics, networks, and agents. Annu Rev Public Health 33(1):357–376. https://doi.org/10.1146/annurev-publhealth-031210-101222
McKinnon MC, Cheng SH, Dupre S, Edmond J, Garside R, Glew L, Holland MB, Levine E, Masuda YJ, Miller DC, Oliveira I, Revenaz J, Roe D, Shamer S, Wilkie D, Wongbusarakum S, Woodhouse E (2016) What are the effects of nature conservation on human well-being? A systematic map of empirical evidence from developing countries. Environ Evidence 5(1):1–25. https://doi.org/10.1186/s13750-016-0058-7
Meadows DH (2008) Thinking in systems: a primer (D. Wright (ed.)). Chelsea Green Pulbishing, White River Junction (VT)
Mollentze N, Streicker DG (2020) Viral zoonotic risk is homogenous among taxonomic orders of mammalian and avian reservoir hosts. Proc Natl Acad Sci 117(17):9423–9430. https://doi.org/10.1073/pnas.1919176117
Morse SS, Mazet JA, Woolhouse M, Parrish CR, Carroll D, Karesh WB, Zambrana-Torrelio C, Lipkin WI, Daszak P (2012) Prediction and prevention of the next pandemic zoonosis. The Lancet 380(9857):1956–1965. https://doi.org/10.1016/S0140-6736(12)61684-5
Moxnes E (2000) Not only the tragedy of the commons: misperceptions of feedback and policies for sustainable development. Syst Dyn Rev 16(4):325–348. https://doi.org/10.1002/sdr.201
Moxnes E, Saysel AK (2009) Misperceptions of global climate change: information policies. Clim Change 93(1–2):15–37. https://doi.org/10.1007/s10584-008-9465-2
Newell B, Siri J (2016) A role for low-order system dynamics models in urban health policy making. Environ Int 95:93–97. https://doi.org/10.1016/j.envint.2016.08.003
Ollivier G, Magda D, Mazé A, Plumecocq G, Lamine C (2018) Agroecological transitions: what can sustainability transition frameworks teach us? An ontological and empirical analysis. Ecol Soc 23(2):5. https://doi.org/10.5751/ES-09952-230205
Ostrom E (2009) A general framework for analyzing sustainability of social-ecological systems. Science 325(5939):419–422. https://doi.org/10.1126/science.1172133
Parkes MW, Bienen L, Breilh J, Hsu LN, McDonald M, Patz JA, Rosenthal JP, Sahani M, Sleigh A, Waltner-Toews D, Yassi A (2005) All hands on deck: transdisciplinary approaches to emerging infectious disease. EcoHealth 2(4):258–272. https://doi.org/10.1007/s10393-005-8387-y
Pongsiri MJ, Gatzweiler FW, Bassi AM, Haines A, Demassieux F (2017) The need for a systems approach to planetary health. The Lancet Planetary Health 1(7):e257–e259. https://doi.org/10.1016/S2542-5196(17)30116-X
Rabinowitz PMG, Pappaioanou M, Bardosh KL, Conti L (2018) A planetary vision for One Health. BMJ Global Health 3(5):1–6. https://doi.org/10.1136/bmjgh-2018-001137
Redding DW, Atkinson PM, Cunningham AA, Lo Iacono G, Moses LM, Wood JLN, Jones KE (2019) Impacts of environmental and socio-economic factors on emergence and epidemic potential of Ebola in Africa. Nat Commun 10(1):4531. https://doi.org/10.1038/s41467-019-12499-6
Ripple WJ, Abernethy K, Betts MG, Chapron G, Dirzo R, Galetti M, Levi T, Lindsey PA, Macdonald DW, Machovina B, Newsome TM, Peres CA, Wallach AD, Wolf C, Young H (2016) Bushmeat hunting and extinction risk to the world's mammals. R Soc Open Sci 3(10):160498. https://doi.org/10.1098/rsos.160498

Rockström J, Steffen W (2009) Planetary boundaries: exploring the safe operating space for humanity. Ecol Soc 14(2):32. https://www.ecologyandsociety.org/vol14/iss2/art32/main.html

Scoones I, Jones K, Lo Iacono G, Redding DW, Wilkinson A, Wood JLN (2017) Integrative modelling for one health: pattern, process and participation. Philos Trans R Soc B Biol Sci 372(1725). https://doi.org/10.1098/rstb.2016.0164

Shereen MA, Khan S, Kazmi A, Bashir N, Siddique R (2020) COVID-19 infection: origin, transmission, and characteristics of human coronaviruses. J Adv Res 24:91–98. https://doi.org/10.1016/j.jare.2020.03.005

Singh SK (2013) The socio-ecology of viral zoontic transfer. In: Singh SK (ed) Viral infections and global change. Wiley Blackwell, Hoboken (NJ), pp 133–150

Sterman JD (2000) Business dynamics: systems thinking and modeling for a complex world. Irwin, Boston (MA)

van Vliet N (2018) “Bushmeat crisis” and “cultural imperialism” in wildlife management? Taking value orientations into account for a more sustainable and culturally acceptable wildmeat sector. Front Ecol Evolution 6:112. https://doi.org/10.3389/fevo.2018.00112

Vennix JAM (1999) Group model-building: tackling messy problems. Syst Dyn Rev 15(4):379–401. https://doi.org/10.1002/(SICI)1099-1727(199924)15:4%3c379::AID-SDR179%3e3.0.CO;2-E

van Vliet N, Mbazza P (2011) Recognising the multiple reasons for bushmeat consumption in urban areas: a necessary step toward the sustainable use of wildlife for food in Central Africa. Hum Dimens Wildl 16(1):45–54. https://doi.org/10.1080/10871209.2010.523924

Walker B, Holling CS, Carpenter S, Kinzig AP (2005) Resilience, adaptability and transformability in social–ecological systems. Ecol Soc 9(2):5. https://doi.org/10.5751/ES-00650-090205

Walker B, Salt D (2006) The system rules: creating a mind space for resilience thinking. In: Resilience thinking: sustaining ecosystems and people in a changing world. Island Press, Washington, pp 28–52

Waltner-Toews D (2017) Zoonoses, one health and complexity: wicked problems and constructive conflict. Philos Trans R Soc B Biol Sci 372(1725):20160171. https://doi.org/10.1098/rstb.2016.0171

Wolstenholme EF, Coyle RG (1983) The development of system dynamics as a methodology for system description and qualitative analysis. J Oper Res Soc 34(7):569–581. https://doi.org/10.1057/jors.1983.137

Wood JLN, Leach M, Waldman L, MacGregor H, Fooks AR, Jones KE, Restif O, Dechmann D, Hayman DTS, Baker KS, Peel AJ, Kamins AO, Fahr J, Ntiamoa-Baidu Y, Suu-Ire R, Breiman RF, Epstein JH, Field HE, Cunningham AA (2012) A framework for the study of zoonotic disease emergence and its drivers: spillover of bat pathogens as a case study. Philos Trans R Soc B Biol Sci 367(1604):2881–2892. https://doi.org/10.1098/rstb.2012.0228

Environmental Education as a Tool to Improve Sustainability and Promote Global Health: Lessons from the COVID-19 to Avoid Other Pandemics

Thiago Leite Cruz, Pedro Israel Mota Pinto, and Anderson Hélio Juwer Ferreira

Abstract Human impacts such as industrial activities and uncontrolled urbanization have intensified environmental degradation, harming global ecosystems and public health. In this context, several studies highlight the emergence of infectious zoonotic diseases which potentially lead to global pandemics. The current COVID-19 pandemic illustrates such a trend, as its origins are related to the mismanagement of natural habitats and wildlife. Thus, it is fundamental to effectively address the interacting factors that stimulate the outbreak of pandemics. The present research assesses how environmental degradation contributed to the COVID-19 emergence, and present Environmental Education as an essential tool to raise awareness on the interaction of environmental protection, ecology and public health. We analyze successful international initiatives that encourage Environmental Education, such as the United Nations's Earth School, the Unesco's Global Education Coalition and the Brazilian initiative *Ame o Tucunduba*. We conclude that promoting Environmental Education may play a vital role in tackling challenges such as global pandemics by improving sustainability, and thus may avoid future environmental and health threats.

Keywords Environmental degradation · Global pandemics · COVID-19 · Environmental Education

---

T. L. Cruz (✉)
Faculdade de Direito, Universidade Federal Do Pará, Rua Augusto Corrêa, 01—Guamá, 66075-110 Belém, Pará, Brasil
e-mail: thiago.james@live.com

P. I. M. Pinto
Faculdade de Geografia, Universidade Do Estado Do Pará, Rua Do Una, 156—Telégrafo, 66050-540 Belém, Pará, Brasil

A. H. J. Ferreira
Curso de Relações Internacionais, Universidade da Amazônia, Av. Alcindo Cacela, 287—Umarizal, 66060-000 Belém, Pará, Brasil

W. Leal Filho (ed.), *COVID-19: Paving the Way for a More Sustainable World*, World Sustainability Series, https://doi.org/10.1007/978-3-030-69284-1_17

1 Introduction

Human activities such as intense industrialization and urbanization have severe consequences on the environment and on society as well, such as deforestation, pollution and habitats destruction. Zhody et al. (2019) claim that such factors significantly interfere in ecological patterns and contribute to the global emergence of zoonotic infectious diseases. In 2007, Cheng et al. (2007) analyzed the patterns of previous infectious diseases and predicted the outbreak of global pandemics in the following years, due to environmental degradation and wildlife mismanagement.

In fact, bats are the main suspects to be the ecological reservoirs of the novel coronavirus, SARS-CoV-19 (WHO 2020a). This acronym stands for Severe Acute Respiratory Syndrome Coronavirus 2, the virus that causes the coronavirus disease 19 (COVID-19). It is a mutation of previously existent coronaviruses such as SARS-COV and MERS-COV. Its symptoms range from asymptomatic or mild respiratory infection to fulminant pneumonia, and may eventually lead to death (Chen et al. 2020; Yeo et al. 2020).

The first patients infected with COVID-19 had visited a livestock market in Wuhan, China (Jandu 2020). Although freshly slaughtered livestock is preferred in Chinese cuisine, animals in wet markets are usually kept under poor sanitary conditions, including exotic and wild animals such as pangolins, civets and bats (Woo et al. 2006). The infection quickly expanded to other Chinese provinces and countries around the world (Huang et al. 2020). By June 2020, around 9 million people have been infected and more than 450 thousand have died in 216 countries (WHO 2020b).

Indeed, it has zoonotic origins, but the COVID-19 is easily transmitted by social contact, as respiratory droplets expelled from an infected person can spread to other people and surfaces. Thus, the World Health Organization strongly recommends social distancing as one of the most effective means to combat person-to-person contamination (WHO 2020b).

The international press quickly noticed significant decreases in pollution levels and changes in wildlife behavior patterns, in response to the sudden changes in industrial production and urban activities due to social isolation measures. Reports from NASA, the European Space Agency, the Center for Research on Energy and Clean Air and the American Energy Department indicated a significant reduction in the emission of air pollutants in Asian, European and American towns, due to reduced traffic and industrial production (Watts and Kommenda 2020; Wright 2020; Natter 2020). Wild animals were also attracted to previously busy city streets and marine routes in Europe, South America and other parts of the world. (BBC 2020a; Dias 2020).

However, negative side effects of the COVID-19 have been much more prominent. Besides the high death rates worldwide, the UNEP 2016 Report and the World Health Organization highlighted that the pandemics harm both mental health and economic development, with direct costs of more than US$ 100 billion due to measures such as social isolation and dismissions. Such a loss could evolve to several trillion dollars in case of prolonged global pandemics (WHO 2020c; UNEP 2016).

In fact, many nations worldwide have faced severe economic challenges due to lockdown to control the COVID-19 pandemics. Factories, churches, stores, shopping and entertainment centers had to remain closed and their revenue fell significantly during the social isolation period, even though some stores reported increases on online sales (Jones 2020; IGB 2020; BBC 2020b).

Global Health Academy (2019) indicated that poor and marginalized social groups are the most vulnerable to zoonotic infectious diseases due to lack of access to sanitation, healthcare services, economic vulnerability and poor educational background. Nonetheless, the whole planet and mankind itself suffers the catastrophic effects of the COVID-19 pandemics, which is a consequence of human mismanagement of natural resources and wildlife. Thus, it is a matter of ultimate importance to restore the balance between society and nature, effectively addressing the factors that contribute to the outbreak of global pandemics.

The SARS-CoV-2 origins and outbreak relate to the interaction of environmental, social and economic factors. Modern human actions promote negative changes in natural ecological patterns, leading to habitat losses and fragmentation due to pollution, deforestation, illegal trade and climate change. Such issues have also contributed to the emergence of global pandemics of infectious zoonotic diseases (UN Environment 2020a).

Therefore, a better understanding of ecological dynamics and the role of humankind in this context can potentially mitigate possibilities of future global environmental and health crises. The UN Environment (2020b) reckoned that Environmental Education may contribute to such a relevant issue, by improving quality education and environmental awareness. It is a pedagogical field that promotes the balance between mankind and nature, through interdisciplinary studies and hands-on practices that should be present in all educational levels, to generate sustainable benefits for society (Orr 1992; UN Environment 2020c).

As an interdisciplinary field, it seeks to understand the multiple connections between human existence, economic affairs and the environment. We consider that Environmental Education raises awareness on important issues such as sustainability and ecology in all educational levels, thus contributing to improve environmental management and control zoonotic diseases, benefiting public health (Badjanova; Ilisko 2015; UNEP 2016; UN Environment 2020c).

Our methodology consisted of bibliographical reviews and secondary data analysis. Using GoogleScholar (Harzing and vanderWal 2008), we selected relevant papers that analyzed the nexus between environmental degradation and the emergence of zoonotic infectious diseases, focusing mainly in the SARS-CoV-2. We also gathered studies that assessed the characteristics, different approaches and the importance of Environmental Education for reaching balance between society and nature, and we presented their main findings and contributions.

Finally, we selected the more prominent initiatives worldwide that promote Environmental Education in schools and universities during the COVID-19 pandemic. The Brazilian project *Ame o Tucunduba* was selected for a slightly deeper discussion because it was created and conducted by actual students, and aims at other students to actively share knowledge about their environmental surroundings. Besides, the authors are engaged in this project. We describe the main characteristics of each initiative and their contributions during pandemic times.

Our research confirms that Environmental Education plays a vital role in raising awareness of how the environmental system works, under influences of social and economic factors. Thus, it may prevent future outbreaks of novel global pandemics as social awareness grows and environmental degradation is discouraged.

The present paper has 4 chapters, including this Introduction. Chapter 2 discusses the mutual influences of environmental impacts and the COVID-19 pandemics, analyzing its origins as a consequence of the current environmental management practices. Chapter 3 analyzes Environmental Education as an effective tool to raise social awareness on environmental matters. Some successful international initiatives are presented and discussed, such as UN's Earth School and a Brazilian project named *Ame o Tucunduba.*

Chapter 4 presents our main conclusions, acknowledging the importance of Environmental Education to improve the relationship of humankind with nature, thus enhancing sustainability, public health and quality of life. Stimulating better practices in environmental management through improved education has the potential to build resilience on the long-term, avoiding future environmental and health threats.

2 Environmental Mismanagement and Global Health Crises

There is strong evidence that SARS-CoV-2 has zoonotic origins, as its emergence is intrinsically related to the handling and consumption of animals, most likely bats (WHO 2020a, Yang 2020). Fan et al. (2019) indicated that such animals can transmit coronaviruses both for humans and livestock, despite their important ecological role. They naturally promote pollination and pest control in nature—but its inadequate consumption and commercialization has led to a major global pandemic (WHO 2020a; Yang 2020). Bats tend to host higher concentrations of zoonotic viruses in comparison to other animals, due to their ecological characteristics: migratory patterns, high density population and metabolic protection against febrile illness (Zhody et al. 2019).

It is known that the COVID-19 pandemics started at the Huanan seafood market in Wuhan City, Hubei Province, China. There, many different species of living animals, including exotic species such as civets and bats, are confined in small cages and mishandled by the dealers under poor sanitary conditions to be sold for human consumption (Woo et al. 2006). Thus, livestock could easily scratch and bite handlers and other animals, allowing the mixing and exchange of bodily fluids between different hosts (Jandu 2020). The cultural preference for fresh food and the habit of consuming wild animals contribute to this kind of situation (Woo et al. 2006).

Indeed, human actions significantly impact the environment and animal behaviors. The mandatory social isolation and lockdown measures worldwide have promoted significant changes in urban and natural spaces. The National Aeronautics and Space Administration (NASA), the European Space Agency (ESA) and the Center for Research on Energy and Clean Air (CREA) released analyses that revealed significant reduction in nitrogen dioxide emissions in Chinese towns during the first months of social isolation. Nitrogen dioxide originates from car engines and industrial processes, and harms respiratory conditions such as asthma and others (Watts and Kommenda 2020). Carbon dioxide emissions were also reduced by around 25%, between January and February, as some public facilities remained closed, traffic was reduced and procedures for remote work were introduced (Wright 2020).

A similar trend was also identified in countries such as South Korea, Italy and the United Kingdom, as demonstrated by satellite images from CREA (Watts and Kommenda 2020). In the USA, forecasts from the Energy Department predict similar trends in decreases of carbon dioxide emissions, by around 7.5% due to social isolation measures (Natter 2020). The ease in air pollution in India has allowed the sighting of the Himalayan mountain range for the first time in decades. In fact, India's Central Pollution Control Board reported a 44% reduction in air pollution levels during the first months of lockdown measures (Picheta 2020).

Besides significant changes in pollution levels, social isolation has also affected wildlife behavior throughout the world. Wild animals have been spotted wandering in emptied streets, waterways and tourist areas. Dolphins in Istanbul are swimming and jumping much closer to the shores in the Bosphorus, as the traffic of cargo ships and passenger boats have significantly reduced. Similar conditions attracted herds of dugongs to the Hat Chao Mai National Park in Thailand, as tourism level has decreased exponentially in the region. Wild boars have been photographed foraging for food in downtown Haifa–Israel (BBC 2020a).

Even more dangerous animals have been found exploring urban areas, such as cougars and pumas in the streets of Santiago–Chile (Dapcevitch 2020). A Brazilian ounce has also been seen twice in tourist pathways in an ecotourism complex in Mato Grosso do Sul–Brazil, due to the lack of visitors (Dias 2020). Those animals look for new places to forage for food, and wander into areas they have not explored previously due to intense urban traffic and noises.

In fact, Beena and Saikumar (2019) and Lorusso et al. (2020) indicated that processes of deforestation and uncontrolled urbanization significantly reduce the habitat of wildlife, especially vectors of infectious diseases. Zhody et al. (2019) noted that the habitat fragmentation isolate specimens and accelerate mutation processes, quickly increasing the diversity of disease-causing microbes. It also increases the encounter rates of such animals with human beings, thus facilitating the spill-over of zoonotic agents (Decaro; Lorusso 2020).

The global intensification of climate change favors vector-borne diseases by proliferating vectors and promoting their occupation of new ecological niches (Decaro; Lorusso, 2020). The process known as tropicalization expands the range of tropical diseases to temperate climate regions such as Europe, as around one-third of infectious pathogens are climate-sensitive (McIntyre et al. 2017; McMahon et al. 2018). Furthermore, Wilson et al. (2010) also found that zoonotic pathogens tend to be more sensitive to climate changes than human- or animal-exclusives.

Decaro and Lorusso (2020) demonstrated that previously known coronaviruses had zoonotic origins as well. The HCoV-229E originated in bats, and was transmitted to humans through alpacas as vectors, while HCoV-OC43 commuted from rats to cattle to humans Corman et al. (2015, 2018). The HCoV-NL63 originated in bats, and the HCoV-HKU1, in rodents. Both SARS-CoV and MERS-CoV originated in bats, but the former was transmitted to humans by wild carnivores and the latter, by dromedary camels (Guarner 2020).

In fact, back in 2007, Cheng et al. (2007) analyzed the origins and emergence of the SARS-CoV in Chinese wet markets, identifying very similar conditions that led to the emergence of the SARS-CoV-2 twelve years later: overcrowded cages, poor sanitary conditions and lack of biosecurity measures that allowed the transmission of the virus from animal hosts to humans. Their research concluded that bats were natural reservoirs of SARS-CoV-like viruses, which were prone to genetic mutations. Nonetheless, many people in Southern China still had the habit of consuming exotic mammals such as bats. Thus, Cheng et al. (2007) identified major dangers in the future, warning to the emergence of novel epidemics if habits and environmental impacts did not change.

Olival et al. (2017) also emphasized that determining the significant factors for effective cross-species transmission of zoonotic diseases was crucial for successful pandemic surveillance programs. It depends upon investments on science, technology, research and development. But it all starts with Environmental Education, which may raise awareness on such important matters from very early school years. Therefore, promoting the importance of Environmental Education is essential for Science to be valued and environmental matters to be better managed (Athman; Monroe 2001).

3 Environmental Education: The Path to a Healthier Planet

3.1 Characteristics and Approaches

The intense and long-lasting industrial production and urbanization processes made society realize, in the twentieth century, that the historical damages to nature and society needed to be repaired by the adoption of new perspectives and changes in social practices (Sterling 2001). The United Nations Conference on Environment and Development, held in Rio de Janeiro in 1992, was crucial to strengthen international commitment with environmental matters. Eventually, it led to the launching of the Sustainable Development Goals (SDGs) in 2015 (UN Environment 2020c).

The paradigm of Environmental Education emerged as a driver for sustainability, proposing different approaches to ecological issues (Layrargues; Lima 2011).

Generally speaking, Environmental Education can be currently understood as the processes by which the individual and the community build social values, knowledge, skills, attitudes and competencies focused on the conservation of the environment, just as the well-being of people and necessity to a healthy quality of life and sustainability (MMA 2020). It goes beyond mere ecological topics, as it encompasses political, social, cultural and economic issues which influence and are influenced by environmental matters (Spironello et al. 2012).

Stapp (1969) firstly defined it as a pedagogical approach, in which mankind should be considered as an indissoluble part of a system that includes culture and the biophysical environment. Through cultural practices, mankind changes nature to improve social welfare and, consequently, shapes the whole system. Environmental Education aims at improving such a system, motivating and providing knowledge, techniques and tools for mankind to identify problems associated with the environment and society and to learn how to solve them. Leal Filho and Pace (2016) highlight that the Tbilisi Conference, held in 1977, expanded the discussion and established basic principles of Environmental Education, encouraging the adoption of the concept worldwide.

Indeed, some of its features needed to be expanded and adapted to better fit and work in different educational contexts in a changing world (Tilbury 1995). Athman and Monroe (2001) discussed how some valuable procedures significantly improved quality environmental education in schools worldwide, such as holistic and constructivist pedagogical approaches.

The holistic methodology demands interdisciplinarity, as environmental issues involve more than single and isolated subjects such as pollution or specific natural resources. A broader set of factors and their interactions must be considered when dealing with environmental matters: historical, economic, political and cultural processes influence each other and the whole environmental system. Tilbury (1995) highlights that the purpose of a holistic approach is not to replace any of such subjects, but rather to combine their potentials in an effective pedagogical tool for improving environmental education.

On the other hand, Constructivism implies moving from a traditional passive learning experience to an active one, with hands-on experiments and actual engagement of the pupils in the learning process (Lerman 1989; Klein and Merritt 1994). In this context, real-life environmental issues are presented to the students and they interact with each other, discussing its relevance in their lives and looking for solutions, while the teacher acts as a mediator of the interactions, assessing and communicating their progress (Klein and Merritt 1994). This process emphasizes the crucial role of individual actions on the collectivity, as each student is actively engaged in the construction of environmental knowledge.

Considering the crucial role of human beings in the ecological system, it is clear the holistic and constructivist approaches significantly improve quality environmental education and ultimately contribute to reduce the impacts of human activities on nature and society. After all, the main goal of Environmental Education is to disseminate knowledge about the environment, stimulate and naturalize positive attitudes towards it, promote competency in citizen action skills regarding the environment and empower people for self-management when dealing with nature (Athman; Monroe 2001).

3.2 *Environmental Education in Schools Worldwide*

It is clear that ecological imbalance has deleterious effects not just in nature itself but in the whole system, harming society in unpredictable manners. Recent data from the non-governmental organization World Wide Fund (WWF 2018) reveal that the human consumption of natural resources exceeded the capacity to renew the biosphere by 42.5%, what demonstrates how irresponsible the current consumption patterns are. The COVID-19 is another dreadful consequence of the mismanagement of natural resources and the poor knowledge of ecological cycles.

The harms in natural habitats favored the dissemination of the novel coronavirus among wildlife, which was consumed and led to contaminations. This serious situation clearly illustrates the urgency of integrating and promoting proper knowledge of the interactions between ecological dynamics and mankind, as health crises tend to become increasingly more common due to zoonotic diseases that are stimulated by environmental mismanagement (Cheng et al. 2007; Zhody et al. 2019).

Therefore, it only gets clearer the importance of developing Environmental Education as a formal subject in all educational levels, strongly based on multidisciplinary and multicultural approaches to achieve sustainable results globally (Orr 1992). Athman and Monroe (2011) noticed that the conduction of environmental studies in schools encouraged students to adopt more conscient and responsible attitudes towards the environment.

In fact, Disinger (1985), Svalfors (2017) and Zsóka et al. (2013) discussed some successful initiatives that attempted at strengthening the importance of environmental education in the school curriculum. Disinger (1985) identified that factors such as social class, age and sex significantly influence students' perceptions of the environment and the outcomes of Environmental Education. Thus, properly managing such variables, based on a Constructivist approach, is crucial to adequately shape student's attitudes and values towards the environment as each student has peculiar background experiences and personal histories that influence how the environment is perceived.

The importance of the holistic approach was stressed by Svalfors (2017), who investigated the curriculum and syllabuses of Swedish upper secondary schools. Even though Swedish schools tend to value the three dimensions of sustainable development (social, economic and environmental), there was a much stronger emphasis on the ecological aspects of Environment Education. The factors are discussed in classes, but such aspects do not play central roles in the national curriculum and syllabuses.

Layrargues and Lima (2011) analyzed the characteristics of Environmental Education in Brazil and identified a similar trend. At first, such subject was under a strong conservationist influence due to its focus on ecology and the vision of mankind merely as guilty and victim of the environmental crises. In the 1990's, researchers from Humanities and Social Sciences increasingly engaged in environmental matters and it reflected in education. Although theoretical studies include social and economic issues, there is still a strong emphasis on Ecology to the detriment of other factors, in practical approaches.

In such a context, Badjanova and Ilisko (2015) and Zygmunt (2016) highlight that each teacher should conduct their Environmental Education classes based on Holistic approaches, to promote social-cultural competence which is absent in the curriculum and syllabuses. Thus, stressing the essential role of economy and society in environmental studies is crucial to understand Environmental Education as an actual social phenomenon, promoting an effective understanding of people's influences in environmental processes. Studies in the United Kingdom, Germany, Romania, India and Mexico identified similar trends that demonstrate the hegemony of ecological discourses to the detriment of social and economic dimensions in the school curriculum (Chatzifotiou 2006; Bagoly-Simó 2014; Iyengar and Bajaj 2011).

Nevertheless, both the Constructivist and Holistic approach to Environmental Education were found to be very relevant in Hungary. Zsóka et al. (2013) aimed at characterizing high school and university students' engagement with environmental issues and their level of knowledge about the subject, identifying which factors may encourage eco-friendly behaviors. Besides the complexity of the matter and the difficulty in measuring a causal relationship, it was identified that Environmental Education significantly shaped how high school and university students perceived their impacts on the environment through daily activities. Thus, it promoted positive changes on consumption-related lifestyles and other behaviors.

But Zsóka et al. (2013) also noticed that merely theoretical classes were not enough to stimulate pro-environmental attitudes among students who were not interested previously in ecological issues. Students' actual engagement with the matter tended to be intensified by cognitive and affective vectors involved in environmental lessons. It means that teachers must go beyond environmental knowledge, basing their classes on relatable real-life situations in which familiar social, economic and environmental factors intertwin, in order to reach wider audiences. Such perspective is also found in the strategies indicated by the United Nations Economic Commission for Europe (UNECE) regarding Education for Sustainable Development (UNECE 2005).

The COVID-19 pandemics is a great example of a real-life situation that could inspire Environmental Education classes in the near future, due to the complex ecological factors involved in the outbreak of the SARS-CoV-2 and the need to avoid similar situations (Woo et al. 2006; Yang 2020; UN Environment 2020a). However, most schools remained closed worldwide during the COVID-19 pandemics, deeply affecting people's behaviors, attitudes and feelings. Social distancing, lockdowns and other restrictions are somehow essential measures to control the transmission of the novel coronavirus during pandemic times. The lack of activities such as strolling in the parks or gathering in social events can be harmful to health and even cause anxiety disorders in young people and adults (WHO 2020b, c). During such times, Environmental Education must expand and reach wider audiences beyond schools and universities.

3.3 Succesful Initiatives During Pandemic Times

Raikumar (2020) emphasizes that the current major health crisis is also a crisis for education. The current changes are both triggers and motivations to promote Environmental Education beyond school boundaries, reaching homes, daily activities and leisure times. In fact, Leal Filho and Pace (2016) stress that Environmental Education should not be merely an academic subject. It should rather be a daily participative process, applying the knowledge and experiences acquired during academic formation into professional life and everyday attitudes. Therefore, some successful programs involving Environmental Education during pandemic times can illustrate the importance of engaging citizens with the subject through theory and hands-on experiences.

The *Center for Open Science* (COS) is a great example. Located in Virginia-USA, its mission is to increase openness, integrity, and reproducibility of research, valuing transparency and accessibility. COS aims to make visible and accessible the content and process of all research to facilitate reuse and self-correcting to accelerate discovery. Its ultimate goal is to foster the sustainable thinking in a global range, to benefit all potential contributors and consumers (COS 2020).

Another well succeeded project is the United Nation's *Earth School*, an online platform created and sponsored by UN Education Department that provides free, high-quality educational content to help students, parents and teachers in isolation around the world. It provides videos, texts and activities that cover various topics such as animals, climate change and underwater farms, to help young people and children to understand the environment and their role in the ecosystem. The materials are offered in ten languages, and it constitutes the most significant online learning initiative ever created in the educational-sustainable field (UN Environment 2020c).

The Cornell University, through its high-tech ornithology lab, launched the *Citizen Science Initiative*, conducting some smaller projects to increase social awareness regarding ecology in the USA. One of its most popular initiatives during the COVID-19 outbreak and social isolation times promotes the preservation of birds' species by encouraging bird watching and photographing. Participants watch the birds around their homes, create lists of the observed species and take pictures which are shared in online platforms. It can be both a way for people to connect with the outside world and, simultaneously, contribute with Science through data collection (Cornell Lab of Ornithology 2020).

Such data contribute to ecological studies involving birds in the country, as scientists use it to investigate how birds are affected by habitat loss, pollution, disease, climate, and other environmental changes. It is a great opportunity for bored citizens in social isolation to engage in fun activities that impact positively the nature. Thus, the projects work as a powerful tool that takes Environmental Education to people's homes, increasing awareness of ecological matters through hands-on activities. This initiative could be implemented in other countries as well, encouraging people's engagement with Science (Cornell Lab of Ornithology 2020).

In the Brazilian Amazon region, the initiative *Ame o Tucunduba* (Love Tucunduba) was named after Tucunduba, the second largest river basin in the city of Belém (Pará, Brazil). The river is located in low-income region, where families face challenges regarding education and quality of life (Pereira and Vasconcelos Sobrinho 2020). The initiative aims at promoting environmental education among children and young people, so they can act as multiplicators of that knowledge in the future, improving local environmental conditions in a sustainable manner (Brazil Foundation 2020).

The inhabitants' used to dispose household waste in the river frequently, so it was heavily polluted. Therefore, some college students who lived in the area joined forces to create the *Ame o Tucunduba* and teach children and young people about environmental issues, focusing mainly on water and ecology, and the important role of the Tucunduba river for local environment and quality of life (Ribeiro 2018).

This initiative is peculiar because it was launched by actual students and aimed at other students and people from local communities. Its founders studied different undergraduate degrees, such as Geography, Oceanography and Architecture. One of their projects took students to sail down the river while having classes about biodiversity, ecology and geography to raise awareness on the important role of rivers for local environment and quality of life, beyond locomotion, washing clothes and consuming. Other projects include workshops about recycling, sanitation and social entrepreneurship (Ribeiro 2018). Due to the social isolation measures during the COVID-19 pandemics, the projects have been conducted using social medias.

In a global scale, UNESCO has recently launched what is considered to be the most important and daring program ever implemented worldwide during pandemic times. It is named Global Education Coalition. Due to social isolation measures and lockdowns worldwide, around 87% of the world's student population has been negatively affected as schools closed (around 1.5 billion students in 165 countries). Therefore, this initiative aims at supporting countries to improve their distance learning techniques and infrastructure, focusing especially on the most vulnerable children and youth. The coalition mobilize different forces to gather financial and technological resources to achieve its goals. Many different partners from both public and private sectors collaborate with the Global Education Coalition, including famous companies such as Facebook, KPMG and Google, and philanthropic organizations (UNESCO 2020).

All the aforementioned initiatives reflect the undeniable crucial role of Environmental Education in shaping the human interferences on the ecosystem, as society, economics and nature are essential parts of the same system. The greatest challenge for environmental education is to effectively motivate students to consciously adopt pro-environmental attitudes and commit to eco-friendly lifestyles, for the sake of nature and society. Fortunately, some countries such as Hungary have realized the importance of valuing Environmental Education, and the many successful initiatives worldwide during pandemic times contribute in the process, even in a period when many people are isolated at home.

4 Conclusion

This paper investigated how the COVID-19 pandemics is intrinsically related to the human mismanagement of environment and wildlife, and what is the role of Environmental Education in this context. The bibliographical review demonstrated how complex the COVID-19 pandemic is, because its origins and consequences involve social practices and economic factors, intertwined with ecological elements. It was also demonstrated that Environmental Education is a good tool to raise awareness of ecological issues, because it promotes a better understanding of the mutual influences between social, economic and environmental factors that affect humankind and nature.

Environmental degradation destroys habitats and negatively affects the behavior of wild animals, which are also inadvertently consumed by humans due to low-cost and cultural influences. Previous studies analyzed the historical emergence of previous coronaviruses and indicated high probabilities of new pandemics in case patterns of environmental mismanagement did not change (Cheng et al. 2007). However, environmental degradation and animal mishandling continued without effective mitigation measures and the SARS-CoV-2 emerged and spread throughout the world.

Thus, strengthening Environmental Education should raise awareness of important ecological cycles and guide human actions towards healthier and more sustainable practices. It shall also improve the management of the factors that led to the emergence of the COVID-19 pandemic, and thus may avoid future global health crises.

Effective Environmental Education requires a multidimensional approach to properly address modern day issues. However, we found out that the ecological dimension is more emphasized than others in some countries, such as Sweden, Brazil and United Kingdom. On the other hand, Hungary places equal emphasis on the three key-aspects of the environmental system: ecological, economic and social, valuing hands-on experiences that involve the students holistically.

We have also identified different initiatives worldwide that promote Environmental Education outside of schools and universities and enroll people in remote environmental learning processes, what is crucial to control the spread of the COVID-19. The COS, UN's Earth School, the Cornell's Citizen Science and the Brazilian project *Ame o Tucunduba* take environmental knowledge to families' homes and raise awareness of the importance of science and the role played by humankind the global ecosystem. UNESCO's Global Education Coalition mobilizes different agents worldwide to cooperate in improving global access to distance learning tools, to gather funds to bring technology and infrastructure to vulnerable communities.

But our study is definitely not exhaustive. No primary data was collected, as we only reviewed secondary data from other studies in a bibliographical review involving Environmental Education and related initiatives. Future studies should assess students, teachers and families' actual perceptions of the subject and its influences on their behaviors and opinions regarding ecological and social issues.

Additionally, prospective research should investigate the COVID-19 impacts on education during social isolation times. Initiatives such as UN's Earth School and UNESCO's Global Education Coalition should be analyzed in the future to assess their impacts on the long term. We also encourage further research to assess the implementation of Environmental Education in other countries and different educational levels, as it may reveal distinct rationalities.

We concluded that the greatest challenge for Environmental Education is to use adequate approaches that effectively motivate and engage students with ecological issues. Theoretical lessons must be combined with constructivist and holistic approaches, through hands-on experiences and real-life case studies. The COVID-19 pandemic, due to its characteristics and emergence, is a great real-life example to illustrate the dangers of the current human actions on natural resources and its consequences for global health.

References

Athman JA, Monroe MC (2001) Elements of effective environmental education programs. In: Fedler AJ (ed) Defining best practices in boating, fishing, and stewardship education. recreational boating & fishing foundation. Alexandria, VA

Badjanova J, Ilisko D (2015) Holistic approach as viewed by the basic school teachers in Latvia. Discourse Commun Sustain Educ 6:132–140. https://doi.org/10.1515/dcse-2015-0010

Bagoly-Simó P (2014) Tracing sustainability: Education for sustainable development in the lower secondary geography curricula of Germany, Romania, and Mexico. Int Res Geogr Environ Educ 23(2):126–141. https://doi.org/10.1080/10382046.2014.908525

BBC – British Broadcast Corporation (2020a) Coronavirus: Retail sales crash in April as lockdown hits shops. Available in: https://www.bbc.com/news/business-52766856. Access: 07/18/2020

BBC – British Broadcast Corporation (2020b) Coronavirus: wild animals enjoy freedom of a quieter world. Available in: https://www.bbc.com/news/world-52459487. Access: 07/18/2020

Beena V, Saikumar G (2019) Emerging horizon for bat borne viral zoonoses. Virus Dis 30(321):328. https://doi.org/10.1007/s13337-019-00548-z

Brazil Foundation (2020) Ame o Tucunduba. Available in: https://brazilfoundation.org/project/ame-o-tucunduba/?lang=pt-br. Access: 07/18/2020

Chatzifotiou A (2006) Environmental education, national curriculum and primary school teachers. Findings of a research study in England and possible implications upon education for sustainable development. Curriculum J 17(4):367–381. https://doi.org/10.1080/09585170601072478

Chen Y, Chen L, Deng Q et al (2020) The presence of SARS-CoV-2 RNA in feces of COVID-19 patients. J Med Virol. 92(7):833–840. https://doi.org/10.1002/jmv.25825

Cheng VCC, Lau SKP, Woo PCY, Yuen KY (2007) Severe acute respiratory syndrome coronavirus as an agent of emerging and reemerging infection. Clin Microbiol Rev 2007:660–694. https://doi.org/10.1128/CMR.00023-07

Corman V, Baldwin H, Tateno A et al (2015) Evidence for an ancestral association of human coronavirus 229E with bats. J Virol 89:11858–11870. https://doi.org/10.1128/JVI.01755-15

Corman V, Muth D, Niemeyer D, Drosten C (2018) Hosts and sources of endemic human coronaviruses. Adv Virus Res 100:163–188. https://doi.org/10.1016/bs.aivir.2018.01.001

Cornell Lab Of Ornithology (2020) We believe in the power of birds to ignite discovery and inspire action. (2020) Available in: https://www.birds.cornell.edu/home. Access: 07/10/2020

COS—Center For Open Science (2020) Show your work, share your work, Advance your science (2020). Available in: https://www.cos.io/. Access: 07/18/2020

Dapcevitch M (2020) Multiple cougars seen wandering the streets of Santiago amid coronavirus lockdown. Available in: https://www.ecowatch.com/cougar-santiago-coronavirus-2645673513.html?rebelltitem=1#rebelltitem1. Access: 07/18/2020

Decaro N, Lorusso A (2020) Novel human coronavirus (SARS-CoV-2): a lesson from animal coronaviruses. Veterinary Biol 244:108693. https://doi.org/10.1016/j.vetmic.2020.108693

Dias F (2020) Sem turistas, onça se acostuma a 'desfilar' por passarela no Pantanal: duas vezes em cinco dias; VÍDEO. Available in: https://g1.globo.com/ms/mato-grosso-do-sul/noticia/2020/06/22/sem-turistas-onca-se-acostuma-a-desfilar-por-passarela-no-pantanal-duas-vezes-em-cinco-dias-video.ghtml?utm_source=facebook&utm_medium=social&utm_campaign=g1&fbclid=IwAR1okXHmyhBzC20NCVR8_iSYLN7LImZsmKZvF5rebkauFeKJVdl2nGkt2mE. Access: 05/20/2020

Disinger J (1985) Environmental education research news. Environmentalist 6(2):85–88. https://doi.org/10.1007/BF02277231

Fan Y, Zhao K, Shi Z-L, Zhou P (2019) Bat Coronaviruses in China. Viruses 11:210. https://doi.org/10.3390/v11030210

Global Health Academy (2019) ADVANZ: advocacy for neglected zoonotic diseases. Available in: https://www.ed.ac.uk/global-health/research/research-programmes/advanz. Access: 05/20/2020

Guarner J (2020) Three emerging coronaviruses in two decades. Am J Clin Pathol 153:420–421. https://doi.org/10.1093/ajcp/aqaa029

Harzing A, vanderWal R (2008) Google scholar as a new source for citation analysis. Ethics Sci Environ Polit 8:61–73. https://doi.org/10.3354/esep00076

Huang C, Wang Y, Li X, Ren L, Zhao J, Hu Y, Cao B (2020) Clinical features of patients infected with 2019 novel coronavirus in Wuhan, China. The Lancet 395(10223):497–506. https://doi.org/10.1016/s0140-6736(20)30183-5

IGB—iGaming Business (2020) Italian Covid-19 lockdown forces Snaitech shops to close. Available in: https://www.igamingbusiness.com/news/italian-covid-19-lockdown-forces-snaitech-shops-close. Access: 05/20/2020

Iyengar R, Bajaj M (2011) After the smoke clears: toward education for sustainable development in Bhopal India. Comparative Educ Rev 55(3):424–456. https://doi.org/10.1086/660680

Jandu N (2020) Human activities are responsible for viruses crossing over from bats and causing pandemics like coronavirus. Available in: https://theconversation.com/human-activities-are-responsible-for-viruses-crossing-over-from-bats-and-causing-pandemics-like-coronavirus-134226. Access: 05/20/2020

Jones E (2020) Poland in COVID-19 lockdown! PM orders bars, restaurants, shopping centres and borders closed—and cancels ALL flights. Available in: https://www.thefirstnews.com/article/poland-in-covid-19-lockdown-pm-orders-bars-restaurants-shopping-centres-and-borders-closed---and-cancels-all-flights-11142. Access: 07/10/2020

Klein E, Merritt E (1994) Environment education as a model for constructivist teaching. J Environ Educ 25:14–19. https://doi.org/10.1080/00958964.1994.9941953

Layrargues P, Lima G (2011) Mapeando as macro-tendências político-pedagógicas da educação ambiental contemporânea no Brasil. In: ENCONTRO PESQUISA EM EDUCAÇÃO AMBIENTAL, 6. Anais. Universidade de São Paulo—Campus Ribeirão Preto, Ribeirão Preto, São Paulo

Leal Filho W, Pace P (2016) Teaching education for sustainable development: implications on learning programmes at higher education. In: Leal Filho W, Pace P (eds) Teaching Education for Sustainable Development at University Level. World Sustainability Series. Springer, Cham. https://doi.org/10.1007/978-3-319-32928-4_1

Lerman S (1989) Constructivism, mathematics, and mathematics education. Educ Stud Math 20(21):1–223. https://doi.org/10.1007/BF00579463

Lorusso A, Calistri P, Petrini A, Savini G, Decaro N (2020) Novel coronavirus (SARSCoV-2) epidemic: a veterinary perspective. Vet Ital 56(1):5–10. https://doi.org/10.12834/VetIt.2173.11599.1

Mcintyre KM, Setzkorn C, Hepworth PJ, Morand S, Morse AP, Baylis M (2017) Systematic assessment of the climate sensitivity of important human and domestic animals pathogens in Europe. Sci Rep 7:7134. https://doi.org/10.1038/s41598-017-06948-9

McMahon BJ, Morand S, Gray JS (2018) Ecosystem change and zoonoses in the Anthropocene. Zoonoses Public Health 65:755–765. https://doi.org/10.1111/zph.12489

MMA—MINISTÉRIO DO MEIO AMBIENTE (2020). Conceitos de Educação Ambiental (2020). Available in: https://www.mma.gov.br/educacao-ambiental/pol%C3%ADtica-nacional-de-educa%C3%A7%C3%A3o-ambiental.html Access: 07/12/2020

Natter A (2020) U.S. Slashes Carbon Emissions Forecast as People Stay Home. Available in: https://www.bloomberg.com/news/articles/2020-04-07/u-s-slashes-carbon-emissions-forecast-as-people-stay-home. Access: 05/20/2020

Olival KJ, Hosseini PR, Zambrana-Torrelio C, Ross N, Bogich TL, Daszak P (2017) Host and viral traits predict zoonotic spillover from mammals. Nature 546:645–650. https://doi.org/10.1038/nature22975

Orr D (1992) Ecological literacy: education and the transition to a postmoderm world. State University of New York Press, Albany, New York. ISBN:1438415214

Pereira CP, Vasconcellos Sobrinho M (2020) Direito à cidade na perspectiva paroara: concepções dos agentes do sistema de justiça do estado do Pará. Rev Direito Práx, Rio de Janeiro 11(1):493–534. https://doi.org/10.1590/2179-8966/2020/48192

Picheta R (2020) Himalayas for the first time in 'decades,' as the lockdown eases air pollution. Available in: https://edition.cnn.com/travel/article/himalayas-visible-lockdown-india-scli-intl/index.html. Access: 05/10/2020

Raikumar RP (2020) COVID-19 and mental health: a review of the existing literature. Asian J Psychiatry 52:102066. https://doi.org/10.1016/j.ajp.2020.102066

Ribeiro MGC (2018) Expedição Tucunduba: Plano de comunicação mobilizadora para a reconexão com os rios da cidade. Trabalho de Conclusão de Curso (Comunicação Social)—Faculdade de Comunicação. Universidade Federal do Pará. Belém, Pará, Brasil. 98p

Spironello R, Tavares F, Silva E (2012) Educação Ambiental: Da teoria à prática, em busca da sensibilização e conscientização ambiental. Revista Geonorte 3(4):140–152. https://doi.org/10.31692/2526-7701

Stapp WB (1969) The concept of environmental education. Environ Educ 1(1):30–31. https://doi.org/10.1080/00139254.1969.10801479

Sterling S (2001) Sustainable education: re-visioning learning and change. Green Books, Bristol, UK. ISBN:1870098994

Svalfors U (2017) Education for sustainable development and multidimensional implementation. A study of implementations of sustainable development in education with the curriculum of upper secondary school in Sweden as an example. Discourse Commun Sustain Educ 8(2):114–126. https://doi.org/10.1515/dcse-2017-0020

Tilbury D (1995) Environmental education for sustainability: defining the new focus of environmental education in the 1990s. Environ Educ Res 1(2):195–212. https://doi.org/10.1080/1350462950010206

UN Environment (2020a) Coronavirus outbreak highlights need to address threats to ecosystems and wildlife. Available in: https://www.unenvironment.org/news-and-stories/story/coronavirus-outbreak-highlights-need-address-threats-ecosystems-and-wildlife. Access: 07/10/2020

UN Environment (2020b) Goal 4: quality education. Available in: https://www.unenvironment.org/explore-topics/sustainable-development-goals/why-do-sustainable-development-goals-matter/goal-4. Access: 07/10/2020

UN Environment (2020c) Earth School (2020). Available in: https://www.unenvironment.org/explore-topics/education-environment/what-we-do/earth-school. Access: 07/10/2020

UNECE (2005) UNECE Strategy for Education for Sustainable Development. Adopted at the High-Level Meeting of Environment and Education Ministries (Vilnius, 17–18 Mar 2005). Available in: https://www.unece.org/fileadmin/DAM/env/documents/2005/cep/ac.13/cep.ac.13.2005.3.rev.1.e.pdf. Acess: 07/10/2020

UNEP (2016) UNEP frontiers 2016 report: emerging issues of environmental concern. United Nations Environment Programme, Nairobi, Kenya. ISBN:978-92-807-3553-6

UNESCO - United Nations Educational, Scientific and Cultural Organization (2020). Available in: https://en.unesco.org/covid19/educationresponse/globalcoalition. Access: 07/10/2020

Watts J, Kommenda N (2020) Coronavirus pandemic leading to huge drop in air pollution. Available in: https://www.theguardian.com/environment/2020/mar/23/coronavirus-pandemic-leading-to-huge-drop-in-air-pollution. Access: 05/20/2020

WHO - World Health Organization (2020a) Coronavirus Disease 2019 (COVID-2019): Situation Report 32. Available in: https://www.who.int/docs/default-source/coronaviruse/situation-reports/20200221-sitrep-32-covid-19.pdf?sfvrsn=4802d089_2. Access: 04/05/2020

WHO – World Health Organization (2020b) Coronavirus disease (COVID-19) pandemic. Available in: https://www.who.int/emergencies/diseases/novel-coronavirus-2019. Access: 05/20/2020

WHO—World Health Organization (2020c) Looking after our mental health. Available in: https://www.who.int/campaigns/connecting-the-world-to-combat-coronavirus/healthyathome/healthyathome-mental-health. Access: 05/20/2020

Wilson N, Lush D, Baker MG (2010) Meteorological and climate change themes at the 2010 international conference on emerging infectious diseases. Eurosurveillance 15(30):1471–1481. PMID: 20684812

Woo PC, Lau SK, Yuen K (2006) Infectious diseases emerging from Chinese wet-markets: zoonotic origins of severe respiratory viral infections. Curr Opin Infect Dis 19(5):401–407. https://doi.org/10.1097/01.qco.0000244043.08264.fc

World Wide Fund – WWF (2018) Living Planet Report 2018 (2018) Available in: https://www.worldwildlife.org/pages/living-planet-report-2018. Access: 07/102020

Wright R (2020) There's an unlikely beneficiary of coronavirus: the planet. Available in: https://edition.cnn.com/2020/03/16/asia/china-pollution-coronavirus-hnk-intl/index.html. Access: 05/20/2020

Yang L (2020) Don't blame bats for COVID-19. Blame Humans. Available in: https://earth.org/bats-importance-dont-blame-bats-for-coronavirus-blame-humans/. Access: 05/20/2020

Yeo C, Kaushal S, Yeo D (2020) Enteric involvement of coronaviruses: is faecal–oral transmission of SARS-CoV-2 possible? The Lancet Gastroenterol Hepatol 5(4):335–337. https://doi.org/10.1016/s2468-1253(20)30048-0

Zhody S, Schwartz T, Oaks J (2019) The coevolution effect as a driver of spillover. Trends Parasitol 35(6):399–408. https://doi.org/10.1016/j.pt.2019.03.010

Zsóka A, Szerényi Z, Széchy A, Kocsis T (2013) Greening due to environmental education? Environmental knowledge, attitudes, consumer behavior and everyday pro-environmental activities of Hungarian high school and university students. J Clean Prod 48(2013):126–138. https://doi.org/10.1016/j.jclepro.2012.11.030

Zygmunt T (2016) Language education for sustainable development. Discourse Commun Sustain Educ 7(1):112–124. https://doi.org/10.1515/dcse-2016-0008

Thiago Leite Cruz has Bachelor's degrees in International Relations and Biology from Universidade da Amazônia and Universidade Federal do Pará — Brazil, and a Professional Master's degree in Sustainable Development. He has a postgraduate certificate in Business Management (London School of Business and Finance - UK). Nowadays, he is an Independent Researcher on Sustainability and Corporate Social Responsibility. He is also an advocate for Sustainability and Human Rights in United Nations conferences.

Pedro Israel Mota Pinto studies Geography at Universidade do Estado do Pará - Brazil. He teaches children and young people in developing neighborhoods of the municipality of Belém - Pará - Brazil, and engages in social works to improve local quality of living.

Anderson Hélio Juwer Ferreira has a Bachelor's degree in International Relations from Universidade da Amazônia - Brazil. He currently works at Cogna Educação, the largest private educational company in the world.

Relationships Matter. New Paths for Tourism Beyond COVID-19 Pandemic. An Exploratory Research from Italy

Concetta Ferrara, Sabrina Tomasi, Chiara Aleffi, Annapia Ferrara, Giovanna Bertella, Gigliola Paviotti, and Alessio Cavicchi

Abstract This chapter aims to understand how and to what extent relational tourism, heavily related to the direct contact among guests, hosts and local communities, can be pursued in the post-COVID-19 scenario and which role could be played by new technologies. According to the multidimensional model of relational tourism (Ruggieri 2007) and the network relationality framework (Marques and Gondim Matos 2020), this research analyses the experience of an online gastronomic "relay race", the *Staffetta della Cucina Ciocheciò,* organised during the COVID-19 lock-down. The role played by technology in maintaining existing relations, creating new ones and promoting relational tourism has been analysed through semi-structured interviews and a questionnaire.

Keywords Relational tourism · New technologies · Post COVID-19 recovery · Local food · Place branding

---

C. Ferrara (✉) · S. Tomasi · C. Aleffi · A. Ferrara · G. Paviotti · A. Cavicchi
Department of Education, Cultural Heritage and Tourism, University of Macerata, 1, P.le Bertelli, 62100 Macerata, Italy
e-mail: c.ferrara3@unimc.it

S. Tomasi
e-mail: s.tomasi@unimc.it

C. Aleffi
e-mail: c.aleffi@unimc.it

A. Ferrara
e-mail: a.ferrara5@unimc.it

G. Paviotti
e-mail: gigliola.paviotti@unimc.it

A. Cavicchi
e-mail: alessio.cavicchi@unimc.it

G. Bertella
School of Business and Economics, UiT The Arctic University of Norway, NO-9037 Tromsø, Norway
e-mail: giovanna.bertella@uit.no

W. Leal Filho (ed.), *COVID-19: Paving the Way for a More Sustainable World*,
World Sustainability Series, https://doi.org/10.1007/978-3-030-69284-1_18

1 Introduction

In the last decade, the tourism demand for more sustainable experience-based authentic interactions with locals (Beverland and Farrelly 2010; Pine and Gilmore 2011; Paulauskaite et al. 2017) has increased. Recent tourism trends show that tourists and travellers, when visiting a destination, are increasingly looking for unique and once-in-a-lifetime experiences and choosing to become more immersed in the daily local life (Booking.com 2019; Mittiga et al. 2019). In this perspective, relational tourism, which puts the emphasis on personal relationships, exchanges, individualised and unique experiences has become popular as a research topic (Purpura et al. 2007; Bertella et al. 2018; Kastenholz et al. 2020; Lin and Fu 2020; Marques and Gondim Matos 2020). The outbreak of COVID-19 pandemic brought the tourism industry to a standstill (ILO 2020; UNWTO 2020; WTTC 2020), changing tourists' behaviours and habits (Del Chiappa 2020) and compromising the social and relational nature of tourism (Higgins-Desbiolles 2020; Qiu et al. 2020). This situation stimulated scholars and researchers to investigate how this industry will recover after COVID-19 and how can be sustainable in a dramatically changed world (Chang et al. 2020; Jamal and Budke 2020; Lapointe 2020; Zenker and Kock 2020). In this context, technology and the relations created by web resources (Gretzel et al. 2020; Marques and Gondim Matos 2020), played a central role in building or maintaining relationships in tourism.

This study aims to understand if relational tourism can be pursued in the post-COVID-19 tourism recovery and how and to what extent new technologies can contribute to promoting authentic tourism experiences during and after a crisis. To this aim, an exploratory case study from Italy presents the experience of the *Staffetta della Cucina Ciocheciò*, ideated during the first national COVID-19 lockdown (March–May 2020) and consisting in an online "relay race" in which participants were asked to post, in a private Facebook group, easy-to-make recipes. This research analyses the role played by technology within the *Staffetta* in maintaining existing relations, creating new ones and promoting a relational tourism destination, through local food and traditions. These aspects have been analysed through qualitative and quantitative methods: a semi-structured interview has been conducted with the 5 organisers, and 71 online questionnaires addressed to participants were collected.

The chapter is structured as follows: after a literature review on relational tourism, the role of gastronomy and local food for place branding and tourism is stressed, considering the opportunities related to web and technologies in supporting relationality in the post-COVID-19 scenario; then the methodology is presented, and results are discussed, by paying attention to the potential integration between the relational tourism model and the network relationality framework. Conclusions highlight that relationality in tourism can play a relevant role also in the context of a crisis thanks to the technology that, far from being a substitute for reality, can facilitate face-to-face interactions and stimulate the visit to places known only virtually.

2 Literature Review

2.1 Relational Tourism

The tourism sector has changed over the years, by producing new forms of tourism and hospitality (Purpura et al. 2007): tourists are increasingly looking for immersive experiences in the culture and traditions of places (Richards 2013b). Travel is perceived as a source of knowledge; tourists want to live like locals (Richards, 2013a; Paulauskaite et al. 2017) and discover the territory, by also preferring less known destinations and inland areas with rich folklore and local culture. In this perspective, the generation of relationships with the place becomes relevant: exchanges and personal relationships characterise the uniqueness and individuality of the tourist's experience (Kastenholz et al. 2020; Lin and Fu, 2020; Marques and Gondim Matos, 2020). Repeated visits are strongly influenced by the tourist's satisfaction with relationality during the tourism experience (Valls et al. 2004).

The concept of relational tourism refers to a relationship established between those who spend time in a destination as tourists and those who live there, as locals. This relationship is perceived as a value and an element of differentiation which takes place spontaneously (Purpura et al. 2007; Bertella et al. 2018; Kastenholz et al. 2020). As argued by Ruggieri (2007, p.54), relational tourism requires the subsistence of at least four conditions:

1. a territory with relational characteristics, such as attractions related to the territory characterised by reduced size, if compared to mass tourism destinations (e.g. small villages, farms, local handicraft companies, etc.);
2. a supply system with elements and conditions facilitating these forms of tourism and hospitality (e.g. agritourism's, historical residences, historical houses, etc.);
3. a type of traveller inclined to interactions and exchanges with the main players in the relational tourism supply chain (e.g., services providers', local community, etc.);
4. interaction, represented by that set of relationships and exchanges that take place between the main players in the relational tourism supply chain (e.g. entrepreneurs, local community, tourist information offices staff, other tourists, etc.).

The author provided a multidimensional model (Fig. 1) to define relational tourism. He describes it as a combination of relationships in which hosts approach tourists in a friendly way, in order to let them discover the beauty and the peculiarity of their own historical, artistic, folkloristic, culinary and human heritage. The tourist becomes a protagonist, a generator of value, completing the tourism offer itself (Ruisi 2004). The productive tissue is also integrated into this system (Purpura et al. 2007).

Advantages related to this relational approach are many. First, this type of tourism can represent a mechanism able to avoid a serial reproduction and to focus instead on the authenticity and uniqueness of the place (Richards and Wilson 2006). Secondly, relational tourism represents a stimulus for the local economy, especially for small

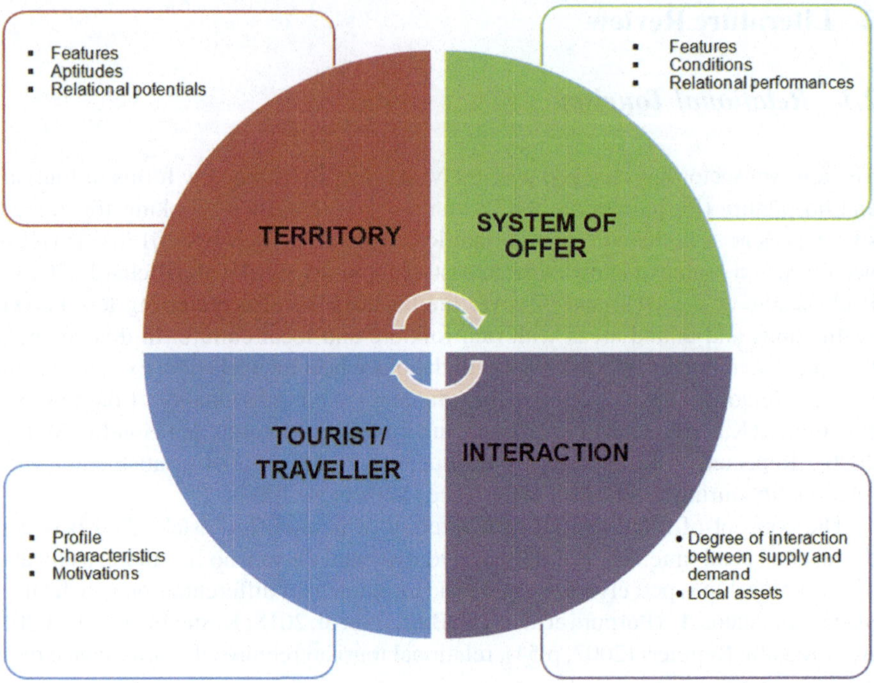

Fig. 1 A multidimensional model for relational tourism. *Source* our elaboration on Ruggieri (2007)

and medium enterprises starting from the agri-food, productive-craft and historical-cultural sectors (Naselli 2005). Thirdly, this kind of approach, encouraging community engagement (Okazaki 2008) and involving the daily lifestyle of the local community (Purpura et al. 2007), can increase residents' awareness about local culture and promote positive relationships between tourists and locals (Sherlock 2001; Teye et al. 2002; Bimonte and Punzo 2016; Lee and Jan 2019).

2.2 *The Role of Gastronomy and Local Food for Place Branding and Tourism*

Gastronomy and local food play an important role in the development of relationships within the tourism sector. Being an expression of local culture and reflecting regional identities and values (Hjalager and Richards 2003; Gyimóthy and Mykletun 2009; Rinaldi 2017), they can differentiate a place from another, thus increasing its attractiveness and competitiveness. This differentiation builds on the idea of an identity-based sense of place, also represented by the bundle of products and services that make up a tourist experience (Harrington and Ottenbacher 2010, p. 17). In this sense, the UNWTO *Global Report on Food Tourism* (2012) emphasises the need for food tourism for its potential to convert food and gastronomy, as heritage, into elements of tourism attraction.

According to Richards (2012, p. 19), food can provide the development of tourism experiences in many ways:

- *linking culture and tourism*: local food could act as a bridge to bring tourists and locals together in a shared cultural experience;
- *developing the meal experience*: meals based on local food represent a central part of the tourism experience, which can be memorable and meaningful;
- *producing distinctive foods*: local foods can act as distinctive elements for place branding and in the marketplace;
- *developing the critical infrastructure for food production and consumption*: local food can stimulate networking among many actors (e.g. producers, chefs, critics, other culinary trendsetters, journalists, bloggers, etc.);
- *supporting local culture*: food experiences can provide the cultural capital necessary to sustain the development of local culture.

In this context, consumers progressively ask to be involved in the production and preparation of food during their tourism experiences. This approach embraces the concept of creative tourism, which includes participation in food experiences and knowledge of food and gastronomy (Richards 2011). Tourism networks can stimulate the establishment of relationships between food producers and tourists, thus giving value to regional products (Rinaldi 2017). It can happen by transmitting the local know-how to tourists and sharing it with them (Bessière 1998): strategic tools can make link quality, diversity and uniqueness of local food products and dishes emerge and link it to the place to support both the image and the brand of a destination (Rinaldi 2017, p. 14).

2.3 The Role of the Web (Network) to Support Relationality in Tourism: Network Relationality

Recently, the reduced spatial distance, the increased physical mobility and virtual contacts due to the extensive use of the internet have strongly influenced the provision of tourism services (Marques and Gondim Matos 2020) and especially the way relationships take place. Here, the concept of hospitality is relevant and characterised by a feeling of empathy between hosts and guests (Bialski 2012). In this relationship, technology is a bridge to interactions (Bawens 2010) and a facilitator of new forms of sociality (Marques and Gondim Matos 2020).

To explain social changes related to technological advancements, Wittel (2001) used the term network sociality, considering five key elements: (1) the level of integration/disintegration with the community; (2) the intensity of social relations; (3) the contents of relations; (4) the boundaries between work and leisure; (5) the integration of technology. Molz (2014) adapted this framework to the hospitality sector, employing the concept of network hospitality, who is based on five aspects: (1) the sharing of private places with strangers; (2) the transformation of strangers into guests; (3) the random nature of guest's choices; (4) the availability of different types of temporary spaces; (5) the fact that guests behave as if they are at home.

Since the relational tourism experience, of which hospitality is an essential part, emphasises relational elements, according to Vázquez and Ruggieri (2011), in order to evoke relationality, the sense of physical encounter and personalised contact with the host community is fundamental. According to Porter (2004), in fact, the distinction between online and offline interactions does not exist, as they are only different means of interaction; moreover, considering the increasing importance of tourist's embeddedness in the local culture (Richards 2013a, 2014), both sociality and relationality are strongly connected to locality (Wittel 2001; Molz 2013, 2014). For these reasons, building on network sociality and network hospitality frameworks, Marques and Gondim Matos (2020) elaborated the network relationality model, which focuses on the relationship between host and guest and, particularly, on how the host influences the tourism experience in a local setting. This model is based on four key principles (Fig. 2):

1. *temporary belongingness*: temporary attachment to a place, providing the conditions to recreate a community and stimulating a sense of places, usually missing in virtual communities;
2. a priori*empathy*: virtual empathy between hosts and guests that starts before the direct encounter. From the hosts' perspective, it is the basis for first positive

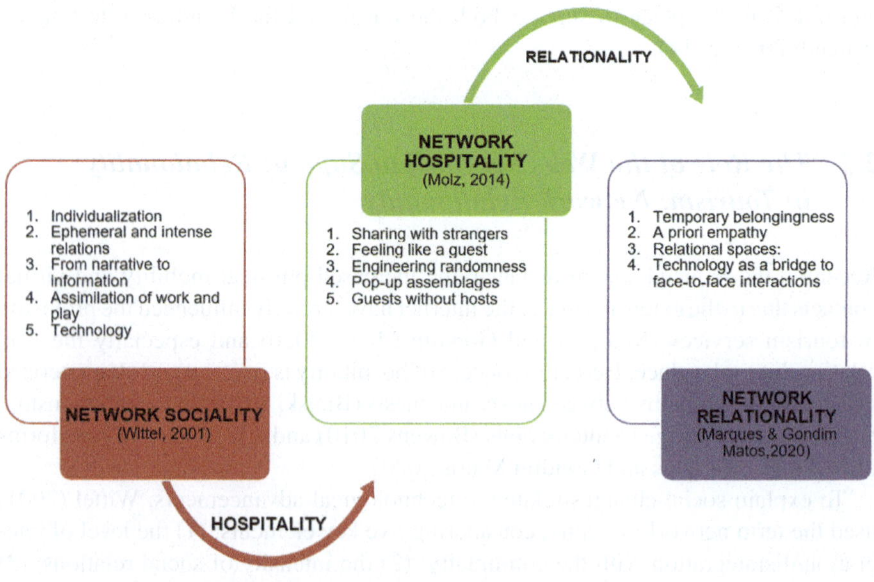

Fig. 2 Hospitality and relationality implications in network sociality. *Source* our elaboration on Wittel (2001), Molz (2014), Marques and Gondim Matos (2020)

contact, and it marks the beginning of an effective two-sided relationship (both online and offline);

3. *relational spaces*: both geographical and virtual spaces, corresponding to different moments of the host–guest relationship and representing a central node to the tourist experience;
4. *technology as a bridge to face-to-face interactions*: a set of tools from which relationships begin and take shape, but not central to the relational experience, as the emphasis is on face-to-face interactions.

2.4 The Outbreak of COVID-19 Pandemic: The Role of Technology in the Tourism Sector

The COVID-19 pandemic has caused economic, social and political damages, still not precisely identifiable. Tourism, hospitality and events sectors have been paralysed by governments efforts to control the pandemic, thus causing a collapse of the whole sector (Higgins-Desbiolles 2020).

The crisis has raised new questions, especially about the strategies for the tourism industry recovery and its evolution. There are no answers to these questions yet, but technology certainly plays a central role in all this. During this pandemic, ICT "*has been widely used, adapted and developed to address some of the pressing problems in people daily life, including work, travel, leisure, business as well as governance*" (Gretzel et al. 2020, p. 2). Technology has become a major factor in addressing specific problems (e.g. traveller screening, case and contact monitoring, online education and entertainment during isolation, to name but a few) and in fostering resilience in tourism (Gretzel et al. 2020; Hall et al. 2020).

In this perspective, the COVID-19 pandemic crisis may offer "a rare and invaluable opportunity to rethink and reset tourism toward a better pathway for the future" (Higgins-Desbiolles 2020, p. 11).

2.5 COVID-19 Challenges: Tourism Trends, Global and Local Changes

One of the most immediate economic effects of the crisis associated with COVID-19 has been the blocking of tourist flows. In response to the generalised measures of social distancing, all tourism activities, at the beginning of March, were reduced to zero (Gössling et al. 2020). At the end of March 2020, UNWTO (2020) estimated the pandemic would have caused international tourist arrivals to decline 20–30% (compared to 2019). Data from the hospitality sector confirm this estimation: for the week of the 21st of March, in comparison to the same week in 2019, in all countries, guest numbers have declined significantly, by 50% or more (STR 2020).

In Italy, travel restrictions have reset to zero an activity that in the quarter of March–May is used to live a seasonal relaunch (ISTAT 2020). Indeed, in the same period in 2019, the expenditure of foreign travellers amounted to 9.4 billion euros (Bank of Italy 2020). COVID-19 also impacted on travel behaviours and perception. A recent survey carried out by the University of Sassari, administered to 5.556 persons, investigated the changes in the way Italians would have travelled during and after the pandemic (Del Chiappa 2020). A good level of cleanliness and sanitisation of public spaces (e.g.: streets, beaches, etc.) (85%); outdoor activities (85.9%) and attractions (e.g.: archaeological sites) (74.3%) are the most important aspects considered by respondents to feel safe during their vacations. The survey also highlighted a relevant propensity to give up some relational aspects of the holiday, such as conviviality during the meals: to avoid overcrowding, 56.5% of respondents would prefer having room-serviced lunch and dinner or meals served in prearranged shifts (73.4%) The research also revealed a higher predisposition to proximity tourism: 67.8% of respondents would travel within the residence region and only the 22.5% would probably travel abroad in the next 12 months.

3 Methodology

In this chapter an exploratory case study from Italy, focused on the experience of the *Staffetta della Cucina Ciocheciò* is presented. The selected case study is relevant as researchers have long investigated the face-to-face relational approach applied to the experiential and relational tourism offer by rural local networks promoted by some of the organizers of the *Staffetta* (Bertella and Cavicchi 2015, 2017; Bertella et al. 2018). From the authors' perspective, it was interesting to understand whether this approach has changed, due to the COVID-19 and to which extent technology has helped in maintaining it during the pandemic. An already existing relationship of trust and openness facilitated the data collection and helped in shedding light on the investigated phenomenon.

The research examined the multidimensional model for relational tourism (Ruggieri 2007) and the role that food can play for the development of tourism experiences (Richards 2012), by also considering the elements of the network relationality framework (Marques and Gondim Matos 2020). The latter was applied with the scope to explore the role played by technology, locality and gastronomic traditions in maintaining existing relations and creating new ones within a relational tourism system of offer.

Both qualitative and quantitative approaches have been chosen for the investigation of the *Staffetta* case (Eisenhardt 1989; Yin 2003): semi-structured interviews were conducted to all *Staffetta*'s organizers (5) and a survey was administered to participants (July 2020). Interviews were performed online due to COVID-19 travel restrictions and lasted approximately 1 hour each. Regardingthe questionnaire, a multilanguage online form (Italian, English and French) was prepared and posted on the Facebook group of the *Staffetta*. Among the 229 members of this Facebook

Table 1 Structure of the interview and questionnaire

<table>
<tr><th></th><th>Interview (organizers)</th><th>Questionnaire (participants)</th></tr>
<tr><td rowspan="4">Cluster 1
(Richards 2012; Ruggieri 2007)</td><td>Relational Tourism</td><td>The Idea of the Staffetta</td></tr>
<tr><td>Knowledge and perception of the relational tourism model</td><td>Perception about the initiative; motivations for participation</td></tr>
<tr><td>Role of gastronomy and locality to support relational tourism</td><td rowspan="2"></td></tr>
<tr><td>The impact of COVID-19 on relational tourism</td></tr>
<tr><td rowspan="3">Cluster 2
(Ruggieri, 2007)</td><td>Premises to the Organization</td><td>Premises to the Participation</td></tr>
<tr><td>Nature of relationships between organizers</td><td>Previous experiences</td></tr>
<tr><td colspan="2">Nature of relationships between organizers and participants before the Staffetta</td></tr>
<tr><td rowspan="3">Cluster 3
(Marques and Gondim Matos 2020; Richards 2012; Ruggieri 2007)</td><td colspan="2">The StaffettaExperience</td></tr>
<tr><td colspan="2">Level of engagement, nature of relationships, interactions and information exchanged during the initiative</td></tr>
<tr><td colspan="2">Network relationality: temporary belongingness, a priori empathy, technology, relational spaces</td></tr>
<tr><td rowspan="3">Cluster 4
(exploratory)</td><td colspan="2">Tourism Implications</td></tr>
<tr><td colspan="2">The role of the Staffetta for post-COVID-19 recovery</td></tr>
<tr><td colspan="2">The role of the Staffetta in promoting destinations</td></tr>
</table>

page, 71 answered, of which 52 actively participated in the initiative; 19 only acted as audience. Both the questionnaire and the interview were organized into four main clusters of questions aimed at investigating issues showed in Table 1.

The data analysis was performed by three members of the research team: two of them, separately, operated the interviews' coding according to a common approach. A third member operated the calculations on the questionnaires' data. These were then checked by the other members.

An interviewer and a rapporteur conducted semi-structured interviews which have been recorded, transcribed and analyzed by highlighting similarities and differences in the five organizers' answers. The emerging aspects were first identified, then categorized on the basis of similarities and synthesized in sentences summarizing the main points. Then, a comparison between the analysis performed by the two was made and the data organized in themes mostly according to the research question. These have been reported in the findings. A phase of interpretation and integration, then followed (Mayan 2009).

Data resulting from the survey have been analyzed by another author to outline the main descriptive statistics.

Both the information emerging from the interviews and from the survey have been reported in the findings, following the initial structure in clusters. In this way, it has been possible to make a comparison between the organizers and participants' perceptions, by reconducting them to the same themes considered through the lens of the models chosen for the analysis. The main themes emerged are the following: relevance of the relational component; the role played by locality and gastronomy within the *Staffetta*; the importance given to the network relationality dimensions; virtual versus real social contacts in a long-term perspective.

4 Findings

4.1 A "relational" Answer to COVID-19 Crisis: The Staffetta Della Cucina Ciocheciò

The *Staffetta della Cucina Ciocheciò* (literally "The relay race of the *Ciocheciò* cooking style") is an initiative promoted during the COVID-19 lockdown (1st May–30th June 2020), using a private Facebook group. Its scope was to face the difficulties provoked by the social distancing during the lockdown, by bringing people together people from several countries, in order to improve and maintain existing relationships virtually and to create new ones.

The *Staffetta* was ideated and organised by three rural hospitality facilities' owners, a journalist, and an extra-virgin olive oil taster. Two of them come from Marche Region (Roberto Ferretti and Anna Maria Monaldi), one from Liguria (Claudio Porchia), one from Veneto (Marisa Saggiotto) and one from Japan (Yoko Moriyama). Each organiser invited participants to enter the Facebook group and eventually present a recipe and had a specific role within the organization. All the organisers are related to each other by a long-lasting friendship and by exchanges (sometimes only virtual) based on three main network experiences, who represented important conditions for the development and organization of the *Staffetta*: the *Ciocheciò* philosophy, the use of spontaneous herbs in the kitchen and relational tourism.

The *Ciocheciò* philosophy (the word is invented and stands for "what is actually available") promotes a form of hospitality in which spending time, sharing and preparing meals together is very important for creating spontaneous, positive relationships between hosts and guests visiting a territory. According to the *Ciocheciò* philosophy, preparing meals means cooking simple and easy-to-make recipes with seasonal, 0 km, healthy, typical products that also sometimes spontaneously grow in a territory. Local knowledge in the use of these ingredients is also relevant. The *Ciocheciò* concept was invented in 2008 by some of the organisers of the *Staffetta* during a conference about the use of spontaneous herbs in the kitchen, organised by the World *Wigwam* Circuit. In that occasion, the name *Ciocheciò* was first used to talk about a dinner prepared by using the available ingredients. The idea of writing a blog to tell about other similar experiences was born, and the *Circuit of the Cucina Ciocheciò* was then created.

The use of spontaneous herbs in the kitchen has been inspired by the figure of Libereso Guglielmi—botanist, and expert in recognising and using spontaneous herbs, who worked as a gardener for the family of the Italian writer Italo Calvino[1]—and by the values of the World *Wigwam* Circuit, joined by one of the organizers through her local association. The *Wigwam* Circuit is a social Promotional Association, which has its headquarter in northern Italy and manages a network of more than 300 clubs in 15 countries. Wigwam clubs aim at re-discovering, protecting and promoting local resources through tourism, leisure and didactic-educational activities (Bertella and Cavicchi 2017).

The *Staffetta* also has to do with the concept of relational tourism. Three of the five organisers are active in promoting relational tourism through their own hospitality facilities. Two of them are also engaged in a relational tourism network composed by 22 members, among which rural hospitality facilities: the *Agritur-Aso association*, established in 2007 in Marche Region (Italy), is aimed at promoting experiential, relational and community-based tourism (Bertella et al. 2018, 2019) for revitalising rural areas and guaranteeing a better quality of life for local communities. Since 2009, the association has also been organising *Le Marche in Valigia* (literally: *Le Marche in your suitcase*), aimed at promoting Marche Region abroad through cultural events and dinners, by re-creating a friendly atmosphere (Bertella and Cavicchi 2017). Due to the COVID-19 restrictions and to national hospitality policies, after the end of the lockdown not all the members of the association re-opened: the ones whose primary income depends from tourism opened their facilities to the public; two of them ideated and organised the *Staffetta*; some others took part to the initiative.

The *Staffetta* involved 229 people (the number of users registered to the Facebook group), coming from several countries. Every participant had to post the recipe according to a weekly schedule. The recipe had to follow the *Ciocheciò* principles. After presenting its recipe in the post, with a combination of text and pictures, the participant had to "pass the baton" to another participant. 77 members actively took part in the initiative by presenting a recipe (47 recipes come from 14 regions of Italy; 30 recipes from 17 different countries in the world). At the end of the *Staffetta*, all the recipes were supposed to be collected and published in the *Ciocheciò* blog.

[1] Italo Calvino (1923–1985) was an Italian journalist and novelist considered one of the most important Italian fiction writers. His best known works include the *Our Ancestors* trilogy (1952–1959), and the novels *Invisible Cities* (1972) and *If on a winter's night a traveler* (1979). (Mondello1990).

4.2 The Experience of Staffetta *According to the Organisers*

4.2.1 Relational Tourism

For all the organisers, the added value of relational tourism consists in an opportunity to live an *immersive experience* in a place through direct *involvement* and active participation in an informal and friendly atmosphere (*doing together*). The emotional component plays a decisive role: sensations and feelings contribute to strengthen the experience, generate reciprocal personal enrichment and wellbeing and, thus, create an ongoing relationship (*loyalty*). The host is a crucial figure (*active and proactive role*), whose task is to put guests at ease (*hospitality*) and to act as the first point of contact with the destination, by sharing personal contacts, information and knowledge about local culture (*pivot and territorial information point*). On the other hand, the relational tourist has an *aptitude for relationships* and direct experiences in the territory.

The time shared by host and guests is essential in the construction of the relationships ("*It is the use of the time that strengthens the relationship"*—R.F.; "*Relational tourism means dedicating time; a time that cannot be monetised"*—A.M.). Hosts dedicate time to guide guests in the discovery of the territory and of the people, acting as *facilitators* ("*The community is a testimony of the local culture, so the relational experience can be conceived within a territorial relationship*"—A.M.; *If a guest, visiting a village, meets friendly and hospitable people, he feels at home and perceives that he is living a story in a welcoming and not hostile territory"*—R.F.). Food and wine traditions support the tale about the identity of a territory (*linking culture and tourism*; *supporting local culture*), stimulate conviviality and experiential aspect of doing together (*developing the meal experience*) and also create a sensory link with the territory, and the experiences lived ("*In relational tourism the 5 senses are important: taste and smell are important to memorise the place where one has travelled"*—Y.M.). All these relationships can be maintained over time, also with the distance and beyond the tourism experience itself.

About the impact of COVID-19 pandemic, respondents agreed that it had some negative consequences as it led to the impossibility to travel and to the need to maintain distances thus compromising the direct human contact, which is a pivotal aspect of relational tourism. On the other hand, it seems to have created new stimuli for domestic tourism and enhanced the search for authentic, hands-on and outdoor experiences.

4.2.2 Premises to the Organization

Before the *Staffetta*, the organisers were linked by a long-lasting friendship, based on shared interests: the *Staffetta* was conceived as a way to keep alive these relationships, share contacts and spread the values of *Ciocheciò*. Organizers invited people with whom they share common values. Indeed, especially the ones who run rural hospitality facilities (3 out of 5 organisers) declared that they met most of the participants they involved, thanks to their relational tourism activity.

4.2.3 The *Staffetta* Experience

All the interviewees affirmed that, concerning the involved participants, this experience enriched (not changed) the nature of the existing relationships: the shared information increased personal knowledge (*sharing common values and visions*) and supported the creation of new contacts, with opportunities, in some cases, for future exchanges and real encounters (*relationships repeated in time*, both *offline and online relationships*).

In terms of contents, as defined in the *Staffetta*'s rules, most of the exchanges concerned information related to the recipes presented (*knowledge, traditions and habits*). Still, there were also moments for sharing private aspects (*intimate and personal stories and moments*), when describing a recipe, participants also decided to share anecdotal details related to their stories. In some cases, some of the participants re-proposed their version of a recipe posted by others, sometimes by re-adapting it with ingredients found locally.

The virtual *temporary belongingness* to the places was stimulated by elements of locality, communicated through the use of products and food and wine traditions in the presented recipes (*locality as identity*). The *Staffetta* also contributed to the definition of a good level of a *priori empathy*. In particular, the choice of a closed group helped to create a pleasant atmosphere of enthusiasm and reciprocal encouragement ("*This empathy emerged from the typology of comments: they were mainly messages of appreciation for the recipes presented and expressions of curiosity for the places visited*"—C.P.). However, the interviewees pointed out that it was a virtual form of empathy: real empathy can also be created by actual human contact. Technology (in particular social media) played a fundamental role both in maintaining existing relationships (*bridge to promote human interactions at a distance*) and in building new relationships (*facilitator of new forms of sociality*). However, all the interviewees reiterated that, although the technology was an essential tool, without previous interpersonal relationships the initiative would not have taken place ("*The Staffetta would not have been possible if there had not been a deep knowledge between us organisers*"—C.P.). All the interviewees agreed that, in the context of the *Staffetta*, physical space and co-presence were not necessary elements, because the conditions imposed by COVID-19 did not allow otherwise. Nevertheless, online space is perceived by all respondents as an additional element, but not as a substitute for physical space ("*Online and physical space are two complementary spaces. When this is not possible, only one space may be sufficient. But for a complete experience both spaces are needed*"—R.F.).

4.2.4 Tourism Implications

Most of the interviewees consider this initiative as a long-term solution for post-COVID-19 recovery. They are planning a second edition and published a book collecting the recipes presented in the first edition. Some interviewees, however, expressed their hope for transforming the online relationships into real ones through a live edition (*going from virtual to real*).

Concerning the role of the *Staffetta* for tourism promotion, even if the objective of the initiative was not clearly related to tourism, the organisers recognised that sharing elements of one's own culture arises interest and curiosity (Staffetta *as knowledge and sharing of mutual identities starting from the gastronomic vehicle*), encourages the creation of new contacts and friendships (Staffetta *as a creator of plots*) and stimulates the desire to deepen this knowledge through real meetings on the respective territories of the participants (Staffetta *as a bridge to face-to-face interactions*; Staffetta *as an attractor)*. This already happened to one of the organisers, which was invited by one of the participants, a new acquaintance, and travelled after the lockdown to visit her region and make direct experience of the local gastronomy. Some of the other participants were also invited to visit other regions and countries.

4.3 The Experience of Staffetta *According to Participants*

4.3.1 The Idea of the *Staffetta*

A total of 71 questionnaires was collected, of which 52 actively participated in the initiative by presenting a recipe and 19 just acted as an audience.

All participants showed enthusiasm for the *Staffetta* ("*I like the topic; original and very useful*"). Among the motivations for participation, the initiative was perceived as a way to practice a personal interest for cooking ("*I am a fan of cooking*"; "*I love both regional cuisine and cuisines from all over the world*"). It was also felt like an occasion for *sharing* and *conviviality* ("*I participated to enhance the value of relationships and for the pleasure of sharing how amazing it is to make food together"; "It seemed an amazing example of conviviality*"), and as a way (*facilitator*) to tell the territory (*locality*) and the local traditions ("*I wanted to introduce to the others my place of origin*"). Some respondents also referred to the *Staffetta* as an opportunity to experience different places and cultures ("*Since it is a good way to get to know different traditions*"; "*...far different from yours*"). Thus, *hospitality* turned into *hosting* since the participants themselves became the hosts for their territory. Moreover, for some of the respondents, it was also a way to promote a sustainable lifestyle ("*it teaches people to live with simplicity and the importance of connecting with nature...*"; "*it promotes a healthy and sustainable way of cooking"; "it encourages de-consumption*").

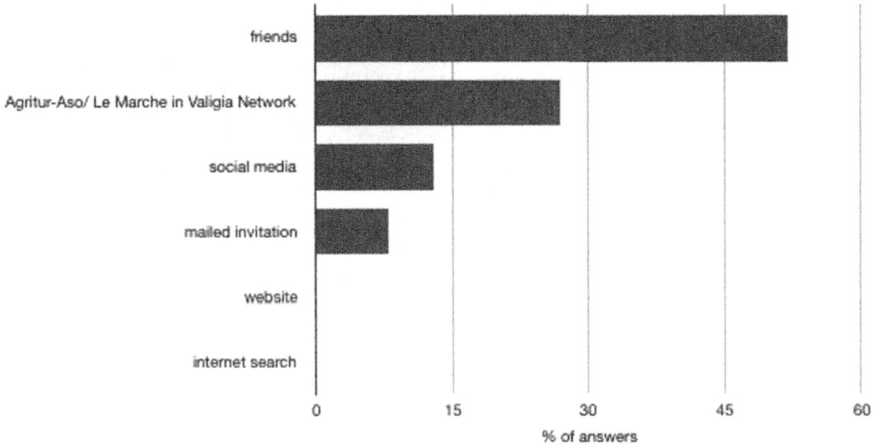

Graph 1 Q1: How did you learn about the *Staffetta* experience?

The impact of COVID-19 became a recurrent element within the answers, being perceived as a restriction to human relationships. As a result, the *Staffetta* was experienced as an occasion to bring social contacts back ("*As a reaction to the unpleasant moments of the lockdown*"; "*it was a wonderful way to connect with others, especially during the COVID-19 lockdown when none of us could meet in person*"). The element acquaintance or *friendship* was of primary importance for participants to know about the *Staffetta*: 52% claimed they got to know the event through their acquaintances, while 27% *through Agritur-Aso /Le Marche in Valigia* network (Graph 1).

4.3.2 Premises to Participation

61% of respondents had already visited the areas and the facilities involved in the initiative before the *Staffetta* and established a *friendly* (41%), and *long-lasting* (17%) *relationship*, where the main reciprocally shared information was about knowledge, traditions and habits. Mutual trust and reciprocity have been shared as well ("I *know the organisers and the quality of their work*"; "*I know who promoted the initiative and his philosophy of relational tourism*"), but also a feeling of solidarity emerged to support the cause ("*I believe in the project"; "I share this way of life*"; "*I agreed with the idea of Ciocheciò cuisine from the very beginning*").

4.3.3 The Staffetta Experience

Thanks to the *Staffetta*, 73% of respondents who actively participated by sharing a recipe, declared to have had the chance to build up new relationships (72%), mostly friendships (58%). Once again, the most shared information concerns knowledge,

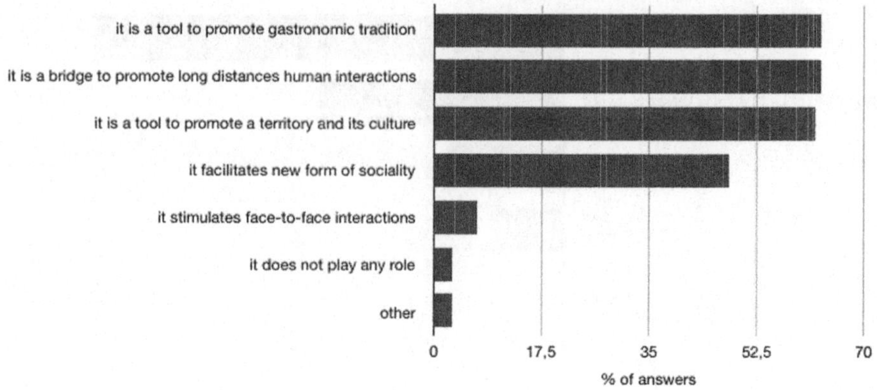

Graph 2 Q.9: Which role did technology play in nurturing relationships in the context of *Staffetta*?

traditions and habits besides tips about local lifestyle, reciprocity and mutual trust. 83% of respondents developed a *sense of belonging* to the group especially by *developing a family feeling with participants* and by *feeling part of a virtual community* where sharing tales about their territory and some daily habits.

The online format helped in consolidating *a-priori empathy* with other participants (83%) who already knew each other and further enhancing the relationships mainly through the virtual community. Technology (Graph 2) was primarily conceived as *a bridge to promote long-distance human interactions*, tool *to promote a territory and its culture* and also as *a tool to encourage gastronomic tradition.*

5 Tourism Implications

Considering the *Staffetta*'s implications, the online relational dimension of the initiative has been seen by participants as a long-term solution for the post-COVID-19 recovery (79%), since it let participants know about new places and traditions (90%). However, it might not completely replace the physical space for interactions. Indeed, as a complementary tool, it can be a way to promote it (24%) and to invite people to visit physical places in the first place (56%).

As a matter of fact, after the event, the 82% (58) of respondents would like to visit (or come back) to the areas virtually acquainted through the experience of food traditions.

6 Discussion

This research has shown that the relational component can play a decisive role in the knowledge of a place and its territorial and cultural peculiarities (Rifkin 2000), even in the contest of a crisis such as the COVID-19 pandemic.

Even if the *Staffetta* is not strictly a tourist experience, the first element that contributed to its success lies in the previous relationships among the organisers and between organisers and participants. This aspect is confirmed by the fact that more than half of the participants became aware of the initiative through their acquaintances, or through *Agritur-Aso/Le Marche in Valigia* network (Graph 1). The data about the 61% of the respondents declaring to have already visited a place or facility related to the *Agritur-Aso* network reveals the centrality of the network as an instrument, associated to the relational approach, for bringing together people sharing common values (interest in cooking; sharing with others upon values related to the *Ciocheciò* principles and the concept of relational tourism; promotion of a sustainable and healthy lifestyle). In this regard, it is relevant that the nature of the relations established before the *Staffetta* between organisers and participants is perceived by the organisers and by the 41% of respondents as a friendship relationship, which for 17% of the respondents is considered to be long-lasting. The other two elements that played an essential role within the *Staffetta* were the territorial roots of the initiative (*locality*) and the role of food and gastronomic traditions (gastronomy). The *Staffetta* was perceived, both by the organisers and the participants, as a vehicle through which narrating a territory and its traditions (Wittel 2001; Molz 2013, 2014). Within the *Staffetta,* gastronomy acted as a bridge able to link local communities to non-local people (*linking culture and tourism*) and also a tool to live memorable and meaningful experiences (*developing the meal experience*) of cultural exchange (e.g. participants reproduce their own version of other participants' recipes) (Richards 2012, p. 19).

The connecting element among these three factors was the online format of the *Staffetta.* According to the network relationality dimensions (Marques and Gondim Matos 2020), both in the organisers and participants' perspective, the *Staffetta* favoured a good level of *temporary belongingness*, stimulating a sense of places especially by developing a family feeling with participants and by feeling part of a virtual community; the online format also helped in creating or consolidating *a-priori empathy* with organisers and other participants, marking the beginning of a two-sided relationship. Despite the positive role recognised to the online format of the *Staffetta* for the creation of new relationships or strengthen existing ones, technology played an instrumental role (Bawens 2010; Porter 2004): in the absence of the other three elements (relationality, locality and gastronomy) the online format would have been an end in itself. This is confirmed by the fact that technology was mainly conceived by respondents as a tool to promote long-distance human interactions, a territory and its culture and gastronomic traditions (Fig. 4). Moreover, the online relational space of the *Staffetta* has been perceived as a way to invite people to visit the physical space (56%).

In terms of implications, the *Staffetta* was an opportunity to create new relationships and/or to strengthen existing ones and to stimulate the desire to deepening knowledge to real meetings on the respective territories of the participants. In this sense, the *Staffetta* can be conceived as a bridge to stimulate face-to-face interactions and a potential tourism attractor. This latter aspect emerged from the organisers' wish to go from virtual to real and to meet each other in a second on-site edition; in the case of the participants, to travel the places known virtually (online) and indirectly (through the recipes) during the *Staffetta*. This aspect is confirmed by the fact that places of origin of the organisers are the ones that participants would most likely visit in the future (Marche: 34%; Japan: 28%; Veneto: 12%; Liguria: 7%). This element could somehow be linked to the role played by previous relationships established between organisers and participants but also to the natural friendly attitude of the organisers in establishing new relationships.

7 Conclusions

Starting from the impact of COVID-19 pandemic in tourism sector in terms of travel restrictions, decrease in the demand and changes in the tourists' behaviour, this contribution has analysed the opportunities given by relational tourism in the post-pandemic scenario with a focus on how and to what extent technologies can contribute to promoting authentic tourism experiences during and after a crisis.

The case of the *Staffetta della Cucina Ciocheciò* has been presented. This initiative, proposed during the lockdown, aimed at maintaining existing relationships and create new ones, by involving organisers and participants in an immersive virtual cooking experience based on *Ciocheciò* shared values and the benefits of relationality in terms of engagement and wellbeing.

Results showed that, even if virtually, elements of relational tourism are included in this experience. Thus, relational tourism can be pursued in the post-COVID-19 tourism recovery, when connected with locality and gastronomy, as in this case. The role played by technology is relevant: far from being a substitute of reality, it can act as a bridge to facilitate face-to-face interactions and stimulate real visits to places known only virtually. In the investigated case, it is the interplay of real and virtual social interactions that has emerged as the key factor for a kind of tourism that can face challenges and crisis such as the COVID-19.

As far as lessons learnt from the *Staffetta* experience and this exploratory research, some suggestions and recommendations emerged. These, taking into account the charter for tourism, travel, and hospitality after COVID-19 proposed by Chang et al. (2020), could benefit practitioners at local, regional and global level, in managing relationships with tourists. On the one hand, this study suggests that social direct interactions are essential elements for the creation of authentic tourism experiences. On the other hand, online interactions can play a decisive role in maintaining stable and long-lasting relationships and in creating new ones that from virtual can turn to real. This is possible during and also after a crisis.

Considering future perspectives, practitioners willing to maintain existing relationships and/or to create new ones could:

- consider the territory and the local communities as *key elements* of the bond they would like to create through their online and offline interactions with guests.
- consider food and immersive experiences, virtual or real, as *supporting elements* to enhance the level of engagement of tourists and improve the relationship itself.

This study presents some limitations. Further research could focus on a quantitative and qualitative analysis of the evolution of the demand for relational tourism after the lockdown and the *Staffetta* experience, based on data on the tourism flows to the destinations and facilities involved in the initiative, to understand if it somehow had an impact in terms of tourism promotion.

Acknowledgements The authors gratefully acknowledge the organisers and participants of the *Staffetta della Cucina Ciocheciò* for contributing to the research and Irene Toscana for her support in carrying out the empirical investigation.

References

Bank of Italy (2020) Indagine sul turismo internazionale, 5 giugno 2020. Retrieved from: https://www.bancaditalia.it/pubblicazioni/indagine-turismo-internazionale/2020-indagine-turismo-internazionale/statistiche_ITI_05062020.pdf (29.07.2020)

Bawens M (2010) Peer-to-peer relationality. The City and Anonymity. Retrieved from: https://www.barcelona.cat/metropolis/ca (18.08.2020)

Bertella G, Cavicchi A (2015) Marchigiane families open their homes to tourists: sharing food and stories at the dinner table. J Gastronomy Tourism 1(1):69–70

Bertella G, Cavicchi A (2017) From sharecroppers to "flying farmers": new forms of tourism entrepreneurship in rural areas. e-Review Tourism Res 14(3/4):133–148. https://hdl.handle.net/10037/14233

Bertella G, Cavicchi A, Bentini T (2018) The reciprocal aspect of the experience value: tourists and residents celebrating weddings in the rural village of Petritoli (Italy). Anatolia 29(1):52–62. https://doi.org/10.1080/13032917.2017.1381854

Bessière J (1998) Local development and heritage: traditional food and cuisine as tourist attractions in rural areas. Sociol Rural 38(1):21–34. https://doi.org/10.1111/1467-9523.00061

Beverland MB, Farrelly FJ (2010) The quest for authenticity in consumption: consumers' purposive choice of authentic cues to shape experienced outcomes. J Consumer Res 36(5):838–856. https://doi.org/10.1086/615047

Bialski P (2012) Technologies of hospitality: how planned encounters develop between strangers. Hospitality Soc 1(3):245–260. https://doi.org/10.1386/hosp.1.3.245_1

Bimonte S, Punzo LF (2016) Tourist development and host-guest interaction: an economic exchange theory. Ann Tour Res 58:128–139. https://doi.org/10.1016/j.annals.2016.03.004

Booking.com (2019) Booking.com reveals key findings from its 2019 sustainable travel report. Amsterdam, 17th April, 2019. Retrieved from: https://globalnews.booking.com/bookingcom-reveals-key-findings-from-its-2019-sustainable-travel-report/ (29/07/2020)

Chang CL, McAleer M, Ramos V (2020) A charter for sustainable tourism after COVID-19. Sustainability 12(9):3671, 1.4. https://doi.org/10.3390/su12093671

Del Chiappa G (2020) How tourist behavior is being changed and transformed by the COVID-19 outbreak? Report. University of Sassari (in press)
Eisenhardt KM (1989) Building theories from case study research. Acad Manage Rev 14(4):532–550. https://doi.org/10.5465/amr.1989.4308385
Gössling S, Scott D, Hall CM (2020) Pandemics, tourism and global change: a rapid assessment of COVID-19. J Sustain Tourism 1-20. https://doi.org/10.1080/09669582.2020.1758708
Gretzel U, Fuchs M, Baggio R, Hoepken W, Law R, Neidhardt J, Pesonen J, Zanker M, Xiang Z (2020) E-Tourism beyond COVID-19: a call for transformative research. Information Technol Tourism 1:1–17. https://doi.org/10.1007/s40558-020-00172-4
Gyimóthy S, Mykletun RJ (2009) Scary food: commodifying culinary heritage as meal adventures in tourism. J Vacation Marketing 15(3):259–273. https://doi.org/10.1177/1356766709104271
Hall CM, Scott D, Gössling S (2020) Pandemics, transformations and tourism be careful what you wish for. Tourism Geographies 22(3):577–598. https://doi.org/10.1080/14616688.2020.1759131
Harrington RJ, Ottenbacher MC (2010) Culinary tourism—a case study of the gastronomic capital. J Culinary Sci Technol 8(1):14–32. https://doi.org/10.1080/15428052.2010.490765
Higgins-Desbiolles F (2020) Socializing tourism for social and ecological justice after COVID-19. Tourism Geographies 22(3):610–623. https://doi.org/10.1080/14616688.2020.1757748
Hjalager AM, Richards G (eds) (2003) Tourism and gastronomy. Routledge, London
Istat (2020) Una stagione mancata: impatto del Covid-19 sul turismo, 29 aprile 2020. Retrieved from: https://www.istat.it/it/files//2020/04/STATISTICATODAY_TURISMO.pdf (29.07.2020)
ILO (International Labour Organization) (2020) COVID-19 and the Tourism Sector. ILO Sectoral Brief. Retrieved from: https://www.ilo.org/sector/Resources/publications/WCMS_741468/lang--en/index.htm (18.08.2020)
Lapointe D (2020) Reconnecting tourism after COVID-19: the paradox of alterity in tourism areas. Tourism Geographies 22(3):633–638. https://doi.org/10.1080/14616688.2020.1762115
Lee TH, Jan FH (2019) Can community-based tourism contribute to sustainable development? Evidence from residents' perceptions of the sustainability. Tourism Manage 70:368–380. https://doi.org/10.1016/j.tourman.2018.09.003
Lin B, Fu X (2020) Gaze and tourist-host relationship–state of the art. Tourism Review (in press). https://doi.org/https://doi.org/10.1108/TR-11-2019-0459
Jamal T, Budke C (2020) Tourism in a world with pandemics: local-global responsibility and action. J Tourism Futures 6(2):181–188. https://doi.org/10.1108/JTF-02-2020-0014
Kastenholz E, Carneiro MJ, Eusébio C, Figueiredo E (2020) 14 Host–guest relationships in rural tourism. In: Artal-Tur A, Kozak M (eds) Culture and cultures in tourism: exploring new trends. Routledge, New York, pp 177–190
Mayan MJ (2009) Essentials of qualitative inquiry. Leaf Coast Press, Walnut Creek, CA
Marques L, Gondim Matos B (2020) Network relationality in the tourism experience: staging sociality in homestays. Curr Issues Tourism 23(9):1153–1165. https://doi.org/10.1080/13683500.2019.1594722
Mittiga A, Kow N, Silva B, Kutschera S, Wernet F (2019) Travel trends report 2019. Retrieved March, 1, 2019. Retrieved from: https://www.trekksoft.com/en/resources/ebooks/travel-trends-report-2019 (18.08.2020)
Molz JG (2013) Social networking technologies and the moral economy of alternative tourism: the case of couchsurfing.org. Ann Tour Res 43:210–230. https://doi.org/10.1016/j.annals.2013.08.001
Molz JG (2014) Toward a network hospitality. First Monday 19(3). https://doi.org/10.5210/fm.v19i3.4824
Mondello E (1990) Italo Calvino. Edizioni Studio Tesi, Pordenone
Naselli F (2005) Integrated tourism as a resource for the development of Mediterranean lands and of strategies for tourism. Gulotta Editore, Palermo
Okazaki E (2008) A community-based tourism model: its conception and use. J Sustain Tourism 16(5):511–529. https://doi.org/10.1080/09669580802159594

Paulauskaite D, Powell R, Coca-Stefaniak JA, Morrison AM (2017) Living like a local: authentic tourism experiences and the sharing economy. Int J Tour Res 19(6):619–628. https://doi.org/10.1002/jtr.2134
Pine J, Gilmore J (2011) The experience economy. Harvard Business Review Press, Boston
Porter CE (2004) A typology of virtual communities: a multi-disciplinary foundation for future research. J Computer-Mediated Commun 10(1). https://doi.org/10.1111/j.1083-6101.2004.tb00228.x
Purpura A, Naselli F, Ruggieri G (2007) La componente relazionale nell'analisi sistemica del turismo. Palumbo, Palermo
Qiu RT, Park J, Li S, Song H (2020) Social costs of tourism during the COVID-19 pandemic. Ann Tour Res 84:1–14. https://doi.org/10.1016/j.annals.2020.102994
Richards GW (2011) Creativity and tourism: the state of the art. Ann Tour Res 38(4):1225–1253. https://doi.org/10.1016/j.annals.2011.07.008
Richards G (2012) Food and the tourism experience. Major findings and policy orientations. In OECD (2012), Food and the Tourism Experience: The OECD-Korea Workshop, OECD Studies on Tourism, OECD Publishing. https://doi.org/10.1787/9789264171923-en
Richards GW (2013a) Creative and relational tourism in Barcelona. In ATLAS cultural tourism group expert meeting on alternative and creative tourism, Barcelona, June 13–14, 2013
Richards G (2013b) Creating relational tourism through exchange. In: ATLAS Annual Conference, Malta, November 6–8, 2013
Richards G, Wilson J (2006) Developing creativity in tourist experiences: a solution to the serial reproduction of culture? Tourism Manage 27(6):1209–1223. https://doi.org/10.1016/j.tourman.2005.06.002
Rifkin J (2000) La rivoluzione della new economy. Mondadori, Milano
Rinaldi C (2017) Food and gastronomy for sustainable place development: a multidisciplinary analysis of different theoretical approaches. Sustainability 9(10):1748, 1–25. https://doi.org/10.3390/su9101748
Ruggieri G (2007) Un modello per l'analisi della relazionalità. In: Purpura A, Naselli A, Ruggieri G (eds) La componente relazionale nell'analisi sistemica del turismo. Palumbo, Palermo, pp 47–59
Ruisi M (2004) Turismo relazionale. Logiche di sviluppo reticolare ed etica dell'ospitalità per le aziende turistiche di piccola dimensione. Giuffrè, Milano
Sherlock K (2001) Revisiting the concepts of host and guests. Tourist Stud 1(3):271–295. https://doi.org/10.1177/146879760100100304
STR (2020) COVID-19: Hotel industry impact. Retrieved from: https://str.com/data-insights-blog/coronavirus-hotel-industry-data-news (29.07.2020)
Teye V, Sonmez SF, Sirakaya E (2002) Residents' attitudes towards tourism development. Ann Tour Res 29:668–688. https://doi.org/10.1016/S0160-7383(01)00074-3
UNWTO (2012) Global report on food tourism. Retrieved from: https://www.e-unwto.org/doi/epdf/10.18111/9789284414819 (28–07.2020)
UNWTO (2020) International tourist arrivals could fall by 20–30% in 2020. Retrieved from: https://ww.unwto.org/news/international-tourism-arrivals-could-fall-in-2020 (29.07.2020)
Valls JF, Bustamante X, Guzmán F, Vila M (2004) Gestión de destinos turísticos sostenibles. Gestión 2000, Barcelona
Vázquez FJC, Ruggieri G (2011) Turismo relacional: desafios y potencialidades. Turydes. Revista de Investigación En Turismo Y Desarrollo Local 4(9):1–14. https://hdl.handle.net/10447/79510
WTTC (2020) Corona Virus Brief: April 14 2020. Retrieved from: https://wttc.org/Portals/0/Documents/WTTC%20Coronavirus%20Brief%20External%2014_04.pdf?ver=2020-04-15-081805-253 (18.08.2020)

Wittel A (2001) Toward a network sociality. Theor Culture Soc 18(6):51–76. https://doi.org/10.1177/026327601018006003
Yin RK (2003) Case study research: design and methods, 3rd edn. Sage, London
Zenker S, Kock F (2020) The coronavirus pandemic—a critical discussion of a tourism research agenda. Tourism Manage 81:1–4. https://doi.org/10.1016/j.tourman.2020.104164

Concetta Ferrara is an industrial PhD student in Human Sciences at the Department of Education, Cultural Heritage and Tourism at University of Macerata. In last years she gained experience in stakeholder's engagement through participatory approaches and European and regional projects on rural development and place branding. Her main fields of interest and research are university-business collaboration for local development and experiential and relational tourism in rural and inner areas.

Sabrina Tomasi has an industrial PhD in Human Sciences. Currently she is a postdoctoral researcher at the Department of Education, Cultural Heritage and Tourism at University of Macerata. Her main current research topic is the Universities' contribution in the regional innovation ecosystem. In recent years she investigated tourism-related topics and took part in several European and regional projects on rural development and place branding, also in the framework of RIS3 and gained experience in stakeholder's engagement and innovation brokerage approaches.

Chiara Aleffi has an industrial PhD in Human Sciences. Currently she is a postdoctoral researcher on the topic University and self-reflection tools for regional action at the Department of Education, Cultural Heritage and Tourism at University of Macerata. In recent years she gained experience in European projects, stakeholder's engagement and innovation brokerage. Her main fields of studies are tourism, agribusiness, rural development and place branding in the framework of RIS3

Annapia Ferrara is an industrial PhD student in Education, Cultural Heritage and Territories at the Department of Education, Cultural Heritage and Tourism at University of Macerata. Her main fields of interest and research are tourism, cultural routes, rural development, and social farming for educational and community based tourism

Giovanna Bertella received her PhD about learning and networking in tourism from the Department of sociology, Political Science and Community Planning at UiT. Currently she is Associate Professor at the School of Business and Economics, UiT The Arctic University of Norway, Tromsø. Her main research interests are management, marketing, innovation/entrepreneurship, tourism and leisure studies and food studies.

Gigliola Paviotti holds a PhD in Education and works as research fellow at the University of Macerata, Italy. She has worked more than twenty years in the education and training field, focusing on employability topic and transitions from education to work. In recent years, her research work has focused on place-based approaches to support graduates' employment in local labour markets.

Alessio Cavicchi (PhD, MSc) is Full Professor at the Department of Education, Cultural Heritage and Tourism at University of Macerata. His main fields of interest and research are consumer food choice, economics of food quality and safety, sustainable tourism and innovation in the agro-food sector.

Managing the Impact of COVID-19 on the Education Plans and Activities of South African Schools

Bongani Sibusiso Mchunu, Annatoria Zanele Ngcobo, and Thamsanqa Thulani Bhengu

Abstract The chapter reviews the education sector's COVID-19 policy implementation for managing the impact of the coronavirus on school plans in South Africa. Complex and problematic policy issues emanate from the effects of the coronavirus and the quick-fix strategies used to implement policy without consideration of the contextual factors in underprivileged schools. Closure of schools as a short-term measure was not sustainable and had unintended consequences. On paper, the COVID-19 policy looks good, but it is devoid of the contextual realities of the places where it needs to be implemented. Gaps identified in the policy gave rise to protracted resistance by stakeholders due to lack of ownership. These challenges led to shifting of goals regarding the early reopening of schools. Systems thinking offers tools to improve on the implementation challenges faced by policy makers, while a systems dynamics approach seeks to bring understanding of the interconnection among stakeholders and the views they hold. The Soft Systems Methodology (SSM) provides a perspective for modelling the underlying problematic issues in structured manner. It shows the complex relationship that exists starting from the conception of the policy to its implementation. The SSM model addresses the challenges faced by policy makers in communicating the policy and engaging the stakeholders, and conceives the root cause of problematic issues in the implementation of the policy by proposing an ideal conceptual model.

B. S. Mchunu (✉)
Chief Education Specialist, Infrastructure Planning Directorate, Kwazulu-Natal Department of Education, 5 Scott Street Bowden House Pietermaritzburg, P.O. Box 135, Edendale 3217, South Africa
e-mail: mchunudoctor2016@gmail.com

A. Z. Ngcobo
Mathematics Education Lecturer, Department of Mathematics Education, University of KwaZulu-Natal, Private Bag X05, Ashwood 3605, South Africa
e-mail: ngcoboA2@ukzn.ac.za

T. T. Bhengu
School of Education, University of KwaZulu-Natal, Private Bag X03, Ashwood 3605, South Africa
e-mail: bhengutt@ukzn.ac.za

W. Leal Filho (ed.), *COVID-19: Paving the Way for a More Sustainable World*, World Sustainability Series, https://doi.org/10.1007/978-3-030-69284-1_19

Keywords School closure · Complexity · Systems thinking · Sustainable · Soft systems methodology

1 Background to the Problem and Contextual Factors

We present some of the initial conversations in South Africa (SA) regarding the COVID-19 pandemic, its devastating effects on education, and the response of the Department of Basic Education (DBE). The chapter is conceptual and draws largely from current scholarship, news reports, and reviews of documents from the DBE regarding its state of readiness to reopen schools as COVID-19 increases its footprint in SA and the world. The discussion begins with the background to the problem and contextual factors, followed by a literature review and theoretical framework. Discussion of the findings, and the recommendations and conclusion bring the chapter to a close.

The whole world is reeling under the devastating effects of the spread of coronavirus, which started in the city of Wuhan, Hubei province, China. COVID-19 has caught education systems off-guard and unprepared to provide alternative plans for sustaining teaching and learning (Djalante et al. 2020; Mahaye 2020). The United Nations Educational, Scientific and Cultural Organization (2020) reported that 188 countries have imposed countrywide school closures, affecting more than 1.5 billion children and youth. The widespread effect is illustrated in Fig. 1.

It is evident that closure of schools is a global phenomenon which has negative effects on schooling, bringing to a halt the learning of an estimated 1.5 billion children and youth (World Bank 2020). Many national examinations were put on hold (DBE 2020a; World Bank 2020). The unprecedented spread of the coronavirus and its devastating effects on the education sector makes us wonder how countries will cope with the issue of school reopening. Also, the scale and scope of the pandemic puts the ability of education departments globally and locally to the test, in responding effectively at school level and ensuring that schooling happens.

The scale of the impact is unprecedented, and studies suggest that the goals and targets of the Sustainable Development Agenda 2030 might be greatly compromised (Daschle and Frist 2020). According the World Health Organization (2020) adaptive leadership is needed and has heeded the call to arrest the spread of the virus in its initial stages. At the beginning of March it was observed that the whole of the African continent had been significantly affected by the coronavirus, with SA the most affected country in Africa (Mahaye 2020). Disaster management measures have been declared in most countries affected by the spread of the coronavirus, including the abrupt closure of schools and universities (Tam and El-Azar 2020). Scholars expressed their concern about the negative effects of the prolonged closure of schools (UNESCO 2020; Viner et al. 2020), thus suggesting that schools need to reopen. While the wisdom of reopening schools so soon is subject to debate, schooling faces numerous challenges and complexities, particularly in the developing economies.

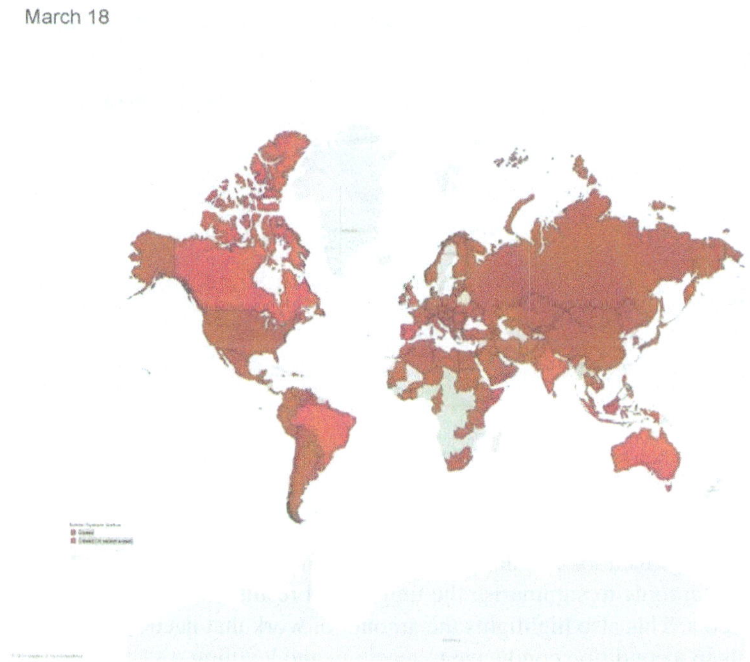

Fig. 1 Red areas indicate countries affected by school closure (World Bank 2020)

In the South African context, for instance, one of the negative consequences of school closure due to the pandemic included 1783 burglaries in schools across the nine provinces of the country (DBE 2020a). These happened during the first few weeks of the national lockdown. Furthermore, challenges facing schools are compounded by the shortage and lack of water supply, broken and unhygienic toilets, and broken windows and doors, particularly in rural areas (C-19 People's Coalition 2020; DBE 2020g). The DBE made a commitment to attend to urgent repair and renovation of the damage to schools before the reopening of schools on 8 June 2020 (DBE 2020a, g). However, collated data from the nine provinces indicated a lack of preparedness for the reopening of schools (C-19 People's Coalition 2020; DBE 2020b).

According to various teacher unions, the lack of preparedness is closely linked to the lack of water provision and hygienic toilets in some rural and township schools. This is critical, since hygiene plays a vital role in preventing infections and controlling the spread of COVID-19 (Álvarez et al. 2020; UNESCO 2020). The WHO has also advised countries to embark on early interventions, including the sudden closure of schools, but with consideration for underprivileged learners (Djalante et al. 2020). On 16 March 2020 the Minister of Basic Education, Angie Motshekga, officially announced the closure of schools in order to avoid spread of COVID-19.

Various scholars (Brooks et al. 2020; Viner et al. 2020; Sintema 2020) highlight the numerous consequences of continued school closure, which include physical,

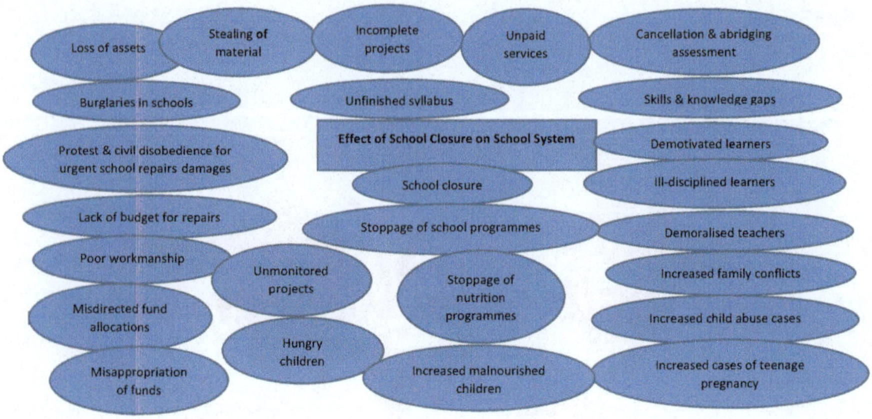

Fig. 2 Multicause diagram showing the unintended consequences of school closures due to COVID-19

psychological and educational damage to the school system. Figure 2 uses Soft Systems approach tools to summarise the unintended results of the disruption and closure of schools. This also highlights the amount of work that needs to be done to restore schools to a condition conducive to teaching and learning.

The main aim of this chapter is to highlight some key decisions implemented in SA and events that unfolded as a response to the COVID-19 pandemic. It is hoped that lessons will be learnt from what was done and not done by the DBE.

2 Related Literature and Theoretical Framework

Scholars like Petrie et al. (2020) decry the lack of preparedness for dealing with pandemics, as many countries were caught off-guard by the sudden outbreak of COVID-19. While there is general agreement about the positive effect of the drastic measures taken to arrest the spread of COVID-19, their sustainability for the education system has been questioned. For instance, scholars decry the loss of learning time and other negative educational effects of prolonged school closures (Álvarez et al. 2020; Esposito and Principi 2020; Kekić and Miladinović 2016). Tam and El-Azar (2020) advocate that education systems need to build resilience in times of emergencies such as the COVID-19 pandemic and ensure sustainability. Sustainable development relates to the principle of meeting human development goals while at the same time sustaining the ability of natural systems to provide the natural resources and ecosystem services which the economy and society depend on (Emas 2015). Therefore, educational leaders are faced with the new challenge of providing a healthy environment for teaching to resume.

Health guidelines and policies have been provided and designed to fit into the school environment (DBE 2020c, d; WHO 2020). Unfortunately, the literature reveals a range of challenges that hinder the effective and sustainable implementation of policies (Pons 2015; Repenning and Sterman 2002). Key among these challenges is the lack of appropriate skills among the school leaders (Repenning and Sterman 2002).

In this regard, it is important that the response of the DBE and school-based leaders is sustainable, and that school leaders have the requisite skills to meet the needs of the knowledge society (Pons 2015).

There is some acknowledgement that stakeholders tend to be dissatisfied with the outcomes of policies, due to the lack of sustainability in their implementation, and hold policy makers accountable for this (C-19 People's Coalition 2020; Repenning and Sterman 2002). In the South African education system, the nine provinces and education district offices are at the coalface of implementation of COVID-19 policies (DBE 2020a, b, c, d, e, f, g, h). Scholars like Esposito and Principi (2020) and Petrie et al. (2020) acknowledge the need to focus more on implementation processes where there is likely to be resistance. Repenning and Sterman (2002) acknowledge that education policy makers need to understand that the policy-making and implementation process is complex, owing to the multiplicity of actors with different interests. Policy implementation often has many consequences which were not unintended by the original policy (Repenning and Sterman 2002). Therefore, the need for a model that deals with stakeholders effectively cannot be overemphasised.

We propose a Soft System Methodology (SSM), as advocated by Checkland and Haynes (1994), as a tool to frame our analysis of how stakeholders can be involved in preparing an effective COVID-19 response. Using the SSM we deal with the soft issues, where there are negotiations between stakeholders from different interests and perspectives. The SSM facilitates the participation of a variety of stakeholders in working on issues where there are divergent views. The SSM has seven stages that reflect how engagements with the complex and problematic issues at hand occur. For this book, we have used five (5) of the seven (7) steps in the SSM process, namely: (i) root definition, (ii) customers, (iii) actors, (iv) transformation process, and (v) Weltanschauung (worldview). The seven phases are captured in Fig. 3.

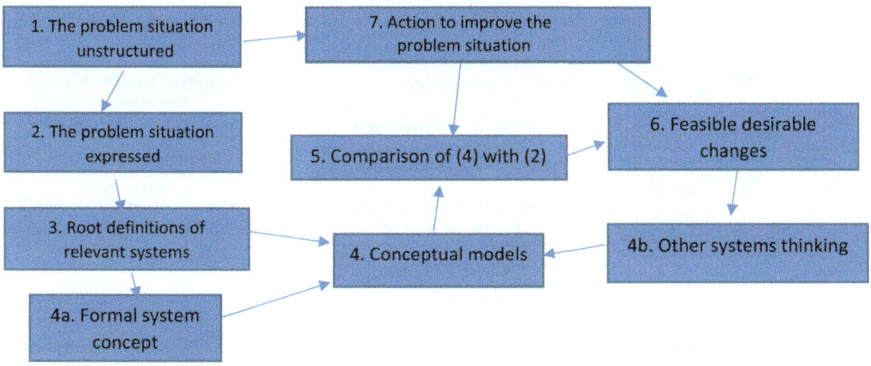

Fig. 3 The seven steps of the SSM model

3 Framework of the Department of Basic Education COVID-19 Sector Plan

Minister Basic Education Angie Motshekga involved all key stakeholders in the process of developing the DBE's COVID-19 Sector Plan. The teacher unions, Congress of South African Students (COSAS) and South African Governing Body Associations (SGBA, i.e. parents' representatives) made inputs into the draft DBE COVID-19 Sector Plan before it was finalised as a policy document. Figure 4 represents all stakeholders with an interest in the formulation and implementation of the policy at school level. The stakeholders that are listed below are formed of Teacher Unions and students associations and parents. The abbreviations stand for the following in full viz.; Suid Afrikaanse Onderwyse Unie—SAOU; National Teachers Union—NATU; South African Democratic Teachers Union—SADTU; National Professional Teachers Union of South Africa—NAPTOSA; and Congress of South African Students—COSAS.

The COVID-19 Sector Plan policy made provision for Standard Operating Procedures (DBE 2020b, c) which provide guidance to schools and education offices on how to prepare for the opening and management of schools during the COVID-19 pandemic. The DBE worked with National Treasury to supply schools with a minimum health package of personal protective equipment (PPE). The DBE policy, at least on the surface, includes the inputs of the Joint Teacher Unions, SADTU, NAPTOSA, SAOU, PEU and NATU which contributed 14 points as non-negotiables for schools to reopen (Joint Media Statement by Education Teacher Unions and National Governing Body Associations 2020). The DBE COVID-19 Sector Plan outlines the guidelines for managing hygiene, social distancing, provision of PPE and screening processes during arrival of teachers and learners at schools (DBE 2020c). These guidelines include considerations to help administrators to plan for the continuity of teaching and learning if there is a need to quarantine learners or staff within schools or to close schools as a result of spread of the coronavirus.

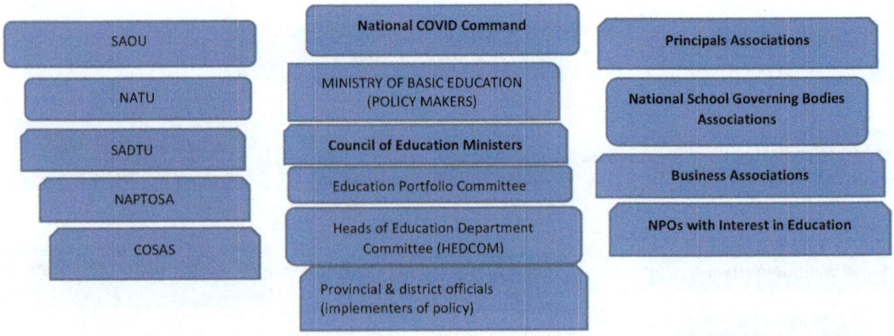

Fig. 4 Multicause diagram showing multiple stakeholders in COVID-19 policy implementation in the school sector in SA

Principals have a critical role in implementing the policy, by ensuring daily screening and testing of learners and teachers. School principals are required to monitor the health routines and social distancing and to refer teachers and learners suspected of having COVID-19 to the nearest health facilities (DBE 2020d). Every school needs to reserve a room for isolation purposes. Teachers and learners who may have been in contact with a person considered to be positive for COVID-19 need to quarantine themselves in their homes for a period of 14 days (DBE 2020c). The policy provides procedures to be followed and actions to be taken in the case of learners or members of staff testing positive for COVID-19. In the event of confirmed cases of coronavirus, the nearby health officials need to be contacted.

Principals need to prepare for possible early school closure (Brooks et al. 2020). This should only be considered after recommendations from a public health official. School principals are not expected to make decisions about dismissal or cancelling of school events without contacting the District or Circuit Office (DBE 2020c). In the event of the closure of schools, principals need to take steps to temporarily cancel all extracurricular group activities and large events; cancel or postpone field trips and sporting events; and discourage learners and staff from socialising and gatherings, as they try to ensure the continuity of teaching and learning through alternative means (DBE 2020c). These may include digital and distance-learning options where feasible, and provision of workbooks for learners. Principals need to consult school District Officials or other relevant partners on using online lessons and how to train educators to do so.

An agreement was reached in the Education Labour Relations Council Chamber on procedures for handling the issue of teachers with reported comorbidities. Firstly, teachers who are aged 60 years and above and those with a particular set of comorbidities are allowed to apply to work from home (ELRC, Circular No1 of 2020; DBE 2020c). Secondly, all teachers with a proven record of the underlying illnesses listed in the policy were also allowed to produce proof of such and to apply for exemption from working at the school site.

All of the educational establishments need to be cleaned prior to their use by learners and teachers. The DBE SOPS guidelines outlines that learners and teachers who had been in contact with those suspected of being positive need to be kept in isolation (DBE 2020c). The cleaners allocated to the school are required to disinfect the institution. All surfaces and objects with suspected contaminated body fluids need to disinfected and cleaned with the required detergents provided by the DBE (2020c).

4 Designation of Schools and Health Care Requirements

The DBE COVID-19 Sector Plan gives guidance for the management of school facilities by the principals. Principals have the added responsibility of ensuring that the health guidelines are followed in order to minimise disruption to teaching and learning, while also taking care of the lives of the children at school (DBE 2020c, d).

The COVID-19 Plan provides guidelines for preventing social stigma and discrimination. Principals are given the space to be innovative in handling cases on a daily basis, as envisaged in the DBE policy guidelines (DBE 2020c, d). Principals have to ensure that learners and teachers adhere to hygiene protocols; for example, by washing hands prior to entry into the school grounds. There also needs to be strict monitoring of the routine activities of children within the school premises. Daily reports on learner attendance and absences have to be collected (DBE 2020c). The Plan prescribes how to conduct tests and screenings in schools, including guidelines for social distancing in schools and within the classrooms, provision of basic sanitation and hygiene packages for classrooms, and psychosocial support services for everyone at schools (DBE 2020c, d).

While all of the above are based on good intentions, the lack of capital and human investment may jeopardise the achievement of these ideals. The DBE Plan assumes that people will learn on the spot as new challenges emerge.

Personal Protective Equipment Packages (DBE 2020c)

Table 1 indicates the norms and standards for the supplying of PPE to all schools.

Other conditions that were laid down for schools to be considered functional included water supply, sanitation, disinfection and cleaning of schools, and the orientation and training of principals, teachers and support staff on these guidelines. Schools were allocated persons responsible for screening and cleaning according to the number of learners. Learner transport was also required to be sanitised every day and learners had to observe social distancing. New norms were introduced for seating arrangements for learners in classrooms in terms of observing social distancing (DBE 2020c). This meant revisiting timetables and forecast planning for the phasing-in of grades. The observation of 1.5 m social distancing in classrooms has serious consequences for the allocation of learners at schools.

Table 1 Distribution of personal protective equipment packages (DBE 2020c, d)

Principal and support staff	Teachers	Learners	Teachers and other staff
• 1 × antiseptic bar of soap • 2 × 3-layer cloth face masks (supplemented with surgical masks) • 500 ml hand sanitiser • 1 × 25 L hand sanitiser for refilling of all at school • 2 × temperature scanners for school	• 1 × antiseptic bar of soap • 2 × 3-layer cloth face masks (supplemented with surgical masks) • 500 ml hand sanitiser • 1 × 25 L hand sanitiser for refilling of all at school • 2 × temperature scanners for school • 10 × biohazard bags for school	• Antiseptic bar of soap per month • 2 × 3-layer cloth face masks (supplemented with surgical masks) • Face shield for deaf learners	• 2 × 3-layer cloth face masks (supplemented with surgical masks) • 500 ml hand sanitiser • Face shield (teachers only) • 2 sets of heavy-duty gloves per assistant • 16 × sets of surgical gloves for teachers per school

Table 2 Phased-in return of learner grades (DBE 2020g)

No	Phased-in return	Grades	Dates
1	Phase 1	Grades 12 and 7	8 June
2	Phase 2	Grades 11	6 July
3	Phase 2	Grades 6	6 July
4	Phase 3	Grades 8 and 9	3 August
5	Phase 3	Grades R, 1, 2, 3, 4 and 5	3 August

After protracted negotiations with stakeholders, the Minister of Basic Education finally pronounced the scheduled dares for the phasing of other grades into the schools (DBE 2020g). A revised school calendar plan was also issued, indicating the new dates for school terms and short vacations. As we write this chapter, schools are managing with the few learners in the Grades that have reopened, i.e. Grade 12 and Grade 7 in the secondary and primary schools respectively. To observe social distancing, classes accommodate only 20–25 learners, depending on the size of the classroom. Complications in terms of accommodating the number of learners may arise when the next group and cohort of classes reopens, as indicated in Table 2.

These guidelines are not sustainable for including all learners in the return to schools, as the issue of space in classrooms will be compromised since social distancing has to be maintained. Schools have been advised to consider adopting any of four models, namely: bi-weekly (1 week attendance with rotation of grades); alternate-day (some grades attending some days); platoon model (some learners attending in the morning, others in the afternoon) and maintenance of the status quo (small schools). Schools with large enrolments will start experiencing shortages of space and furniture as soon as the third cohort of learners arrives on 3 August 2020 (KZN Circular No 48 of 2020). This also impacts negatively on changes to the timetables. The worst scenario—and it does not seem to have been thought through—is that there will be shortages of teacher supply due to those applying for exemption owing to comorbidities. Teacher unions have negotiated postponement of the return of Grades R, 1, 2, 3 and 10, due to the lack of space in classrooms and the sudden outbreak of cases testing positive for COVID-19 in a number of schools (Macupe 2020).

The latest renewed mounting pressure from Joint Teacher Unions (SADTU, NATU and NAPTOSA) led to the RSA President Cyril Ramaphosa announcing on the 26th July 2020 the early school closure starting from 27 July to 24th August 2020 (Bolowana 2020; DBE 2020h; Maromo 2020b; Ngqakamba 2020). The latest revised schedule leaves only Grade 7 and 12 to be at school (KZN Circular No. 63 of 2020). As we write there is still uncertainty about when will the other grades be phased into the schooling system.

5 Findings and Recommendations for Soft System Methodology Application as Theoretical Framework

The SSM model is that preferred to be applied in implementation of the DBE COVID-19 Sector Plan. As highlighted elsewhere in this chapter, SSM gives policy makers the capacity to deal with complex issues and handle multi-stakeholder engagements effectively. All of the SSM stages are thought out and tested for review in action with the actors and owners of the policy (Checkland and Poulter 2006, 2020). SSM discourages quick-fix solutions that may lead to failure. It stands a better chance of succeeding because it enables stakeholders to take ownership of the policy implementation processes (Hynes et al. 2020; Checkland and Haynes 1994).

However, implementation of the DBE COVID-19 Sector Plan encountered some resistance from grassroots stakeholders and Joint Teacher Unions (C-19 People's Coalition 2020; Maromo 2020a; Mahlati 2020). This was probably due to less efficient co-ordination of information from the schools to the national policy makers. For instance, there was vehement resistance from teacher unions, SGBAs and COSAS (student movement), who argued against what they saw as rushed implementation of the policy while there were outstanding issues in terms of delivery of PPE, toilet repairs and provision of clean water at some schools. Such became a bargaining tool for teacher unions to use to defy calls from the Minister of Basic Education for teachers to return to their classrooms (Mahlati 2020; Maromo 2020a; Naki 2020).

Multiple stakeholders (as indicated in Fig. 4) were consulted by policy makers on some key decisions (DBE 2020b, c). However, the teacher unions used the 14 points which they regarded as non-negotiable, as a bargaining tool for resisting the early reopening of schools (Cele 2020; Naki 2020; Nkanjeni 2020). These challenges and contextual factors could have been taken into consideration and dealt with honestly by interaction between the Ministry of Basic Education and the stakeholders (C-19 People's Coalition 2020).

Our contention is that the issue of school reopening required a combination of adaptive leadership and systemic thinker leadership on the part of the policy makers at DBE. The Ministry of Basic Education used adaptive leadership in closing down the schools as a contingency measure to prevent spread of the coronavirus, without consideration of a detailed analysis of a reopening strategy. However, the lack of resources for learners in disadvantaged communities means they will always be left behind in their learning. Lack of provision of digital learning tools for learners in disadvantaged schools left a void that showed a lack of preparedness by the DBE to provide equity for all learners (C-19 People's Coalition 2020). Resources that were provided online for learners to continue with their learning assisted only those in affluent sections of South African society. The DBE policy framework is good on paper, but due to the lack of urgency and data collection tools in both the provinces and districts, implementation was resisted by teacher unions owing to lack of readiness on the ground. Teacher unions were justified in blocking the reopening of schools, as the situation was not yet conducive for learners and teachers to return.

A complex and unequal South African schooling system needs a combination of both adaptive leadership and systems thinker leadership to consider all of the dynamics and the inequality of resources that exist between schools in affluent urban areas, those in townships and small rural schools. The COVID-19 policy is based on a one-size-fits-all approach, thereby lacking in sustainability and adaptability to handle issues pertaining to the diverse schooling system that exists in SA.

Teacher unions objected to what they consider at the gambling with the lives of learners and teachers in a rushed reopening of schools to save the academic year (C-19 People's Coalition 2020; Maromo 2020a).

On the flip side, the DBE policy makers can leverage this as a pilot for conducting data gathering for reviews and scenario planning for the next learner cohorts. The lack of a diverse and data-driven monitoring and evaluation system meant that provincial and district officials were left on their own to face the ensuing hurdles and challenges. The SSM provides tools for policy makers to conduct scenario planning in view of envisaging phasing-in of the next learner cohorts (Checkland 2000).

Most of the stakeholders were totally opposed to the idea of reopening, which was considered as 'testing the waters' with Grade 12 and Grade 7 learners and teachers (C-19 People's Coalition 2020). Some risks and other unintended consequences could have been avoided if there had not been prolonged resistance by stakeholders and long delays in implementation of the COVID-19 Sector Plan (C-19 People's Coalition 2020; Maromo 2020a; Mahlati 2020). The lack of on-time instruments and co-ordinated monitoring led to disparities between the information provided by teacher unions and that collated by provincial and district officials. Incapacity in the collation of statistical information created cause for concern for teacher unions and SGBA about supporting an uninformed school reopening. Shifting of the goalposts by the DBE in terms of reopening of schools caused tensions, frustration and anxiety among learners and teachers (C-19 People's Coalition 2020; DBE 2020b; Mlambo 2020). The lack of systems for accountability among provincial and district officials empowered the teacher unions to be more vocal in their stance that the situation at school level lacked readiness (C-19 People's Coalition 2020; Macupe 2020; Mlambo 2020).

The DBE communication strategy lacked urgency in allaying the fears and anxiety of learners, teachers and parents. Teacher unions took advantage of this gap and fanned the flames, by manipulating all available media with negative communication about the DBE COVID-19 Sector Plan (Macupe 2020; Mlambo 2020).

In so doing, teacher unions won the hearts and sympathy of the public and parents (Joint Media Statement by Education Teacher Unions and National Governing Body Associations 2020). The DBE communication officials were also not readily available to communicate the latest information and developments, and hence this situation was exploited with a lot of fake news and misinformation. On the part of the DBE, however, there was a lack of clear co-ordination in terms of ensuring that the correct information was provided timeously from districts and provinces up to national level (DBE 2020d).

The COVID-19 Sector Plan looks watertight on paper, but fails to grasp the complex situations that exist in schools located in poverty-stricken communities of SA where there is poor sanitation. Systems thinkers develop a variety of tools for the purpose of taking stakeholders along with the process, by mapping the issues at hand during consultations (Repenning and Sterman 2002). We have noted the ongoing debate between the Minister of Basic Education, Joint Teacher Unions and School Governing Body Associations. At the core of the debate is that most of the non-negotiables have not been delivered by the DBE, yet it pushed for the schools to reopen (Cele 2020; Naki 2020; Nkanjeni 2020).

Failure to agree on and fix the date for reopening of schools by the DBE caused uncertainties in the system, and invited unnecessary criticism, with teacher unions calling such actions reckless (Cele 2020; Naki 2020; Nkanjeni 2020). Therefore, the implementation of the COVID-19 policy by the DBE and provincial Departments of Education can be characterised as treading on thin ice. It has become an issue of life and death, and has been highly contested. To compound the problem, the DBE has not handled the diverse interests of the stakeholders efficiently (C-19 People's Coalition 2020; Mahlati 2020). We also need to acknowledge that the education system in SA is experiencing a pandemic of this magnitude for the first time. As a result, the huge gap between the DBE policy makers and stakeholders has not been narrowed, for example, by using SSM that has the ability to engage stakeholders with diverse interests effectively (Adam and de Savigny 2012).

This has resulted in mistrust and misrepresentation of the facts during reporting to the various constituencies. The protracted negotiations between the Minister of Basic Education and teacher unions plunged the education system into uncertainty, with more anxiety as to what the exact date for the reopening of schools was. The situation become more of a mess than expected, as days and weeks went by (Mahlati 2020; Nkanjeni 2020). There was a sense of naming and shaming and finger pointing among officials at various levels, relating to the state of readiness of schools to reopen (Mahlati 2020; Naki, 2020).

The linear and narrow approach in drawing up the policy, which excluded key stakeholders, emboldened the latter—who were quick to identify various loopholes when they ultimately became involved in the consultation process. In terms of SSM, a root definition could have initially been outlined to all stakeholders in order to get them on board (Carey et al. 2015).

With a root definition adopted, all of the stakeholders could have been brought on board and taken ownership of the process (Checkland 2000). The process needed to take into consideration the diversity of interests of the variety of stakeholders, as mapped out in Fig. 4. For instance, COSAS, as a student body, was left out of key decisions. The policy could have accommodated the transformation process that it envisages, so that all stakeholders would know what they wanted to achieve. This is a critical stage that cannot be divorced from the others, in the sense that ongoing engagements need to be the order of the day. This could have been done, for example, by the establishment of effective and sustainable structures at the school, circuit, district and provincial level. The policy never made provision for such—hence the perception that the policy makers were forcing the COVID-19 policy upon the stakeholders, without due consideration for those on the ground. However, if SSM had been adopted, many of the problems and tensions among the various stakeholders could have been avoided. A discussion of how each element of the SSM could be applied follows.

How soft systems methodology could have been adopted and implemented

The discussion below assumes that the DBE is using SSM in engaging with various stakeholders, and thus reflects the seven stages of that model.

Root Definition: A DBE COVID-19 Sector Plan is a product that is integrated, holistic, fully owned by all the stakeholders (teachers, learners, and parents) and aims to empower School Management Teams at the schools to manage school plans and activities effectively for sustainable progress.

Customers: The learners are the primary beneficiaries of an effectively and sustainably implemented COVID-19 policy. Teachers are also primary beneficiaries if all of the non-negotiables presented by the teacher unions are implemented at school level. Parents also benefit if the learners are provided with a healthy and safe school environment. Representative Councils of Learners (RCLs) in all schools, where applicable, are also beneficiaries. Any hiccups and/or blockages could have been attended to at the level of the established and legitimate structures. These School Health Management Structures could be established at the level of the school, circuit, district and province, and would have given legitimacy to the DBE COVID-19 Sector Plan at all of these levels.

Actors: The Actors are DBE officials as policy makers, teacher unions, SGBAs (representing parents), principals' associations, teachers, and learners. In public transport they include drivers of taxis transporting children to school. RCLs can be empowered to take responsibility for the monitoring of learner transport routes where learners board buses and taxis, especially the DBE-funded learner transport.

Transformation process: This refers to the core of the purposeful activity that transforms an input into an output. The transformation of the DBE COVID-19 policy implementation and management of school plans needs to be a programme that is sustainable, developmental, outcomes-based, with ownership, that is participatory, supported, monitored and evaluated, with accountability measures.

Weltanschauung (Worldview): This refers to the view of the world that makes the root definition meaningful in context. The Joint Unions statement captures the views of teachers, who are totally against the haphazard reopening of schools. The SGBA documents represent the views of parents, who are against the reopening of schools. One of the structures of COSAS wrote a letter to the Minister of Basic Education representing the views of learners on the reopening of schools.

Owners: In the case of implementation of the COVID-19 policy, this refers to all the teacher unions, the SGBA, COSAS and parents. In the above discussion we have noted how these three structures have wielded their weight against implementation of the COVID-19 Sector Plan. It seems as if they were never part of the stakeholders who were consulted, whereas the Education Minister always alleges that she has consulted all the stakeholders.

Environmental constraints: This refers to those factors affecting the situation. This environment is full of mistrust, fear, anxiety, indecision, negative perceptions, strife, under-resourced schools, institutional incapacity, a disintegrated system, infighting; unionism, dysfunctional schools, and political opportunism. Some of these constraints could have been managed at the level of the recognised and established structures before negative statements were made in the news media.

The RCLs needed to be involved in the role of monitoring implementation of some of the envisaged tasks, for example, the formation of classroom and school ground rules to monitor wearing of masks in and out of school. COSAS is a recognised student structure that could have been mobilised to engage in debates on the pros and cons of prolonged school closure versus the merits of working within the constraints of social distancing, wearing of masks and other hygiene measures. Another way would have involved assigning tasks in Life Orientation for Grade 12 learners to conduct research on how other countries managed to open their schools during the COVID-19 pandemic.

The COVID-19 policy framework that has been developed was a step in the right direction. Education officials in the provinces and districts can only build and improve on this framework. Education officials in the districts need to be capacitated to have core competency in health policy management, as they are closer to the schools. Policy development as a key strategic response to crisis needs to be taught to those who are responsible for education improvement at provincial and district level. Those officials working in systems data management need to support the districts in developing data collection tools.

The process of reopening the schools needs to be managed in collaboration with the key stakeholders. The SSM model provides an opportunity for looking at problematic issues from a holistic and integrated perspective. In times of emergency we need more partnerships between government, teacher unions and all of the relevant stakeholders. We suggest that district and provincial officials be trained in research and data collection tools especially Statistical Package for the Social Sciences (SPSS) and Monkey Survey as part of capacity development in the sector. For the future there is a need to conduct action research using the SMM as part of capacity development especially in areas of policy implementation.

6 Conclusion

The COVID-19 pandemic caught the education sector off-guard and exposed the shortcomings and weaknesses of the system in handling epidemics. For instance, it exposed the extent to which basic infrastructure such as clean water and sanitation is made available in schools, as well as the levels of hygiene practices in the schools. Secondly, the pandemic exposed the manner in which the DBE espouses and applies the notion of stakeholder involvement in dealing with critical issues affecting the lives of learners and staff.

Drawing from the process, we can conclude that stakeholder participation was not taken as seriously as it could have been, and hence no convergence of views on decisions could be attained. Closely linked to this reality is the lack of a widely inclusive process in which all key stakeholders were taken on board. That is why we propose the use of the SSM in this regard.

References

Adam T, de Savigny D (2012) Systems thinking for strengthening health systems in LMICs: need for a paradigm shift. Health Policy Plann (27):iv1–iv3. Accessed 20 June 2020

Álvarez C, Cancino A, Castillo C, de Wolff T, Gajardo P, Lecaros R, Marıa S (2020) Report# 5: scenarios for the opening schools during the Chilean COVID-19 outbreak. https://covid-19.cmm.uchile.cl/wpcontent/uploads/2020/04/Reporte5CMMAM2VCEPS-VF.pdf. Accessed 27 July 2020

Bolowana A (2020) SADTU rubbishes suggestions that teachers shouldn't get paid if schools are closed. @SABCNews. https://www.sabcnews.com/sabcnews/sadtu-rubbishes-suggestions-that-teachers-shouldnt-get-paid-if-schools-are-closed/. Accessed 27 July 2020

Brooks SK, Smith LE, Webster RK, Weston D, Woodland L, Hall I, Rubin GJ (2020) The impact of unplanned school closure on children's social contact: rapid evidence review. Euro Surveill 25(13):2000188

Carey G, Malbon E, Carey N, Joyce A, Crammond BR, Carey A (2015) Systems science and systems thinking for public health: a systematic review of the field. BMJ Open 5:e009002. https://doi.org/10.1136/bmjopen-2015-009002. Accessed 22 July 2020

Cele S (2020) 'I'm not stubborn': Angie Motshekga amid criticism from MPs. Sunday Times. Retrieved from: https://www.timeslive.co.za/politics/2020-07-22-im-not-stubborn-angie-motshekga-amid-criticism-from-mps/. Accessed 23 July 2020

Checkland P (2000) Soft systems methodology: a 30 year retrospective. Syst Res Behav Sci 17(S1):S11–S58

Checkland PB, Haynes MG (1994) Varieties of systems thinking: the case of soft systems methodology. Syst Dyn Rev 10(2–3):189–197

Checkland P, Poulter J (2006) Learning for action: a short definitive account of soft systems methodology and its use for practitioner, teachers, and students. Wiley, Chichester

Checkland P, Poulter J (2020) Soft systems methodology. In: Reynolds M, Holwell S (eds) Systems approaches to making change: a practical guide. Springer, London

C-19 People's Coalition (2020) Society is not ready to safely reopen schools and education centres. University of the Witwatersrand COVID-19 News. Retrieved from: https://www.wits.ac.za/covid19/covid19-news/latest/society-is-not-ready-to-safely-reopen-schools-and-education-centres.html. Accessed 20 June 2020

Daschle T, Frist B (2020) In the covid-19 pandemic response: global health is U.S. health. April 2020 https://bipartisanpolicy.org/wpcontent/uploads/2020/04/BPC_Health_COVID19-Pandemic-Response.pdf. Accessed 13 May 2020

Department of Basic Education (2020a) Basic education department postpones May/June rewrite matric examinations. https://kfmulaudzi.files.wordpress.com/2020/04/basic-education-department-postpones-may-june-rewrite-matric-examinations.pdf. Accessed 15 May 2020

Department of Basic Education (2020b) Basic Education Minister in consultation process on sector recovery plans ahead of the reopening of schools. https://kfmulaudzi.files.wordpress.com/2020/05/basic-education-minister-in-consultation-process-on-sector-recovery-plans-ahead-of-the-reopening-of-schools.pdf. Accessed 20 May 2020

Department of Basic Education (2020c) DBE policy guidelines for schools on maintaining hygiene during COVID-19 pandemic. Available online: https://www.dbe.gov.za. Accessed 30 May 2020

Department of Basic Education (2020d) Standard operating procedure for the prevention, containment and management of Covid-19 in schools and school communities. https://org.za/wp-content/uploads/2020/05/dbe-standard-operating-procedure-for-covid-19.pdf. Accessed 28 May 2020

Department of Basic Education (2020e) Basic education department condemns vandalism of 397 schools since Covid-19 lockdown. Retrieved from: https://kfmulaudzi.files.wordpress.com/2020/04/basic-education-department-condemns-vandalism-of-397-schools-1.pdf. Accessed 10 May 2020

Department of Basic Education (2020f) Minister Angie Motshekga: basic education sector recovery plans for the reopening of schools, following the coronavirus COVID-19 lockdown adjustment of regulations. Retrieved from: https://www.gov.za/speeches/minister-angie-motshekga-basic-education-sector-recovery-plans-reopening-schools-following. Accessed 20 May 2020

Department of Basic Education (2020g) Statement by the Minister of Basic Education, Mrs Angie Motshekga on the state of readiness for the return of the second cohort of grades back to school. Accessed 7 July 2020

Department of Basic Education (2020h) Basic Education Minister welcomes Cabinet decision, calls for communities to protect schools against vandalism. https://kfmulaudzi.files.wordpress.com/2020/07/basic-education-minister-welcomes-cabinet-decision.pdf. Accessed 27 July 2020

Djalante R, Lassa J, Setiamarga D, Rafliana I (2020) Review and analysis of current responses to COVID-19 in Indonesia: period of January to March 2020. https://www.ingsa.org/covidtag/academic-papers/djalante-indonesia/. Accessed 10 May 2020

Education Labour Relations Council Collective Agreement No 1 of 2020 concessions process to flow from employees with comorbiditie

Emas R (2015) The concept of sustainable development: definition and defining principles. https://sustainabledevelopment.un.org/content/documents5839GSDR%202015_SDconcept_definiton_rev.pdf. Accessed 15 May 2020

Esposito S, Principi N (2020) School closure during the coronavirus disease 2019 (COVID-19) pandemic: an effective intervention at the global level? JAMA Pediatr https://doi.org/10.1001/jamapediatrics.2020.1892. Accessed 16 May 2020

Hynes NRJ., Kumar JS, Kamyab H, Sujana JAJ, Al-Khashman OA, Kuslu, Y, … & Suresh B (2020). Modern enabling techniques and adsorbents based dye removal with sustainability concerns in textile industrial sector-A comprehensive review. J cleaner production 122636

Joint Media Statement by Education Teacher Unions and National Governing Body Associations on the Readiness of Schools to Re-Open (2020) https://www.sadtu.org.za/content/joint-media-statement-education-teacher-unions-and-national-governing-body-associations. Accessed 2 June 2020

Kekić D, Miladinović S (2016) Functioning of educational system during an outbreak of acute infectious diseases. Retrieved from: https://www.researchgate.net/publication/309728224. Accessed 24 May 2020

KZN Circular No. 48 of 2020. Re-engineering if the schooling system

KZN Circular No. 63 of 2020. Closure of schools until the 23 August 2020

Macupe B (2020) Teacher unions warn Motshekga not to jump the gun on reopening schools. Mail & Guardian. https://mg.co.za/coronavirus-essentials/2020-05-13-teacher-unions-warn-motshekga-not-to-jump-the-gun-on-reopening-schools/. Accessed 19 May 2020

Mahaye NE (2020) The impact of COVID 19 pandemic on education: navigating forward the pedagogy of blended learning. https://www.researchgate.net/project/The-Impact-of-COVID-19-Pandemic-on-South-African-Education-Navigating-Forward-the-Pedagogy-of-Blended-Learning-dagogy. Accessed 5 May 2020

Mahlati Z (2020) Angie Motshekga blames uneven state of readiness for further delays in reopening schools. IOL. https://www.iol.co.za/news/politics/angie-motshekga-blames-uneven-state-of-readiness-for-further-delays-in-reopening-schools-48798539. Accessed 22 July 2020

Maromo J (2020a) Teacher union resists push to reopen schools under level 4 of lockdown. IOL. https://www.iol.co.za/news/politics/teacher-union-resists-push-to-reopen-schools-under-level-4-of-lockdown-47357383. Accessed 10 May 2020

Maromo J (2020b) Sadtu calls for schools across South Africa to close as Covid-19 infections rise. ANA. https://www.africannewsagency.com/coronavirus/Sadtu-calls-for-schools-across-South-Africa-to-close-as-Covid-19-infections-rise-26298637. Accessed 25 July 2020

Mlambo S (2020) Teacher unions slam DBE's June 8 date for reopening schools to pupils. IOL. Retrieved from: https://www.iol.co.za/news/politics/teacher-unions-slam-dbes-june-8-date-for-reopening-schools-to-pupils-48794286. Accessed 8 June 2020

Naki E (2020) SADTU calls on teachers to stay away until it's convinced schools are safe. The Citizen. https://www.oudtshoorncourant.com/News/Article/National/SADTU-calls-on-teachers-to-stay-away-until-it-s-convinced-schools-are-safe-202005111244. Accessed 25 May 2020

Ngqakamba S (2020) School closures: unions 'cautiously' welcome decision but private school concerns remain. News 24. https://www.news24.com/news24/southafrica/news/school-closures-unions-cautiously-welcome-decision-but-private-school-concerns-remain-20200724. Accessed 28 July 2020

Nkanjeni U (2020) Department of basic education claps back at school opening criticism. Sunday Times. https://www.timeslive.co.za/politics/2020-07-15-department-of-basic-education-claps-back-at-school-opening-criticism/. Accessed 20 July 2020

Petrie C, Aladin K, Ranjan P, Javangwe R, Gilliland D, Tuominen S, Lasse L (2020) Spotlight: quality education for all during COVID-19 crisis. Retrieved from: https://www.hundred.org/en/research. Accessed 10 May 2020

Pons X (2015) Educational policies. The implementation, Claude Lessard, Anylène Carpentier. Paris, PUF, 208 p. Int Rev Educ Sèvres (69):37–39

Repenning NP, Sterman JD (2002) Capability traps and self-confirming attribution errors in the dynamics of process improvement. Adm Sci Q 47(2):265–295

Sintema EJ (2020) Effect of COVID-19 on the performance of grade 12 students: implications for STEM education. EURASIA J Math Sci Technol Educ 16(7). https://doi.org/10.29333/ejmste/7893. Accessed 20 July 2020

Tam G, El-Azar D (2020) Resilience must be built into our educational systems. Minerva Projects. https://www.weforum.org/agenda/2020/03/3-ways-coronavirus-is-reshaping-education-and-what-changes-might-be-here-to-stay/. Accessed 16 June 2020

United Nations Educational, Scientific and Cultural Organization (2020) COVID-19 educational disruption and response. https://en.unesco.org/themes/education-emergencies/coronavirus-schoolclosures. Accessed 25 May 2020

Viner RM, Russell SJ, Croker H, Packer J, Ward J, Stansfield C, Booy R (2020) School closure and management practices during coronavirus outbreaks including COVID-19: a rapid systematic review. Lancet Child Adolesc Health. https://doi.org/10.1016/S2352-4642(20)30095-X. Accessed 19 May 2020. p:/0.1016/52350095-X33

World Bank (2020) COVID-19 strategic preparedness and response program and proposed 25 projects under phase 1: using the multiphase programmatic approach. Human Development Practice Group, New York. https://documents1.worldbank.org/curated/en/993371585947965984/pdf/World-COVID-19-Strategic-Preparedness-and-Response-Project.pdf. Accessed 10 May 2020

World Health Organization (2020) Director-General's opening remarks at the mission briefing on COVID-19. 2020. https://www.who.int/dg/speeches/detail/who-director-general-s-opening-remarks-at-the-missionbriefing-on-COVID-19. Accessed 12 May 2020

Dr. Bongani Sibusiso Mchunu received his Ph.D in 2016 at the University of Kwazulu-Natal. His field of specialisation is (School development) (theories on systems thinking with focus on improvement of the practice for holistic school development). To date he has 1 publication in on journal. However, there are four pending papers based on the COVID-19 disruptive impact on education and churches sent to other journals, awaiting publication.

Dr. Annatoria Zanele Ngcobo received her Ph.D in 2015 at the University of KwaZulu Natal. Her Field of Specialisation: Mathematics and Mathematics Education (learning theories with special focus to knowledge construction and knowledge for teaching). She is the Full bright Alumni Fulbright. To date she has 14 publications made up of eleven journal articles, two conference proceedings paper and one book chapter all in the field of Education.

Thamsanqa Thulani Bhengu is an Associate Professor in the Discipline of Educational Leadership, Management and Policy in the School of Education, University of KwaZulu-Natal. He teaches modules at BEdHonours and Master of Education levels, and supervises PhD students. He also serves as Academic Leader: Doctoral Studies.

COVID-19 and Education for Sustainable Development Role: An Assessment of Latin American Schools

Vanessa Tibola da Rocha, Bárbara Fritzen, and Luciana Londero Brandli

Abstract The COVID-19 pandemic is not only a health crisis but also a crisis for all humanity. In early 2020 and the "Decade of Action" to accelerate progress on the Sustainable Development Goals, humanity needed to change its ways of relating, communicating, consuming, and teaching to reduce the spread of COVID-19. However, this pandemic may mark a transformational period for all society around the world, if the right actions are taken, since behavioral changes have already been forced on humanity, a fundamental requirement for sustainable practical actions. Education for Sustainable Development (ESD) allows students to be empowered to face present and future global challenges, creating more resilient and sustainable societies. This approach signals and emphasizes the transforming role of ESD increasingly in the current global context. The objective of this article is to present the impacts of the COVID-19 pandemic in schools' ESD, as well as to which extent ESD may prepare resilient citizens to face a future global crisis. The methodology is classified as a descriptive and qualitative critical assessment based on UNESCO reports and government websites. Schools from the Latin America region are the scope of the study. The results show how this region is conducting the primary and secondary education process during the pandemic, how these methods are evaluated, highlighting how quality education, in addition to encouraging students to promote virus prevention, prepares future citizens for the promotion of a society committed to present and future generations.

Keywords Education · Sustainability · COVID-19

V. T. da Rocha · B. Fritzen (✉) · L. L. Brandli
University of Passo Fundo, BR 285, São José, Passo Fundo, Rio Grande do Sul 99052-900, Brazil
e-mail: barbaramaria.fritzengomes@haw-hamburg.de

L. L. Brandli
e-mail: brandli@upf.br

W. Leal Filho (ed.), *COVID-19: Paving the Way for a More Sustainable World*,
World Sustainability Series, https://doi.org/10.1007/978-3-030-69284-1_20

1 COVID-19 and Quality Education: Figures, Facts and Challenges

The COVID-19 crisis is more than a health crisis. Education is one of the social areas most affected by this pandemic, declared on March 12, 2020, by the World Health Organization (WHO 2020). According to the UN Educational, Scientific and Cultural Organization (2020a, b, c, d, e) since then, 22% of the worldwide enrolled students were affected by the shutdown. Following this declaration, immediately the closure in schools around the world got double size (44% on March 16, 2020) considering 110 country-wide closures. In April 2020, more than 1.5 billion children and young people had been affected (91.3% of the global students' population in 194 countries).

In the countries with the weakest health infrastructures, the duration of these measures has been extended until a vaccine becomes available (Reimers et al. 2020). In Latin American region, COVID-19 triggered an unprecedented educational crisis, and the situation is even harder than other regions in the world: more than 95% of children were out of school at the end of March 2020, around 150 million children from early childhood centers and pre-primary, primary and secondary schools (UNICEF 2020a).

Educational institutions' shutdown was driven by the need for lockdowns and social-distancing policies (Sintema 2020; Zhang et al. 2020). Physical distancing is one of the most effective measures to limit and minimize the spread of viruses (Jones 2020; Kamal et al. 2020; Jackson et al. 2016; Eames et al. 2010; Sintema 2020), considering the estimated global impact may range between 40 million deaths in case of no prevention measures and it can be reduced 50% with interventions (Walker et al. 2020). In each country or region, the adopted prevention measures change, for instance, Reimers et al. (2020) research reveals Argentinean schools could be open, however, the teachers were working in rotating shifts. This measure allowed the delivery of teaching resources and food for students. In Costa Rica, other study resources have been used together with online reading.

Viner et al. (2020) argue that 'evidence of COVID-19 transmission has an important role in the outbreak'. However, studies carried considering other outbreaks and viruses' diseases argue that the effectiveness of mass school closure is often low (Nafisah et al. 2018) and perhaps this disruption will be drastic for the education process, causing severe learning losings for students (Reimers and Schleicher 2020; Sintema 2020; Zhang et al. 2020; Leal Filho et al. 2020), significantly higher for marginalized students (Viner et al. 2020; UNESCO et al. 2020; Ferdig et al. 2020; Zhang et al. 2020; Leal Filho et al. 2020). UNESCO (2020b), Reimers et al. (2020), Sintema (2020), and Saavera (2020) state adverse consequences regarding school closure such as: interrupted learning and development; compromised nutrition; confusion and stress for teachers and unprepared parents; gaps in childcare; rise in dropout rates; social isolation considering schools' hubs of social activity and human interaction.

As cited, the COVID-19 represents an adaptive and transformative challenge for students, teachers, and parents (Reimers et al. 2020; Zhang et al. 2020). Moreover, many countries have unequal education systems (Saavera 2020), reporting different human, pedagogical, and technical challenges regarding the schools' closure and the most cited are the availability of tools, technological devices, channels, and good quality internet (UNICEF 2020a; Reimers et al. 2020). The gaps and challenges are amplified by the unavailability of technological devices and internet access, differences among students' parents support, differences in schools' capacity to support remote learning, differences among students' resilience, motivation and skills to learn independently, and a lack regarding guidelines to orient the teaching and learning process (Reimers et al. 2020; Zhang et al. 2020; Ferdig et al. 2020; Leal Filho et al. 2020).

At the beginning of twenty-first century, UNICEF emphasized that quality education would be required, ensuring five aspects: (1) learning what they need to learn, for learning throughout life; (2) being healthy, well-nourished, and free from exploitation, violence, and labor; (3) being aware of their rights and have opportunities to realize them; (4) being able to participate in decisions that affect their lives in accordance with their evolving capacities; (5) being able to respect diversity, practice equality, and solve differences without violence. Two decades later the publication of the UNICEF document, these guidelines are currently encouraged by the United Nations Sustainable Development Goals (SDG), with a special focus on SDG 4 and its targets. However, the COVID-19 crisis has expanded the challenges for SDG 4, which is part of the UN Agenda 2030. Leal Filho et al. (2020) states that "the pandemic will therefore harm education in all spheres (SDG 4—Quality education)".

Besides adverse effects in students' performance (Sintema 2020) and cited challenges for all education stakeholders, this breakdown may exacerbate the risk of permanent dropout of vulnerable students in disadvantaged communities (UNICEF 2020a; UNESCO 2020a) and the risk of teenage pregnancy, sexual exploitation, child marriage, violence, and other threats may increase (UNESCO et al. 2020).

Considering these gaps, challenges and risks, governments, organizations, and researchers worldwide have carried studies and created materials to guide effective responses to the COVID-19 pandemic and future crises. The Organisation for Economic Co-operation and Development (OECD 2020) has developed a series of modules 'to facilitate the rapid design process and implementation of adaptive responses to the emerging education challenges and to protect young people's educational opportunities during and following the pandemic' (Reimers et al. 2020). UNESCO et al. (2020) worked together to provide a framework to guide governments in facilitating the reopening of schools in a safe way for students and educational staff. Table 1 summarizes some actions on the process of basic education in the world in order to minimize COVID-19 spread, considering positive and negative impacts emphasized by the global scenario.

Focusing on the Latin American schools' scenario, this study aims to present the main challenges and gaps triggered by COVID-19 pandemic and strategies, actions, and initiatives adopted to address these issues that countries have faced. These initiatives were classified regarding quality education (SDG 4) aspects and Education for

Table 1 The impact of COVID-19 on basic education in the world

Action	Positive	Negative	Source
Use of ICTs	Insertion of Information and Communication Technologies (ICTs), in the context of schools; ICTs Storage of archives, documents and content made available virtually by the clouds Joint learning of teachers, students and relatives for the use of ICTs	Students without access to the Internet and equipment (mobile phone, computer, printer) suitable for monitoring online learning In the early grades, children have greater difficulties in managing the educational use of ICT and need the help of parents and/or legal guardians to access online education Risk and exposure of human rights, due to social networks and access to personal data of people while using the networks	UNESCO Institute for Statistics Blog 2020 Education International (2020)
Open Access Media: TV and Radio	They enable the continuity of education (in remote conditions), information and access to the local, national and global context—in the face of the current crisis stimulated by COVID-19	Families in socially or geographically vulnerable conditions located in remote access regions, remain without access to education because they do not have the appliances to access the information; without power installed in the homes and/or in difficult access condition for the transition signal and connection to such media (TV and radio)	UNESCO Institute for Statistics Blog 2020

(continued)

Table 1 (continued)

Action	Positive	Negative	Source
Remote learning	Remote teaching is a disease prevention strategy It stimulates new learning and skills in the school community (teachers, students and staff) and in families	There has been a significant increase in the number of children and women victims of domestic violence Increase in the number of pregnant girls in adolescence High risk of abdoning education and education, due to the difficult access and reaching, mainly, the most vulnerable people in society Scenario of instability in public education and risks of teacher unemployment	UNESCO (2020d)
School snack	Undescribable[a]	Children without access to school meals and exposed to poor nutrition. For the most vulnerable, school lunch means the main and only meal of the day	UNESCO (2020d)
Social contact and group learning	Undescribable[a]	Difficulty in communication relations and inclusion among all. The school environment represents for the children their first experience with life in community/society	UNESCO (2020d)
Children and youth with disabilities and affected by trauma or mental health issues	Undescribable[a]	Low efficiency in monitoring the teaching and learning of children and youth with disabilities and affected by trauma or mental health issues	Education Cannot Wait (2020)

[a]Positive aspects were not presented by the search sources until the time of the investigation
Source Prepared by the authors based on the references

Sustainable Development principles in order to identify their impact in the future education. The study answers the following questions: (1) How does ESD relate to COVID-19? and (2) How does the pandemic impact on SGD 4, regarding the quality education?

2 COVID-19 and Education for Sustainable Development in Schools

Education is the key to sustainable development (UNESCO 2005). For Agbedahin (2019) education is also a potential tool for the transformation of contemporary society. Vare and Scott (2006) already emphasized that education for sustainable development would be a challenge for contemporary society. According to Hill et al. (2020), the pandemic has accelerated the process of rethinking the global education model and it has been necessary to develop teaching strategies considering social distance, difficulties in accessing remote education, social inequalities and prioritizing the health and well-being of people.

Currently the disciplines and educational institutions at different levels present a commitment to Education for Sustainable Development (ESD). Notwithstanding, there are still difficulties in relation to practical issues related to the subject. There is a need to change habits and behaviors in favor of a better future for all. For Korobar and Siljanoska (2016), teaching focused on sustainable development approaches is focused, in most cases, on knowledge transfer between the groups involved in the educational process (teachers and students) and there is a need to go further.

For Jorge (2020), rethinking the education model in the twenty-first century is an emerging need to ensure gender equality, social inclusion, quality education and all other principles related to ESD. The ESD stimulates the development of human skills and competences to answer emergency questions; to manage their emotions; to develop resilience in the face of uncertainties; to positively impact the community where the person is inserted; among other important characteristics for a better life, with more quality and sustainability (UNESCO 2020c).

In contemporary times, it is clear that global problems need solutions thought out together, COVID-19 is an example of this reflection (UNESCO 2020c). Lockdown and quarantine were inserted to prevent and to reduce the number of infected people by the new coronavirus. As a consequence, a significant reduction in CO_2 emissions was globally identified (UNESCO 2020c). COVID-19 has awakened in governments, decision-makers, public and civil entities greater attention to public health and the quality of the built environment. According to UNICEF (2020b), in response to the fight against the COVID-19 disease, the WASH Programme has become priority. UNICEF's water, sanitation and hygiene (WASH) supports affected, at-risk, low-capacity and fragile countries to develop and implement infrastructure to access of drinking water, basic sanitation and adequate sanitation. There has been an expansion of WASH networks across the planet. Unfortunately, a high number of people do not have access to WASH Programme, in most cases, people in situations of widespread social vulnerability. Though, some questions still remain: what will it be like when the routines return to the "new normal"? Will happens a change in behavior to further reduce emission levels? Or will levels continue to rise significantly until future crises?

UNESCO (2020c) empathizes incentives for ESD is as urgent as responding to COVID-19 and its challenges. It is still necessary to broaden the incentive for ESD at the different levels of education so that these transformations are indeed observed on a global scale. Even as the new coronavirus pandemic, ESD is an effective strategy for combating global problems, based on eight basic skills, such as systemic thinking, anticipatory, normative, strategic, collaborative competence, critical thinking, self-awareness, and integrated problem-solving competences (Hoffmann and Siege 2020; UNESCO 2017), encouraging the development of sustainable behaviour and transforming people and their realities in a positive way. The coronavirus pandemic has required the same skills and competences.

According to the Australian Research Institute for Environment and Sustainability (ARIES 2009), ESD is based on seven principles: Transformation and change; Life-long learning for all; Systems thinking; A better future; Critical thinking; Participation; Partnerships for change. Each principle contributes to the human development of capabilities and competences towards sustainable development.

The commitment to ESD and Quality Education (SDG4) must ensure learning for all contexts and learners, regardless of gender, geographical location, socio-economic status, availability of ICT, and intellectual capacity (UNESCO 2005). Cebrián et al. (2020) emphasize that COVID-19 presents challenges and opportunities for global education and the school environment is providing answers for the transformation and updating of the physical and virtual school context. The teaching processes directed to ESD, treating the classroom space as smart, and using active teaching methodologies should be considered. However, Wolff (2020) emphasizes that online education made possible by technological resources such as computers and cell phones needs immediate response as to the proper end destination of these devices from the moment they are discarded or become electronic waste. ESD is associated with the entire circle and must analyse the benefits and challenges of the use of electronics and the final destination considering the different global realities.

Ensuring inclusion and the right of everyone to access online education is paramount for global communities to follow a successful path towards sustainable development. However, there are numerous factors that need to be considered throughout the process if ESD and Quality Education are to occur (UNESCO 2005).

3 Methodology

The focus of this study is schools from the Latin America region, which is composed of 20 countries from South America, Central America, and North America. According to the United Nations, the Latin American countries are Argentina, Bolivia, Brazil, Chile, Colombia, Costa Rica, Cuba, Dominican Republic, Ecuador,

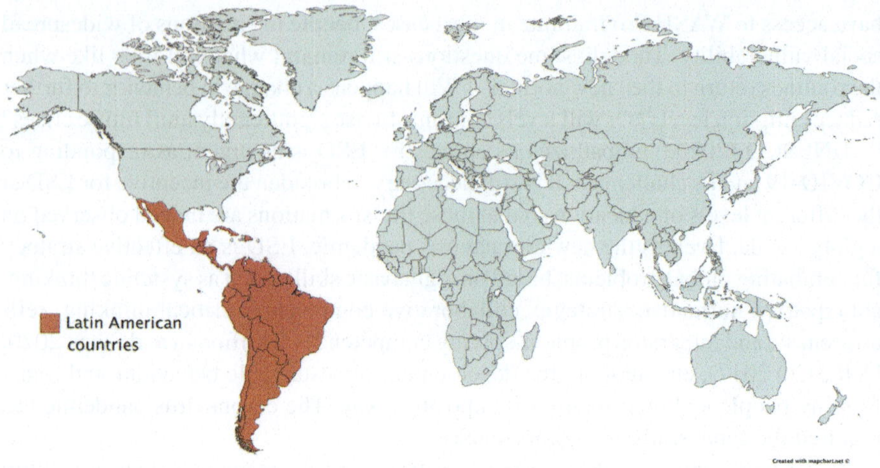

Fig. 1 Latin America region

El Salvador, Guatemala, Haiti, Honduras, Mexico, Nicaragua, Panama, Paraguay, Peru, Uruguay, and Venezuela. The Caribbean countries were not considered in this study. Figure 1 shows the Latin American countries.

The adopted methodology is classified as a descriptive and qualitative critical assessment based on reports published mainly by the United Nations Educational, Scientific and Cultural Organization (UNESCO) and countries' government reports and websites. After a literature review on how the COVID-19 crisis has affected education and how ESD has been impacted by this pandemic, an identification of educational measures, strategies, and actions taken to mitigate this educational crisis in each Latin American country was developed. These measures were classified as suggested by the Economic Commission for Latin America and the Caribbean (ECLAC 2020) in the COVID-19 Observatory in Latin America and the Caribbean.

After listing and categorizing the initiatives from each country, their impacts were discussed and they were classified according to the seven principles of Education for Sustainable Development (ARIES 2009) and the five aspects of Quality Education—SDG 4 (UNESCO 2020a). Figure 2 presents the seven principles of ESD and the five aspects of quality education considered for the analysis.

Figure 3 presents the methodology overview and the research stages, according to the paper aim.

4 Results

The results are divided into three sections, presenting a scenario pre-COVID-19, the main challenges and gaps the Latin American educational stakeholders are facing

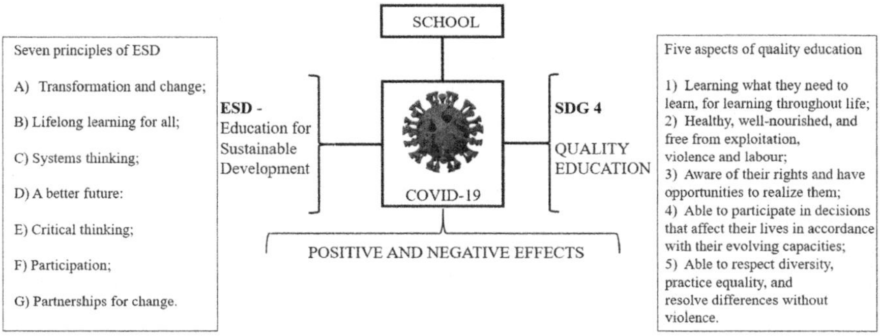

Fig. 2 COVID-19 and effects on ESD and SDG 4. *Source* Developed by the authors based on: UNICEF (2000) and ARIES (2009)

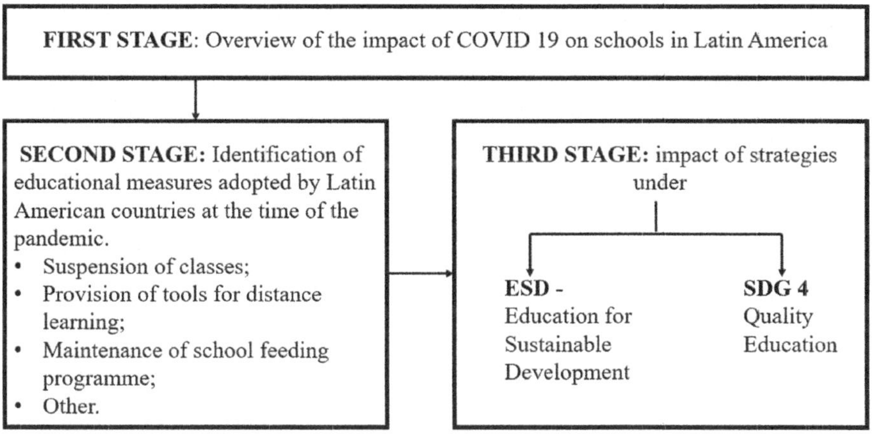

Fig. 3 Methodology overview and research stages

during the pandemic, and which strategies and actions these twenty countries are taken to address and to minimize the impacts.

4.1 *Latin American: Pre-pandemic Situation*

Latin America has faced socioeconomic problems prior to the COVID-19 crisis. In 2016, 186 million people were living in poverty, and 61 million in extreme poverty in this region (ECLAC 2020). Unfortunately, poverty exacerbates social inequality and social-economic segregation is persistent.

Fiszbein and Stanton (2018) state that in the past two decades Latin America had reached great outcomes in education. In 2000, only six Latin American countries spent more than 5% of GDP on public education expenditures. In 2015, this number increased to fifteen countries. Never before, too many children and teenagers were enrolled in schools. Currently, Latin America has more than 150 million students (World Bank 2020b), nevertheless, almost 12 million children and youth were excluded from education in Latin America and the Caribbean even before the pandemic.

Therefore, Latin America's greatest challenge is expanding equitable access to the education system (Fiszbein and Stanton 2018). There are around 10.5 million children and adolescents in child labor, compromising SDG 8 targets—"ending child labor in all its forms by 2025" (ILO 2017). For instance, even after compulsory basic education laws, Paraguay and Peru are examples of countries that permit child labor at age fourteen (Global Partnership for Education 2020).

Inclusive education also addresses students with disabilities. In Latin America, only 29% of schools have adapted infrastructure and materials. Ethnic origin segregation is another hard issue for the region. According to the Global Partnership for Education (2020), in Brazil, Mexico, and Peru, the indigenous background still increases social inequality, especially regarding education. As well as, gender and race are other focuses. Global Partnership for Education (2020) reveals that over 50% of Brazilian, Colombian, and Mexican teachers reported needs on teaching students with special needs.

4.2 Covid-19 and Latin American Schools: Main Challenges and Gaps

Fiszbein and Stanton (2018) cite an important challenge in implementing a quality educational system in the Latin America: lack of a strategic vision or plan to guide decision-making. Addressing the challenges never before was so relevant. COVID-19 pandemic promptly required countries to put in place equitable and inclusive initiatives to mitigate the several impacts of the crisis in the educational systems (Global Partnership for Education 2020).

The Economic Commission for Latin America and the Caribbean (2020) estimates the number of people in extreme poverty will severely increase to 83 million due to the impacts of the COVID-19 pandemic (ECLAC 2020). This crisis Latin America is facing compromises inclusive social development, defined as the access to basic infrastructure services, such as education (ECLAC 2020). Living in the poorest families and discriminated against by society, these children and teenagers usually are not prioritized in terms of education and the risk of exclusion widens during the pandemic (Global Partnership for Education 2020; UNESCO 2020b). Only a quarter of Latin American and the Caribbean countries has laws and regulations for inclusive education covering all learners (Global Partnership for Education 2020).

Deep pre-existing inequalities within and across Latin American countries have increased and, for education, the most significant impact of COVID-19 is the school closure. It is supposed to have potential negative impacts and to worsen the educational situation outcomes, such as losses in learning, increased dropout rates, and children's mission their most important meal of the day (World Bank 2020b; Leal Filho et al. 2020). Education impacts can exacerbate existing vulnerabilities in the Latin America scenario: child marriage, teenage pregnancy, and child labor (ECLAC 2020). In the Latin America scenario, in some cases, distance learning is hypothetical. It is noticed that learners from low-income families, and in remote rural areas have difficulties attending the classes and follow the learning, especially because of the lack of technological devices and internet, and sometimes not even the access to TV and radio (Global Partnership for Education 2020; Leal Filho et al. 2020).

4.3 Covid-19 and Latin American Schools: Strategies and Actions

The role of the education system may be strategic for social issues during the pandemic time. Considering its own scenario, each country is working and taking innovative and flexible measures to minor the impacts on education (World Bank 2020b). The COVID-19 Observatory in Latin America and The Caribbean reveals how the countries are adapting ways during the crisis. They are using classification for the measures, such as suspension of classes, provision of tools distance learning, maintenance of school feeding program, and others. Figure 4 shows the number of measures/actions adopted by each Latin American country and Fig. 5 presents the percentage of the twenty Latin American countries is taking the main adopted strategies and actions.

Most of the Latin American countries are adopting many strategies and actions to face the pandemic restrictions. As mentioned above, 95% of the countries in Latin America have adopted school closure. At the end of July 2020, among the 20 countries, only, in Uruguay and Nicaragua, schools are open, however with limitations (World Bank 2020b). Figure 6 shows the schools' situation and the number of students enrolled in each Latin American country.

Uruguay has the lowest rates of COVID-19 cases and deaths, therefore, at the end of April 2020, schools reopened in remote rural areas, mitigating the impact on vulnerable children and adolescents. At the end of June 2020, schools in the Uruguayan main cities also reopened (UNESCO 2020c). Nicaragua is another Latin American country that students are having classes in the physical classrooms. This country has faced various challenges in education even before the pandemic. A high number of out-of-school children and low quality of preschool education are the hardest challenges (Global Partnership for Education 2020). At the end of April 2020, more than 1.6 million students returned to schools in Nicaragua (UNESCO 2020c).

In Brazil, the country with the highest number of students in Latin America, basic education was supposed to distribute the classes in 200 days. After the pandemic was

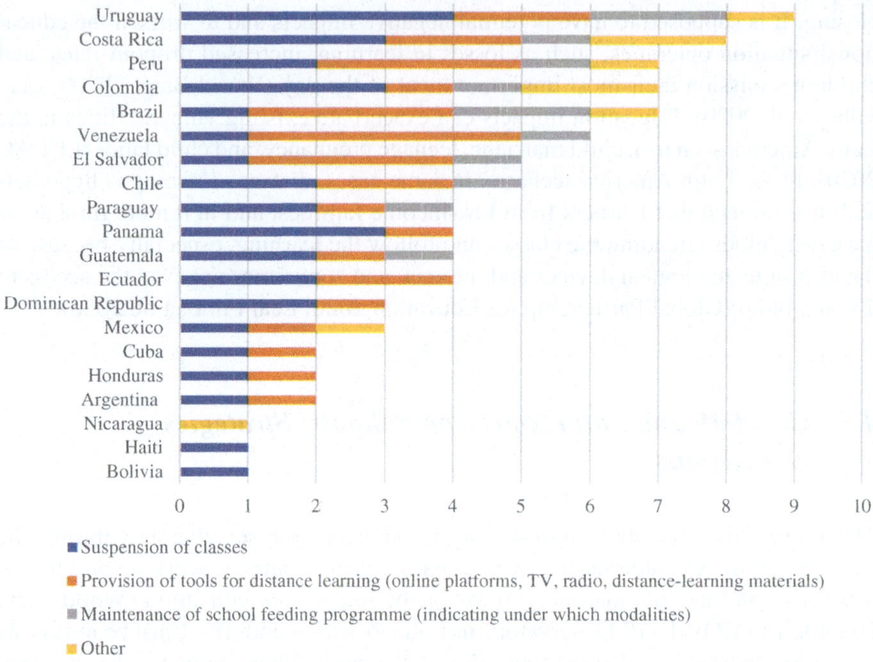

Fig. 4 Number of measures/actions adopted by each country. *Source* Authors based on the COVID-19 Observatory in Latin America and the Caribbean (ECLAC 2020)

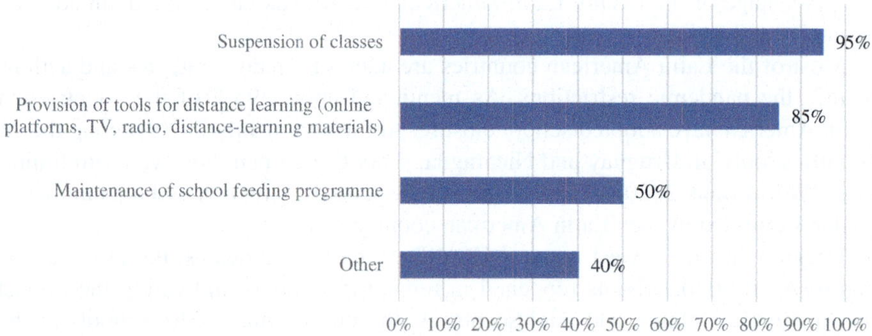

Fig. 5 Percentage of countries implementing the main measures. *Source* Authors based on the COVID-19 Observatory in Latin America and the Caribbean (ECLAC 2020)

announced by the World Health Organization, the Brazilian Ministry of Education authorizes the replacement of in-person classes to distance learning. This could be adopted for thirty days. In June 2020, this period was extended until the end of 2020 (31st of December) (MEC 2020). Considering the vast territory of Brazil, each

	Country	Number of students
Closed	Argentina	11,061,186
	Bolivia	2,966,735
	Brazil	51,326,929
	Chile	3,652,100
	Colombia	10,434,248
	Costa Rica	1,100,782
	Dominican Republic	2,450,277
	Ecuador	4,462,460
	El Salvador	1,414,326
	Guatemala	4,192,944
	Haiti	1,650,480
	Honduras	2,024,045
	Mexico	33,159,364
	Panama	837,246
	Paraguay	1,519,678
	Peru	9,015,606
	Venezuela	6,866,822
	Total	**148,135,228**
Open with limitations	Nicaragua	1,607,008
	Uruguay	793,841
	Total	**2,400,849**
*	Cuba	1,876,156

Fig. 6 Schools' situation due COVID-19 shutdown and the number of students in Latin America. *Source* Developed with Mapchar.net by authors based on World Bank (2020b). *no information

Brazilian state has had the autonomy to decide about the strategies to face COVID-19 pandemic.

Considering the needs triggered by distance learning, technological actions have been used by countries to respond to the crisis, incorporating different channels and media to facilitate teaching and learning. A national repository of digital resources and learning materials has been offered for the students. Communications channels such as WhatsApp, phone callings, and social media were adopted to provide pedagogical guidance and support for teachers and parents. Radio and TV are also part of the strategy program since the internet is not accessible for all stakeholders (World Bank 2020a). Partnerships for providing free Wi-Fi and technological devices (computers, laptops, smartphones, and tablets) for students are also measures to minimize the digital lack in the Latin American countries (UNESCO 2020c).

To avoid babies, children and teenagers face a lack in their nutrition, some countries have implemented strategies to deliver or distribute scholar feeding during the schools' closure. Such measures include the delivery of milk for younger children, kits of food, and/or monthly financial contribution (UNESCO 2020c). Other measures are psychological support for students, teachers and families, and special activities for people with disabilities or in a vulnerable situation.

The adopted measures aim to avoid the COVID-19 spread and mitigate the negative impacts in the education sector triggered by the suspension of the in-person classes. Quality education is a goal for the Latin America, even before the pandemic, since this is a premise of SDG 4 targets. During the COVID-19 pandemic, the measures taken by these countries have addressed quality education guidelines. Considering ESD can stimulate the development of competences to deal with emergency and uncertain questions (UNESCO 2020c), these measures can also go to meet with Education for Sustainable Development principles. Table 2 shows the main measures adopted by Latin American countries, the countries that have taken them, and guidelines for quality education and ESD principles they address.

Each country is developing different actions for this moment. However, some countries are not considering their own social and economic scenario. Therefore, efficient short, medium and long-term solutions have been failing, since the main initiatives are not accessible to the majority of the learners (Global Partnership for Education 2020), and this may also compromise quality education aspects and also the Education for Sustainable Development.

5 Conclusion Remarks

COVID-19 represents a global challenge that affects all areas of knowledge and amplifies the challenges for a Quality Education (SDG 4) and the Education for Sustainable Development. The main measure adopted by countries, not only in the Latin American region, but also all over the world was schools' closure, forcing a remote education strategy. However, in many cases, governments are not considering the social vulnerability of the population, and pre-existing inequalities may be exacerbated. In some countries' scenario, the access to needed resources for remote teaching and learning is precarious or inexistent. Moreover, for many children in poverty or extreme poverty, schools' closure goes beyond compromising their education, increasing risks of poor nutrition, violence, and sexual abuse, as well as opens a lack of scholar dropouts.

Nevertheless, if the right social actions are taken by governments, the COVID-19 crisis may also be seen as a possibility for society's transformation. This pandemic is evidencing how human has the abilities to respond to emergencies and collaborate in finding the best solutions (UNESCO 2020c), prerequisites for good staring in sustainable development achievement. In this sense, this crisis' experience may contribute to sustainability implementation, since human behavior has been requested to change.

As a social aspect, education through the teaching and learning process had to undergo adaptations in the face of the new scenario (COVID-19). Even remote education faces difficulties as a lack of technological devices, internet access, and, in some cases, access to communication media, positive effects were triggered by this transformation, considering a global context that involves schools, children, their families, and teachers. Thus, it is possible to highlight: (1) the use of ICT; (2) the distribution of school meals; (3) the use of free access media (TV and radio), providing more training content; (4) and the increased possibility of preventing the disease caused by the new coronavirus.

Table 2 Educational measures for COVID-19 pandemic and guidelines for quality

Strategy	Guidelines for quality education	ESD principles	Country
Suspension of classes			
School closure	2	A, B	Argentina; Bolivia; Brazil; Chile; Colombia; Costa Rica; Cuba; Dominican Republic; Ecuador; El Salvador; Guatemala; Haiti; Honduras; Mexico; Nicaragua; Panama; Paraguay; Peru; Uruguay; Venezuela
Anticipation of winter/summer holidays	2	–	Brazil; Chile; Colombia; Costa Rica; Guatemala; Mexico; Peru; Uruguay
Changes in the school year	3, 5	B, F	Brazil
Special activities for holidays	2	B, D, F	Mexico
Provision of tools for distance learning			
Online platform	1, 2, 3	A, B, C, D, E, F, G	Argentina; Brazil; Chile; Colombia; Costa Rica; Cuba; Dominican Republic; El Salvador; Guatemala; Haiti; Honduras; Mexico; Panama; Paraguay; Peru; Uruguay
Online live channel	1, 2, 3	A, B, C, D, E, F, G	Ecuador
Radio and TV	1, 2, 3, 5	A, B, C, D, E, F, G	Argentina; Brazil; Chile; Colombia; Cuba; Ecuador; Panama; Peru; Venezuela
YouTube channel	1, 2, 3	A, B, C, D, E, F, G	Argentina; Brazil; Ecuador; El Salvador; Honduras; Mexico; Panama
Communication channels and social medias (WhatsApp, telephone calls, Facebook, etc.)	2, 3, 4, 5	A, B, D, E, F, G	Brazil; Honduras; Mexico; Panama;

(continued)

Table 2 (continued)

Strategy	Guidelines for quality education	ESD principles	Country
Training and support tools for teachers	2, 3, 4, 5	A, B, D, E, F, G	Argentina; Bolivia; Brazil; Chile; Colombia; Costa Rica; Dominican Republic; Honduras; Mexico; Panama; Peru; Uruguay; Venezuela
Support and orientation for families	2, 4, 5	A, B, D, E, F	Colombia; Costa Rica; Honduras; Panama; Uruguay; Venezuela
Provision of devices and internet for vulnerable students	1, 2, 3, 4, 5	A, B, C, D, E, F, G	Argentina; Colombia; El Salvador; Mexico; Paraguay; Peru;
Free download of didactic material and exercises	2, 3, 5	B, D, E, F, G	Argentina; Brazil; Chile; Cuba; El Salvador; Peru; Uruguay
Delivery of printed materials	2, 3, 5	B, C, D, E	Brazil; Chile; Colombia; Dominican Republic;
Virtual library and museums	1, 2, 3, 5	A, B, C, E, G	Chile; El Salvador; Guatemala; Mexico; Uruguay; Venezuela
Own tool for videoconference	2, 3, 4	B, D, E, F, G	Uruguay
Software, Apps and games	1, 2, 3, 5	A, B, C, D, F, G	Chile; Colombia, Mexico
Maintenance of school feeding programme			
Home gardening, focus on growing flowers, vegetables and vegetables	2, 4	D, E, F	Argentina
Check the functioning of the School Feeding Program	1, 2	C, D, G	Colombia
Food delivery and distribution	2, 3	B, C, D	Colombia; Costa Rica; Dominican Republic; El Salvador; Guatemala; Haiti; Paraguay
Other			
Incentives for good relations in the family context for quality education at home	1, 3, 4, 5	A, B, D, E, F, G	Colombia

(continued)

Table 2 (continued)

Strategy	Guidelines for quality education	ESD principles	Country
Psychological assistance for students and their families	1, 3, 4	C, D, E, G	Costa Rica; México
School health and personal hygiene	1, 2, 3	B, D, G	Cuba; Panamá; Nicaragua
People with disabilities in vulnerable situations receive government support	1, 3, 4, 5	B, F, G	Ecuador; Chile
School insurance (prevention COVID-19 telephone guidance guidelines)	1, 3, 4	B, D, F, G	Guatemala
Basic biosafety standards course	1, 2, 4	B, C, D, F, G	Honduras
Assistance and advice to migrants on educational issues	1, 3, 4, 5	D, F, G	Uruguay
Financial support for schools	1, 2	A, B, D, G	Dominican Republic

Source Developed by the authors, based on official websites of the ministries and presidents of the countries described in the table, general search source https://en.unesco.org/fieldoffice/santiago/covid-19-education-alc/monitoring

The use of ICT in the teaching and learning process has been discussed for a long time, however, before the COVID-19 pandemic, part of society had resistance in adopting or inserting technologies in the education system. As well as, programs focused on the implementation of technologies and internet quality access were not prioritized by governments. In this regard, this crisis has changed this reality, going in a pathway of significant changes in the future of education.

Latin American governments are shedding light on education and seeking alternatives for higher quality education with a focus on sustainable development. Many important initiatives were identified in this research, demonstrating the future of education requires the use of ESD as a roadmap, strengthening human capabilities to tackle challenges (UNESCO 2020c). Through this study, it was possible to identify how the taken measures could address Education for Sustainable Development, considering its seven principles. Every measure could be related to at least one ESD principle and may (a) transform and change, (b) bring lifelong learning, (c) promote systems thinking, (d) bring a better future, (e) promote critical thinking, (f) stimulate participation, and (g) incentive partnerships.

Regarding the aspects of quality education, contemplating the Sustainable Development Goal 4 targets, the taken initiatives could be associated with not less than one aspect. In this sense, these mitigation actions may stimulate (a) learning throughout life, (b) health and children's non-exploration, (c) awareness of the rights and opportunities, (d) participation in decisions, and (e) diversity respect and equality.

Due to differences in Latin American general context, each country is managing on its own, facing difficulties considering the availability of financial resources and accentuated by the social inequalities present in this territory even before the pandemic. It was also possible to identify some countries are not considering their own social and economic scenario. ESD can help to translate responses into concrete solutions for local scope. Education that goes beyond providing basic knowledge and skills is required for a better future (UNESCO 2020c). In the same way, governments, policymakers, and leaders are learning how to deal, manage, and respond to the challenges triggered by COVID-19. This know-how and experience may be used for future issues regarding global sustainable development promotion.

References

Agbedahin AV (2019) Sustainable development, education for sustainable development, and the 2030 agenda for sustainable development: emergence, efficacy, eminence, and future. Sustain Dev 27:669–680. https://doi.org/10.1002/sd.1931

ARIES (2009) Education for sustainability: the role of education in engaging and equipping people for change. Published by the Australian Research Institute in Education for Sustainability, Macquarie University, Parkes, Australia. Retrieved from: https://aries.mq.edu.au/publications/aries/efs_brochure/pdf/efs_brochure.pdf. Accessed 10 July 2020

Cebrián G, Ramon P, Mogas J (2020) The smart classroom as a means to the development of ESD methodologies. J Sustain 12:3010. https://doi.org/10.3390/su12073010

Eames KT, Tilston NL, White PJ, Adams E, Edmunds WJ (2010) The impact of illness and the impact of school closure on social contact patterns. Health Technol Assess 14:267–312

ECLAC (2020) Preventing the COVID-19 crisis from becoming a food crisis urgent measures against hunger in Latin America and the Caribbean. Published by United Nations. Retrieved from: https://repositorio.cepal.org/bitstream/handle/11362/45726/1/S2000392_en.pdf. Accessed 20 July 2020

Education Cannot Wait (2020) COVID-19 and education in emergencies. Published by Education Cannot Wait. Retrieved from: https://www.educationcannotwait.org/covid-19/

Education International (2020) Protecting the human rights of teachers and students and education in the digital age. Published by Education International. Retrieved from: https://www.ei-ie.org/en/detail/16870/protecting-the-human-rights-of-teachers-and-students-and-education-in-the-digital-age

Ferdig RE, Baumgartner E, Hartshorne R, Kaplan-Rakowski R, Mouza C (eds) (2020) Teaching, technology, and teacher education during the COVID-19 pandemic: stories from the field. Association for the Advancement of Computing in Education (AACE). Retrieved from https://www.learntechlib.org/p/216903/. Accessed 11 Aug 2020

Fiszbein A, Stanton S (2018) The future of education in Latin America and the Caribbean. Published by The Dialogue, Washington, DC. Retrieved from: https://www.observatorioeducacion.org/sites/default/files/usaid-layout-6.12.2018-final_pdf.pdf. Accessed 11 April 2020

Global Partnership for Education (2020) Education during and post COVID-19: the role of civil society. Published by Global Campaign for Education Secretariat, France.

Retrieved from: https://www.globalpartnership.org/blog/education-during-and-post-covid-19-role-civil-society. Accessed 10 July 2020
Hill C, Rosehart P, St Helene J, Sadhra S (2020) What kind of educator does the world need today? Reimagining teacher education in post-pandemic Canada. J Educ Teach. https://doi.org/10.1080/02607476.2020.1797439
Hoffmann T, Siege H (2020) What is education for sustainable development (ESD)? Published by ESD. Retrieved from: https://www.esd-expert.net/files/ESD-Expert/pdf/Was_wir_tun/Lehr-%20und%20Lernmaterialien/What_is_Education_for_Sustainable_Development.pdf
ILO (2017) Estimaciones mundiales sobre el trabajo infantil. Published by Oficina Internacional del Trabajo. Retrieved from: https://www.ilo.org/wcmsp5/groups/public/@ed_norm/@ipec/documents/publication/wcms_596481.pdf. Accessed 30 June 2020
Jackson C, Vynnycky E, Mangtani P (2016) The relationship between school holidays and transmission of influenza in England and Wales. J Epidemiol 184(2016):644–651
Jones N (2020) How coronavirus lockdowns stopped flu in its tracks. Published by Nature. Retrieved from: https://www.nature.com/articles/d41586-020-01538-8. Accessed 10 June 2020
Jorge M (2020) Teacher education for the twenty-first century (and a post-pandemic world). Revista Brasileira de Linguística Aplicada 20(2). Epub July 03, 2020. https://doi.org/10.1590/1984-6398202016853
Kamal R, Mahtani CH, Jeffrey K (2020) What is the evidence for social distancing during global pandemics? Published by the Centre for Evidence-Based Medicine, University of Oxford, Oxford, UK. Retrieved from: https://www.cebm.net/covid-19/what-is-the-evidence-for-social-distancing-during-global-pandemics/. Accessed 20 June 2020
Korobar VP, Siljanoska J (2016) Challenges of teaching sustainable urbanism. Energy Build 115:121–130. https://doi.org/10.1016/j.enbuild.2015.04.049
Leal Filho W, Brandli L, Salvia A, Rayman-Bacchus L, Platje J (2020) COVID-19 and the UN sustainable development goals: threat to solidarity or an opportunity? Sustainability 12(5343). https://doi.org/10.3390/su12135343
MEC (2020) Educação e Coronavírus. Published by Brazilian Ministry of Education, Brasilia, Brazil. Retrieved from: https://portal.mec.gov.br/component/content/article?id=86791. Accessed 28 May 2020
Nafisah SB, Alamery AH, Al Nafesa A, Aleid B, Brazanji NA (2018) School closure during novel influenza: a systematic review. J Infect Public Health 11(5):657–661
OECD (2020) Supporting the continuation of teaching and learning during the COVID-19 pandemic. Published by OECD. Retrieved from: https://www.oecd.org/education/Supporting-the-continuation-of-teaching-and-learning-during-the-COVID-19-pandemic.pdf. Accessed 13 June 2020
Reimers FM, Schleicher A (2020) A framework to guide an education response to the COVID-19 pandemic of 2020. Published by OECD. Retrieved from: https://www.hm.ee/sites/default/files/framework_guide_v1_002_harward.pdf. Accessed 12 June 2020
Reimers FM, Schleicher A, Saavedra J, Tuominen S (2020) Supporting the continuation of teaching and learning during the COVID-19 pandemic. Published by OECD. Retrieved from: https://www.oecd.org/education/Supporting-the-continuation-of-teaching-and-learning-during-the-COVID-19-pandemic.pdf. Accessed 12 June 2020
Saavera J (2020) Educational challenges and opportunities of the coronavirus (COVID-19) pandemic. Published by World Bank. Retrieved from: https://blogs.worldbank.org/education/educational-challenges-and-opportunities-covid-19-pandemic. Accessed 20 June 2020
Sintema EJ (2020) Effect of COVID-19 on the performance of grade 12 students: implications for STEM education. J Math Sci Technol Educ 16(7):1–6. https://doi.org/10.29333/ejmste/7893
UNESCO (2005) Contributing to a more sustainable future: quality education, life skills and education for sustainable development. Published by UNESCO, Paris. Retrieved from: https://unesdoc.unesco.org/ark:/48223/pf0000141019. Accessed 19 June 2020
UNESCO (2017) Education for sustainable development goals: learning objectives. Published by UNESCO, Paris. Retrieved from: https://unesdoc.unesco.org/ark:/48223/pf0000141019. Accessed 19 June 2020

UNESCO (2020a) Data for sustainable development. Published by UNESCO, Paris. Retrieved from: https://sdg.uis.unesco.org/2020/05/15/the-importance-of-monitoring-and-improving-ict-use-in-education-post-confinement/. Accessed 19 June 2020

UNESCO (2020b) COVID-19 educational disruption and response. Published by UNESCO, Paris. Retrieved from: https://en.unesco.org/themes/education-emergencies/coronavirus-school-closures. Accessed 19 June 2020

UNESCO (2020c) Education: from disruption to recovery. Published by UNESCO, Paris. Retrieved from: https://en.unesco.org/covid19/educationresponse. Accessed 19 June 2020

UNESCO (2020d) Adverse consequences of school closures. UNESCO, Paris. Retrieved from: https://en.unesco.org/covid19/educationresponse/consequences. Accessed 19 June 2020

UNESCO (2020e) COVID-19 and higher education: today and tomorrow. Impact analysis, policy responses and recommendations. Retrieved from: https://www.iesalc.unesco.org/en/wp-content/uploads/2020/04/COVID-19-EN-090420-2.pdf. Accessed 19 June 2020

UNESCO, UNICEF, the World Bank, WFP, UNHCR (2020) Framework for reopening schools. Published by UNESCO, UNICEF, the World Bank, WFP, & UNHCR. Retrieved from: https://www.unicef.org/sites/default/files/2020-06/Framework-for-reopening-schools-2020.pdf. Accessed 19 June 2020

UNICEF (2000) Defining quality in education. Published by United Nations Children's Fund, New York, NY, USA. Retrieved from: https://www.right-to-education.org/sites/right-to-education.org/files/resource-attachments/UNICEF_Defining_Quality_Education_2000.PDF. Accessed 19 June 2020

UNICEF (2020a) COVID-19: more than 95 per cent of children are out of school in Latin America and the Caribbean. Published by UNICEF. Retrieved from: https://www.unicef.org/press-releases/covid-19-more-95-cent-children-are-out-school-latin-america-and-caribbean. Accessed 20 2020

UNICEF (2020b) UNICEF WASH programme contribution to COVID-19 prevention and response. Published by UNICEF. Retrieved from: https://www.unicef.org/media/66091/file/UNICEF-WASH-COVID-19-prevention-response-overarching.pdf. Accessed 19 June 2020

Vare P, Scott W (2006) Learning for a change: exploring the relationship between education and sustainable development. J Educ Sustain Dev 1(2). https://doi.org/191-19810.1177/097340820700100209

Viner MR et al (2020) School closure and management practices during coronavirus outbreaks including COVID-19: a rapid systematic review. Lancet 4(5):397–404. https://doi.org/10.1016/S2352-4642(20)30095-X

Walker PGT, Whittaker C, Watson O et al (2020) The global impact of COVID-19 and strategies for mitigation and suppression. WHO Collaborating Centre for Infectious Disease Modelling, MRC Centre for Global. Retrieved https://www.imperial.ac.uk/media/imperial-college/medicine/sph/ide/gida-fellowships/Imperial-College-COVID19-Global-Impact-26-03-2020v2.pdf. Accessed 30 June 2020

World Health Organization (2020) WHO announces COVID-19 outbreak a pandemic. Retrieved https://www.euro.who.int/en/health-topics/healthemergencies/coronavirus-covid-19/news/news/2020/3/who-announces-covid-19-outbreak-a-pandemic. Accessed 30 June 2020

Wolff L-A (2020) Sustainability education in risks and crises: lessons from covid-19. J Sustain 12:5205. https://doi.org/10.3390/su12125205

World Bank (2020a) World Bank education and COVID-19. Published by World Bank. Retrieved from: https://www.worldbank.org/en/data/interactive/2020/03/24/world-bank-education-and-covid-19. Accessed 30 June 2020

World Bank (2020b) Education systems' response to COVID-19. Published by World Bank. Retrieved from: https://pubdocs.worldbank.org/en/729801593526005043/Education-Sector-Brief-June-26.pdf. Accessed 30 June 2020

Zhang W, Wang Y, Yang L, Wang C (2020) Suspending classes without stopping learning: China's education emergency management policy in the COVID-19 outbreak. J Risk Financ Manage 13(55):1–6. https://doi.org/10.3390/jrfm13030055

Vanessa Tibola da Rocha is an architect and holds a master's degree and a PhD in Engineering at University of Passo Fundo. Currently she teaches for graduation.

Bárbara Fritzen is an environmental and safety engineer. She holds a master's degree in Engineering at University of Passo Fundo. Currently she is a PhD researcher in Civil and Environmental Engineering.

Luciana Londero Brandli is a civil engineer from Federal University of Santa Maria. She holds a master's degree in Civil Engineering, a PhD in Production Engineering at Federal University of Santa Catarina and a post-doctorate at Hamburg University of Applied Sciences. She is Deputy Editor of the Encyclopedia of Sustainability and Higher Education and of the Encyclopedia of Sustainable Development Goals: Transforming the World We Want. Holds a master and a PhD degrees in Engineering. Currently she teaches for graduation. She acts on the editorial board of several magazines, supervises several masters and doctoral students in engineering, environment and sustainability issues and has more than 300 publications, including books, book chapters and articles in those magazines. Her current research interests include sustainability in higher education, the 2030 Agenda for Sustainable Development, Climate Change, Education for Sustainability and Sustainable Cities. She is currently carrying out the international project Transforming Universities for the Climate Change, involving the United Kingdom, Kenya, Mozambique and Fiji.

Using COVID-19 to Teach Sustainability Futures Thinking

Courtney E. Quinn and Matthew Cohen

Abstract Students of Sustainability Science have a unique lens through which to engage current events and envision a sustainable future that may arise from current and future hardships. In our undergraduate Sustainability Science courses, students learn and practice key sustainability competencies (Wiek et al. in Sustain Sci 6:203–218, 2011) including systems thinking, strategic thinking, and futures thinking. In response to COVID-19, in the latter half of the Spring 2020 semester, we re-tooled our courses to help our students (1) apply competencies to understand Coupled Human and Natural Systems drivers behind COVID-19 spread as well as individual, societal, and natural responses, and (2) to envision how our future can be deliberately different if people choose to push past previously accepted paradigms as we emerge from the immediate crisis. In this chapter, we first overview how our curriculum lends itself to a deep systems understanding of COVID-19 by providing specific pedagogical examples of how we have utilized this pandemic to explain key sustainability theories, concepts, and competencies. Second, we offer a novel qualitative analysis of future sustainability-focused visions from student assignments from two courses that occurred between March and May 2020. Together these two lines of evidence provide information on how today's college students envision and plan to shape our shared future.

Keywords Sustainability · COVID-19 · Undergraduate · Case studies

1 Introduction

Students studying sustainability science are equipped with a unique set of tools with which to analyze the COVID-19 global pandemic. Sustainability science students

C. E. Quinn (✉) · M. Cohen
Department of Earth, Environmental, and Sustainability Sciences, Furman University, 3300 Poinsett Highway, Greenville, SC 29613, USA
e-mail: Courtney.quinn@furman.edu

M. Cohen
e-mail: Matthew.cohen@furman.edu

W. Leal Filho (ed.), *COVID-19: Paving the Way for a More Sustainable World*,
World Sustainability Series, https://doi.org/10.1007/978-3-030-69284-1_21

can consider our current situation from multiple lenses including through a systems lens, along with the skills of strategic, values, and futures thinking. This is due to sustainability science education's framing through the lens of training students in key competencies required for solving complex sustainability problems. Although there has been an array of disparate and overlapping proposals (see for instance: Barth et al. 2007; Wiek et al. 2011; Habron 2012; Besong and Holland 2015; Vare et al. 2019), the literature has mostly coalesced around five key competencies: systems thinking, values thinking, futures thinking, strategic thinking, and collaborative thinking (Wiek et al. 2011, 2016), with numerous studies building off of this foundation (for instance, Remington-Doucette and Musgrove 2015; Giangrande et al. 2019; Brundiers et al. 2020).

Sustainability challenges are complex, wicked problems that require systems thinking to understand underlying root causes and drivers of current problems as well as the web of feedback loops relevant to the initial problems and the unintended consequences of superficial solutions. Grounded in core values, including equity and justice as well as ecological integrity, sustainability is a normative science (Gibson 2006; Ziegler and Ott 2011; Schlaile et al. 2017) in which values-based judgments are rendered on systems' current states, and normative goals are designed for the future. These futures are imagined through visioning processes to describe desirable outcomes in which sustainability has been realized. Once a vision is set, solution strategies are needed to lead from the unsustainable present towards the sustainable future. The process of understanding sustainability challenges, imagining their solutions, and designing and implementing strategies are meant to be inclusive and collaborative (Wiek et al. 2011, 2016).

While we acknowledge the integrative nature of these key competencies and the need to avoid prioritizing one over the others, in this paper, we focus primarily on students demonstrating futures thinking competence, while connecting student demonstrations of futures thinking to systems and strategic thinking as well. When reading across the literature on sustainability education, there are numerous competencies on which to draw, and both futures and systems thinking commonly appear in diverse competency-based frameworks (Barth et al. 2007; Habron et al. 2012; Wiek et al. 2012; Giangrande et al. 2019). As such, we present case studies from our own teaching during the Spring 2020 semester to demonstrate how we adapted our teaching to the COVID-19 pandemic and how undergraduate Sustainability Science majors applied their learning to interpret the crisis and envision more sustainable futures.

These case studies provide an overview how our curriculum lends itself to a deep-systems understanding of COVID-19 and how sustainability science students might project these systems into the future. As illustration, we provide specific pedagogical examples, from two classes, of how we have utilized this pandemic to teach key sustainability theories, concepts, and competencies. We present our two classes as case studies in teaching sustainability concepts, theories, and competencies using COVID-19, with a specific focus on futures thinking. Second, we offer a novel qualitative analysis of sustainability-focused visions created by students in the two courses. Together these two lines of evidence provide information on how today's college students envision and plan to shape our shared future.

2 Methods

This paper utilizes case study research and qualitative content analysis methods to generate evidence and draw conclusions regarding student learning. This first series of subsections presents two case study learning experiences. Following the case studies, we present our analytical methods, which include a qualitative content analysis of student products generated in the two courses.

3 Case Study of Two Sustainability Classes

Both authors teach in the Sustainability Science (SUS) B.S. major that is housed within the Department of Earth, Environmental, and Sustainability Sciences at Furman University, a liberal arts university in the United States. The department has approximately 100 majors and typical classes consist of 12–24 students. Due to health concerns and social-distancing measures, students did not return to in-person classes after their March spring break and all learning for the rest of the spring semester occurred virtually. Here we present our two classes as case studies in teaching sustainability concepts, theories, and competencies using COVID-19, with a specific focus on futures thinking.

3.1 Introduction to Sustainability Science (SUS 120)

Introduction to Sustainability Science is the first course students encounter in the major. The course is open to freshmen students, most of whom have yet to declare a major. The course is designed to introduce key theories and models such as Planetary Boundaries (Rockström et al. 2009; Steffen et al. 2015), the Doughnut Model by economist Kate Raworth (2017), and Coupled Human and Natural Systems (CHANS) (Liu et al. 2007a, b, Quinn and Quinn 2020). The course also introduces each sustainability competency and allows students entry-level practice in applying the skills to relevant sustainability topics such as; food systems, consumption and waste, biodiversity and ecosystem services, climate change, and environmental justice. The course builds on each competency until students can use their knowledge to create sustainability visions for the future. The course always references timely events that help to illustrate concepts. In the past few years such topics included; Hurricane Harvey in the United States, the spread of the Zika virus, and regionally relevant examples such as intense and frequent flooding in Charleston, South Carolina.

When COVID-19 was first in United States newspapers in January 2020, the class was working on the competency of systems thinking, learning basic systems concepts such as stocks and flows and the importance of scale (Meadows 2008; Gordon et al. 2008). Zoonotic diseases stemming from human-animal interactions became a key example to illustrate CHANS. Students practiced drawing CHANS diagrams with elements, drivers, feedbacks, and connections related to COVID-19 (NSF CHANS) (Fig. 1).

The next competency covered in the introductory class is Values Thinking. Here the class once again looked at the interconnections between humans and nature to examine the ethical considerations of deforestation, wet markets, and broader patterns of consumption that have indirect effects on the natural world as well as people around the globe. COVID-19 provided a poignant example to debate the ethical considerations we owe to other species or even to ecosystems. It also allowed for lively debate about valuing individual rights versus social obligations, and what happens when those two value systems are seemingly in conflict. The immediacy

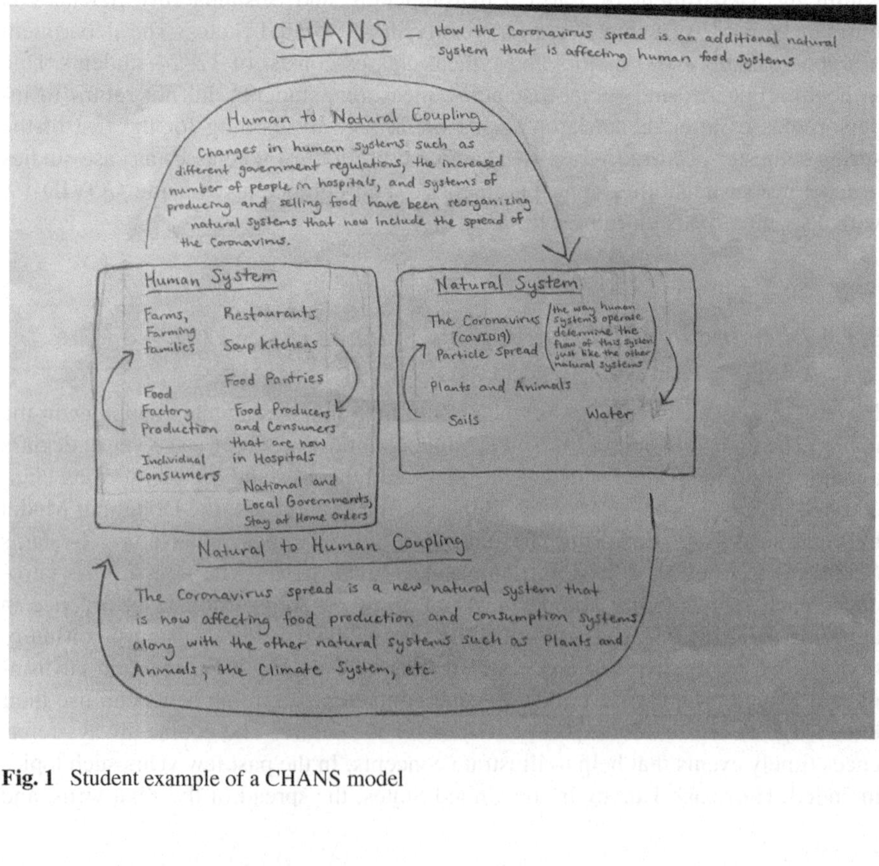

Fig. 1 Student example of a CHANS model

Table 1 Student proposed use of leverage points (from shallow to deep) during COVID-19 pandemic

Leverage point		Student proposed intervention
Shallow	Numbers, parameters, constants	*"Increase the number of ventilators available to hospitals"*
	Delays	*"Social distancing and quarantine act to delay the spread of the virus"*
	Balancing feedback loops	*"Stores should limit purchases per customer of items such as bleach and toilet paper to create a balancing feedback loop to the reinforcing feedback loop of panic shopping"*
Deep	Rules and norms	*"Use social norms to influence mask wearing"*

of the pandemic was juxtaposed during class discussions with the seemingly slower emerging ethical conflicts around climate change.

When covering strategic thinking, students return to, and build upon, systems concepts by learning about leverage points (Meadows 2008). In past semesters, interventions in food systems and megacities were primary topics for lectures and small group work. This semester, the original topics were retained but responses to COVID-19 was a popular addition to the curriculum. It was noted, anecdotally, that students, experiencing the immediate and deeply personal effects of COVID-19, were especially interested in the topic of leverage points, above and beyond students' interest in the past. Examples of leverage points interventions proposed by students can be found in Table 1.

The next competency covered in the introductory class is futures thinking. Two methods of futures thinking are covered: (1) scenario creation and (2) vision creation. This semester, students created scenarios for the future utilizing the novel driver of COVID-19 along with a sub-topic such as food systems, consumption and waste, or biodiversity. Last, students explored the sub-set of scenarios that are positive visions for the future. For the exceptional circumstances of this semester, students started by reading others' visions for moving forward after this crisis and then proceeded to create their own. The course also focused on sustainability storytelling (Bernier 2020). Students wrote a story as if they were already in a post-COVID future that brought positive changes. One story is offered below.

> An excerpt of a speech given to the Local Farmers Association, as the speaker addresses the changes implemented in farming since the pandemic, and how these changes have been beneficial for the earth:
>
> Thirty years from now, we have healed from the major impacts of the coronavirus pandemic. This major event that took months from our lives is now being incorporated into history books. Those who were old enough to remember that time learned an important lesson from the experience, as the world was forced to slow down for a bit. We learned that the environment can heal itself, with less pollution and a slower lifestyle. We learned that smaller scale farms and neighborhood gardens were more reliable and adaptable to unforeseen events.

> We learned how to distribute our food in a smarter way, so that no one has to go hungry while another person is throwing away their excess food. We learned that food can be grown hydroponically in cities, recycling water instead of wasting it, and taking in excess carbon emissions. The pandemic was not fun, but it was a learning experience that allowed us to improve in a way that we never would have otherwise.

4 Social Systems*(SUS 241)*

Social systems is a 200-level core requirement for the Sustainability Science (SUS) major. The majority of students that enroll in this course have formally declared the SUS major and taken introductory sustainability science and environmental science courses. Many have also taken an additional core requirement or upper-level elective from the major. This class is predominantly sophomore and junior students.

To frame the course, we define social systems as "political, economic, industrial, and other structures created by human beings that provide the societal base for a fulfilling human existence (Komiyama and Takeuchi 2006)." We focus on studying institutions within political and economic systems, to help students better understand the social dynamics and structures that create sustainability challenges and that need to be leveraged or transformed to create sustainability solutions. Because nearly all students enrolled in the course have already taken Introduction to Sustainability Science, the course assumes students have had exposure to the key competencies.

The first half of the Spring 2020 semester followed the course's usual plan, defining sustainability problems as collective action problems, exploring relevant social systems and institutions, conducting institutional and stakeholder analyses, and exploring the paradigms that underlie the systems and institutions that guide our individual and collective behaviors. As such, students were well-versed in political-economic realities of a neoliberal world order, the tensions between liberal economics and minimal vs. affirmative states (Fung and Wright 2001) and power dynamics in sustainability cases (Hayward and Lukes 2008).

Following spring break and the University closure due to COVID-19, the instructor redesigned the course for remote instruction. At this point, the instructor also shifted the focus of the remaining weeks of the course to leverage student learning from the first half of the semester. The goal was to understand the current events of the COVID-19 crisis (systems and values thinking) as well as anticipate plausible futures (future thinking and strategic thinking) that might result from the pandemic disrupting contemporary systems and institutions (systems thinking and strategic thinking).

Readings and course content in the first weeks of the redesigned course focused on four decades of neoliberalism eroding government capacity to respond to the public health crisis, the role of valuing exponential economic growth in driving the spread of the virus and weakening our collective ability to slow it, and the relationships between COVID-19 and our responses to other sustainability challenges, including climate change. To engage students on imagining a post-COVID-19 world, we read an article in Politico magazine (2020) in which scholars from diverse disciplines applied their expertise to explain where society is and how systems, institutions, and life might shift as a consequence of the pandemic. Students were able to draw on much of their training in institutional analysis, political economic thought, and sustainability science to interpret these scholars' ideas. Additional lessons included exploring green stimulus proposals for a post-crisis economic recovery, and ways that the economic collapse around the world shift perceptions of plausibility around sustainability transition agendas and degrowth principles.

For the rest of the semester, we explored the originally-planned course content, but with the lens of how these systems and institutions might look and function differently once global society emerges from the coronavirus lockdown.

5 Qualitative Analysis of Student Work on Futures After COVID

For the second objective of this chapter, we present the results of a qualitative content analysis of relevant student products across the two courses presented above. We asked the research question, 'How do Sustainability Science students consider, create, and present future-thinking visions that arise from the COVID-19 pandemic?' To collect our data, we deposited student work from four online assignments (listed below) into MaxQDA. We followed recommended guidelines for qualitative coding (Basit 2003; Saldaña 2015). First, we both read all student entries and coded for emergent themes. During the first round of coding, three broad areas emerged which then became our a priori coding themes for second round coding; (I) Understanding of Current Systems, (II) Visions of a post-COVID world, and (III) Strategies to Achieve Visions. Through an iterative process of co-coding within the three a priori themes alternating with discussion, we identified the most salient examples of student futures-thinking for each theme along with sub-themes such as 'Connection' as a sub-theme within the area of Visions.

Introduction to Sustainability collected assignments:

1. Reflection assignment: Using the guidelines of writing high-quality scenarios and visions, create a titled, two-paragraph, post-COVID sustainability vision regarding a sub-topic (food systems, consumption and waste, biodiversity) of your choice.
2. Final exam question: Using the guidelines of effective sustainability storytelling, write a two or three-paragraph story as if your positive post-COVID vision has come to pass. This may be in the form of a news article, story, speech, etc. You may choose the timeframe of your story, but please specify within your story.

Social Systems collected assignments:

1. Reflection assignment: How do you see this crisis as a logical outcome given our social systems and institutions? While wellbeing is suffering around the planet, we have concurrently seen the natural environment recover in places while the

global capitalist economy has been frozen. How do you feel about this tradeoff between wellbeing and environmental health? Does this tradeoff have to be a feature of our social systems in the future? What do you hope for the world we can create out of this crisis?

2. Final exam question: Read back over the reflection you submitted for Module 2 on the week of March 30th. Please summarize your hope for the future and describe what systems changes would be required for your ideas to be realized.

We focus on presenting examples of student work to allow students' voices to illustrate their views.

5.1 Understanding of Current Systems

Although Sustainability Science is inherently focused on futures-thinking and future-creating, students must have a solid foundation of current systems and paradigms. Students, familiar with the Doughnut Economics model (Raworth 2017), frequently identified areas where we are surpassing planetary boundaries or falling short on providing key social foundations.

Many students saw the pandemic crisis and its toll on society as a logical end to neoliberalism. Students highlighted the globalized economic system and eroded public capacity in the U.S. as key contributors to the spread of the virus and its economic fallout.

> Over the rise of neoliberalism, we have discovered that lack of government regulations only helps a small percentage and hurts the larger majority. The minimalist state ideals our governments have adopted (Fung and Wright 2001) has also wreaked havoc on the environment. We believe that less government can cultivate a higher level of wellbeing, yet this has led to the endless exploitation of the environment and people in the name of profit.

> The flaws in our social systems are not new, but many are being revealed and exacerbated in new ways. I hope that this shock will be sufficient to help people understand the impacts our systems have on our lives as well as the pressing need to transform them in order to achieve a sustainable society.

> Our current system is not very resilient, which we are seeing through the externalities caused by this disease. Globalization has created a connected and interdependent world system that focuses on economic wellbeing over the common good. This global philosophy has enabled the disease to spread quickly and we are seeing the global economy failing as countries effectively shut down their economies to reduce the spread of the disease.

> The coronavirus has spread rapidly because of the state of our social systems and institutions. Our world is increasingly globalized and interconnected, which facilitated the rapid spread of the virus between and within countries. In the United States particularly it became such a crisis because of the systematic underfunding of healthcare systems, medical research, the disbanding of the pandemic response team, and general mistrust in the institutions of science and research. All of these failures have led to misinformation, rapid transmission, overwhelming of healthcare systems, and deaths.

Students noted that currently, "*consumerism is about convenience and overconsumption*" and that "*our society pushes materialistic views upon us*" and therefore "*the patterns of our lifestyles led to this pandemic.*"

Our students, growing up during times of massive social divisions and upheaval, see the "*tension*", "*social unrest*", and "*globalization*" that are "*exacerbating income inequality*".

5.2 Visions of a Post-COVID World

Within the theme of Visions, we identified six important sub-themes: Connection, Cooperation, Resilient Human and Natural Systems, Paradigm Shifts as a Vision, Wellbeing, and Values. Overall, students' visions were positive and many saw long-term changes for economic, political, social, and environmental systems as a result of the COVID-19 pandemic.

Connection: Our students want to see humans recognizing, seeking, and creating greater connections between humans and the natural world and between individuals and human communities.

> People have grown an appreciation for nature and all it offers. Along with this appreciation, there is a gained understanding of how many resources we can actually take without causing strain. We have finally connected with nature and are living more sustainably all over the world.

> One of my hopes for the future is that we lean into our humanity and become more connected to both the choice-based and geographic communities we belong to. Kate Raworth, when tweeting about the implications of COVID-19, that we need to "nurture our human nature" by understanding "our brains are wired for empathy, cooperation and mutual aid" (Raworth 2017). It is my hope, that in a time such as this, that we can begin to revise our self-portrait to base our happiness and gratification on the relationships we have.

Cooperation/Coordinated Action: Our students discussed cooperation between various actors and stakeholders including businesses, governments, and individuals. They see a world where cooperation is the norm and we are applying the lessons of coordinated action learned through the pandemic to deep problems such as climate change and equality.

> Imagine the United States in 2050 where corporations work together to create sustainable options for their products, from clean energy to the morally just production of clothing, that are accessible and not only the rich can afford. This is the United States that I want to live in.

> My vision is one where the ability to get your hands on food is not a privilege but rather a right. Countries have really joined hands during the time of this pandemic and my vision is one where they will continue such.

> In my previous reflection, I discussed my hope for the planet to learn to coordinate better for future worldwide issues. In other words, we must be better at collective action problems. The coronavirus requires a worldwide effort to slow and decrease its spread. Similarly, climate change needs comparable efforts to decrease its impacts.

Our collective values and actions should be toward increasing equality, cooperation, health, education, accommodations, and open dialogue across vertical and horizontal scales.

Resilient Human and Natural Systems: In our classes, students learn about resilience and vulnerability. This theme was a part of many student visions. Students desire to see just and resilient climate systems, health systems, and agricultural systems.

Although system level change is never something that happens without resistance, I hope through this crisis we can see the flaws in our system and work towards developing a more resilient system that promotes wellbeing of people and the Earth.

I hope we will invest into continued short term immediate needs so that people are not hurting due to our previous failures as a system. But most of all, I hope we look long-term to see how we can prevent a catastrophe like this from ever happening again by creating a more resilient future- one that is built on green energy and equality.

Most importantly, I hope there is an effort to become more resilient in the face of crises like the pandemic. In particular, I hope this helps give the climate movement some mobility and more validity.

Another hope I have for the future is an increase in the resiliency of the healthcare system. When everything hit, hospitals desperately struggled to get enough supplies and workers to deal with the pandemic. This vulnerability isn't inherent—it's because the healthcare system has not been prioritized, even though it is the institution that literally saves our lives.

Farmers that focus on one crop will be a thing of the past, and all farmers will learn to operate and thrive on a diverse and healthy stretch of farmland. Biodiversity will enrich the land and will increase the types and amounts of wildlife that are able to live on farms. Once more and more farms become biodiverse, the world as a whole will be more naturally stable and will bounce back more quickly from future worldly or natural disasters.

Paradigm shifts as a vision: In our sustainability classes, students learn about leverage points (Meadows 2008) as places to intervene in a system. One of the most powerful leverage points is to shift paradigms. When this occurs, all other parts of a system including goals, rules, information flows, and feedback loops, also change. Our students discussed paradigm shifts as both strategies for change and something they envision emerging from this pandemic. Students want vast and deep paradigm shifts in our values, attitudes, individual consumption patterns, and system structures.

I want to see a change in our paradigms where we go from anthropocentrism to ethical holism. If we take care of our planet and value what we have we can start a movement and real systematic change that could save ourselves and everything we love.

My hope is that this crisis will instill a widespread desire for change that will fuel a compelling call for change as well. In order for changes to remain well into the future, they must be embedded into social systems, institutions, and structures; the rules, norms, values, and goals must all shift for meaningful, long-lasting change.

I think that this is the time for us to all change our mental model and to reduce our own consumption patterns.

My vision is to change our consumption patterns. Obviously, some items and products will not be able to be geared towards long-term use but my vision for change would be a societal shift away from such rampant short-term consumption. This task would not be easy and

> would require a paradigm shift in how we view products and advertisements, as well as the structure of large portions of our economic model.
>
> Much of the current economic and social system in the west socially pressure or motivate people into widespread consumerism.
>
> I've seen a quote floating around that reads 'in the rush to return to normal, use this time to consider which parts of normal are worth rushing back to', and while that's a kitschy Instagram quote, I think it's exactly what we've been talking about in this class since the pandemic began. What's worth going back to and what should change?

Well-being: Our students articulated a vision for a holistic, overall wellbeing for human and environmental health that moves past current economic-focused measures of a society's health.

> However, there is a balance that needs to be created in which human wellbeing and environmental health can live in unison. It will be interesting to see the action plans post-pandemic to help make this balance happen. I think for our social systems in the future, there can be a more concrete system in which neither human wellbeing or natural capital has to suffer. Both can live together by giving and taking equitable amounts to benefit each other. I hope that out of this the world can create a more just and equal place for both human wellbeing and the environment.
>
> We could begin measuring progress in terms of wellbeing and access to the social foundations instead of in terms of GDP growth.
>
> My hopes for the future following the pandemic include a social system that produces equitable outcomes, promotes human wellbeing, protects the environment, and is sustainable long-term.

Values: Referring back to the students' diagnosis of a crumbling neoliberal order that we highlight above under *Understanding the Current System*, many students framed alternative, desirable futures in contrast to that world order based upon different value systems than those currently creating unsustainable and unjust systems.

> My hope for the future is characterized entirely by change – changing attitudes, changing mental models…the rules, norms, values, and goals must all shift for meaningful, long-lasting change.
>
> Our collective values and actions should be toward increasing equality, cooperation, health, education, accommodations, and open dialogue across vertical and horizontal scales.

5.3 *Strategies to Achieve Visions*

Students had many ideas to put their visions into action ranging from changing individual behaviors up to complete overhaul of government and economic systems. They were forceful in their calls to change. Students also recognized the COVID-19 crisis as an opportunity; a wake-up call where tragedy can be harnessed to ensure a better future for all. Although students stated that they would, of course, want these changes to occur without loss of life through the virus, they also recognized that this moment could be a tipping point for a change to sustainable systems.

Systems-Scale Change: Based on course content, particular reading assignments, and online discussion board activity, it was not surprising to see Social Systems students propose structural changes and policy-scale strategies as solutions. For instance, many students were inspired by Green New Deal-style platforms for restructuring national economies following the crisis.

> Now creates an opportunity for us to shift how we see our economy and instead of worrying about the short term we can shift to post-growth economic principles. We need to focus on the wellbeing of humans and also the ecosystems around us.
>
> New jobs could be created in green industries such as wind and solar energy, the government could invest in green infrastructure installations, public and employee ownership could be expanded, and carbon emissions could be cut rapidly. Because our social systems dictate our behaviors and norms, changing those systems to become more sustainability-oriented would have positive effects in individuals' lives as well.

Further, as the U.S. Congress was preparing a large economic relief package, we read in class proposals for a Green Stimulus (Bozuwa et al. 2020), which were presented as a means for rebuilding national and global economies around wellbeing, public health, and green energy, among other outcomes. As such, students focused their strategies on significant changes to the U.S. economy.

Common student proposals included replacing GDP with wellbeing outcomes to assess economic success, refusing fossil fuel bailouts while investing in renewables to strengthen the energy transition, and rebuilding government capacity in the wake of a declining neoliberal order. One student also connected these larger scale interventions with a reorientation of economic activity around local communities, and others highlighted initiatives from the Transition Town movement as ways to reorient society around community building and local sustainability initiatives.

> The European Union is increasing recycling, decoupling economic growth from resource use, and shifting towards a circular economy. South Korea's Green New Deal is similar to the EU's and includes a carbon tax, investments in renewable energy, and a requirement to end investment in coal. The United States can and should follow suit, although partisan divisions are making it difficult to make progress on the environmental goals. Instead of focusing on short-term relief, policymakers and citizens should consider how our response to the pandemic will shape our society in the coming decades.

Individual Behavior Change: Although students in Introduction to Sustainability Science begin to learn about systems, freshmen students tend to relate better to learning how their individual behavior affects the environment. For example, students learn that over 70% of freshwater withdrawals go to agriculture, but they tend to find it more interesting to compare the average water use of their daily shower to the gallons of water embedded in their hamburger at lunch. Therefore, their strategic recommendations tended to encourage lifestyle changes.

> I see myself in a beautiful world, a better world. One where we do not sit mindlessly in front of a screen watching Chopped or Top Chef. Instead of watching people make beautiful plates of food, we make those plates ourselves. We use local fruits and vegetables, we grow our own spices, and we rely less on meat. During the pandemic we take action to change our food systems.

> It is the year 2075, finally. We have seen the many ups and downs in this century. This includes the memorable quarantine that lasted several months, keeping people in their homes. During this quarantine the great switch happened. People began to read more on the earth and how to sustain it. The newest trends became all about eliminating waste and becoming more friendly with our mother earth. Because of this boom, our waste slowly began to diminish. Our landfills became smaller. Sewing classes became the biggest craze, as almost all sewing machines were sold out on amazon for months. Famous designers made competitions to see who could design the coolest looks, and runway shows were a fun way to show off handmade clothes. More people bought thrifted clothes than ever before. That was the year the world slowly began to heal, and the people with it.

Crisis as Opportunity: Our students realize that this crisis provides us with a chance to make deliberate changes in our individual behaviors and in our systems. They can put this current situation into a larger historical and future-thinking context to call for changes large and small.

> While this crisis does present many short term challenges, it also can be used as an intervention point for long term solutions that can address both social and environmental problems.

> Historically, crises have brought about dramatic societal or political change and this could be an opportunity for a shift in values, norms, and goals. Ideally, changing from the idea of constant growth and progress to valuing the natural world while also maintaining a stable economy.

> Sometimes, you need a life changing event to change your mind. We have seen this many times in history but now we have to live through history.

> Maybe with this opportunity to reorganize society and the economy we can prioritize resilience and environmental protection while still maintaining a stable economy.

6 Reflection on Student Work

Through our teaching, we have seen that young people are ready to support big, sweeping systems-scale reform and the willingness to make changes in their individual daily lives to create a sustainable future. This desire became even more profound, and students more adamant in their calls for change, during the COVID-19 pandemic. Students in our introductory class easily identified many aspects of our current systems that harm the planet and people. While they advocated for more individual behavior change than more advanced students, they were willing to make changes to their diets and their broad consumption patterns. Students who have declared Sustainability Science as their major (i.e. those in Social Systems), were more able to express the need for radical transformation, degrowth principles, and systems-level transitions. Yet, all of our students envisioned a more just and safe planet emerging from this crisis.

7 Conclusion

The rapid change in daily lives and large-scale system functioning during this time of COVID-19 has been profoundly challenging for our students who had to abruptly leave campus and not return for the remaining of the spring semester. And yet, we feel lucky to have the opportunity to work with our students and provide them with the tools to have a deeper understanding of this life, and world, altering situation. This pandemic offers all people a chance to reevaluate their individual choices and the systems we support and create, but for young people it offers them a chance to reshape system paradigms, their own lives, and the lives of those near to them now and in the future.

From an analysis of our own students' assignments, we can see examples of undergraduate sustainability science students making concerted efforts to think about desirable futures and connect both personal commitments and transformational changes to lead us from a global crisis to a more sustainable world. While it is encouraging to see that as students progress through the major, they are able to describe the importance of systems and institutional change, we also recognize shortcomings in their abilities to articulate clear, actionable steps to enact transformation. This shouldn't surprise us, as paradigm shifts are bigger than any one individual, and we as educators must strive to build sustainability education programs that can equip students at all levels with the tools to enact meaningful change across scales. While there is abundant research on sustainability education as a transformative enterprise, we would argue that transforming students to transform the world is a large undertaking, and this area would benefit from additional empirical research.

We know that our students for at least the next decade will have had intense personal experiences of the pandemic that we will continue to leverage for relevant sustainability lessons. Perhaps, within that time, we will also be able to teach about the positive system changes that emerged after the crisis that made our world a more safe and just space for humanity. We hope to one day teach students who are already living in the world created by our students of today, who took to heart the lessons we are learning now about the need to protect the environment and each other.

References

Barth M, Godemann J, Rieckmann M, Stoltenberg U (2007) Developing key competencies for sustainable development in higher education. Int J Sustain High Educ 8(4):416–430. https://doi.org/10.1108/14676370710823582

Basit T (2003) Manual or electronic? The role of coding in qualitative data analysis. Educ Res 45(2):143–154

Bernier A (2020) Sustainability storytelling is not just telling stories about sustainability

Besong F, Holland C (2015) The dispositions, abilities and behaviours (DAB) framework for profiling learners' sustainability competencies in higher education. J Teach Educ Sustain 17(1):5–22

Bozuwa et al (2020) A green stimulus to rebuild our economy. https://medium.com/@green_stimulus_now/a-green-stimulus-to-rebuild-our-economy-1e7030a1d9ee
Brundiers K, Barth M, Cebrián G et al (2020) Key competencies in sustainability in higher education—toward an agreed-upon reference framework. Sustain Sci. https://doi.org/10.1007/s11625-020-00838-2
Fung A, Wright EO (2001) Deepening democracy: innovations in empowered participatory governance. Politics Soc 29(1):5–41
Giangrande N, White RM, East M, Jackson R, Clarke T, Saloff Coste M, Penha-Lopes G (2019) A competency framework to assess and activate education for sustainable development: addressing the UN sustainable development goals 4.7 challenge. Sustainability 11(10):2832
Gibson RB (2006) Sustainability assessment: basic components of a practical approach. Impact Assess Proj Appraisal 24(3):170–182
Gordon LJ, Peterson GD, Bennett EM (2008) Agricultural modifications of hydrological flows create ecological surprises. Trends Ecol Evol 23(4):211–219
Habron G (2012) Competency-based sustainability specialization at Michigan State University. Sustain J Rec 5(6):379–385
Habron G, Goralnik L, Thorp L (2012) Embracing the learning paradigm to foster systems thinking. Int J Sustain High Educ 13(4):378
Hayward C, Lukes S (2008) Nobody to shoot? Power, structure, and agency: a dialogue. J Power 1(1):5–20
Komiyama H, Takeuchi K (2006) Sustainability science: building a new discipline. Sustain Sci 1(1):1–6. https://doi.org/10.1007/s11625-006-0007-4
Liu J, Dietz T, Carpenter SR, Alberti M, Folke C, Moran E, Pell AN, Deadman P, Kratz T, Lubchenco J, Ostrom E (2007a) Complexity of coupled human and natural systems. Science 317(5844):1513–1516
Liu J, Dietz T, Carpenter SR, Folke C, Alberti M, Redman CL, Schneider SH, Ostrom E, Pell AN, Lubchenco J, Taylor WW (2007b) Coupled human and natural systems. AMBIO A J Human Environ 36(8):639–650
Meadows DH (2008) Thinking in systems: a primer. Chelsea Green Publishing, Vermont
Politico Magazine (2020) https://www.politico.com/news/magazine/2020/03/19/coronavirus-effect-economy-life-society-analysis-covid-135579
Quinn JE, Quinn CE (2020) Coupled human and natural systems: a review and anthrome case study. Encyclopedia of the world biomes, vol 10. Reference module in earth systems and environmental sciences. Elsevier. https://doi.org/10.1016/B978-0-12-409548-9.12435-2
Raworth K (2017) Doughnut economics: seven ways to think like a 21st-century economist. Chelsea Green Publishing, Vermont
Remington-Doucette S, Musgrove S (2015) Variation in sustainability competency development according to age, gender, and disciplinary affiliation. Int J Sustain High Educ
Rockström J, Steffen W, Noone K, Persson Å, Chapin III FS, Lambin E, Lenton TM, Scheffer M, Folke C, Schellnhuber HJ, Nykvist B (2009) Planetary boundaries: exploring the safe operating space for humanity. Ecol Soc 14(2)
Saldaña J (2015) The coding manual for qualitative researchers. Sage Publications, New York
Schlaile MP, Urmetzer S, Blok V, Andersen AD, Timmermans J, Mueller M, Pyka A (2017) Innovation systems for transformations towards sustainability? Taking the normative dimension seriously. Sustainability 9(12):2253
Steffen W, Richardson K, Rockström J, Cornell SE, Fetzer I, Bennett EM, Biggs R, Carpenter SR, De Vries W, De Wit CA, Folke C (2015) Planetary boundaries: guiding human development on a changing planet. Science 347(6223)
Vare P, Arro G, De Hamer A, Del Gobbo G, De Vries G, Farioli F et al (2019) Devising a competence-based training program for educators of sustainable development: lessons learned. Sustainability 11(7):1890
Wiek A, Lang DJ (2016) Transformational sustainability research methodology. Sustainability science. Springer, Dordrecht, pp 31–41

Wiek A, Withycombe L, Redman CL (2011) Key competencies in sustainability: a reference framework for academic program development. Sustain Sci 6(2):203–218. https://doi.org/10.1007/s11625-011-0132-6
Wiek A, Farioli F, Fukushi K, Yarime M (2012) Sustainability science: bridging the gap between science and society. Sustainability Sci 7(1):1–4
Ziegler R, Ott K (2011) The quality of sustainability science: a philosophical perspective. Sustain Sci Pract Policy 7(1):31–44

Courtney Quinn is a Visiting Research Associate of Sustainability Science in the Department of Earth, Environmental, and Sustainability Sciences at Furman University, USA. Her research focuses on social networks within regional food systems.

Matthew Cohen is Assistant Professor of Sustainability Science in the Department of Earth, Environmental, and Sustainability Sciences at Furman University, USA. His research focuses on sustainability education and urban sustainability.

Re-thinking Sustainability Teaching

Qudsia Kalsoom and Ayesha Jalil Shah

Abstract The Covid-19 pandemic has highlighted two key issues in our education. (1) Educational disruption of millions of students across the world due to the lockdown for Covid-19 indicates that the current model of education in most parts of the world is not sustainable. (2) Extensive sharing of social media information without checking its authenticity indicates that our education is not empowering students as critical beings. The chapter discusses the need of re-thinking education and sustainability teaching in the light of the aforementioned issues.

Keywords Covid-19 · Sustainability education · Sustainability competencies · Online teaching and learning · Critical thinking · Epistemological development

1 Introduction

The field of sustainability has evolved over the years with an aim to address the existing and anticipated complex anthropogenic challenges of desertification, climate change, pandemics, war, and poverty (Wiek et al. 2011). As sustainability-related problems are complex and different from the problems in other fields, they have no obvious and simple solutions. The role of education becomes critical in such a context. Education is expected to build students' capabilities to anticipate sustainability challenges and solve the problems creatively. Wiek et al. (2011, p. 204) argue that "sustainability education should enable students to analyze and solve sustainability problems, to anticipate and prepare for future sustainability challenges, as well as to create and seize opportunities for sustainability". Similarly, Schank and Rieckmann (2019) contend that sustainability education is supposed to empower people to reflect on their own actions critically and adopt the actions that can contribute to sustainable development. Leal Filho et al. (2019) view the role of education as of catalyst in achieving sustainable development goals (SDGs).

Q. Kalsoom (✉) · A. J. Shah
School of Education, Beaconhouse National University, Tarogil, Off Raiwind Road, 13 km Thokar Niazbaig, Lahore, Pakistan
e-mail: qudsia.kalsoom@bnu.edu.pk

W. Leal Filho (ed.), *COVID-19: Paving the Way for a More Sustainable World*, World Sustainability Series, https://doi.org/10.1007/978-3-030-69284-1_22

Considering the key role of education for sustainable development (ESD) or sustainability education in realizing sustainable development, scholars have discussed or studied different aspects of ESD e.g. ESD outcomes and ESD processes (Tilbury 2011); ESD pedagogies (Eilam and Trop 2010); ESD competencies (Vare et al. 2019); sustainability teaching (Reid and Petocz 2006), etc. The pandemic of Covid-19 has affected education worldwide and has highlighted the problems in education across the world. This viewpoint chapter discusses the need of re-thinking sustainability education and sustainability teaching in the light of the literature on the effects of Covid 19; literature on sustainability education; and[1] a survey of teachers and students about their epistemological orientation and online teaching during lockdown for Covid 19 (Shah and Kalsoom 2020). The literature search for sustainability education was done through Google Scholar and the databases of the international journal of sustainability in higher education; journal of cleaner production; sustainability; and environmental education research. The following keywords were used for the search: sustainability teaching; ESD; sustainability education; ESD pedagogy and critical thinking. Literature related to the effects of Covid 19 on students was searched through Google Scholar.

The chapter has been organized under the following headings:

- Covid 19 and Educational Problems
- Effects of Covid 19 on Economy, Society, and Environment
- Covid 19 and Fake News
- Covid 19 and Online Teaching & Learning
- Covid 19, Education for Sustainable Development and Sustainability Teaching
- Conclusions and Way Forward.

2 Covid 19 and Educational Problems

The Covid 19 pandemic has affected education severely. Researchers have reported the negative effects of school and university closures on students. The most visible effects are related to their mental health or well-being. Giannini (2020) reported that the pandemic affected the mental health of 70% of the children and youth in Thailand. Similarly, Tiwari et al. (2020) found that restricted social interaction led to more negative emotions (sadness, aggression, quarrelsome behaviours, disturbed routine, and boredom) among 9–11 year old children. Kapasia et al. (2020) reported anxiety and depression among the undergraduate students of India during the lockdown for Covid 19. Kalsoom and Hasan (2020a) found that students who lacked the disposition of self-regulatory learning were impacted more psychologically. The effects of Covid 19 on education may be understood in the light of some key issues in global education systems. Some of them have been discussed in this section.

[1]Shah and Kalsoom (2020) conducted a survey with 116 students and 88 teachers in Pakistan to know about their perceptions of social media and online classes during the lockdown for Covid 19. This survey has been referred to frequently because of its relevance to the chapter's theme.

The foremost problem in education systems across the world is that they have not been developed corresponding to the pace of other services such as banking, world libraries, online purchase, health diagnostic services, online ticketing, and stock exchange businesses. As a result, the lockdown during Covid 19 affected the educational service the most as compared to any other service. Banking or other online work mostly did not stop during the lockdown anywhere in the world, but the education of many students was disrupted due to the non-availability of learning management systems or internet connectivity. Similarly, large scale examinations were cancelled in many parts of the world due to the non-availability of alternate or online examination mechanisms. According to a report, "58 out of 84 countries had postponed or rescheduled exams, 23 introduced alternative methods such as online or home-based testing, 22 maintained exams while in 11 countries, they were cancelled altogether" (UNESCO 2020b).

Another major issue is related to unequal access to educational resources. Students living in developed countries have more access to digital and online resources whereas students in the developing countries have limited access. In Pakistan, there is further discrimination on the basis of the socio-economic status of the students. Students studying in the elite private schools have better access to educational resources as compared to the students studying in low-income private schools or public sector schools. In addition, students living in urban areas have better access to learning resources as compared to the students of rural areas. The students in the rural areas of some developing countries such as Pakistan face frequent electricity failure in their houses. Moreover, a large number of students do not have access to digital devices. UNESCO (2020a) mentions that in many countries, university students do have digital devices. However, their devices do not have essential educational applications and cannot be used for educational purposes.

The non-availability of digital resources has affected students' class attendance, participation, and performance in online classes. Even 'A grader' students could not attend the classes due to no or poor access to digital resources. In this situation, educational assessment becomes an important ethical issue. The students having no or limited access to digital resources may fail to perform well on the assessment day. At small scales, organizations made alternate arrangements. For example, Higher Education Commission of Pakistan suggested the universities to give a choice to the students to select a pass/fail grade instead of a grade or GPA. In this way, their overall CGPA would not suffer (HEC 2020a). Mr Chang, Chief of Section of Education Policy UNESCO highlighted another issue related to assessment by questioning the validity of assessments. He maintained that "not all subjects and competencies can be assessed online or by phone" (UNESCO 2020b).

Teachers' capacity to conduct online classes has emerged as another issue. Although many organizations have tried to train teachers for online teaching, the mode of training was found problematic by some teachers (Shah and Kalsoom 2020). Teachers might have learned about the use of online teaching applications easily if they were trained in face-to-face training sessions. This reflects in teachers' perceptions regarding the effectiveness of online classes. Shah and Kalsoom (2020) reported that only 18% of the students claimed that online classes were more effective as compared to on-campus classes. It is relatively a higher percentage in comparison to the teachers. Only 3% of the teachers mentioned in the survey that online classes were more useful.

3 Effects of Covid 19 on Economy, Society, and Environment

The experience of Covid 19 indicates that the world's socio-economic system is not sustainable and resilient. Laborde et al. (2020) refer to the report of the International Labour Organization which states that 220 million people across the globe would lose their jobs. Ahmed et al. (2020) predict that COVID-19 could cost the world more than $10 trillion and the poor will be affected the most. On the other hand, many internet-based businesses boomed. Iyengar (2020) reported that Zoom's revenue skyrocketed 169% up till April 2020. Global Market Insights projected that the video conferencing market will grow to $50 billion by 2026, nearly 3.6 times more than its market in the year 2019 (Iyengar 2020). Covid 19 has affected health services as well. There is an alarming decline in the number of children receiving life-saving vaccines (WHO 2020). The WHO report mentions that the world is seeing a reduction in the immunization coverage of DTP3 (diphtheria, tetanus, and pertussis) for the first time in the past 28 years (WHO 2020).

Contrary to the negative socio-economic effects, the lockdown for Covid 19 has impacted Nature positively in some respects. Henriques (2020) reported that greenhouse gas emissions and pollution had fallen across all continents when countries cancelled flight operations to curtail the spread of Covid 19. Daly (2020) reported an abundance of insects and birds in the cities as a result of lesser traffic. Kasriel (2020) noted that pandemic led to promoting more pro-environmental behaviours among people as a spillover effect. Similarly, Amos (2020) reported a clear fall in UK air pollution. These reports indicate a clear relationship between economic activities and the environment. Socio-economic and environmental impacts of the Covid 19 lockdown collectively suggest reviewing the current model of development and replacing it with a more sustainable and resilient model (Marinov 2020).

4 Covid 19 and Fake News

The Covid 19 pandemic has highlighted that educational institutions in many parts of the world have not developed people's capacity to question social media information and claims. People shared real as well as fake news extensively without questioning their authenticity. BBC (2020) reported that there was an increase of 27% in bad traffic or fake news in the month of February 2020. This increase may be associated with fears related to Covid 19. BBC (2020) also compiled a short list of fake news that became viral during the Covid 19 pandemic. Such an undirected influx of misinformation not only added to the stress level of people but also made people believe in a variety of myths regarding food-related remedies and herbal/alternative medication for Covid 19. Usually, it is believed that less educated people do not question the authenticity of information. However, it has been observed that Covid 19 related myths were believed and shared by highly educated people, too. Shah and Kalsoom (2020) reported that only 23% of students and 31% of teachers (in Pakistan) did not trust social media information about Covid 19. The rest of the students and teachers either fully or partially trusted social media information. Fifty percent of the students and teachers shared social media information occasionally. It is also interesting to note that the teachers (36.36%) shared social media information more often as compared to the students (23.27%). These figures suggest revisiting and revising the overall education to help people learn to question the claims and information.

5 Covid 19 and Online Teaching and Learning

Education across all levels has been badly affected by the lockdown for Covid 19. According to UNESCO (2020c), approximately 1.2 billion students and youth across the world have suffered due to the closure of schools and universities during the lockdown. As an alternative to campus-based education, online classes are being held all over the world. In Pakistan, many universities (e.g. Beaconhouse National University, Forman Christian College, Lahore University of Management Sciences) shifted to online classes as soon as the lockdown was announced in the month of March 2020. Most of the Pakistani universities announced a summer break for a period of 10 weeks. The school sector took time (3–10 weeks) to prepare for the online classes. The government of Pakistan developed educational videos and started educational programmes on national television for the students who did not have access to the internet. All these steps were taken to help the students continue their education without any interruption.

It is also important to note that Covid 19 pandemic has made the teachers learn the use of online teaching. With each passing day the teachers, who were once mostly adapted to the physical classroom, are exploring new methods of e-teaching and e-learning. The pandemic has acted as a catalyst and teachers are developing learning materials by using interactive apps and online tools like Google Classroom, MS Teams, synchronous face-to-face audio and video interaction, and learning tools like ZOOM, GoToMeeting, WEBEX, Skype, etc. (Ahmed 2020). In addition, teachers have started employing the concept of flipped teaching more frequently (Singh and Arya 2020). Flipped teaching allows students to take ownership of their learning and feel independent instead of feeling dependent and over-shadowed by the few high achievers of their class.

Another interesting development (during the Covid 19 lockdown) is the frequent organization of webinars and e-conferences by educational institutions. A recent example is organization of a three day event entitled "School of Tomorrow" by a school chain in Pakistan (BSS 2020). A large number of people are joining these webinars. It is unlike the situation before Covid 19. It may be claimed that Covid 19 has enhanced the sharing of ideas. Sharing of ideas through webinars should continue in the future, too, as webinars are more inclusive and allow the people from any part of the globe to participate, which is not possible if a campus-based seminar is organized.

6 Covid 19, Education for Sustainable Development and Sustainability Teaching

6.1 Education for Sustainable Development

Ever since 1992, with the birth of the idea of Education for Sustainable Development (ESD), there are consistent debates about the aims and focus of sustainability education. In all different debates, the underlying assumption is that the world is an integrated whole, and sustainability education should help the students to understand the connections between economy, environment, and society and develop pro-sustainability lifestyles. UNESCO (2005) maintained that ESD aims at:

> … an understanding of social institutions and their role in change and development, as well as the democratic and participatory systems … an awareness of the resources and fragility of the physical environment and the effects on it of human activity and decisions ... a sensitivity to the limits and potential of economic growth and their impact on society and on the environment, with a commitment to assess personal and societal levels of consumption out of concern for the environment and for social justice (p. 5).

UNESCO's (2005) delineated three dimensions of sustainable development (economy, environment, and society) into fifteen strategic perspectives. These perspectives may constitute content for ESD. These perspectives are:

> Human rights, peace & human security, gender equality, cultural diversity, and intercultural understanding, health, HIV/AIDS, and governance (socio-cultural perspectives)
>
> Natural resources (water, energy, agriculture, biodiversity), climate change, rural development, sustainable urbanization, disaster prevention and mitigation (environmental perspectives)
>
> Poverty reduction, corporate responsibility and accountability, and market economy (economic perspectives)

In addition to raising students' awareness about the complexity of the world and development issues, scholars have identified the capabilities, competencies, behaviours, and attitudes necessary for sustainability. Wiek et al. (2011) developed competencies—framework. According to this framework, there are five broader key competencies i.e. strategic competence, anticipatory competence, systems-thinking competence, normative competence, and interpersonal competence. There is a range of sub-competencies under each key competency. Lozano et al. (2017) identified a set of twelve sustainability competencies. They are: systems thinking; interdisciplinary work; anticipatory thinking; justice responsibility and ethics; critical thinking and analysis; interpersonal relations and collaboration; empathy and change of perspective; communication and use of media; strategic action; personal involvement; assessment and evaluation; and tolerance for ambiguity and uncertainty. These competencies serve as expected learning outcomes of ESD (Wiek et al. 2011).

Scholars (e.g. Vare and Scott 2007; Sterling 2011) argue that ESD is much more than raising students' awareness about unsustainability and its consequences. ESD should essentially aim at building students' capacity "to think critically about [and beyond] what experts say and to test sustainable development ideas" (Vare and Scott 2007, p. 193). Gough and Scott (2003) argue that the future is uncertain and what is known in the present is not adequate to address future problems. Therefore desired 'end-states' cannot be specified in ESD. Similarly, Jickling and Wals (2008) insist that identifying exact end-states is mis-educative and ESD needs to be open-ended. According to Vare and Scott (2007), ESD may be classified as ESD 1 and ESD 2. According to this classification, ESD 1 is about teaching sustainability content and promoting informed, skilled behaviours and ways of thinking among the students. Contrarily, ESD 2 is about empowering the students to think critically, question the accepted truths and experts' opinions. In other words, the aim of ESD 2 is to transform students' thinking and worldviews. Sterling (2011) explains this perspective of ESD as "experience of seeing our worldview rather than seeing with our worldview" (p. 22).

6.2 *Sustainability Teaching*

Sustainability teaching is a complex and under-theorized concept. Most of the scholars have described sustainability teaching in terms of pedagogies employed by teachers. Constructivist pedagogies are generally considered sustainability pedagogies by the scholars from the field of sustainability education (Wiek et al. 2014). Constructivist pedagogies are the pedagogies that allow construction of knowledge through active, individual and/or social engagement. In ESD, this engagement should centre upon the real-world or sustainability issues (Kalsoom 2019). Some of the constructivist pedagogies found useful by the sustainability researchers include problem and project-based learning (Du et al. 2013; Lehmann et al. 2008; Wiek et al. 2014); undergraduate research (Kalsoom and Khanam 2017); service-learning (Lasen et al. 2015); transdisciplinary problem-oriented learning involving dialogical learning as a means of knowledge generation (Dlouha and Burandt 2015); inquiry-based learning (Pretorius et al. 2016); project-based learning (Leal Filho et al. 2016); action research (Kemmis et al. 2014); and drama (McNaughton 2010). The underlying elements of constructivist pedagogies are ESD processes i.e. collaboration, reflection, dialogue and systems thinking (Tilbury 2011).

Jabreen (2012) described sustainability teaching in terms of its aims. Jabareen (2012) maintains that the aim of sustainability teaching is to help the students acquire various skills including critical and creative thinking, problem-solving, conflict management, and communication. According to this view, any teaching that aims at developing the afore-mentioned skills is sustainability teaching. Vare et al. (2019) viewed sustainability teaching in terms of teachers' competencies to teach for sustainability. Their proposed list of competencies include the competencies of transdisciplinarity; criticality; futures; empathy; innovation; responsibility; participation; engagement; action; attentiveness; systems and decisiveness. According to this thought, a teaching process where teachers act or demonstrate these competencies may be regarded as sustainability teaching.

Keeping ESD 1 in view, sustainability teaching can be described as a set of content and processes that allow students to understand the issues related to environment, economy, and society and their complex relationship; question environmental and socio-economic injustice; and transform their attitudes and behaviours in favour of sustainability. With a transformative perspective of ESD, sustainability teaching may be understood as a process of building students' capacity as critical beings who could question claims; construct knowledge through critical dialogue; anticipate the unpredictable future; and be able to adapt and survive in the uncertain future. In other words, sustainability teaching is a process of transforming the worldviews and making students independent learners.

6.3 Rethinking Sustainability Teaching in Time of the Covid 19 Pandemic

Considering the effects of Covid 19 on society, economy, and environment, there is a need of rethinking the concepts of ESD and sustainability teaching. The occurrence of Covid 19 and large scale lockdowns to curb the spread of Covid 19 support the argument of Gough and Scott (2003) that the future is largely unpredictable and uncertain. According to Gill (2020) our behaviour—particularly deforestation and our encroachment on diverse wildlife habitats—is helping diseases to spread from animals into humans more frequently. The scale and impact of these diseases are not known. Similarly, many other outcomes of unsustainable, human activities are unknown at this stage. Therefore, ESD needs to be open-ended and sustainability teaching should essentially focus on empowering students as independent learners and critical beings who could live in an uncertain and unpredictable future in harmony with other humans and non-humans.

Though the development of critical thinking is emphasized by many educational institutes across the world, the pandemic of Covid 19 indicates that millions of people across the world did not exercise critical thinking as far as information related to Covid 19 was concerned. The Covid 19 situation signifies that critical thinking is a key competence to address emergency and unpredictable situations. In addition to the development of critical thinking, problem-solving and adaptability skills of learners are also questionable particularly in the time of Covid 19. Recent studies

(Kapasia et al. 2020; Tiwari et al. 2020) found that restricted social interaction led to mental health issues among a large number of students during the lockdown. One of the causes of poor mental health during the lockdown seems lack of metacognition or self-regulatory learning among the children and adolescents (Kalsoom and Hasan 2020b). This indicates that education systems are not helping the students to develop metacognition or learn adaptability and problem-solving. These observations have implications for sustainability educators.

6.4 *Epistemological Development*

Trusting and sharing the information (true or fake or misguiding) by the people (less educated as well as highly educated) raises questions about the quality of education in terms of the epistemological development of the people. Epistemological development refers to the transformation of epistemological beliefs from viewing knowledge as fixed and unchanging to viewing knowledge as uncertain and changing. A critical thinker views knowledge as uncertain and changing. Epistemological development is an important goal of education, particularly of higher education. The importance of epistemological development reflects in research studies in the context of higher education. Scholars have investigated the epistemological beliefs of university students extensively (Baxter Magolda 1987/2004; Belenky et al. 1986; Perry 1970). They have also classified epistemological beliefs/ epistemological positions or epistemological perspectives. A brief description of epistemological beliefs is as follows.

Schraw (2013) defines epistemological beliefs as specific beliefs that are about "some aspect of knowledge". Sandoval (2005) and Schommer-Aikins (2004) view epistemological beliefs as 'a set of beliefs' that individuals hold about the nature of knowledge and its acquisition or production i.e. ways of knowing. Baxter Magolda (1987/2004) classified epistemological beliefs under four categories i.e. absolute knowing, transitional knowing, independent knowing, and contextual knowing. Baxter Magolda's classification describes the nature of knowledge and the relationship between assumptions about knowledge and the nature of learning. According to Baxter Magolda, an absolute knower views knowledge as 'certain' thus tries to gain knowledge/information from teachers or other sources and memorizes the information. Transitional knowers view some knowledge as 'uncertain' and try to understand that knowledge. They shift from the stage of acquiring knowledge to understanding knowledge. They collaborate with peers to explore different interpretations. Independent knowers see most of the knowledge uncertain and focus "on thinking for themselves, sharing views with peers to expand their thinking, and expecting teachers to promote independent thinking and avoid judging students' opinions" (Baxter Magolda 1987/2004, p. 37). Contextual knowers view knowledge as uncertain and evolving. To them, knowledge is "continually reconstructed on the basis of new evidence and new contexts" (Baxter Magolda 1992 cited in Hofer and Pintrich 1997). In other words, they operate through 'internal authority' (Kegan 2009) which allows them to construct and reconstruct knowledge in the light of changing contextual evidence.

Schommer (1994) labels epistemological beliefs as sophisticated and naïve beliefs. She argues that learners holding sophisticated epistemological beliefs think that a "vast amount of knowledge is evolving, some knowledge is yet to be discovered, and a very small amount of knowledge is unchanging. With this belief distribution, individuals would be critical readers" (p. 301). To naive learners, on the other hand, a vast amount of knowledge is certain and a very small amount of knowledge is changing. Learners holding naïve epistemological beliefs are uncritical readers (Schommer 1994). Baxter Magolda's (1987/2004) categories of 'absolute knowing' and 'transitional knowing' may be placed under naïve epistemological beliefs while 'independent knowing' and 'contextual knowing' may be labeled as sophisticated epistemological beliefs. Epistemological transformation involves shifting away from being 'made up by' the values of community, culture, family, and friends towards being 'made up by' one's internal authority or self-authored belief system" (Kegan 2009). Similarly, Mezirow (1978,1991), Lange (2004), and Gravett (2004) describe the transformation in terms of changes in the 'frame of references' or habits of thinking. Epistemological development is in line with the agenda of ESD (Vare and Scott 2007; Sterling 2011) i.e. capacity building of the students to think critically about [and beyond] what experts say.

7 Conclusions and Way Forward

The pandemic of Covid 19 has exposed the weaknesses of global educational systems in terms of inclusive educational provision, readiness to educate in emergencies, and development of critical thinking and resilience among people. This suggests changing the centuries-old, current model of education as it is not sustainable. There is a need for environment-friendly, just, inclusive, and holistic education and pedagogies. Such education would require a lifelong learning attitude of the teachers. They would need to learn new skills regularly and act as role models of lifelong learning. In addition, they would need to revisit their teaching content and integrate environmental, social, and economic issues in their teaching. They would organize ESD processes of dialogue, critical reflection, and cooperation. The underlying aim of their teaching would essentially be the capacity building of the students to address unpredictability through creative solutions.

Covid 19 pandemic supports the claim that "the world is a complex, interconnected, finite, ecological-social-psychological-economic system" (Meadows 1982 cited in Sterling 2010). Sustainability teaching should essentially help students to see the inter-connectedness of the world. The pandemic situation further suggests that sustainability teaching should build students' capacities to anticipate the uncertain future, think strategically, perform systems-thinking, solve unforeseen problems, be resilient, and question the knowledge claims (Fig. 1).

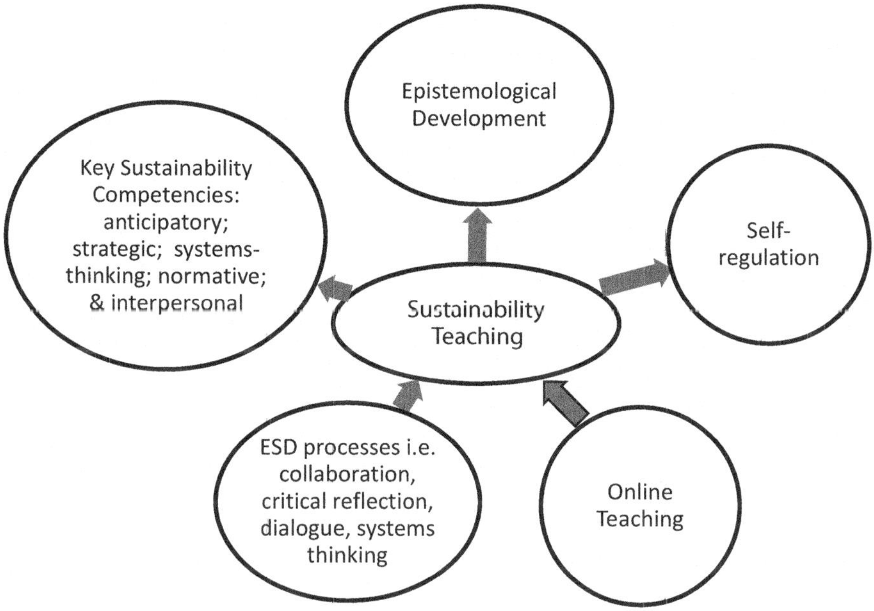

Fig. 1 Sustainability teaching

The use of technology in education has long been advocated by the researchers. O'Donoghue et al. (2002) argued that technology in education had reformed learning and teaching processes and created new opportunities for learning and assessment. Considering the benefits of technology in education, blended learning became a characteristic feature in some programmes (Bonk and Graham 2012; Means et al. 2013). Despite all the advancements in technology-based education, switching to online classes during the lockdown did not happen in a uniform manner across most of the countries and their institutions. The major reason is that majority of the educational institutions (in developing countries) had not invested in digital resources. There is a need of developing digital infrastructure to support online teaching at large scale to ensure educational access of all the learners. Moreover, teachers should keep exploring new educational and teaching technologies after the lockdown too because online teaching seems to be an integral component of future education.

Teacher preparation programmes should train teachers for online teaching methods. At the moment, curricula of many teacher preparation programmes do not include any content related to online teaching. The Higher Education Commission of Pakistan revised the curricula of B.Ed. Honours in 2012. A course on Information Communications Technologies is a part of this programme. However, its contents do not include anything related to online teaching (HEC 2013). It is also important to note that virtual teaching has evolved tremendously in the past 10 years. It is no more limited to the teacher's lectures or videos. Now students can directly communicate with their teachers and classmates during online sessions, study groups, and breakout room apps. Teachers employ different features of these apps to keep students engaged. It may be claimed that online classes in 2020 include many elements of teaching and learning that are found in face-to-face classes.

Although online classes were started as an alternative arrangement for face-to-face education during the lockdown for Covid 19, they seem more in line with the agenda of sustainability education for three major reasons: (1) Online classes reduced petrol consumption because of no traveling; (2) Reduced the use of paper; (3) Allowed the students to access education from remote places. Considering these benefits, it is suggested that online classes should be made a permanent feature of university education and secondary education in schools. There could be a provision of a blended mode of teaching (face-to-face and online teaching). Social development is an important focus of early childhood and elementary education. This can happen through social interaction. Therefore, in elementary grade levels, there could be more face-to-face classes as compared to online classes. It is also important to note that the success of sustainability teaching would require structural support from the institutions and governments.

Acknowledgements We thank Sibte Hasan, Jahan Aara Syed and Mehr-un-Nisa Syed for their valuable suggestions.

References

Ahmed A (2020) Synchronizing pedagogy and technology in post COVID Scenario. ResearchGate. Retrieved from https://www.researchgate.net/publication/341988034_Synchronizing_Pedagogy_and_Technology_in_Post_COVID_Scenario

Ahmed F, Ahmed NE, Pissarides C, Stiglitz J (2020) Why inequality could spread COVID-19. The Lancet Public Health 5(5):E240

Amos J (2020) Coronavirus: lockdown prompts clear fall in UK air pollution. Retrieved from https://www.bbc.com/news/science-environment-52113695

Baxter Magolda MB (1987/2004) The affective dimension of learning: faculty-student relationships that enhance intellectual development. Coll Stud J 21:46–58

BBC (2020) What's real? What's distortion? Retrieved from https://www.bbc.com/news/topics/cjxv13v27dyt/fake-news

Belenky MF, Clinchy BM, Goldberger NR, Tarule JM (1986) Women's ways of knowing: the development of self voice and mind. Basic Books, New York

Bonk CJ, Graham CR (2012) The handbook of blended learning: global perspectives, local designs. Wiley, San Francisco

BSS (2020) School of Tomorrow. Retrieved from https://www.sotevents.com/

Daly C (2020) Coronavirus: how nature is reacting to the lockdown. Retrieved from https://www.bbc.com/news/uk-northern-ireland-52382956

Dlouhá J, Burandt S (2015) Design and evaluation of learning processes in an international sustainability oriented study programme. In search of a new educational quality and assessment method. J Clean Prod 106:247–258

Du X, Su L, Liu J (2013) Developing sustainability curricula using the PBL method in a Chinese context. J Clean Prod 61:80–88

Editorial (May, 2020) Include the true value of nature when rebuilding economies after coronavirus. Nature 581(119). https://doi.org/10.1038/d41586-020-01390-w

Eilam E, Trop T (2010) ESD pedagogy: a guide for the perplexed. J Environ Edu 42(1):43–64

Giannini S (2020) Prioritize health and well-being now and when schools reopen. Retrieved from https://en.unesco.org/news/prioritize-health-and-well-being-now-and-when-schools-reopen

Gill V (2020) Coronavirus: this is not the last pandemic. Retrieved from https://www.bbc.com/news/science-environment-52775386

Gough S, Scott W (2003) Sustainable development and learning: framing the issues. Routledge, London

Gravett S (2004) Action research and transformative learning in teaching development. Edu Action Res 12(2):259–272

HEC (2013) Information and communication technologies in education. Retrieved from https://hec.gov.pk/english/services/universities/RevisedCurricula/Documents/2011-2012/Education/ICTsEdu_Sept13.pdf

HEC (2020a) HEC policy guidance series on Covid-19. Guidance on assessments and examinations. Retrieved from https://hec.gov.pk/english/HECAnnouncements/Documents/nCoVirus/Assessments-Examinations.pdf

HEC (2020b) HEC COVID-19 Policy papers policy guidance Note 5: Online readiness. Retrieved from https://www.hec.gov.pk/english/HECAnnouncements/Documents/nCoVirus/Covid-19-Policy-Guidance-No.5-Online%20Readiness.pdf

Henriques M (2020) Will Covid-19 have a lasting impact on the environment? Retrieved from https://www.bbc.com/future/article/20200326-covid-19-the-impact-of-coronavirus-on-the-environment

Hofer BK, Pintrich PR (1997) The development of epistemological theories: beliefs about knowledge and knowing and their relation to learning. Rev Edu Res 67(1):88–140

Iyengar R (2020) Zoom's revenue soars 169% as people flock to service during pandemic. Retrieved from https://edition.cnn.com/2020/06/02/tech/zoom-earnings-coronavirus/index.html

Jabareen Y (2012) Towards a sustainability education framework: challenges, concepts and strategies—the contribution from urban planning perspectives. Sustainability 4(9):2247–2269

Jickling B, Wals AE (2008) Globalization and environmental education: looking beyond sustainable development. J Curric Stud 40(1):1–21

Kalsoom Q, Hasan S (2020a) Children and Adolescents' Affect during Lock-down for COVID-19: case of Pakistan. Unpublished Report

Kalsoom Q, Hasan S (2020b) Students' affect during Lock-down. Mendeley Data. https://doi.org/10.17632/xsv4tfmgh7.1

Kalsoom Q (2019) Constructivism and sustainable development. In: Leal Filho W (eds) Encyclopedia of sustainability in higher education. Springer, Cham

Kalsoom Q, Khanam A (2017) Inquiry into sustainability issues by preservice teachers: a pedagogy to enhance sustainability consciousness. J Clean Prod 164:1301–1311

Kapasia N, Paul P, Roy A, Saha J, Zaveri A, Mallick R, Barman B, Das P, Chouhan P (2020) Impact of lock-down on learning status of undergraduate and postgraduate students during COVID-19 pandemic in West Bengal, India. Children and Youth Services Review, 105194

Kasriel, E.(2020). A 'mass experiment' for the climate. Retrieved from https://www.bbc.com/future/article/20200624-has-covid-19-brought-us-closer-to-stopping-climate-change

Kegan R (2009) What "form" transforms? A constructive-developmental approach to transformative learning. In: Illeris K (ed) Contemporary theories of learning: learning theorists …in their own words. Routledge, New York, pp 35–52

Kemmis S, McTaggart R, Nixon R (2014) Introducing critical participatory action research. The action research planner. Springer, Singapore, pp 1–31

Laborde D, Martin W, Vos R (2020) Poverty and food insecurity could grow dramatically as COVID-19 spreads. IFPRI Blog: Research Post, 16. Retrieved from https://www.ifpri.org/blog/poverty-and-food-insecurity-could-grow-dramatically-covid-19-spreads

Lange EA (2004) Transformative and restorative learning: a vital dialectic for sustainable societies. Adult Educ Q 54(2):121–139

Lasen M, Tomas L, Hill A (2015) Potential of service-learning to promote sustainability competencies in pre-service teachers: A case study. Teach Edu 26(4):341–365
Leal Filho W, Shiel C, Paco A (2016) Implementing and operationalising integrative approaches to sustainability in higher education: the role of project-oriented learning. J Clean Prod 133:126–135
Leal Filho W, Tripathi SK, Andrade Guerra JBSOD, Giné-Garriga R, Orlovic Lovren V, Willats J (2019) Using the sustainable development goals towards a better understanding of sustainability challenges. Int J Sustain Dev World Ecol 26(2):179–190
Lehmann M, Christensen P, Du X, Thrane M (2008) Problem-oriented and project-based learning (POPBL) as an innovative learning strategy for sustainable development in engineering education. Eur J Eng Educ 33(3):283–295
Lozano R, Merrill MY, Sammalisto K, Ceulemans K, Lozano FJ (2017) Connecting competences and pedagogical approaches for sustainable development in higher education: a literature review and framework proposal. Sustainability 9(10):1889
Marinov GK (2020) Reboot the economy for resilience. Nature 581:262 (2020). https://doi.org/10.1038/d41586-020-01501-7
McKeown R, Hopkins CA, Rizi R, Chrystalbridge M (2002) Education for sustainable development toolkit. Retrieved from https://www.esdtoolkit.org/about.htm
McNaughton MJ (2010) Educational drama in education for sustainable development: ecopedagogy in action. Pedagogy, Cult Soc 18(3):289–308
Means B, Toyama Y, Murphy R, Baki M (2013) The effectiveness of online and blended learning: a meta-analysis of the empirical literature. Teach College Record 115(3):1–47
Mezirow J (1978) Perspective transformation. Adult Educ 28(2):100–110
Mezirow J (1991) How critical transformative learning reflection triggers. In: Mezirow J (ed) Associates, fostering critical reflection in adulthood. Jossey-Bass Publishers, San Francisco, pp 1–20
O'Donoghue J, Singh G, Dorward L (2002) Virtual education in universities: a technological imperative. British J Edu Technol 32(5):511–523
Perry WG (1970) Forms of intellectual and ethical development in the college years: a scheme. Holt Rinehart and Winston, New York
Pretorius R, Lombard A, Khotoo A (2016) Adding value to education for sustainability in Africa with inquiry-based approaches in open and distance learning. Int J Sustain High Educ 17(2):167–187
Reid A, Petocz P (2006) University lecturers' understanding of sustainability. High Educ 51(1):105–123
Sandoval WA (2005) Understanding students' practical epistemologies and their influence on learning through inquiry. Sci Educ 89(4):634–656
Schommer M (1994) Synthesizing epistemological belief research: tentative understandings and provocative confusions. Educ Psychol Rev 6(4):293–319
Schommer-Aikins M (2004) Explaining the epistemological belief system: introducing the embedded systemic model and coordinated research approach. Edu Psychol 39(1):19–29
Schraw G (2013) Conceptual integration and measurement of epistemological and ontological beliefs in educational research. Int Scholarly Res Not. https://doi.org/10.1155/2013/327680
Shah, A, Kalsoom Q (2020) Teachers and students' perceptions about social media news and online teaching & learning. Mendeley Data, V2. https://doi.org/10.17632/m3v6np4tn8.2
Schank C, Rieckmann M (2019) Socio-economically substantiated education for sustainable development: development of competencies and value orientations between individual responsibility and structural transformation. J Edu Sustain Dev 13(1):67–91
Singh S, Arya A (2020) A hybrid flipped-classroom approach for online teaching of biochemistry in developing countries during Covid-19 crisis. Biochem Mol Biol Educ. https://doi.org/10.1002/bmb.21418
Sterling S (2010) Living in the earth: towards an education for our tim. J Edu Sustain Dev 4(2):213–218
Sterling S (2011) Transformative learning and sustainability: sketching the conceptual ground. Learn Teach Higher Educ 5(11):17–33

Tilbury D (2011) Education for sustainable development: an expert review of processes and learning. UNESCO, Paris
Tiwari GK, Singh AK, Parihar P, Pandey R, Sharma DN, Rai PK (2020) Understanding the perceived health outcomes of children during COVID-19 pandemic. Retrieved from https://www.researchgate.net/profile/Gyanesh_Tiwari/publication/341417291_Understanding_the_perceived_health_outcomes_of_children_during_COVID-19_pandemic/links/5ebf823f92851c11a86c3dc2/Understanding-the-perceived-health-outcomes-of-children-during-COVID-19-pandemic.pdf. Accessed 25 May 2020
UNESCO (2020a) COVID-19 and higher education: today and tomorrow: impact analysis, policy responses and recommendations. Retrieved from https://www.iesalc.unesco.org/en/wp-content/uploads/2020/04/COVID-19-EN-090420-2.pdf
UNESCO (2020b) Exams and assessments in COVID-19 crisis: fairness at the centre. Retrieved from https://en.unesco.org/news/exams-and-assessments-covid-19-crisis-fairness-centre
UNESCO (2020c) Global education coalition. Retrieved from https://en.unesco.org/covid19/educationresponse/globalcoalition. Accessed 30 May 2020
UNESCO (2005) Guidelines and recommendations for reorienting teacher education to address sustainability. UNESCO Education for Sustainable in action. UNESCO, Paris
Vare P, Scott W (2007) Learning for a change: exploring the relationship between education and sustainable development. J Edu Sust Dev 1(2):191–198
Vare P, Arro G, De Hamer A, Del Gobbo G, De Vries G, Farioli F, Kadji-Beltran C, Nijdam C (2019) Devising a competence-based training program for educators of sustainable development: lessons learned. Sustainability 11(7):1890
WHO (2020) WHO and UNICEF warn of a decline in vaccinations during COVID-19. Retrieved from https://www.who.int/news-room/detail/15-07-2020-who-and-unicef-warn-of-a-decline-in-vaccinations-during-covid-19
Wiek A, Xiong A, Brundiers K, Van Der Leeuw S (2014) Integrating problem-and project-based learning into sustainability programs. Int J Sustain High Educ 15(4):431–449
Wiek A, Withycombe L, Redman C (2011) Key competencies in sustainability: a reference framework for academic program development. Sustain Sci 6(2):203–218

Dr. Qudsia Kalsoom is currently serving as Head of the Department of Educational Leadership and Management in the School of Education at Beaconhouse National University, Pakistan. Dr Kalsoom is associated with the teaching profession for more than 22 years. As a researcher, she studies sustainability education in higher education and particularly in teacher education.

Ms. Ayesha Jalil Shah has an MPhil in Educational Leadership and Management from Beaconhouse National University (BNU). As an alumna of BNU, Ms Shah actively collaborates with the faculty of the School of Education, BNU, to investigate issues related to ESD. Ayesha Jalil Shah is currently employed at The City School (pvt limited), Pakistan as a partner in the department of school improvement and quality assurance.

Women and Higher Education in COVID-19 Times: Challenges and Possibilities for Sustainable Development'

Tayyaba Tamim

Abstract COVID-19 has not only created a health emergency but also seriously disrupted the functioning of economic and social institutions. Its impact on higher education has been direct, swift and devastating. Higher education institutions (HEIs) form the backbone of progress towards Sustainable Development Goals (SDGs 2030) in their stewardship of research and innovation, engagement with policy debates and commitment to the education and training of professionals. The crucial element that facilitates this trajectory, however, is women's participation in HEIs, premised on principles of equity and inclusivity, widening participation and access that enables individual empowerment and prosperity on the one hand and ensures valuable contributions towards the SDGs 2030 on the other hand. Nevertheless, it is now feared that COVID 19 may reverse the gains made towards Goal 4 of SDGs 2030, as it structures new barriers that make it difficult for girls to stay in education. This is a major worry because the ripple effect from this may not only derail achievement of Goal 5 but other SDGs as well. Given that the World Bank and UNESCO anticipate large number of students, especially girls in poor countries to drop out of education, the question is, would it be the same for females in higher education? As formal education is relocated in homes, where unequal gendered relations of power might play out strongly against them. Women may find it difficult to pursue their higher education. This may be because of their limited access to resources, burden of domestic work, caring responsibilities, domestic violence, mental and physical health or sociocultural norms. Conversely, the question is also, does this virtual space of higher education, offer some unique opportunities for us to facilitate the access of women to higher education and ensure their retention? Is it an opportunity to rethink education in a way that is more inclusive and flexible? These are important questions to contend with to make an informed, gender sensitive response to the challenges faced by women in higher education.

Keywords Women in higher education · COVID-19 · Sustainable development goals · South Asia · Vulnerabilities · Opportunities

T. Tamim (✉)
School of Education, Lahore University of Management Sciences, Sector U, Defence Phase V, Lahore, Pakistan
e-mail: tayyaba.tamim@lums.edu.pk

W. Leal Filho (ed.), *COVID-19: Paving the Way for a More Sustainable World*, World Sustainability Series, https://doi.org/10.1007/978-3-030-69284-1_23

1 The Disruption of COVID-19 and Higher Education

The first COVID-19 case was identified in Wuhan in December 2019 and by March 11, 2020 it had been declared a pandemic by the World Health Organization (WHO 2020). August 23, and the world was still smarting with pain, suffering under strained health systems, crippled economy and disruption of social systems and threat to life itself, as the pandemic struck more than 23 million and left over 8 hundred thousand dead (Covid-19 Dashboard, John Hopkins 2020). Several countries suffered from a second wave of the pandemic after the first had been abated (for example, Hong Kong, China, Vietnam, Fiji, South Korea, Spain, Iran, Australia, UK, Germany and France) (Philipose 2020; Chadwick 2020). Given, its massive scale, the multidimensional impacts of this pandemic can only be expected to persist for quite some time to come (BBC 2020; Nicola et al. 2020). UNESCO (2020a) estimates that COVID-19's impact on the disruption of education alone, will outlive this generation.

Higher education sector has been one of the hardest hit, left to grapple with an unprecedented global situation, with few hard choices (UNESCO 2020a, b; Bassett and Arnhold 2020). According to an estimate by April 8, 2020, higher education institutions had been closed down in 175 countries affecting '220 million post-secondary students,' with their studies either ended or 'significantly disrupted due to COVID-19' (World Bank Group: Education 2020, p. 1). Table 1 captures the global scale of this disruption. While the devastating impact of COVID-19 may be displayed to be more significant at the given moment in upper and lower middle income countries, in lockstep with the spread of the virus, it is anticipated that numbers would rise dramatically for lower middle and low income countries, especially in South Asian and African regions as the pandemic spreads (ibid.).

The response of higher education institutions (HEIs) to COVID-19 can be mapped across a spectrum of inter-and intra-regional variation, between choosing to remain open, if possibility existed, (as in the case of University of Virginia in Australia) to halting all educational delivery (for example in Malaysia). A majority, however, closed campuses and moved their courses online (Crawford et al. 2020). A survey conducted between March 25–April-17 2020, of 424 HEIs in 109 countries, across multiple regions, revealed that 59% of these had their campuses completely shut

Table 1 The scale of disruption in higher education

Income level	Out-of-school tertiary ed students	Total tertiary ed students	%
High income	53,479,089	54,103,566	99
Upper middle income	97,493,490	97,934,594	96
Lower middle income	65,358,490	66,421,264	98
Low income	3,808,691	4,146,072	100
Grand total	220,139,760	222,605,496	99

Source The COVID-19 crisis response: Supporting tertiary education for continuity, adaptation, and innovation, World Bank Group: Education- Table 1, p1, 2020

down (Marinoni and de Wit 2020).These decisions, nevertheless, have been dynamic, responsive to the pandemic spread (Crawford et al. 2020). Even as HEIs stagger to their feet, clutching onto online mode of instruction, they now operate in a different world reality, as the health emergency created by COVID-19 leaves behind 'widespread socioeconomic implications,' with no aspect of life untouched (Nicola et al. 2020 p. 185). This points to the urgent need of taking into stock the challenges by HEIs and planning an informed response, sensitive to at risk students, particularly women to achieve Goal 4.[1] Failure to do so would consequently affect the attainment of goal 5[2] of the sustainable development agenda 2030 (UNESCO 2017), with a domino effect on the achievement of all the other SDGs also, because of their fundamentally interlinked status (ibid.).

The imperative of HEIs to shift to online instruction in the face of COVID-19 was not without its problems. Online courses had to be designed, at a breakneck speed over a matter of days. It was a 'speed staggering Emergency Remote Teaching,' which may be different from regular online courses, which typically take six to eight months of preparation (Hodges et al. 2020). While some universities were already following hybrid or online modes of instruction, for others, for example, several in Afghanistan, India and Pakistan, with limited technological capacity could do very little to support their students (Khan et al. 2020). Nevertheless, it was a major challenge for HEIs all over the globe to simultaneously update their infrastructure, train faculty, develop online resources and support their students, while facing serious financial problems (Bao 2020; Crawford et al. 2020).

The financial challenges for HEIs were a major concern, given the immediate new investments required and the paradoxical drop in the tuition fees, as large numbers of fee paying international and national students headed home with uncertain plans to return, given both health security concerns and loss of family incomes (Friedman et al. 2020; Maslen 2020; McKie 2020; The Japanese Times 2020). The magnitude of the disruption begins to emerge, when one realizes that only in US Chinese make up 33.7% and Indians 18.4% of the international student population, a majority of which had to return home (Times 2020).To make matters worse, with no definitive date for re-opening, HEIs were obliged to make refund fees charged to students for the use of physical resources i.e. hostels, parking spaces etc. (Maslen 2020), while forgoing annual raise in tuition fee (ibid.). This affected universities across the globe. SOAS University of London UK, in the face of financial crisis had to opt for jobs cuts (McKie 2020), while Monash, the largest Australian university, with 80,000 students and 43 campuses and with more than AU$5 billion (US$3.2 billion) assets, reported a major financial shortfall of AU$350 million (US$226 million) in revenue in the current year (Maslen 2020).

Public universities and colleges, which work around the agenda of equity, took the major brunt of the economic down turn, as governments redirected their funding. With the economic recession looming large and failing health systems, funding higher

[1]"Ensure inclusive and equitable quality education and promote lifelong learning opportunities for all." (UN, Transforming our world: The 2030 Agenda for Sustainable Development, 2015).

[2]Achieve gender equality and empower all women and girls (ibid).

education was not the priority. In US, Universities of California and Rutger faced serious shortfalls in their budgets that far exceeded the federal funding they received, as they simultaneously tried to support their students' transition to online instruction and make refunds to them (Maslen 2020). In lower-middle and low-income countries, while little data is accessible in this regard, the situation is bound to be worse, where governments already struggling with fragile economies, and poor health systems, now juggling through a conundrum of competing demands on their resources, find it easier to slash the funding of HEIs. In Pakistan, for example, the funds allocated for higher education in the budget were drastically cut by 5.94 billion for the fiscal year 2020–21, seriously threatening not only the quality of subsidized education these HEIs offered but also their very survival (Imran 2020). To make matters worse, the foreign aid on which HEIs could count on earlier could also not be forthcoming, as the world prepares for a global recession (Nicole et al. 2020). UK's decision to assimilate DFiD under the foreign office, for example, may squeeze the funding it can offer to lower-middle and low-income countries (Landale 2020). Even more worrying is the anticipation that this 'state disinvestment in higher education may become the norm in years to come' (UNESCO 2020c, p. 3; Yuen 2020). This would be a baffling stumbling block for not only the achievement of Goal 4 focusing on the provision of inclusive quality education for all but also for the gender focused Goal 5, affecting all other SDGs 2030. UNESCO (2020c) estimates university students to be most at risk of dropping out, among the 24 million who may not be able to continue their education (half of these will belong to South and West Asia and sub-Saharan Africa ibid.), while others estimate this figure to be close to 1.5 billion (World Bank 2020).

The hardest hit, as always, are the ones already vulnerable and disadvantaged stratified across class, disability or other stigmatized markers of identity, with gender being a salient category cutting across all other dimensions. For women and girls, the worry is not just the anticipated learning losses or disruption due to the closure of educational institutions but that they may drop out of education at all levels (Giannini and Albrectsen 2020). The specific vulnerabilities of women in relation to their participation in higher education in the face of COVID-19 and the resultant implications for SDGs remain under discussed in literature. This paper aims to address this gap, given that gender equality is central to SDGs 2030, addressed specifically in Goal 5, but also 'mainstreamed into numerous others goals' including Goal 4 (Fredman et al. 2016). Hence, the emphasis on higher education in SDGs 2030, given the understanding that while progress might have been made regarding the participation of women in higher educations, the gender gaps remain significant in several low and lower income countries, especially conspicuous when class is brought in as a variable (Ilie and Rose 2016).

Method

The interest in the topic actually triggered from conversations with women students at a university in Pakistan. As some themes became consistent, for example, dropping out or semester freezing, marriages during COVID-19, time squeeze; the burden of children's education and domestic work, caring responsibilities and access to digital resources, we decided to use to explore it using a 'desktop analysis approach' following (Crawford et al. 2020). Specific attention was paid to the reliability of the sources (ibid.). A total of 88 articles were used for this paper, which were filtered from more than 200 papers which were read up and not used either because they became irrelevant as the argument took shape or because of the limitation of space. In the first instance, resources were identified using the key words 'COVID-19 spread.' This was to determine the scale of the pandemic. This data was updated several times when writing the paper. Simultaneously, internet search was done using terms 'gendered impact of COVID-19,' 'COVID-19 impact on higher education,' and COVID-19 women and higher education.' We could find little in relation to the latter. Once the gendered themes began to emerge, we started to explore each theme in relation to women in pre-COVID situations, for example women's participation in higher education, unpaid work, ownership of assets and other forms of discrimination. This allowed us to make connections and interpret how the existing forms of vulnerabilities translate into barriers for their participation in higher education.

This paper now proceeds to discuss the significance of higher education and within it the role of women's participation for SDGs 2030; the third section extrapolates the specific issues faced by women across country contexts, well documented in literature, and now exacerbated on their access to higher education in times of COVID-19; the fourth section discusses the affect it may have on the achievement of SDGs 2030, followed by the possibilities that the current situation offers. The fifth section concludes the argument. In this paper, the term 'women' is applied broadly to females of college/university going age here. The discussion of higher education institutions applies to all post secondary colleges and universities, while the arguments made in the paper are more relevant to lower-middle and low-income countries than others. Similarly, some aspects discussed here may be more salient to some contexts than others, nevertheless, they do cut across borders and affect women in general and their higher education in one form or the other.

2 Higher Education, Sustainable Development Goals and the Participation of Women

Higher education is undeniably the stepping-stone for the achievement of sustainable development, a realization which has come after missing out its significance in Millennium Development Goals 2015 (UNESCO 2015; Ilie and Rose 2016). The target 4.3 of SDG 4 pledges to 'ensure equal access for all women and men to affordable and quality technical, vocational and tertiary education, including university.'(UNESCO n.d.)

Universities as ultimate seats of teaching/learning, knowledge and scholarship not only have the capacity to influence policies for transformative change but also have a role in the training of educators, and nurturing a whole generation that can work towards 'sustainable economic, social and political directions of the globalized interdependent world' (Blessinger et al. 2018). Along with encouraging critical thought and deliberation for sustainable development, higher education institutions (HEIs) also lead to inclusive mindsets and behavior that form the basis of socially just democratic societies; hence, the centrality of HEIs in pushing the agenda of SDGs 2030. UNESCO emphasizes:

> Higher education also forms an important part of other goals related to poverty (SDG1); health and well-being (SDG3); gender equality (SDG5) governance; decent work and economic growth (SDG8); responsible consumption and production (SDG12); climate change (SDG13); and peace, justice and strong institutions (SDG16) *(UNESCO: Higher Education and Sustainable Development Goals,*n.d.*)*

This role of HEIs in achieving SDGs 2030, apart from being contingent of the quality of education is also based on inclusion of women in specific who form almost 49.5% of the world's population (World Bank Data 2019). While quality of higher education has been passionately guarded ever since the first seat of higher education was established in 859 AD, inclusion of women has been a struggle. Historically, females have been excluded from sites of higher education (Morris 2011), Oxford and Cambridge in UK, for example, fiercely held the fort against the entry of women and their access to degrees until last century (Dyhouse 2003). It is only over the past 100 years that women have found their way into universities after a long and painful struggle that challenged both the long established norms of the academia and the social and cultural traditions (Moore 1987). Ever since, women have been making their mark in a number of fields: law, medicine, architecture, politics, business, art, literature and education itself, more often than not made possible through access to higher education. While women's participation has been steadily increasing, it remains problematic in several middle and low-income countries (Klasen 2019).

The emphasis of SDGs 2030 on gender equality and empowerment and the need to address all forms of discrimination against women more specifically in Goal 5 is also a unifying thread that runs through all the other SDSs. This is an acknowledgement of the core importance of addressing gender based inequality for achieving sustainable development and an admission that a lot more needs to be done in relation to the discrimination faced by women. The target 4.3 of SDG4 that aims to 'ensure equal access for all women and men to affordable and quality technical, vocational and tertiary education, including university,' (UNESCO 2017), underscores the importance of the participation of women in higher education. Women's participation in higher education can carve pathways to achieve Goal 5, by increasing women's access to resources and giving them voice and representation; while also ensuring their wider participation in social, economic and political fields and pushing the agenda of SDGS (Szto 2015). Ensuring the access and retention of women in HEIs then may be seen as both as a matter of equity and as a strategic investment to accelerate the process of socioeconomic development, and establishing of democratic and peaceful societies as espoused in SDGs 2030 (UNESCO: Higher education and the Sustainable Development Goals, n.d.). Hence, the need to take into account the challenges faced by women to pursue higher education in these unusual times to ensure their presence in these sites.

3 In the Eye of the Storm: The Challenges of COVID-19 for Women Participation in Higher Education

Women have been facing glaring discrimination in all fields of life and it is until recently that some progress has been made in terms of their inclusivity in all walks of life, and higher education is no different (UNESCO 2017). Although the number of females in higher education has increased, this progress must be seen as fragile, especially in lower-middle and low-income countries, where wider social inequalities are more persistent (ibid.). The Global Gender Gaps Report (2020) based on 107 countries declares that it will take on average 99.5 years to close the gender gap. It estimates that women are worst off in political participation and followed by economic participation. It states that.

> Financial disparities are slightly larger (on average), explaining the step back registered by the Economic Participation and Opportunity subindex this year. On average, only 55% of adult women are in the labour market, versus 78% of men, while over 40% of the wage gap (the ratio of the wage of a woman to that of a man in a similar position) and over 50% of the income gap (the ratio of the total wage and non-wage income of women to that of men) are still to be bridged (Global Gender Gap Report 2020, p. 6)

Although countries did better in educational attainment gender gap, the report argued that:

> Even in countries where education attainment is relatively high, women's skills are not always in line with those required to succeed in the professions of the future. In addition, they encounter barriers to employment in the most dynamic and in-demand occupations (Global Gender Gap Report 2020, p. 5).

Now, the gendered impacts of COVID-19 (Wenham et al. 2020) threaten their presence in HEIs, which offered them transformative possibilities, as these conflate with their low positioning within families and in the society. Five major challenges may be identified (although several others might also exist) that women may confront as they plan to access or remain in HEI in the current situation and which could possibly push them to drop out:

3.1 Shrinking Economies: The Financial Crunch and Shifting Priorities of Households

The economic setback in the COVID-19 has resulted in the loss of millions of jobs and family incomes across the world. It has also drastically affected the funding of public education institutions, which has been contingent upon economic growth (Evans et al. 2020). The situation is worse lower middle and low income countries, who can no longer rely on financial aid from richer countries, unlike earlier disaster situations or during the financial global crisis of 2008–09 (ibid.). This threatens the agenda of both equity and quality of public sector higher education systems, while increasing the pressure on families for paying up for the education of their children, as financial aid and scholarships dry up.

As millions of families suffer from the economic backlash of COVID-19 with loss of jobs and incomes, against rising inflation and real health risks, their priority might not be the higher education of their children but survival. Student dropouts are the likely outcome (World Bank 2020). As part-time jobs become rare and financial support difficult, a large number of students may not be able to continue their education. In Japan, for example, an online survey of 1000 participants in April, 2020 concluded that one in every 13 students was considering to leave university, following economic hardship (The Japane Times 2020). The situation can be assumed to be worse in lower middle and low-income country contexts, where large populations fall below the poverty line (Strauss 2020). Mali, Niger and South Sudan are three countries with lowest levels of enrollment for girls, and the closures have further forced 4 million girls out of schools (Giannini and Albrectsen 2020). With hardly any data, one can only guess the number of women leaving or who may leave higher education in similar contexts, where they are economically dependent on male members of the family to provide for their education. An income which may now be either gone or severely reduced.

Women maybe the first to leave higher education if the financial constraints continue, amid pressures of finding working to support family income, free up family funds, help with domestic work or care for the sick, elderly or the young. The 'choice' may be made by the women themselves, socialized into their role of putting the needs of others before them (Qureshi 2020). In places where it is not uncommon for women to be used as bargaining chips to negotiate truce with enemies, as in the case of 'vani' an age old custom practiced in certain regions in South Asia (Munir and Akhtar 2014), the onus is on the 'good' woman, like in the fairy tale of the Beauty and the Beast, to give up her ambitions- to be imprisoned so others are free. Hence women may just feel obliged to give up their education so that male siblings could continue with theirs when incomes become uncertain.

In several regions, where the positioning of women remains secondary in the family, so does their education. For example, in contexts like Pakistan, India and Vietnam, where male children are traditionally entrusted with caring for aging parents, a simple cost benefit analysis may typically make families conclude that material returns to the family come only with investment in the education of male children (Saha 2013). In addition, the widespread custom of dowry, whereby a family needs to spend a hefty amount to marry their daughters, one way or the other, may also deter families to spend the money on the higher education of their daughters in these times of financial instability. In all probability they may also consider it only fair to spend the sons' share of family funds on their education and save the share of daughters for their prospective marriages, with the latter now in foreseeable future given the age of these women.

3.2 Increase in Domestic Violence and Reduced Support

The social distancing, isolation and lock downs to contain the spread of COVID-19, was soon followed up by surging reports of domestic violence around the world (Taub 2020; Campbell 2020; Usher et al. 2020). This rise in domestic violence has been related to a complex range of factors 'economic stress, disaster-related instability, increased exposure to exploitative relationships, and reduced options for support,' issues that have been observed to follow at the heels of a pandemic (Peterman et al. 2020). While the victims may also be men (Fulu et al. 2013), it is the women and children who are in the eye of the storm (Usher et al. 2020). Even before the pandemic, according to an estimate every one in three women across the world suffered from one kind of abuse or the other in her lifetime (Devries et al. 2013).

Now the magnitude of their vulnerability only increases, as women find themselves locked up in homes (Campbell 2020; Usher et al. 2020; van Gelder et al. 2020; Bradbury and Isham 2020). The quarantines imposed by governments enable the perpetrators of domestic violence abuse to effectively use their classic strategies of isolating victims and keeping them their daily activities under surveillance (Hagan et al. 2019). Away from the public eye, the violence in the private sphere hardly noticed. Violence and abuse may exacerbate as psychological and economic pressures rise with extended lockdowns and 'negative coping mechanisms' for example alcohol or drug use, would now take place at homes, leaving women more vulnerable than ever (ibid.).

For women in HEIs, as the 'critical emotional support and […] opportunity for a "reprieve" from their abusive home environment [offered by HEIs in the form of hostels, libraries and classrooms] disappear (Campbell 2020) and friends move into shadows, it may become very difficult for them to continue with their education. Even asking for help and support may also become more complicated under the watchful eye of the perpetrators of domestic violence (Fielding 2020). All this compounds the dilemma of several women in higher education, who need to cope with both academic pressures and the unbearable situation at home.

In contexts, where domestic violence is normalized, and justified by women themselves (Biswas et al. 2017), the risk of exacerbation of violence becomes manifold. The stigma attached to being a single mom, social pressures to stay in marriages and family homes, financial dependence and a general silence around family based violence or abuse, and the strong likelihood of the blame rebounding to the female victim, may force women to be bear it quietly. While access to higher education is a way out of the distressing situation, with the possibility of financial independence, awareness and finding a voice eventually against this injustice, several may not be able to cope with the existing pressures. They may drop out or perform poorly as higher education move into unpredictable and volatile spaces of their homes.

3.3 Division of and the Demands on Time

Despite the significant achievements in closing the gender gaps in education and labour market where women have entered domains that were historically male dominated fields, culturally embedded inequitable gendered division of labour at home persist. When women contribute to family incomes, working outside homes, this only adds to the burden of unpaid, time intensive domestic work they continue to carry disproportionately more than men (Yavorsky et al. 2015; Mannino and Deutsch 2007; Milkie et al. 2009; Offer and Schneider 2011; Raley et al. 2012; Yeung et al. 2001) Although men are now expected to be involved in the care of their children, especially in western contexts (Pleck 2010), the time they actually invest in domestic work is still much lesser than women (Garey 1999). There is not a single country where men are involved in equal share of unpaid work at home (Global Gender Gap Report 2020).

A justification for this inequitable distribution is often offered from 'time availability perspectives', which means that the person spending most time at home does most of the domestic work (ibid.). What is often forgotten in this seemingly just arrangement is that women are expected to cut down their paid work hours after children, while men are not (Ferrant et al. 2014). The resulting loss of income, along with historically low ownership of assets and the culturally embedded expectations of domestic work, lowers womens' 'bargaining power' to ask for more gender equitable division of labour at home (Sanchez and Kane 1996; Gupta 2006). It may also be argued that this is a part of "doing gender" i.e. reinforcing the gender role performances that are culturally expected of men or women (Sandhya and Jayaraman 2019). Hence, women suffer from a "time squeeze" (Milkie et al. 2009), which becomes even more pressing with the caring of the sick and elderly in a family (ibid.) or a child with disabilities (Traustadottir 1991). Such patterns of gendered time usage in domestic work, more or less is a feature common to all over the world (Hong 2019; Nawaz and McLaren 2016; Strong et al. 2016; McMunn et al. 2020), which can be argued to have intensified during COVID-19 as all the family members stay at home, expecting to be served.

Women in higher education, especially during COVID-19, are on a fragile footing because they are not only NOT contributing to the family income but making demands on it, which dramatically reduces their ability to negotiate the time (Fafchamps and Quisumbing 2003), they need to be freed up to focus on their studies. The pressures become manifold in the case of a large number of mature female students who may be married, divorced or single parenting. Concentrating on academic work from home may be a challenge in itself amid competing demands on their time, by children, elderly, the sick and other male members, who in certain Asian contexts consider it derogatory to help with domestic work. Deflecting these demands may be difficult because they can no more detach themselves physically, as they could earlier in the space offered by higher education institutions. To make matters worse, the closure of schools means that they might have to take on the additional role of educators for their children, or siblings without any chance of other duties being taken off their shoulders (Moreno and Shaw 2018).The consequences, may it be lower course grades or the decision to put their higher education on hold, would have significant negative implications for their lives.

3.4 Health Risks: Mental and Physical

Women are not only more susceptible to the pandemic, by virtue of being primary carers but also to physical and mental health risks in these times that may inevitably affect their participation in higher education. The higher number of women in medical professions (Boniol et al. 2019) automatically feeds into an assumption that more women than men might be engaged in the discipline of medicine in higher education institutions and if closer to the final year may also be engaged in practical work in hospitals. The 'systematic differences' in their positioning in the health sector, across 104 countries, reveal more women than men as nurses, midwives and health workers (ibid.). This means that they do not only have little power for decision making but also face a higher risk to the disease because of more intimate contact with patients (Ramakrishnan et al. 2014), either at home or as part of their higher education.

Although gender segregated data of COVID-19 indicates more deaths in men than women, may be 'because of sex related immunology or lifestyle patterns,' (Wenham et al. 2020), it does not cover the vulnerability to contracting the disease, which may be higher for women (ibid.). If learning from past can shed any light on the health risks for women, in the 2014–16 the spread of Ebola and Zika viruses showed that women were not only more susceptible to catching the virus because they were the 'primary care-givers in families, as well as the front-line health-care workers,' but also because they had little decision making power and had limited access to health care, while government resources for reproductive health for directed elsewhere (ibid.). It has also been reported that when women catch the COVID-19 they are more severely, stigmatized and less cared for (McLaren et al. 2020). Women in higher education may then be coping with these larger issues as try to continue their studies and these pressures may compound their stress related to completion of academic work etc.

3.5 The Digital Divide

While gender-specific quantitative data is limited, particularly in developing country contexts, various studies indicate a strong evidence of the digital gap across genders (Hafkin and Huyer 2007), more across marginalized groups and minorities than others (Tolbert et al. 2007; Mumporeze and Prieler 2017). In low-income countries on average, 30% women are less likely than men to use Internet, while in some countries this gap stands at 80% (Global Development Lab: USAID 2020). Negating the common assumptions that women are 'technophobic,' while men excel in the use of digital services, Hilbert found empirical evidence from Latin America and Africa that the digital divide between the genders, both in terms of usage and access emerged from women's limited access to well paid jobs and education (Hilbert 2011). This is also an important aspect of the well-documented issues of women's poor access to resources within the family (Brännström 2012). Whatever the cause, this means that the digital divide across gender will clearly aggravate the disadvantage of women in HEIs during COVID-19. Poor connectivity of Internet, limited access to computers and other devices will make things increasingly difficult for women as they study from home. While this may also affect men living in rural areas where internet connectivity is poor, the velocity with which it hits women is much more, especially in contexts where their social mobility is constrained not only by COVID-19 but also be sociocultural norms (Parvazian et al. 2017).

These multiple issues disproportionately weigh on women and make it difficult for them to participate in higher education. May it be worry of financing their education, loss of family incomes, pressures of domestic work, caring responsibilities, domestic violence, limited access to technology, health risks, competing demands on time or a general sense of powerlessness and loss of control, these may make them think twice about continuing their higher education. If HEIs turn a blind eye to the multiple issues faced by women who come to them in these times, it may not be possible for these women to continue. This in turn may not only directly affect achievement of Goal 4, that pledges inclusive and equitable education for all, but also severely affect Goal 5 that aims to eradicate all forms of discrimination against women to ensure their equal participation provide and contribution towards all the other sustainable development goals. Participation in higher education is a crucial means for women to be able to challenge inequitable structures, have greater decision making power, acquire financial independence, close gender wage gaps and take on leadership positions. The threat to women's participation in higher education does not only jeopardize their own progress but also wider sustainable development processes, which remains deficient in their impacts without the educated and informed contribution of women. It is then important to deliberate on ways that women can be supported to access and continue their higher education.

4 Enabling Possibilities Through Support Systems for Women in HEIs

The new normal of the higher education in terms of online instruction during COVID-19 times creates unique challenges for women as discussed before but the situation also offers possibilities for women's education, which may be capitalized with well-thought out support.

Many higher education institutions were offering online degrees and courses, before COVID-19. However, these were not considered at par with those offered face to face in classroom settings, with little market value (Hodges et al. 2020). Now as universities around the world have been pushed into the virtual space, there is now an opportunity to access educational content of the best -Ivy league universities from home. In addition, several universities have also slashed their tuition fees, or waived off their annual fee increase. The aim of the HEIs has been to facilitate the access and retention of students in these difficult times, although they themselves crunch under the weight of a slow economy. Furthermore, as online instruction is normalized, many universities also allow transfer of credits from online courses taken elsewhere. This reduction of cost and online access to a large number of universities across the globe is certainly an opportunity for women.

Theoretically speaking, in contexts where women face restrictions on their mobility arising from tradition, religious norms, concerns for their security or expec tations to take up domestic responsibilities, online higher education now ensures a viable option without comprise on the quality and prestige of the education offered. The current times open up the opportunity for them to continue their study from home, with a well-recognized college or university with lesser cost, while being physically at home and navigating their time through multitasking. Similarly, mature women students who are single parenting, working and simultaneously studying may find it easier to access their HEIs from home. This takes the edge off their day which was earlier spent in commuting, pick/drop of children from schools/ day cares and then rushing to their colleges/ universities. Now, all that is a simple click away for an education whose market value is not compromised.

However, for women to actually capitalize the current situation, considering the issues discussed earlier (see Sect. 3), we need to deliberate on support systems specifically for women to ensure that we do not lose them in HEIs. The first and foremost thing in this regard is a recognition of gendered COVID-19 impacts on women and its inevitable impact also on their participation in higher education. In this regard, HEIs need to reach out to their female students to understand their specific issues. While simple surveys might help, some qualitative data collection in the form of focus group or individual interviews may be needed to fully capture the deeper problems. This is best followed up by discussions across HEIs so that some collaborative work can be done in this direction and collective deliberations are undertaken to understand and respond to the problems faced by women participating in higher education.

Second, although the financial crunch facing the higher education's institutions may limit to the number and extent of scholarships and financial aid, they can offer, women cannot do without them. Even if fees are lowered, several of them might find it difficult to expect their family to pay up when jobs have been lost or economic certainty prevails (Maital and Barzani 2020). The universities could initiate discussions to find corporate partners to fund women or atleast offer special loans to women students with minimal or no interest to engage with higher education. Smaller installments spread over a longer period of time, could help. Although banks in several countries offer student loans but it is far from a norm in lower middle and low-income countries and hardly any specific concession is made for women. In addition, NGOs or philanthropic organizations can also be encouraged to do their part in this, for instance, AKHUWAT in Pakistan, has been offering interest free loans for quite some time now for starting off small businesses. Similar models could be used for providing financial support to women in HEIs. In addition, opportunities for part time paid work, for example internships, or TA-ships or other pedagogical partnerships, some especially for women students could help the universities to get their work done, with minimum cost and support their female population to stay on.

In addition, HEIs in coordination with government and non- government actors also needs to plan other forms of support based on the needs of women. For example, webinars around the topic of time management, especially taking into account their domestic responsibilities, seeking help in the case of domestic violence, noticing signs of abuse in peers and reporting it, and staying connected could be helpful. Reaching out to parents, involving and encouraging them to support the women students with farsightedness is again an important dimension that needs to be taken into account. There is also an urgent need for HEIs to be cognizant of the mental health and wellbeing of women and to be prepared to deal with these issues. Possibilities of engaging with NGOs already working in this area may be a good idea, considering the financial constraints that HEIs are facing. This could open opportunity for consultations with counselors at a wider scale if need be. In addition, ensuring small group discussions, and formation of online peer support groups and faculty support can give women the strength to cope with their higher education.

Furthermore, flexibility in terms of the mode of course delivery is a must. For instance both synchronous but also asynchronous modes of instruction need to be creatively applied, keeping in mind that a large number of women may only have intermittent access to internet facility or computers. HEIs may need to begin their planning with this sense of limitation that women may be facing and find ways to address them. One could explore the possibilities of distribution of Internet devices (if possible), USBs with pre-recorded lectures, followed up by small group tutorials on flexible timings might resolve several issues in this regard. The digital access issue may need collaboration among several actors who can play part by addressing the barriers faced by women (UNESCO 2020d). This may include subsidies and tax incentives by the government that encourage the private sector to develop multiple technological options, while the 'academia and civil society' helps in developing capacity for digital use (ibid).

Lastly, HEIs must not just limit themselves to lecture delivery in offline/ online modalities but also reimagine ways to encourage social networking among student populations and build social and psychological social networks that particularly support women students during this difficult time. They should play their part in creating awareness about the increased risk of violence and how to tackle it during COVID-19 (Usher et al. 2020). This has to be with an understanding that those suffering might not be able to give a distress call and the role of their peers in staying connected, looking out for signs of distress and reporting might be critical (Campbell 2020). Hence, 'Social connectedness' while maintaining social distance is an important strategy in these times (Usher et al. 2020). Without identifying and paying attention to at risk women in higher education, we might witness large numbers of dropouts that foreclose opportunities for this vulnerable section of population. The Goal 17 of SDGs 2030 emphasizes the need to develop partnerships. This is the time for partnerships to be built across HEIs both at national and international level and across countries to make a concerted effort to ensure that women are able to access and stay in higher education undeterred by the multiple vulnerabilities they face in COVID-19.

5 Conclusion

The COVID-19 devastation has been multisectoral and pervasive, not only threatening life but all aspects of it. The global economy is estimated to shrink by 5.2% doubling the numbers of those suffering from extreme poverty, and leaving millions jobless in the recession that follows (Buttar 2020). Higher education has been one of the worse hit, affecting millions across the world (UNESCO 2020c). Several HEIs closed down altogether, while a majority tried to move their courses online (Crawford et al. 2020). However, given the digital divide and resource differences across and within countries and regions this has only exacerbated inequalities, even more so for women who have much less access to technology than men (UNESCO 2020d; World Bank 2020). For example 40% of the low-income countries could do very little to support their students during the COVID-19. (Global Education Monitoring Report 2020).In several countries the discrimination and exclusion is deliberate, persistent and now predicted to worsen, especially in relation to women's participation in education (ibid.). UNESCO (2020c) estimates that the largest number of those at highest risk of never returning to higher education are those in HEIs, of which women form a significantly larger part (ibid.). While HEIs struggle to stay afloat, juggling with the multiple demands of the new normal and tough financial constraints, and governments slash the funding of HEIs to pump money into failing health systems and slow economy, the urgency of the issues of women in higher education may be lost, which will directly affect the achievement of Goals 4 and 5 of SDGs 2030 agenda and indirectly make it difficult to achieve all the other SDGs.

This paper highlights the devastating consequences of COVID-19 on women's participation in higher education, as their vulnerabilities become exaggerated. Their relative poverty in relation to men, financial dependency on male family members, lower positioning and valuing in households means that their higher education might no longer be supported as family incomes are lost, reduced or become uncertain. In addition, given the dramatic increase in domestic violence during COVID-19 (UNESCO 2020d; Mazza et al. 2020) poor access to technological resources, demands on their time, expectation of unpaid domestic work, sociocultural norms and learning from history, pressures to marry in uncertain times, are all factors that force women out of higher education. While locality, ethnicity, race, class or disability will configure the nature and extent of issues faced by individual women, these are issues that have been well documented across several regions.

The dropping out of women from HEIs will not only severe their long struggle for inclusion in higher education but also in the wider society—voice, political and economic empowerment. This will be a major setback to the achievement of Goal 4 and Goal 5, of the sustainable development agenda 2030, having a disruptive domino effect on all the other SDGs 2030. Higher education has been a major driving force in the conceptualization of SDGs and will be in their achievement, given its pivotal role in steering research, innovation, critical thinking and finding sustainable answers to the problems of today. Given the pivotal role higher education in the training of professionals, intellectuals, and shaping of education itself would mean that a loss of footing here, would seriously set the clock back for women and restrict their participation in socioeconomic development processes. Failing to sustain their education would lead to the disengagement of women who form 50% of the population, increase intolerance and worsen the discrimination in society.

The paper recommends that there is an urgent need to recognize the 'equity challenges,' within higher education, in relation to women (Marinoni and Wit 2020), and those which will arise in the larger society, if inequities within education is allowed to fester. HEIs need to recognize and adopt a more sensitive approach to the women students, it is not just their education but their 'lives have been uprooted and left unmoored,' (ibid. p. 2). Hence, the need to go beyond provision of education and to engage with families and communities to understand the nature of the COVID-19 impacts for women that may act as barriers to their education. It is time to reframe and rethink education that is more 'inclusive, flexible and resilient, specifically addressing and ensuring the social and emotional welfare' of all students and teachers (UNDP 2020, p. 3; GEMR 2020), but specifically that of women students. Partnerships as emphasized in Goal 17 of SDGs 2030 may be the key to tackle the 'risk of growing inequalities,' across countries, HEIs and student populations. Forging partnerships with local governments, not for profit and for profit organizations locally and globally, engaging with communities and families, tech companies, can help creative planning of interventions and low cost tech solutions and generate support for women's higher education. These partnerships can allow a collective deliberation on equity, equality and democratic participation (Harkavv et al. 2020, p. 6) and the design of 'equity-oriented policies, frameworks, and targeted funding,' (Marinoni 2020). This however will only be fruitful if they are primarily targeted to ensure women's sustained participation higher education.

References

Bao W (2020) COVID-19 and online teaching in higher education: a case study of Peking University. Human Behavior and Emerging Technologies 2(2):113–115

Basset MR, Arnhold N (2020) COVID-19s immense impact on equity in tertiary education, April, 30, 2020, Education for Global Development https://blogs.worldbank.org/education/covid-19s-immense-impact-equity-tertiary-education

BBC News (2020). Coronavirus: worst could be yet to come, WHO warns, June 29, 2020. BBC News: https://www.bbc.com/news/world-53227219

Biswas RK, Rahman N, Kabir E, Raihan F (2017) Women's opinion on the justification of physical spousal violence: a quantitative approach to model the most vulnerable households. PLoS ONE 12(11):e0187884. https://doi.org/10.1371/journal.pone.0187884

Blessinger P, Sengupta E, Makhanya M (2018) Higher education's key role in sustainable development. University World News. https://www.universityworldnews.com/post.php?story=20180905082834986
Boniol M, McIsaac M, Xu L, Wuliji T, Diallo K, Campbell J (2019) Gender equity in the health workforce: analysis of 104 countries (No. WHO/HIS/HWF/Gender/WP1/2019.1). World Health Organization. https://apps.who.int/iris/bitstream/handle/10665/311314/WHO-HIS-HWF-Gender-WP1-2019.1-eng.pdf
Bradbury-Jones C, Isham L (2020) The pandemic paradox: the consequences of COVID-19 on domestic violence. J Clin Nurs 2047–2049. https://doi.org/10.1111/jocn.15296
Brännström I (2012) Gender and digital divide 2000–2008 in two low-income economies in Sub-saharan Africa: Kenya and Somalia in official statistics. Gov Inf Q 29(1):60–67
Buttar J (2020) 7 ways to help understand the world's challenges in 2020—and have hope for future, World Bank Blogs, August 4, 2020. https://blogs.worldbank.org/voices/7-ways-help-understand-worlds-challenges-2020-and-have-hope-future
Campbell AM (2020) An increasing risk of family violence during the Covid-19 pandemic: strengthening community collaborations to save lives. Forens Sci Int: Rep **100089**. https://doi.org/10.1016/j.fsir.2020.100089
Chadwick L (2020) Europe's coronavirus resurgence: Are countries ready to prevent a 'second wave? Euro News, August 4, 2020. https://www.euronews.com/2020/08/04/europe-s-coronavirus-resurgence-are-countries-ready-to-prevent-a-second-wave
Crawford J, Butler-Henderson K, Rudolph J, Glowatz M (2020) COVID-19: 20 Countries' higher education intra-period digital pedagogy responses. J Appl Teach Learn (JALT), 3(1):1–20
Devries KM, Mak JY, García-Moreno C, Petzold M, Child JC, Falder G, Pallito C (2013) Watts Ch. Global health. The global prevalence of intimate partner violence against women. Science **340**(6140), 1527–1528
Dyhouse C (2003) Troubled Identities: gender and status in the history of the mixed college in English universities since 1945. Women's History Rev 12(2):169
Evans D, Hares S, Sandefur J, Steer L (2020) How much will covid cut education budgets? May 8, 2020 Center for Global Development. https://www.cgdev.org/blog/how-much-will-covid-cut-education-budgets
Fafchamps M, Quisumbing AR (2003) Social roles, human capital, and the intrahousehold division of labor: evidence from Pakistan. Oxford Econ Pap 55(1):36–80
Ferrant G, Pesando LM, Nowacka K (2014) Unpaid care work: the missing link in the analysis of gender gaps in labor outcomes. Boulogne Billancourt: OECD Development Center https://www.oecd.org/Dev/Development-Gender/Unpaid_Care_Work.Pdf
Fredman K, Campbell M. (2016) Transformative equality: making the sustainable development goals work for women. Eth Int Affairs **30**(20), 177–187
Friedman S, Hurley T, Fishman T (2020) Covid-19s impact on higher education: strategies for tackling challenges facing colleges and universities. Downloaded on August 9th from https://www2.deloitte.com/content/dam/Deloitte/us/Documents/public-sector/us-gps-covid-19-impact-on-higher-education.pdf
Fulu E, Jewkes R, Roselli T, Garcia-Moreno C (2013) Prevalence of and factors associated with male perpetration of intimate partner violence: findings from the UN Multi-country Cross-sectional Study on Men and Violence in Asia and the Pacific. the Lancet Global Health 1(4):e187–e207
Garey AI (1999) Weaving work and motherhood. Temple University Press, Philadelphia
Giannini S, Albrectsen A (2020) Covid-19 school closures around the world will hit girls hardest. March 31st 2020. UNESCO: https://en.unesco.org/news/covid-19-school-closures-around-world-will-hit-girls-hardest
Global Education Monitoring Report (2020) Inclusion and education-All means all, UNESCO https://en.unesco.org/news/global-education-monitoring-gem-report-2020
Global Gender Gap Report (2020) World Economic Forum, 2019. www.weforum.org
Gupta S (2006) Her money, her time: Women's earnings and their housework hours. Soc Sci Res 35(4):975–999

Hafkin NJ, Huyer S (2007) Women and gender in ICT statistics and indicators for development. Inf Technol Int Dev 4(2):25

Hagan E, Raghavan C, Doychak K (2019) Functional isolation: understanding isolation in trafficking survivors. Sexual Abuse. https://doi.org/10.1177/1079063219889059

Harkavv I, Bergan S, Gallagher T, Land H (2020) Universities must help shape the post-COVID-19 world, University World News: The Global Window to Higher Education, https://www.universityworldnews.com/post.php?story=20200413152542750

Hilbert M (2011) Digital gender divide or technologically empowered women in developing countries? A typical case of lies, damned lies, and statistics. Women's Stud Int Forum 34(6):479–489

Hodges CB, Moore S, Lockee B, Trust T, Bond A (2020) The difference between emergency remote teaching and online learning. Edu Rev 27 https://er.educause.edu/articles/2020/3/the-difference-between-emergency-remote-teaching-and-online-learning

Ilie S, Rose P (2016) Is equal access to higher education in South Asia and Sub-saharan Africa achievable by 2030? High Educ 72(4):435–455

Imran M (2020) HEC denounces cut in higher education budget, June 14, 2020. The News: https://www.thenews.com.pk/print/672533-hec-denounces-cut-in-higher-education-budget

John Hopkins Coronavirus Resource Centre (2020) COVID-19 Dashboard by the Center for Systems Science and Engineering (CSSE) at Johns Hopkins University (JHU). https://coronavirus.jhu.edu/map.html

Khan AA, Niazi S, Saif KS (2020) Universities unprepared to shift to remote learning (March, 26, 2020), University World News. https://www.universityworldnews.com/post.php?story=20200326141547229

Klasen S (2019) What explains uneven female labour force participation levels and trends in developing countries? World Bank Res. Observer 34(2):161–197

Landale J (2020) International development and foreign office to merge. June 16, 2020. BBC News: UK Politics: https://www.bbc.com/news/uk-politics-53062858

Maital S, Barzani E (2020) The global economic impact of COVID-19: a summary of research. Samuel Neaman Inst Nat Policy Res. https://www.neaman.org.il/EN/Files/Global%20Economic%20Impact%20of%20COVID-19_20200322163553.399.pdf

Mannino CA, Deutsch FM (2007) Changing the division of household labor: A negotiated process between partners. J Res 309–324

Marinoni G, de Wit H (2020) A severe risk of growing inequality between universities' University World News, June 8, 2020. https://www.universityworldnews.com/post.php?story=2020060815405140

Maslen G (2020) Saving Australia's biggest university. May 7, 2020, University World News. https://www.universityworldnews.com/post.php?story=20200507090424381

Mazza M, Marano G, Lai C, Janiri L, Sani G (2020) Danger in danger: interpersonal violence during COVID-19 quarantine. Psychiatry Res 289:113046. https://doi.org/10.1016/j.psychres.2020.113046

McKie A (2020) SOAS faces 'viability problems' amid pandemic crisis, director warns. Times Higher Education. https://www.timeshighereducation.com/news/soas-faces-viability-problems-amid-pandemic-crisis-director-warns

McLaren HJ, Wong KR, Nguyen KN, Mahamadachchi KND (2020) Covid-19 and Women's Triple Burden: Vignettes from Sri Lanka, Malaysia Vietnam Australia. Soc Sci 9(5):87. https://doi.org/10.3390/socsci9050087

McMunn A, Bird L, Webb E, Sacker A (2020) Gender divisions of paid and unpaid work in contemporary UK couples. Work Employ Soc 34(2):155–173

Milkie MA, Raley SB, Bianchi SM (2009) Taking on the second shift: Time allocations and time pressures of U.S. parents with preschoolers. Soc Forces 88(2):487–517

Moore KM (1987) Women's access and opportunity in higher education: toward the twenty-first century. Comp Educ 23(1):23–34

Moreno J, Shaw D (2018) Women's empowerment following disaster: a longitudinal study of social change. Nat Hazards 92(1):205–224
Morris LV (2011) Women in higher education: access, success, and the future. Women in higher education: access, success, and the future. Innov High Educ 36:145–147
Mumporeze N, Prieler M (2017) Gender digital divide in Rwanda: a qualitative analysis of socioeconomic factors. Telematics Inform 34(7):1285–1293
Munir A, Akhtar N (2014) A social custom "Vani": Introduction and critical analysis. VFAST Trans Educ Soc Sci 3(2):1–4
Nawaz F, McLaren HJ (2016) Silencing the hardship: Bangladeshi women, microfinance and reproductive work. Soc Altern 35(1):19–25
Nicola M, Alsafi Z, Sohrabi C, Kerwan A, Al-Jabir A, Iosifidis C, Maliha A, Agha R (2020) The socio-economic implications of the coronavirus pandemic (COVID-19): a review. Int J Surg 78:185–193 https://doi.org/10.1016/j.ijsu.2020.04.018
Offer S, Schneider B (2011) Revisiting the gender gap in time-use patterns: multitasking and well-being among mothers and fathers in dual-earner families. Am Sociol Rev 76(6):809–833
Parvazian S, Gill J, Chiera B (2017) Higher education, women, and sociocultural change: a closer look at the statistics. Sage Open 7(2):2158244017700230
Peterman A, Potts A, O'Donnell M, Thompson K, Shah N, Oertelt-Prigione S, Gelder NV (2020). Pandemics and violence against women and children. Center for Global Development Working Paper, 528. https://www.un.org/sexualviolenceinconflict/wp-content/uploads/2020/05/press/pandemics-and-violence-against-women-and-children/pandemi
Philipose R (2020) which countries are seeing a resurgence of Socid-19 cases?, Indian Express, August 1st 2020. https://indianexpress.com/article/world/which-countries-are-seeing-a-res,urgence-of-covid-19-cases-6534369/
Pleck JH (2010) Paternal involvement: revised conceptualization and theoretical. In: Lamb M (ed) The role of the father in child development. Wiley, New Jersy, pp 94–153
Qureshi Z (2020) The analytical angle: Covid-19 and the looming education crisis, May 16, 2020. Harvard Kennedy School: https://epod.cid.harvard.edu/article/analytical-angle-covid-19-and-looming-education-crisis.
Raley S, Bianchi S, Wang W (2012) When do fathers care? Mothers' economic contribution and fathers' involvement in child care. Am J Sociol 117(5):1422–1459
Ramakrishnan A, Sambuco D, Jagsi R (2014) Women's participation in the medical profession: insights from experiences in Japan, Scandinavia, Russia, and Eastern Europe. J Womens Health 23(11):927–934
Saha A (2013) An assessment of gender discrimination in household expenditure on education in India. Oxford Dev Stud 41(2):220–238
Sanchez L, Kane EW (1996) Women's and men's constructions of perceptions of housework fairness. J Fam Issues 17(3):358–387
Sandhya V, Jayaraman H (2019) citationality of gender: judith butler and kamala das. J English Stud 14(2):32–41
Strauss V (2020) Why girls in poor countries will suffer the most from worldwide closing of schools during covid-19, April 14, The Washigton Post. https://www.washingtonpost.com/education/2020/04/14/why-girls-poor-countries-will-suffer-most-worldwide-closing-schools-during-covid-19/
Strong A, Schwartz DA (2016) Sociocultural aspects of risk to pregnant women during the 2013–2015 multinational Ebola virus outbreak in West Africa. Health Care Women Int 37(8):922–942
Szto C (2015) Serving up change? Gender mainstreaming and the UNESCO–WTA partnership for global gender equality. Sport Soc 18(8):895–908
Taub A (2020) A new Covid-19 crisis: domestic abuse rises worldwide. Domestic-Abuse-Rises-Worldwide-New York-Times.pdf. https://chescocf.org/wp-content/uploads/2020/04/Domestic-Abuse-Rises-Worldwide-New-York-Times.p

The Japanese Times (2020) One in 13 college students in Japan considering quitting due to COVID-19, survey finds, April 26, 2020. https://www.japantimes.co.jp/news/2020/04/26/national/one-13-college-students-japan-considering-quitting-due-covid-19-survey-finds/

Tolbert C, Mossberger K, King B, Miller GL (2007) Are all American women making progress online? African-Americans and Latina. Inf Technol Int Dev 4(2):61–88

Traustadottir R (1991) Mothers who care: Gender, disability, and family life. J Fam Issues 12(2):211–228

UNDP (2020) Social justice for women amid COVID-19,UNDP, IDLO, UNODC, World Bank, May 14, 2020

UNESCO (n.d.) Higher education and the sustainable development goals. UNESCO https://en.unesco.org/themes/higher-education/sdgs

UNESCO (2015) Transforming our world: the 2030 agenda for sustainable development. Sustainable development: https://sustainabledevelopment.un.org/post2015/transformingourworld#:~:text=The%20High%2Dlevel%20Political%20Forum,Summit%20on%2025%20September%202015.&text=This%20Agenda%20is%20a%20plan,for%20people%2C%20planet%20and%20prosperity

UNESCO (2017) Ensure quality education for all: sustainable development goal 4; ten targets. https://unesdoc.unesco.org/ark:/48223/pf0000259784

UNESCO (2020a) Educational disruption and response, UNESCO March 20, 2020. https://en.unesco.org/themes/education-emergencies/coronavirus-school-closuresView in article

UNESCO (2020b) Education during COVID-19 and beyond: policy Brief, August 2020. https://www.sdg4education2030.org/education-during-covid-19-and-beyond-un-secretary-general-august-2020

UNESCO (2020c) COVID-19 education response: how many students are at risk of not returning to school:UNESCO [52913]Documentcode:ED/PLS/EDP/2020/07 https://unesdoc.unesco.org/ark:/48223/pf0000373992

UNESCO (2020d) Digital technologies critical in facing COVID-19 pandemic. UNDESA. https://www.un.org/development/desa/en/news/policy/digital-technologies-critical-in-facing-covid-19-pandemic.html

USAID Global Development Lab (2020). COVID-19 and the Gender Digital Divide. https://www.usaid.gov/sites/default/files/documents/15396/COVID-19_and_Gender_Digital_Divide.pdf

Usher K, Bhullar N, Durkin J, Gyamfi N, Jackson D (2020) Family violence and COVID-19: Increased vulnerability and reduced options for support. Int J Ment Health Nurs. https://doi.org/10.1111/inm.12735https://onlinelibrary.wiley.com/doi/abs/10.1111/inm.12735

Van Gelder N, Peterman A, Potts A, O'Donnell M, Thompson K, Shah N, Oertelt-Prigione S (2020) COVID-19: reducing the risk of infection might increase the risk of intimate partner violence. eClinical Med. 1(21).https://www.thelancet.com/journals/lancet/home

Wenham C, Smith J, Morgan R (2020) COVID-19: the gendered impacts of the outbreak. The Lancet 395(10227):846–848

World Bank (2020) Education and COVID-19, April 30, 2020. The World Bank. https://www.worldbank.org/en/data/interactive/2020/03/24/world-bank-education-and-covid-19

World Bank Group: Education (2020) The COVID-19 crisis response: supporting tertiary education for continuity, adaptation, and innovation, April 9, 2020. https://pubdocs.worldbank.org/en/621991586463915490/WB-Tertiary-Ed-and-Covid-19-Crisis-for-public-use-April-9.pdf

World Health Organization (2020) Timeline of WHO's response to COVID-19, June 2020. https://www.who.int/news-room/detail/29-06-2020-covidtimeline

Yavorsky JE, Kamp Dush CM, Schoppe-Sullivan SJ (2015) The production of inequality: the gender division of labor across the transition to parenthood. J. Marriage Fam. 77(3):662–679

Yeung WJ, Sandberg JF, Davis-Kean PE, Hofferth SL (2001) Children's time with fathers in intact families. J. Marriage Fam. 63(1):136–154

Yuen V (2020) Mounting peril for public higher education during the coronavirus pandemic, June 11, Centre for American Progress. Washington, DC. https://www.americanprogress.org. https://eric.ed.gov/?id=ED606582

Tayyaba Tamim is currently Associate Professor & Director Academics at the School of Education, and the faculty lead for Pedagogical Partnership Programme at LUMS. Dr. Tamim has her PhD from University of Cambridge as a fully funded RECOUP scholar and MPhil RSLE (Research in Second Language Education Across Cultures) from Cambridge, UK as a British Council Chevenning scholar. In addition, she also has an MA ELT from Kinnaird College for Women University and MA English from the University of Punjab, Pakistan. Dr Tamim has led several funded research projects with national and international partners, including those with USAID, British council and the World Bank. She has also published and presented research papers at a number of national and international forums. Her work covers issues of social justice, equity and inclusivity in education with specific reference to languages in education, gender and caste.

Evolving Schools in a Post-pandemic Context

Dustin Eirdosh and Susan Hanisch

Abstract The Covid-19 pandemic has required schools and students to radically and immediately adapt teaching and learning processes to more distributed, digital, and home-based learning strategies for an indefinite period of time. This situation provides an unprecedented opportunity for students and teachers around the world to reflect on the purpose and design of schools post COVID-19, particularly on the kinds of school design elements that may best meet the needs of all students and teachers in the future. Before the pandemic emerged, the Community Science Lab at the Max Planck Institute for Evolutionary Anthropology had been developing a student-led effort, the Evolving Schools project, for exploring the everyday cooperation dynamics of teaching and learning within their own school communities. This chapter offers an illustrative case study of this project as a novel approach to student participation in school improvement efforts. By involving a small group of students and teachers in reflecting on their experience of changing schooling practices during the COVID-19 pandemic, as well as in exploring scientific perspectives on human learning and cooperation, a new theoretical foundation for the practical empowerment of students and teachers is framed within the aims of Education for Sustainable Development.

D. Eirdosh (✉) · S. Hanisch
Max Planck Institute for Evolutionary Anthropology, Department of Comparative Cultural Psychology, Leipzig, Germany
e-mail: dustin.eirdosh@eva.mpg.de
URL: http://www.globalesd.org

Global ESD, Elkins Park, USA

S. Hanisch
e-mail: susan.hanisch@uni-leipzig.de
URL: http://www.globalesd.org

S. Hanisch
University of Leipzig, Department of Primary School Science Education, Leipzig, Germany

W. Leal Filho (ed.), *COVID-19: Paving the Way for a More Sustainable World*, World Sustainability Series, https://doi.org/10.1007/978-3-030-69284-1_24

1 Introduction

Teaching and learning was a central driver of the biological and cultural evolutionary emergence of humans as a species ~200,000 years ago, with the origins of such behavioral capacities likely dating to millions of years prior to that (Sterelny 2012). At its core, teaching and learning among humans is about sustaining and innovating adaptive understandings of the world we are inheriting, in order to shape the world as we might imagine it could be (Fuentes 2017). Relative to this context, formal schooling is a remarkably new feature of humanity (Geary and Berch 2016; Gray 2013; see Fig. 1).

How schools can adapt to a rapidly changing world has been a theme in educational development since at least the origins of compulsory education, and has been just as evident in twenty-first century discourse on school improvement (Bryk et al. 2011; Gray 2013). With the onset of the Covid-19 pandemic, this question became all the more urgent to engage with in an effective manner. International school shutdowns, in various forms, have shown a spotlight on educational inequalities, the challenges and opportunities of online learning, and even on broader questions regarding the purpose of schooling and education itself.

Against this background remains the still more complex question of how education research and education science can or should best engage the grand challenges now facing this sector (see extended discussions in Mintrop 2016; LeMahieu et al. 2017). As education systems around the world clamour towards purportedly effective solutions to the myriad challenges wrought by this pandemic, intuition and real or perceived authority may hold as much clout as evidence in the decision making processes at any given level of a school's organizational ecosystem. Educators need immediate solutions to meet the academic and social-emotional needs of students in the present moment, under dynamically uncertain conditions, while education advocates and activists are keen to leverage this moment towards creating a "new

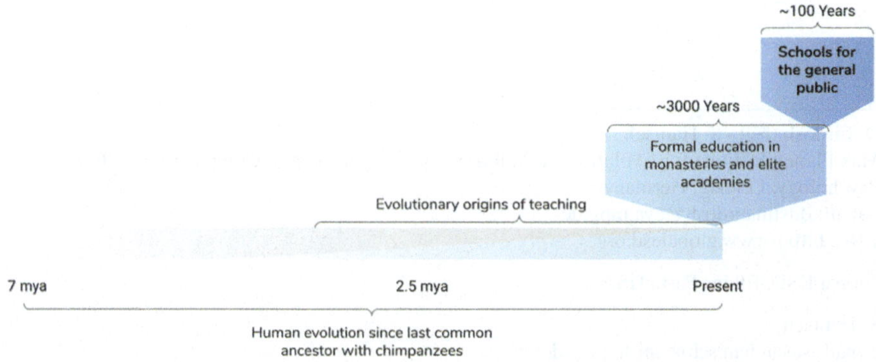

Fig. 1 A schematic timeline of the evolution of teaching and formal education. (Image from the student reading for the Evolving Schools project)

normal", a new understanding of the fragility of past systems and the potential of re-envisioning the potential purposes of education and schooling.

It is remarkably easy to get lost in the noise and needs related to school adaptation processes in the face of this evolving pandemic. Our aim in this chapter is not to offer one more voice advocating for change, nor one more solution purported to address immediate needs. Rather, what we intend to provide here is a minimal yet unique vision for involving school stakeholders in the school improvement process, the *Evolving Schools* project as an example of our *Community Science Lab* model. It is minimal in the sense that it is a project built on highly generalized principles of systems improvement practices, and also in the sense that we can only offer a glimpse at the first four months of this exploratory effort to kickstart a new engine for educational innovation. It is unique in the sense that our approach to stakeholder empowerment is grounded within an emerging interdisciplinary synthesis in the human and educational sciences that is simultaneously pluralistic in the breadth of pedagogical strategies it can engage, and unifying in its foundational perspectives on teaching and learning as a core capacity of the human species.

We begin with a brief overview of the theoretical basis and practical context for the Community Science Lab model developed within our larger *Education for Sustainable Development* project (GlobalESD; www.GlobalESD.org) and the Max Planck Institute for Evolutionary Anthropology in Leipzig, Germany. We then offer an illustrative case study of our exploratory work in developing the Evolving Schools project with a small team of local students within the Community Science Lab context. This work connects to broader educational discourse on the need for *Networked Improvement Communities* (NICs; Bryk et al. 2011) as an adaptive engine for translating educational science into effective innovation through an explicit focus on continuous improvement. Our conclusions highlight the further opportunities for engaging a global discourse towards driving participatory approaches to school improvement that is simultaneously pluralistic and unifying.

2 Community Science Lab Model Development

Human behavior is at the center of our everyday experience, it has been a driver of our evolution as a species, and it remains a driver of our development as individuals and societies. Despite this centrality in our lives, the place of scientific perspectives on human behavior within the general education curriculum remains elusive. In many ways, human behavior is an implicit topic across traditional subject areas, yet it hardly has a place explicitly, save for the occasional introductory psychology electives offered in some secondary school contexts.

To address the challenge of teaching human behavior as an interdisciplinary theme across the general education curriculum, we have advanced a long-term Design-Based Implementation Research (DBIR, Fishman et al. 2013) project, Global ESD (Eirdosh and Hanisch 2019; Hanisch and Eirdosh 2019), which aims to advance a design concept in *Education for Sustainable Development* (ESD) that integrates concepts across evolution, behavior, and sustainability sciences.

The Global ESD design concept contains a wealth of teaching tools and other structural elements to support educators in embracing interdisciplinary perspectives, yet it is based on three relatively straightforward design principles:

1. Focus on Human Behaviors
   Focus on the aspects and everyday experience of human behaviors relevant to human well-being and sustainable development (eg, prosociality, cooperation, sense of belonging, trust, curiosity and creativity, learning and teaching, empathy and compassion, sense of fairness, perspective taking, flexibility, self-control, goals and values, health, prevention). *Focusing on human behaviors helps students relate to and understand the causes of biological and societal phenomena.*
2. Explore Complex Causality
   Explore and reflect on the many causes and consequences of human behavior and on the complex causal relationships in human evolution, behavior, and social-ecological systems: How do immediate internal and external factors, as well as individual development and evolutionary history, function as causes of human behavior? Why do these mechanisms and patterns of behavior exist compared to other possibilities? What consequences do behaviors have for individuals and their environment, in the short-term and in the long-term? Diverse teaching tools such as causal maps and payoff matrices help in reflecting on these questions. *Exploring complex causality helps students understand and relate causal factors in the emergence of human behaviors.*
3. Teach for Transfer
   Ensure students can transfer understandings to novel phenomena, everyday experience and relevant problems of sustainable development across multiple scales and contexts of global society, with the help of analogies, analogy maps, and other teaching tools. *Teaching for transfer requires the iterative exploration of diverse human behaviors and contexts.*

   Since 2015 we have been developing a database of teaching materials created through teacher development programs and teacher-researcher collaborations in real-world classroom contexts, within this educational design concept. These classroom materials provide usable and pedagogically informed approaches to engaging students in understanding human behavior as an interdisciplinary theme, however, they largely fall within the realm of more traditional classroom learning and do not engage students in the real-world issues of sustainable development. In 2019, through our work with the Department of Comparative Cultural Psychology at the Max Planck Institute for Evolutionary Anthropology, we launched the *Community Science Lab* as an exploratory approach to, broadly, engage students and teachers in understanding and influencing the cooperation dynamics that pervade their everyday lives and which impact local to global sustainability issues. Building on the Global ESD concept of "reflecting on the everyday experience of human behavior in the light of evolution and sustainability" (Hanisch and Eirdosh 2019), the inaugural year of the Community Science Lab aimed to follow the minimal design principles set out by Global ESD, while empowering and following student voices in the process.

Starting in September 2019, a team of four students from a local 8th grade English class volunteered to visit the Community Science Lab within the institute, for 1.5 h per week. Sessions began by exploring some conceptual foundations in human behavior across evolutionary, cross-cultural, developmental, and sustainability contexts. Given that the student participants were engaged members of their local *Fridays for Future* climate action school protest group, our team (led by student decision-making) decided to focus on trying to understand the cooperation dynamics in this context.

Building on the international *Prosocial World* research program (Atkins et al. 2019), we engaged students in understanding the *theory of collective action* originally developed by Nobel laureate, Elinor Ostrom, as a democratic framework for understanding the *core design principles* associated with effective group cooperation across diverse contexts and levels of social organization (Ostrom 1998; Poteete et al. 2010; Wilson et al. 2013). Informed by this theoretical framework, students and lab leaders developed a short survey focusing on various stakeholder perceptions regarding the purpose and efficacy of the *Fridays for Future* climate action school protests. While some stakeholders in the *Fridays for Future* movement espoused an "us versus them" mindset of movement supporters versus opponents, the student-led survey tentatively revealed a more complex picture. Students gained insights from this educational project suggesting that many supporters of the movement (students, teachers, and parents) also harbored significant critical views of some aspects of this activism. Likewise, many stakeholders claiming to be critical of the movement often described significant aspects which they supported and valued.

In discussing next steps with our student researchers, we identified challenges in further elaborating research within the *Fridays for Future* movement, namely that the decentralized structure of the movement made it challenging to further elaborate the study with significant community buy-in. Simultaneously, we also explored further the *root causes* of the movement itself. While *Fridays for Future* is ostensibly about action on climate change, our students were keen to elaborate that it is also about the purpose of school itself, and a perception among students that current school practices are not empowering them with the knowledge and skills needed to tackle the grand challenges they will clearly face in their adult lives. Against this background, and with a student vote, we decided to shift emphasis into understanding the cooperation dynamics within schools themselves. This occurred in the first week of February 2020, just prior to a planned ~6 week break, both to allow students time to engage their final exams and enjoy spring break, as well as allow our team to plan the upcoming semester within the context of this emergent focus on school culture.

3 The Evolving Schools Community Science Project

During this ~6 week break in the Community Lab sessions, our team developed a plan for the *Evolving Schools* project, as a means of both going deeper into the evolution science perspectives on teaching, learning, and human cooperation, as well as to elevate student voice within school improvement discussions by empowering them to use more advanced methods from the Prosocial World research program.

We were able to have one inaugural project session at the Max Planck Institute in mid-March prior to the full shutdown of schools and our institute in late March. We present here a brief overview of the project as planned, a summary of the challenges faced during implementation within the pandemic context, and highlights from the collaborative project report produced from this work.

3.1 *Project Overview*

The central aim of the Evolving Schools project was to develop a model for authentic student engagement in the improvement of school culture through project-based conceptual learning about the science of learning and culture. That is, we sought to empower our students to elevate their voice as drivers in the cultural evolution of their school, while using these empowerment processes to drive academic learning in the sciences of evolutionary anthropology itself.

We developed this idea broadly within the Global ESD design concept, and specifically building on the emerging pedagogical knowledge synthesis occurring in the field of *conceptual learning* (Cope and Kalantzis 2015; Stern et al. 2017). Towards these aims, we began by structuring an overarching research question and several student facing questions. Within this direction we then drafted some student readings to engage and elaborate student conceptions in relation to scientific perspectives in the evolution of teaching and learning, and outlined a Project-Based Learning protocol adapted from Stern et al. (2017, p. 80).

3.2 *Research Questions*

The *Evolving Schools* project is conceptualized as an education project aiming to empower students as community scientists, and as such, has been structured around a guiding question for educators and project coordination partners, as well as several student facing questions (Table 1).

Table 1 The essential questions driving the planned *Project-Based Learning* protocol for the Evolving Schools project

Guiding question	What learning potential and challenges exist in engaging students in evolution science perspectives on teaching, learning, and school culture?
Student facing questions	What conceptions do school students have regarding the purpose and autonomy of their learning experiences [esp. during the pandemic as it relates to normal school contexts]?
	What conceptions and questions do students have regarding the diversity of learning theories for the design of school curricula?
	Can the collective conceptions of students on these issues inform educational policy and design discussions?

3.3 What is Meant by "Evolving Schools"?

An important clarification should be made regarding our choice of the project name *Evolving Schools*. The concept of evolution is used in a wide diversity of ways across various scientific disciplines as well as within popular culture. In biology, evolution can variously refer to changes in the frequency of genes within a population, or more generally, to changes in the frequency of heritable variation of traits in a population. In evolutionary anthropology, the conceptualization of evolution has been generalized to include changes to the complex systems that enable the retention or reconstruction of behavioral, cognitive, and cultural traits (see Table 2; Hanisch and Eirdosh 2020 for an elaborated discussion).

How diverse scientists conceptualize evolution, and how scientists relate these diverse conceptualizations to practical implications for school design remains controversial. Indeed, it is exactly this controversial space in which we wanted to engage student reflection and voice.

Given that our 8th grade students had not yet had significant instruction in evolution science, we crafted a four page reading that framed a simple historical context for the origins of modern schooling, and offered three different narrative perspectives. We summarize these perspectives below, however the full student reading can be found within the Evolving Schools (2020) project report.

3.4 Three Perspectives on Evolution and Education

Perspective 1: Students should learn like our ancestors evolved to learn

Humans have been teaching and learning from each other for millions of years, therefore we are genetically adapted to learning in certain ways. By studying how small-scale hunter-gatherer communities engage in teaching and learning we can identify important aspects of how modern education should be designed. For example,

Table 2 An analogy map comparing evolutionary concepts across domains of genetics, learning, and culture

Concept, process, principle	Genetic evolution	Cognitive-behavioral evolution (learning)	Cultural evolution
What is the relevant level of analysis?	Populations of organisms in an ecosystem	Populations of concepts, mental models, and behaviors in an individual	Populations of individuals and groups exchanging information
How is variation of traits caused?	Putation, recombination	Mistakes, recombination of prior learning, trial-and error learning, reactions to new environments, creativity, social learning	Mistakes, recombination of ideas, trial-and error learning, reactions to new environments, creativity, between-group social learning
How does selection of traits occur	Higher chances of survival and reproduction	Selective attention, emotional strength, relation to prior learning, practical consequences	Higher chances of survival and reproduction (*natural selection*); greater reward, appeal or attractiveness of the trait (*cultural selection*)
How are traits inherited, transmitted, or retained?	Biological reproduction, mitosis/meiosis	Encoding into long-term memory for later retrieval	Social learning/imitation, teaching; technologies and infrastructure that endure

In this context, schools are immensely complex evolving ecosystems of individual and social learning processes. To the degree we can clearly define the goals of education as a whole, and for specific schools in particular, we can begin to think about the social conditions that will favor or hinder the characteristics of these systems that we value most

students should have unlimited free time to play with the tools of the culture in a community of helpful supporters, rather than being tested according to a pre-set curriculum.

Perspective 2: Teachers should help students evolve their minds

Humans have been teaching and learning from each other for millions of years, therefore we are genetically adapted to engaging with the tools of the culture we are born into. However, through cultural evolution the world that we live in today is drastically different from the world that our ancestors lived in, therefore we need new tools and knowledge about how to learn best in today's world. Because the tools of modern societies include complex ideas and concepts we would not have encountered in our evolutionary past, it is the role of teachers to use modern teaching practices to evolve new species of thought within the minds of students.

Perspective 3: Students should be empowered to evolve their own education system

There are aspects of Perspectives 1 & 2 that are correct, but neither one is fully adequate to design modern schools that are enjoyable and effective for all students. Instead of choosing one perspective or the other, students themselves, with the help of teachers and scientists, should understand the evolutionary science of learning and be empowered to evolve their own school system to be most enjoyable and effective in their own communities.

3.5 The Project-Based Learning Protocol

In order to structure the investigation, we adapted the Project-Based Learning (PBL) protocol from the *conceptual learning* approach of Stern et al. (2017, p. 80; see Table 3). This model provides a clear roadmap for project development while integrating best practices in teaching for conceptual understanding. Such an approach aims to

Table 3 Planned project-based learning protocol for the evolving schools project

PBL protocol	Evolving schools project plan
1. Project launch	• Discuss project aims • Engage students in student reading on perspectives in the evolution of teaching and learning • Reflect on core concepts and conceptual relationships in the readings
2. Help students plan inquiry and build background knowledge	• Engage students in peer reflection and thematic analysis on their interpretations of the student reading • Identify areas of agreement, disagreement, or interest for further investigation among their peers • Plan community science investigation(s)
3. Monitor student inquiry processes and guide student reflection	• Support students in the development and implementation of their community science investigations
4. Support students as they construct high-quality products through critique and revision	• Use Community Science Lab weekly sessions to engage students in peer reflection and rubric-based analyses of their emerging research products and final project report
5. Organize students to present or publish their final products to a real-world audience	• Support students in the publication and community presentation of the Evolving Schools report
6. Provide opportunities for reflection on the process	• Final reflection session • Planning for project continuation in the following school year • Celebration

weave student learning across a multitude of pedagogical approaches, including conceptualizing, experiencing, critical analysis, and creative design (sensu Cope and Kalantzis 2015).

3.6 Challenges Within the Broader School Adaptation Context

Immediately following the project launch session, the Covid-19 situation in Germany (and across the western world) escalated at a dramatic rate. School closures occurred in Leipzig by the end of March, and the fate of our project (and of student learning as a whole) was suddenly put into an unknowable jeopardy.

To proceed in this landscape of uncertainty, we began to work in closer partnership with the teacher coordinator at our partner school. This teacher had previously been largely supporting the logistics of engaging our student lab members at the Max Planck Institute, but now served as a critical point of information and process reflection on if and how we might proceed. We agreed to let student interest drive any possible continuation of the project. Within the first couple of weeks of school shut down, our students reported significant stressors, both from the general uncertainty of the pandemic situation, but as well, from a relative lack of coordination in the approach and clarity of expectations across teachers in navigating how student learning should best be supported. Teachers, for their part, also reported a state of stressful chaos as they got limited and sometimes conflicting guidance from higher levels of leadership. Despite this stress, our core student group decided they did want to continue to try to engage through virtual collaboration, albeit somewhat reduced.

To account for the multiple limitations imposed by this lack of direct classroom engagement, we crafted a set of adapted research questions and project processes that focused more on capturing and reflecting on student voice, and less on students as direct drivers of community science investigations.

We outlined two simplified research questions:

1. What are student conceptions of current debates in the evolutionary anthropology of modern education systems?
2. What are student conceptions about the possibility of elevating student voice in the ongoing cultural evolution of modern education systems?

The process adaptation included working with our teacher collaborator to also offer the project to the 10th grade class, from which ($n = 29$) students volunteered to engage. Due to the challenges of inconsistent IT access for students, we adopted an asynchronous online learning plan. In this context, a combination of YouTube videos, student essays, and iterative video/questionnaires based on thematic analysis of their essays provided a set of mechanisms to drive the conversation forward based on student voice, while also minimizing additional stressors on student work load. These methods also served as our primary source of data collection.

Our teacher partner has a strong interest in student well-being and the role of student voice in schools. Throughout this process, and during the final project reflection interview, she repeatedly voiced concern about the lack of clear and coordinated guidance leading to cooperation dilemmas among teachers in regards to developing adaptive innovations during this time. That is, in the face of unclear expectations, varying teachers had varying responses. Those teachers who adopted a view that normal school work and grading expectations would simply be ported to online contexts created demands on students' time and skill sets that, in essence, hindered students' ability to engage potentially more valued opportunities regarding social-emotional learning, self-care, and other activities designed to support students in the self-reported challenges of stress and anxiety many of them were facing.

Against this challenging background, our continuous engagement with ($n = 33$) students throughout the remainder of the school year speaks to their motivation and interest in advancing student voice within the scientific framework offered by this project.

3.7 The Evolving Schools Report

While the original project plan suggested the project report would be fully student driven, the adapted version, edited by us, still captures and highlights a wealth of student voices within the core theoretical framework. The full report is available at the project website (https://EvolvingSchools.GlobalESD.org), our intent here is only to highlight three key findings relevant to our discussion on next steps for the project.

First, through student reflections on their conceptions of what is meant by the concept of "Evolving Schools", we were able to identify that, while they collectively hold a diversity of ideas about what this phrase means, there is a wealth of scientifically adequate thinking as it relates to modern perspectives in cultural evolution science. That is, even without instruction, students' intuitive, popular conceptions of "evolving minds" and "evolving schools" contains kernels of scientific conceptions that could be developed with further opportunities for engagement. Importantly, because the dynamics of cultural evolution are both similar to, and in some regards, different than, biological evolution, some of these student conceptions stand in contrast to the understood aims of evolution education within the biology classroom. We discuss these important conceptual overlaps and divergences at length in our article collection on *teaching evolution as an interdisciplinary science* (Hanisch and Eirdosh, in press). Critically however, students tended to have a non-trivial and partially scientific understanding of the complexities of cultural evolution science within this context, and reportedly did not find this view in contrast with their understanding of biological evolution.

Second, students overwhelmingly preferred perspective #3 (students should be empowered to evolve their own school system), explicitly viewing this as an integration of all of the perspectives offered. There was a high diversity of thinking, especially regarding the specifics of school design policy options, but ultimately students seemed to strive for a relatively nuanced balance between student autonomy (which they deeply value and want more of) and the value of structured learning from expert educators (which many of them also deeply value). Also of note, students tended to reject perspective #1 (students should learn like our ancestors evolved to learn) most often due to a perception that this amount of freedom would not prepare them for the modern world. When we discussed this finding with educators in various *Self-Directed Education* movements (which this perspective was aiming to capture), there was concern that the emphasis on ancestral learning in small-scale societies, while being a theoretical core to *Self-Directed Education*, may have biased student responses. Importantly, while students frequently overtly rejected perspective #1 in their direct response to it, their later justifications of support for perspectives #2 and #3 commonly invoked core elements of perspective #1. Thus, these findings should not be interpreted as an overt student rejection of self-directed education practices.

Finally, while student attitudes regarding the potential of this project to actually lead to school improvements ranged from cynical skepticism about the motives of educational policy makers, to more nuanced optimism, students overwhelmingly agreed the process was useful. In a concluding survey, only one student preferred to not engage in the project next school year, four suggested they would like to stay engaged if they could get a grade for the work, and the remainder ($n = 28$) reported an intrinsic interest in continued project engagement.

3.8 Next Steps for the Evolving Schools Project

Given the strong student interest and conceptual richness of their engagement, we are now planning the next steps to improve the project implementation within our partner school, and also to open the methods and materials to allow for a more networked, collaborative, open science approach to this work.

Improve teaching materials. Having found that students intuitively prefer the third perspective on evolving their own school system, largely because it integrates the first two perspectives, we will make several changes to the student reading. First, we plan to offer only the first two perspectives during the initial reading and reflection phase to better capture students' conceptions of this specific problem space in the learning science literature. Second, because we found students tend to reject perspective one on a perception that it is an outdated mode of learning, we will include more explicit examples of modern school environments and research into their efficacy for student adaptation to future academic and work life in modern societies (e.g. Sudbury school models, see Gray 2013). We are working to additionally develop and curate new resources to help students dive deeper into school culture improvement (see our *Community Science Field Guide for School Culture*; Hanisch et al. 2020) as well as evidence-informed teaching and learning environments (see Hattie 2012; Kirschner and Hendrick 2020). In particular, the in-development *Prosocial Schools Inventory* will provide a roadmap for student–teacher teams to sustain strategic discussion and community science investigations towards school culture improvement over the course of a semester or entire school year.

Adapt PBL protocol for the challenges of pandemic school closures. As of this writing, the Covid-19 pandemic situation remains highly unstable around the world. The state of school opening and school strategies for students in the coming year is unknowable at the moment, therefore we aim to "Covid-Proof" the project protocol by planning for not only in person options, but also strategically improved blended and fully online options. Again, given the known inequities of synchronous online learning (e.g. video conferencing), we will adopt asynchronous strategies as the dominant approach. Key to this model will be social annotation technologies, such as the free online platform Perusall (Miller et al. 2018), which enables asynchronous peer-to-peer collaboration in interpreting and clarifying key readings and videos, while providing strategic analytics on student engagement and understanding to educators. Social annotation can also be used within the *Evolving Schools* project context to quantify student consensus or disagreement on a variety of specific propositions about school design. Combined with short questionnaire, essay, and peer-to-peer chat platforms, we believe a robust engagement with the project can be advanced even within school closure contexts.

Create tools for cross-school collaborations. Because we work within an open education science framework (van der Zee and Reich 2018; Makel et al. 2019), adopting strategies to adapt to the potential challenges of the pandemic has a positive side effect of expanding the potential for cross-school collaborations. In the section that follows, we describe the need and potential to evolve just such a network of schools working towards scientifically informed school improvement.

4 Evolving a Networked Improvement Community

Education science is a richly diverse field, filled both with healthy discussions, and sometimes more tribal disagreements. From an evolutionary perspective, a few things are fairly clear. Our species has been shaped over millions of years by the relatively egalitarian small-scale societies of our ancestors, such that the development of folk understandings of the physical, biological, and social world is a biologically primary capacity of all humans. Over at least the past 10,000 years, ultra-social cultural evolution has radically transformed our social organization many times over, and allowed for the acculturation of a massive store of cultural information far beyond the expectations of our biologically primed capacities for social learning. Schools exist, at least in part, to help bridge that gap (Geary and Berch 2016), on this there is little disagreement. What remains up for debate are the specifics of how to design and continuously adapt schools to achieve this aim in a way that also cultivates empowered prosocial citizens and whole communities. The *Evolving Schools* project was developed to engage students and educators in participatory approaches to bridging this gap between theory and practice. Yet this project alone, especially within the isolated context of individual schools, is unlikely to be sufficient. What is further required is for education science to better engage with long-standing critiques about how we conceptualize and structure our inquiry to support knowledge transfer between research and practice (Bryk et al. 2011; Makel et al. 2019). Many schools have been encouraged to adopt new programs and policies on the perceived evidence-basis for efficacy ("What works") across contexts. The drastic changes that the pandemic imposed in various ways across schools highlights that simply waiting for research to provide that evidence-base is not necessarily an adequate approach to improvement.

We suggest that the recent scholarship on *Networked Improvement Communities* (NICs; see LeMahieu et al. 2017, and the work of the Carnegie Foundation for comprehensive resources on the NIC model) provides the most strategic path forward for empowering schools to adapt and learn from the challenges imposed by this pandemic. That is, schools should adopt new programs and policies on the basis of the actual efficacy *in their context* ("What works, for us, now"). The NIC research model suggests that making educational innovations work in local communities is fundamentally a process of intentional improvement of implementation. Schools need to focus on "getting better at getting better" in implementing the complex processes intended to meet identified aims and serve identified values. Furthermore, improvement is fundamentally strengthened through strategic measurement and networks of social learning.

The challenges that our project team faced in implementing the *Evolving Schools* project during the pandemic were merely symptomatic of the challenges many teachers in that school (and countless schools around the world) were facing. Yet this school has, to-date, no systematic process for improving practice or even understanding the range of valued outcomes from current practices in relation to their pandemic adaptation efforts. How many schools globally are in the same position? The improvements to our project planned for the coming school year represent a small step towards institutionalizing a pathway for elevating student and teacher voices within school improvement efforts, but this work needs to be supported by broader networked improvement work.

For this reason, in parallel with the development of this project, we have expanded collaborations within the Prosocial World research community (Atkins et al. 2019) to launch *Prosocial Schools* (www.ProsocialSchools.org) as an international NIC, focused broadly on improving school culture around valued outcomes. This network is now actively developing toolkits and processes to facilitate an applied toolkit for school improvement that is based on scientific perspectives of the human universal aspects of the cultural evolution of cooperation and learning, while emphasizing community empowerment for local adaptation of core design principles. The *Evolving Schools* project is one approach that more explicitly engages students in the evolution science itself, however the broader Prosocial toolkit can be applied without going this deep into the theoretical framework. By serving as a minimal and uniquely interdisciplinary scientific basis for participatory school improvement, we hope Prosocial Schools can support the improved implementation of Evolving Schools, and a diversity of projects with related aims yet employing diverse approaches.

5 Conclusions

The Evolving Schools project is neither a panacea nor a quick fix solution to the daunting challenges wrought by the global pandemic. Instead, it represents a unique approach to engaging student and educator voices in school change processes through scientific reflection on the adaptive cultural evolution of schools, in a time when reflective, inclusive approaches to change are needed perhaps more than ever. Our illustrative case study does not allow for inference to effective or optimal practices, only a suggestion of the possibility for a stronger integration between academic learning and participatory school improvement efforts. Our modest attempt here to involve students in the evolutionary science human learning as a participatory design exercise was met with enthusiasm and competent reflection even under the most challenging of conditions. Further work is required to empower more educators and students around the world to come together to improve these participatory processes of science-informed school improvement pathways. If networks such as Prosocial Schools can help sustain these broad directions, we could surely evolve new cultural adaptations to the challenges this pandemic has brought, as well as to the many other challenges that the future will inevitably bring, to our educational systems.

References

Atkins PW, Wilson DS, Hayes SC (2019) Prosocial: using evolutionary science to build productive, equitable, and collaborative groups. New Harbinger Publications

Bryk AS, Gomez LM, Grunow A (2011) Getting ideas into action: Building networked improvement communities in education. In: Frontiers in sociology of education. Springer, Dordrecht, pp 127–162

Cope B, Kalantzis M (Eds) (2015) A pedagogy of multiliteracies: learning by design. Springer

Eirdosh D, Hanisch S (2019) The Role of Evolutionary Studies in Education for Sustainable Development. In: Geher G, Wilson DS, Head H, Gallup A (eds) Darwin's roadmap to the curriculum: evolutionary studies in higher education. Oxford University Press

Evolving Schools Project (2020) Evolving Schools Project Report. GlobalESD https://EvolvingSchools.GlobalESD.org.

Fishman BJ, Penuel WR, Allen AR, Cheng BH, Sabelli NORA (2013) Design-based implementation research: an emerging model for transforming the relationship of research and practice. Nat Soc Study Edu 112(2):136–156

Fuentes A (2017) The creative spark: how imagination made humans exceptional. Penguin

Geary DC, Berch DB (eds) (2016) Evolutionary perspectives on child development and education. Springer

Gray P (2013) Free to learn: why unleashing the instinct to play will make our children happier, more self-reliant, and better students for life. Hachette UK

Hanisch S, Eirdosh D (2020) Educational potential of teaching evolution as an interdisciplinary science. Evolution: Edu Outreach 13(25). https://doi.org/10.1186/s12052-020-00138-4

Hanisch S, Eirdosh D, Atkins PWB (2020) Community science field guide for school culture. GlobalESD. https://CommunityScience.GlobalESD.org

Hattie J (2012) Visible learning for teachers: maximizing impact on learning. Routledge

Kirschner PA, Hendrick C (2020) How learning happens: seminal works in educational psychology and what they mean in practice. Routledge

LeMahieu PG, Bryk AS, Grunow A, Gomez LM (2017) Working to improve: seven approaches to improvement science in education. Qual. Assurance Edu
Makel MC, Smith KN, McBee M, Peters SJ, Miller EM (2019). Open science 2.0: large-scale collaborative education research. https://doi.org/10.31234/osf.io/ypmjg
Miller K, Lukoff B, King G, Mazur E (2018) Use of a social annotation platform for pre-class reading assignments in a flipped introductory physics class. Front Edu 2018(3)
Mintrop R (2016) Design-based school improvement: a practical guide for education leaders. Harvard Education Press
Ostrom E (1998) The need for civic education: a collective action perspective. In: Workshop in Political Theory and Policy Analysis, pp. 98–26
Poteete AR, Janssen MA, Ostrom E (2010) Working together: collective action, the commons, and multiple methods in practice. Princeton University Press
Sterelny K (2012) The evolved apprentice. MIT press
Stern J, Ferraro K, Mohnkern J (2017) Tools for teaching conceptual understanding, secondary: designing lessons and assessments for deep learning. Corwin Press
van der Zee T, Reich J (2018) Open education science. AERA Open 4(3)
Wilson DS, Ostrom E, Cox ME (2013) Generalizing the core design principles for the efficacy of groups. J Econ Behav Organ 90:S21–S32

Dustin Eirdosh is the education development coordinator at the department of Comparative Cultural Psychology at the Max Planck Institute for Evolutionary Anthropology. He is interested in teaching and learning at the intersection of evolution, behavior, and sustainability science. Working through collaborations across scientific institutions, into rnational non-profits, and local classrooms, he uses human social behavior as a conceptual lens for interdisciplinary education towards deeper understandings of the challenges and solutions to sustainable development at every level of society. He is also co-founder of the educational non-profit organization Global ESD

Dr. Susan Hanisch is an educational researcher, teacher educators and curriculum developer at the working group of primary school science education at the University of Leipzig. She is working to integrate perspective from evolutionary anthropology and human behavioral sciences into modern K-12 curricula and teaching approaches for the development of a deeper understanding of the human condition. She is also co-founder of the educational non-profit organization Global

The manufacturer's authorised representative in the EU is Springer Nature Customer Service Centre GmbH, Europaplatz 3, 69115 Heidelberg, Germany. If you have any concerns regarding our products, please contact ProductSafety@springernature.com

Printed and bound by CPI Group (UK) Ltd, Croydon, CR0 4YY
28/09/2026
02224067-0004